Nanoscience and Nanotechnology

Nanoscience and Nanotechnology

Edited by **Andrew Green**

Willford Press

New York

Published by Willford Press,
118-35 Queens Blvd., Suite 400,
Forest Hills, NY 11375, USA
www.willfordpress.com

Nanoscience and Nanotechnology
Edited by Andrew Green

International Standard Book Number: 978-1-68285-303-0 (Hardback)

Printed in China.

Contents

Preface

It is often said that books are a boon to mankind. They document every progress and pass on the knowledge from one generation to the other. They play a crucial role in our lives. Thus I was both excited and nervous while editing this book. I was pleased by the thought of being able to make a mark but I was also nervous to do it right because the future of students depends upon it. Hence, I took a few months to research further into the discipline, revise my knowledge and also explore some more aspects. Post this process, I began with the editing of this book.

The applications of nanoscience and nanotechnology in the areas of biotechnology, electronics, integrated circuits, chemistry, materials science, etc. are constantly increasing. Thus, the study of nanostructured materials is also becoming a core part of many sciences and engineering disciplines. This book investigates the recent developments and challenges in nanoscience and nanotechnology. It aims to elucidate various experimental tools to understand nanomaterials. The chapters included in this book have undergone a tremendous reviewing process. This book will help students and researchers to better understand the vast field of nanoscience and technology.

I thank my publisher with all my heart for considering me worthy of this unparalleled opportunity and for showing unwavering faith in my skills. I would also like to thank the editorial team who worked closely with me at every step and contributed immensely towards the successful completion of this book. Last but not the least, I wish to thank my friends and colleagues for their support.

Editor

1

Optical interactions in plasmonic nanostructures

Wounjhang Park

Abstract

We present a review of plasmonic nanostructures in which the constituent materials are coupled together by optical interactions. The review first provides a comprehensive coverage of theoretical framework where the optical interactions are described by the multiple scattering among the nanostructures. We then discuss the two limiting cases of weak and strong interactions. The weak interaction regime is described well by the effective medium theory while the strong interaction regime requires theoretical tools that can describe the new eigenmodes delocalized over the entire system. Weakly interacting plasmonic nanostructures have been studied extensively in the metamaterials research, which has been a major research thrust in photonics during the past decade. This review covers some of the latest examples exhibiting perfect absorption and invisibility. Strongly coupled systems started to receive attention recently. As a representative example, plasmonic molecules exhibiting Fano resonance are discussed in detail. Plasmonic nanostructures offer an excellent platform to engineer nanoscale optical fields. With the recent progress in nanofabrication technologies, plasmonic nanostructures offer a highly promising pathway to discovering new phenomena and developing novel optical devices.

1 Introduction

In the Drude theory, metal is modeled as a system of free electron gas undergoing forced oscillation with some damping. Other than some notable exceptions such as the strong interband absorption involving the *d* band in transition metals, the Drude model generally provides good descriptions of the optical properties of metals. It predicts that in the low frequency region the induced polarization oscillates 180° out-of-phase with the external field, resulting in negative permittivity and consequently high reflectivity. At certain frequencies, the electron gas can also undergo collective oscillations whose quantum is called plasmon. The collective oscillation may involve volume charge density (volume plasmon) or surface charge density (surface plasmon). Volume plasmons are longitudinal oscillations and cannot be excited by electromagnetic waves which are normally transverse. Surface plasmons, however, interact strongly with light. By storing a significant portion of its energy in oscillating electron gas, surface plasmon offers an effective means to localize light down to length scales much smaller than wavelength. For this reason, the past decades have seen an explosive growth in research on surface plasmon nanophotonics.

Correspondence: won.park@colorado.edu
Department of Electrical, Computer & Energy Engineering, University of Colorado, Boulder, CO 80309-0425, USA

One of the topics of high interest is nanoscale waveguiding with potential applications in integrated photonics for which a large body of literature exists [1]. Another important class of plasmonic nanostructures more relevant to this review are those supporting localized surface plasmon (LSP) modes. These structures include metal nanoparticles of various shapes, nanoholes in metal films, and some variations of these. The LSP resonances typically manifest themselves with strong scattering and absorption. The characteristic color arising from the LSP resonances has enabled a wide range of applications from the medieval stained glasses to modern optical sensors. For example, colorimetric sensing of biomolecules and ions has been demonstrated [2,3]. Also, strongly scattering nanoparticles have been used to enhance contrast in dark-field microscopy and optical coherence tomography [4] while thermal ablation sing strongly absorbing nanoparticles is being actively developed for targeted therapies of various diseases [5,6].

LSP naturally provides a highly localized and intense optical field which consequently enhances a variety of optical processes. The best-known example is surface enhanced Raman scattering (SERS). Using a rough silver surface, Raman scattering by a single molecule has been observed with enhancements up to a factor of 10^{14} [7]. Much of the enhancement is believed to arise from the local field enhancement due to the highly localized hot spots produced by the silver nanostructure

[8]. Surface plasmon resonance can also be used to enhance luminescence [9] and Förster energy transfer process [10,11]. Since the optical processes depend on the local field strength, strong emphasis has been placed on achieving higher local field enhancement using novel nanostructure geometries. These structures are generally termed nano-antennas analogous to conventional antennas in the microwave frequency region and have been the subject of extensive research in the past decade [12]. Since hot spots are typically formed in the nano-gap between plasmonic nanoparticles, the nano-antenna research naturally directed attention to the interaction between LSP modes, which is the main subject of this review. An important recent development in this field is the concept of plasmon hybridization [13]. In this picture, the individual LSP modes in closely spaced nanoparticles hybridize to form a set of new modes delocalized over the entire structure, in much the same way atomic orbitals hybridize to form molecular orbitals in real molecules. For this reason, these plasmonic nanostructures are also called plasmonic molecules. The hybridized modes in plasmonic molecules have been extensively studied. The simplest form of plasmonic molecule is a dimer or a pair of nanoparticles in which the plasmon resonance of single nanoparticle hybridizes to produce symmetric and anti-symmetric dimer plasmons [14]. Hybridization in more complex structures has also been studied, e.g. trimer [15], quadrumer [16], hexamer [17] and most widely heptamer [18-22]. In these complex structures, symmetry properties of the hybridized plasmon modes should be described by the group representation theory [23]. In this formalism, how each plasmon mode transforms under the symmetry operations is described by the irreducible representations of the symmetry group. This allows for a systematic description of modes with complex field profiles and also their interaction with incident optical fields, providing a complete picture when combined with the numerical techniques that can precisely calculate the hybridized LSP modes. It is noted that the group theoretical approach classifying the symmetry properties of plasmon modes does not require any approximations. For example, in nanostructures with features sizes much smaller than the wavelength, the plasmon modes exhibit predominantly electric dipole nature and are often approximated as point dipoles. The group theoretical approach, however, does not require dipole approximation and is equally applicable to larger and more complex structures exhibiting higher order mulitpole components.

The strength and nature of the optical interaction between plasmonic nanostructures varies with the distance. For long distances, the interaction is mainly radiative and involves propagating electromagnetic waves undergoing multiple scattering among the plasmonic nanostructures. In this case, the original character of the individual LSP modes should largely be preserved. For short distances, evanescently decaying near-field components should be taken into account. In principle, the near-field interaction can be treated the same way as the far-field interaction in the general framework of multiple scattering theory. That is, the evanescent modes undergo multiple scattering between nanostructures as well as the propagating modes. In extreme short distances, where the individual LSP modes have significant overlaps with one another, the near-field interaction will dominate and hybridization resulting in fully delocalized modes would occur. This review is aimed at providing a unified view of plasmonic nanostructures in which the strength of the interaction among individual metallic components is varied. We will first present a theoretical framework followed by examples of nanostructures and their optical properties in the weak and strong coupling regimes.

2 Review

2.1 Theoretical framework

2.1.1 Classical description of surface plasmon resonance

Surface plasmon, while inherently a quantum mechanical entity, is generally well described by the classical theory where surface plasmon presents itself as a surface resonance. There are of course quantum mechanical characteristics that cannot be fully described by the classical theory such as nonlocal response of electrons [24,25], tunneling and screening effect [26], spontaneous and stimulated emission of photons [27] and plasmons [28,29], to name a few. Nevertheless, the classical theory has proved sufficient for a wide range of systems. The classical theory is based on the Maxwell's equations where the boundary conditions are set up by the material properties specified by the macroscopic permittivity and permeability. For the planar geometry, surface wave solutions with fields decaying exponentially away from the interfaces can be found. In the simplest example of a single planar interface between semi-infinite metal and dielectric, the surface wave vector, q, is given by the dispersion relation [30],

$$q = \frac{\omega}{c}\sqrt{\frac{\varepsilon_1\varepsilon_2}{\varepsilon_1+\varepsilon_2}} \tag{1}$$

where ε_1 and ε_2 are the permittivities of metal and dielectric medium, respectively, and often Drude model is used for the metal permittivity, $\varepsilon_1(\omega) = 1 + \omega_p^2/(\omega^2 + i\gamma\omega)$ where ω_p is the plasma frequency and γ is the damping parameter. This surface wave solution specifies the optical mode often called surface plasmon-polariton (SPP). For the lossless case where the permittivities are purely real, the dispersion relation has a singularity when $\varepsilon_1 + \varepsilon_2 = 0$. This condition defines the surface plasmon frequency, ω_s, which is given by $\omega_s = \omega_p/\sqrt{1+\varepsilon_2}$, when using the Drude model for metal permittivity. Here ω_p is the plasma frequency of the electron gas. When there is finite loss, the metal permittivity is complex and the resonance condition should be written as which defines the condition for maximum electric field amplitude. This theory can be straightforwardly extended to more complex geometries involving multiple interfaces to obtain SPP solutions [31-34]. The localized surface plasmon modes (LSPs) generally require numerical calculations as the surface plasmon modes are highly sensitive to the details of the given geometry. An important exception is the highly symmetric geometry of sphere and circular cylinder for which a complete analytical solution can be obtained by using the Mie theory where the scattering and absorption cross-sections are given by the scattering coefficients a_n and b_n corresponding to various multipole terms [35]. For small spheres, the scattering coefficients a_n have vanishing denominators when $m^2 = -(n+1)/n$, where m is the relative refractive index of sphere relative to the embedded medium and n is an integer. This leads to a sharp increase in both scattering and absorption. In the limit of extremely small sphere, the extinction is dominated by the leading term, a_1, resulting in the resonance condition, $m^2 = -2$ or $\varepsilon_1 = -2\varepsilon_2$ when using the Drude model for the metal permittivity. For the lossless case where the permittivities are purely real, the resonance condition is satisfied when $\omega = \omega_p/\sqrt{2+\varepsilon_2}$, which is sometimes called the Fröhlich frequency. Thus, the LSPs supported by metallic spheres and cylinders may be considered special cases of Mie resonances.

2.1.2 Multiple scattering theory

We now consider a cluster of nanoparticles or nanostructures which individually support LSP resonances. The key additional consideration needed to properly describe the cluster is the multiple scattering between the individual components. Since analytical solutions are available for spheres, it is instructive to consider a cluster of nanospheres. In this case, the scattering by the individual spheres can be described by the standard Mie theory. The fundamental difference from the single sphere case is that for each sphere in the cluster the incident field is the combination of the actual incident field impinging on the cluster and the scattered fields by all other spheres in the cluster [36-38]. To briefly provide an overview of this process, we first remind the reader of the standard Mie theory for a single sphere where all fields are described in the coordinate system with the sphere located at the origin. The incident and scattered fields are then expressed in terms of the vector spherical harmonics,

$$\begin{aligned}
\mathbf{E}_{scat} &= \sum_{n=1}^{\infty}\sum_{m=-n}^{n} iE_{mn}\left(a_{mn}\mathbf{N}_{mn}^{(3)} + b_{mn}\mathbf{M}_{mn}^{(3)}\right) \\
\mathbf{E}_{int} &= \sum_{n=1}^{\infty}\sum_{m=-n}^{n} iE_{mn}\left(d_{mn}\mathbf{N}_{mn}^{(1)} + c_{mn}\mathbf{M}_{mn}^{(1)}\right) \\
\mathbf{E}_{inc} &= \sum_{n=1}^{\infty}\sum_{m=-n}^{n} iE_{mn}\left(p_{mn}\mathbf{N}_{mn}^{(1)} + q_{mn}\mathbf{M}_{mn}^{(1)}\right)
\end{aligned} \tag{2}$$

where $\mathbf{E}_{scat}$, $\mathbf{E}_{int}$ and $\mathbf{E}_{inc}$ represent scattered, internal and incident fields, respectively, $\mathbf{M}_{mn}$ and $\mathbf{N}_{mn}$ are the vector spherical harmonics. Similar expressions for the magnetic fields are readily obtained from the Maxwell's equations. The coefficients p_{mn} and q_{mn} are naturally dependent on the exact form of the incident field. For plane wave incidence, all terms with $m \neq 1$ vanish and the full expressions for the coefficients, a_n, b_n, c_n, d_n, p_n and q_n are available in the literature [39]. To describe multiple scattering in a cluster made of L spheres, the coefficients, p_{mn} and q_{mn}, describing the incident field for a sphere needs to be rewritten to include the scattered fields by all other spheres. This in turn requires translating all vector harmonics functions separately obtained in the coordinate systems of individual spheres into the common coordinate system so they may be added. This is achieved by the addition theorem of vector spherical harmonics:

$$\begin{aligned}
\mathbf{M}_{mn} &= \sum_{\nu=0}^{\infty}\sum_{\mu=-\nu}^{\nu}\left(A0_{\mu\nu}^{mn}\mathbf{M}'_{mn} + B0_{\mu\nu}^{mn}\mathbf{N}'_{mn}\right) \\
\mathbf{N}_{mn} &= \sum_{\nu=0}^{\infty}\sum_{\mu=-\nu}^{\nu}\left(B0_{\mu\nu}^{mn}\mathbf{M}'_{mn} + A0_{\mu\nu}^{mn}\mathbf{N}'_{mn}\right)
\end{aligned} \tag{3}$$

where $\mathbf{M}_{mn}$ and $\mathbf{N}_{mn}$ are the vector spherical harmonics about origin O, $\mathbf{M}'_{mn}$ and $\mathbf{N}'_{mn}$ are the vector spherical harmonics about origin O′, and $A0_{\mu\nu}^{mn}$ and are the additional coefficients whose complete expressions are available in the literature [36]. One can then express the scattered field by the lth sphere in the coordinate system of jth sphere using the new coefficients, p_{mn}^{lj} and q_{mn}^{lj},

$$\begin{aligned}
p_{mn}^{lj} &= -\sum_{\nu=0}^{\infty}\sum_{\mu=-\nu}^{\nu}\left(a_{\mu\nu}^{l}A_{mn}^{\mu\nu}(l,j) + b_{\mu\nu}^{l}B_{mn}^{\mu\nu}(l,j)\right) \\
q_{mn}^{lj} &= -\sum_{\nu=0}^{\infty}\sum_{\mu=-\nu}^{\nu}\left(a_{\mu\nu}^{l}B_{mn}^{\mu\nu}(l,j) + b_{\mu\nu}^{l}A_{mn}^{\mu\nu}(l,j)\right)
\end{aligned} \tag{4}$$

where $A_{\mu\nu}^{mn}$ and $B_{\mu\nu}^{mn}$ are related to $A0_{\mu\nu}^{mn}$ and $B0_{\mu\nu}^{mn}$ as given in Ref. [36]. Imposing the boundary conditions at the surfaces of the spheres finally leads to a set of linear equations

$$\begin{aligned}
a_{mn}^{j} &= a_{n}^{j}\left\{p_{mn}^{jj} - \sum_{l\neq j}\sum_{\nu=0}^{\infty}\sum_{\mu=-\nu}^{\nu}\left(a_{\mu\nu}^{l}A_{mn}^{\mu\nu}(l,j) + b_{\mu\nu}^{l}B_{mn}^{\mu\nu}(l,j)\right)\right\} \\
b_{mn}^{j} &= b_{n}^{j}\left\{q_{mn}^{jj} - \sum_{l\neq j}\sum_{\nu=0}^{\infty}\sum_{\mu=-\nu}^{\nu}\left(a_{\mu\nu}^{l}B_{mn}^{\mu\nu}(l,j) + b_{\mu\nu}^{l}A_{mn}^{\mu\nu}(l,j)\right)\right\}
\end{aligned} \tag{5}$$

Once the scattering coefficients, and, are found from the above equations, then the total scattered field by the entire cluster can be obtained by using the addition theorem one more time to express all fields in the common coordinate system of j_0th sphere,

$$\mathbf{E}_{scat} = \sum_{n=1}^{\infty}\sum_{m=-n}^{n} iE_{mn}\left(a_{mn}\mathbf{N}_{mn}^{(3)} + b_{mn}\mathbf{M}_{mn}^{(3)}\right) \tag{6}$$

where

$$a_{mn} = \sum_{l=1}^{L}\sum_{\nu=0}^{\infty}\sum_{\mu=-\nu}^{\nu}\left(a_{\mu\nu}^{l}A_{mn}^{\mu\nu}(l,j_0) + b_{\mu\nu}^{l}B_{mn}^{\mu\nu}(l,j_0)\right)$$
$$b_{mn} = \sum_{l=1}^{L}\sum_{\nu=0}^{\infty}\sum_{\mu=-\nu}^{\nu}\left(a_{\mu\nu}^{l}B_{mn}^{\mu\nu}(l,j_0) + b_{\mu\nu}^{l}A_{mn}^{\mu\nu}(l,j_0)\right) \tag{7}$$

The scattering cross section can be written in a similar fashion to the single sphere Mie scattering case,

$$C_{scat} = \frac{4\pi}{k^2}\sum_{n=0}^{\infty}\sum_{m=-n}^{n} n(n+1)(2n+1)\frac{(n-m)!}{(n+m)!}\left(|a_{mn}|^2 + |b_{mn}|^2\right) \tag{8}$$

A similar multiple scattering theory has also been developed for an infinitely periodic array of spheres [40-42]. In this case, since the system has infinite extent, the main concern is the propagation of light inside the periodic array. Due to the similarity to the Kohn-Korringa-Rostocker (KKR) method widely used for electronic energy band structure studies, this technique is sometimes called the vector KKR method. Once again, vector spherical harmonics are used to take advantage of the spherical geometry. Following the original development by Stefanou et al. [40,43], we begin with the general expression for spherical wave expansion given in equation (6) for a two-dimensional (2D) periodic array of spheres. When a plane wave is incident on the 2D array, the scattered wave must be expressed as a sum of the waves scattered by all spheres in the 2D plane. Due to the periodicity, the wave scattered by a sphere at $\mathbf{r} = \mathbf{R}_n$ differs only by a simple phase factor, $\exp(i\mathbf{k}_{||} \cdot \mathbf{R}_n)$, from the wave scattered by a sphere at the origin. Here, $\mathbf{k}_{||}$ represents the tangential component of k-vector parallel to the plane of spheres. The scattered field can then be written as,

$$E_{scat} = \sum_{n=1}^{\infty}\sum_{m=-n}^{n} \frac{i}{k}b_{mn}^{E}\nabla\times\sum_{\mathrm{R}_n}\exp(i\mathbf{k}_{\|}\cdot\mathbf{R}_n)h_n(kr_n)\mathbf{X}_{mn}(\theta,\phi)$$
$$\left. + b_{mn}^{H}\sum_{\mathrm{R}_n}\exp(i\mathbf{k}_{\|}\cdot\mathbf{R}_n)h_n(kr_n)\mathbf{X}_{mn}(\theta,\phi)\right) \tag{9}$$

where b_{mn} are coefficients to be determined, $\mathbf{r}_n = \mathbf{r} - \mathbf{R}_n$, h_n is the spherical Hankel function and $\mathbf{X}_{mn}$ is defined as $\sqrt{n(n+1)}\mathbf{X}_{mn} = -i\mathbf{r} \times \nabla Y_{mn}$ where Y_{mn} is the spherical harmonics function. While we try to keep the notation of original authors as much as possible, we note that the above equation is essentially in the same form as equation (6) where b_{mn}^{E} and b_{mn}^{H} in equation (9) are analogous to a_{mn} and b_{mn} coefficients in equation (6). The phase factor, exp ($i\mathbf{k}_{||} \cdot \mathbf{R}_n$), in the above equation allows us to express the scattered field in terms of plane waves with wave vectors,

$$K_g^{\pm} = \left(\mathbf{k}_{\|} + \mathbf{g}, \pm\sqrt{k^2 - (\mathbf{k}_{\|} + \mathbf{g})^2}\right) \tag{10}$$

Here $\mathbf{k}_{||}$ represents the reduced wave vector defined as $\mathbf{k}_{tangential} = \mathbf{k}_{||} + \mathbf{g}$, where $\mathbf{g}$ is the reciprocal lattice vectors of the 2D lattice, so that $\mathbf{k}_{||}$ now represents a vector within the unit cell of the 2D reciprocal lattice or the surface Brillouin zone. The sign in the superscript indicates forward and backward propagating waves in the direction perpendicular to the 2D plane. Now we construct the same plane wave expansion for the incident wave as well and obtain a matrix equation relating the coefficients of the incident wave and the total scattered wave,

$$[E_{scat}]_{g,i}^{s} \equiv \sum_{g',i'} M_{gi,g'i'}^{ss'}[E_{inc}]_{g',i'}^{s'} \tag{11}$$

where the new subscripts i, i' indicate the Cartesian components. The explicit expression for the matrix $\mathbf{M}'$ may be found in Ref. [44]. Equation (11) fully describes the interaction of a 2D periodic array of spheres with an incident light but, since the summation is infinite, actual numerical evaluations will involve truncation which introduces numerical errors. Now we extend the theory to three-dimensional (3D) array composed of multiple layers of 2D arrays by constructing a transfer matrix describing the multiple scattering effects between the adjacent 2D planes. That is, the field between the nth and $(n + 1)$th layers are determined by the scattering by the two neighboring planes of spheres. More specifically, the backward propagating wave in the region between nth and $(n + 1)$th layer is the sum of the forward propagating wave in the same region back-scattered by $(n + 1)$th layer and the backward propagating wave between the $(n + 1)$th and $(n + 2)$th layers transmitted (or forward-scattered) by the $(n + 1)$th layer. The forward propagating wave is similarly defined. The transmission and reflection matrices are obtained from the matrix $\mathbf{M}$ in equation (11) modified to account for the shift of origin between the two adjacent planes. This way, one obtains equations relating fields at adjacent layers which must also satisfy the Bloch theorem because the system is periodic in the 3rd dimension as well. Combining the multiple scattering equations and the Bloch theorem, one obtains an eigenvalue equation,

$$\begin{pmatrix} \mathbf{Q}^{++} & \mathbf{Q}^{+-} \\ -[\mathbf{Q}^{--}]^{-1}\mathbf{Q}^{-+}\mathbf{Q}^{++} & [\mathbf{Q}^{--}]^{-1}[\mathbf{Q}^{-+}\mathbf{Q}^{+-}] \end{pmatrix}\begin{pmatrix} \mathbf{E}^{+}(n) \\ \mathbf{E}^{-}(n+1) \end{pmatrix}$$
$$= \exp\left(i\mathbf{k}\times\mathbf{a}_3\right)\begin{pmatrix} \mathbf{E}^{+}(n) \\ \mathbf{E}^{-}(n+1) \end{pmatrix} \tag{12}$$

Here the matrices **Q** are the scattering matrices defined as $Q^{ss'}_{gg'} = M^{ss'}_{gg'} \exp\left[i\left(s\mathbf{K}^s_g\cdot\mathbf{d}_r + s'\mathbf{K}^{s'}_{g'}\cdot\mathbf{d}_l\right)\right]$ where s and s′ are + or − signs, g and g′ index the 2D reciprocal lattice vectors, $\mathbf{d}_r$ and $\mathbf{d}_l$ are right and left translation vectors, the matrix **M** is given in equation (11), and $\mathbf{a}_3$ is the primitive unit vector representing the periodicity in the 3rd dimension. It is noted that this last part is the standard transfer matrix formalism and can be used for non-spherical scatterers as well [45]. For given ω and $\mathbf{k}_{||}$ in the surface Brillouin zone, the matrix in equation (12) is fully specified and thus can be solved using the standard matrix inversion techniques. The plane wave expansion over the reciprocal lattice vectors of course will need to be truncated for numerical evaluations.

2.1.3 Effective medium theory

Now we explore the limiting cases where the multiple scattering theory presented above can either be simplified significantly or is inadequate. An obvious possibility of simplification is found when the interaction among the scatterers is weak. In this case, the multiple scattering effect will be small and may be ignored in the extreme case. A commonly employed approximation for such a case is the effective medium theory which attempts to obtain an average response of a non-uniform medium composed of multiple constituents. The most widely used effective medium theory is the Maxwell Garnett theory [46], which is derived from the Lorentz local field relation covered in many textbooks. Briefly, Lorentz considered an imaginary spherical cavity in a dielectric medium and stated that the local field is simply given by the sum of the dielectric response of the medium and the additional fields due to the surface charge on the spherical cavity and due to the atoms inside the cavity. The lattermost contribution cancels out to zero in a highly symmetric crystal structure. This leads to the famous expression,

$$\mathbf{E}_{local} = \mathbf{E} + \frac{4\pi\mathbf{P}}{3} \tag{13}$$

where **E** and **P** are the macroscopic field and polarization of the medium. Relating the local field to the induced atomic dipole moment, $\mathbf{p} = \alpha\mathbf{E}_{local}$, we arrive at the Clausius-Mossotti equation,

$$\frac{\varepsilon-1}{\varepsilon+2} = \frac{4\pi N\alpha}{3} \tag{14}$$

where ε is the permittivity of the medium and N and α are atomic density and polarizability, respectively. Maxwell Garnett effective medium theory simply replaces the atomic polarizability in the Clausius-Mossotti equation with the polarizability of a spherical object. In the electrostatic limit, the polarizability of a sphere in vacuum has a simple form,

$$\alpha = \left(\frac{\varepsilon_s-1}{\varepsilon_s+2}\right)a^3 \tag{15}$$

where a is the radius of the spherical object and ε_s is the permittivity of the sphere. The Maxwell Garnett mixing rule is obtained by substituting equation (15) into (14),

$$\frac{\varepsilon-1}{\varepsilon+2} = f_s\cdot\frac{\varepsilon_s-1}{\varepsilon_s+2} \tag{16}$$

where f_s is the volume fraction of the sphere. For spheres embedded in a dielectric matrix with permittivity ε_m, we simply replace 1 and 2 in equation (16) with ε_m and $2\varepsilon_m$, respectively.

We now make a connection with the more rigorous multiple scattering theory presented earlier. In the Mie theory for a single spherical scatterer, the scattering coefficients a_n and b_n are generally expressed as infinite series of spherical Hankel functions. Using the power series expansions of the spherical Hankel functions, it can be shown that the leading terms of these coefficients are $a_1 \sim x^3$, $b_1 \sim x^5$, $a_2 \sim x^5$, $b_2 \sim x^7$, etc., where $x = 2\pi a/\lambda$. Therefore, for small spheres, we may retain only the leading term in a_1. Since the scattering and absorption cross sections are proportional to $|a_1|^2$ and $\mathbf{Im}(a_1)$ (imaginary part of a_1), respectively, the scattering and absorption efficiencies defined as the ratio of scattering and absorption cross sections to the physical cross section depend on x^4 and x, respectively, yielding the well-known Rayleigh scattering results. More specifically, the leading term of coefficient a_1 is

$$a_1 = -\frac{i2x^3}{3}\cdot\frac{\varepsilon_s-\varepsilon_m}{\varepsilon_s+2\varepsilon_m} \tag{17}$$

The relationship between the a_1 coefficient and the polarizability is apparent,

$$\alpha = \frac{3ia^3}{2x^3}a_1 \tag{18}$$

This relationship reveals one of the key approximations involved in the Maxwell Garnet effective medium theory: retaining only the leading term of the leading coefficient a_1 in the general solution of the Mie theory. This is essentially the electrostatic approximation in which the polarizability of small spheres is modeled by the static polarizability given in equation (15). A modest improvement on this approximation may be achieved by the extended Maxwell Garnett theory in which the full expansion of a_1 is used to calculate the polarizability [47]. The extended Maxwell Garnett theory has been used successfully for metamaterials [48,49]. However, if higher order terms in coefficient a_1 are to be included, it is logical to include b_1 and a_2

terms as well. These terms represent the contributions of the magnetic dipole and electric quadrupole terms, respectively. The inclusion of b_1 can be done straightforwardly by defining effective permeability in exactly the same way effective permittivity is calculated except that a_1 is replaced by b_1 to define magnetic polarizability. It has been shown that the optical properties of small particle composites may exhibit magnetic permeability even when the constituent materials are all non-magnetic [50]. Thus, the use of extended Maxwell Garnett theory is more appropriate when both effective permittivity and permeability are considered. It has also been attempted to improve the Maxwell Garnett theory using dressed polarizability to better account for the resonant effect [51]. In addition to neglecting higher multipole contributions to the light scattering and absorption, which is appropriate for small particles, the Maxwell Garnett effective medium theory also ignores multiple scattering among particles. This can be seen from the facts that the polarizability is defined by the single sphere Mie coefficient and that the effective permittivity is constructed based on the Lorentz local field concept. There are other issues related to the Maxwell Garnett effective medium theory such as the Lorentz catastrophe and non-symmetry and there exist other effective medium theories that improve on those shortcomings [52]. However, it can be said in general that the effective medium theory is appropriate for small particles with low volume fraction where both the higher order multipole contributions and inter-particle interaction by multiple scattering can be ignored.

2.1.4 Plasmon hybridization

We now us consider the opposite extreme where the particles are closely spaced and thus strongly interacting with one another. In this case, especially near plasmon resonances, strongly overlapping evanescent fields exist, leading to significant modification in the nature of the resonance. In principle, this near-field interaction can be described by the multiple scattering theory presented earlier. However, there is a practical difficulty in that it requires the inclusion of a large number partial waves and thus makes the numerical computation extremely challenging. Instead it is more profitable to look for new eigenmode solutions described by the collective modes delocalized over the entire structure. Prodan et al. proposed the plasmon hybridization model [13]. To be sure, the concept of plasmon hybridization itself is not new. A classic example is coupling of two surface plasmon modes in a metal thin film and the resultant formation of symmetric and anti-symmetric modes [53,54]. But Prodan's model is highly useful as it provides a complete set of eigenmode solutions in complex nanostructures. The plasmon hybridization model considers the electron gas as incompressible fluid subject to Coulombic interaction. The plasmon resonance is then represented by self-sustained oscillations of the electron fluid which may be obtained from the Euler-Lagrange equation for the Lagrangian,

$$L = \frac{n_0 m}{2}\int \eta\nabla\eta dS - \frac{1}{2}\int \frac{\sigma(\mathbf{r})\sigma(\mathbf{r}')}{\mathbf{r}-\mathbf{r}'} dSdS' \tag{19}$$

where n_0 and m are the electron density and mass, respectively, and σ is the surface charge density. This technique was shown to agree well with the full electrodynamic simulation results in the electrostatic limit where all of the involved length scales are small compared to the wavelength. It was successful in describing the plasmonic resonances in complex structures including nanoshell, nanoegg and nanorice [55]. The model has also been applied to multiparticle systems such as dimer, trimer, quadrumer, hexamer and heptamer [14,20,56]. A similar method has been proposed by Mayergoyz et al. who derived a boundary integral eigenvalue equation using the electrostatic approximation. In this approach, the surface plasmon resonance is described as an eigenstate of the homogeneous boundary integral equation [57],

$$\sigma(\mathbf{r}) = \frac{\gamma}{2\pi}\int \sigma(r') \frac{(\mathbf{r}-\mathbf{r}')}{|\mathbf{r}-\mathbf{r}'|^3} \cdot \hat{\mathbf{n}} dS \tag{20}$$

For an arbitrary geometry, the boundary integral equation (20) can be solved numerically to obtain the eigenvalue γ and eigenfunction $\sigma(\mathbf{r})$. The eigenvalue γ is related to the permittivity $\varepsilon(\omega)$ of the metal as

$$\mathrm{Re}\varepsilon(\omega) = \varepsilon_b \left(\frac{1+\gamma}{1-\gamma}\right) \tag{21}$$

where ε_b is the permittivity of the background material. It is thus possible to determine the resonance frequency from the permittivity of the metal. The eigenfunction $\sigma(\mathbf{r})$ describes the self-sustained surface charge distribution. This technique has been used to describe collective plasmon resonance supported by hexamers and heptamers made of gold nanoparticles and nanorods [17,21,22].

It should be noted that both of the two techniques described above employ electrostatic approximation and thus cannot describe the retardation effect. Their applicability is therefore limited to extremely small particles or low frequencies. Nevertheless, these methods proved highly useful in the exploration of coupled plasmon modes in complex nanostructures consisting of multiple nanoparticles. The main reason is the applicability of elegant group theoretical method in a fashion similar to the molecular orbital theory. Just as the molecular orbital can be constructed from the symmetry adapted linear combination of atomic orbitals, the hybridized plasmon modes can be constructed from the symmetry adapted linear combination of individual nanoparticle plasmon modes. This enables a highly effective and systematic exploration of plasmon modes in a complex, strongly coupled nanostructure. Here we briefly describe this approach using the boundary integral method for a collection of nanoparticles. First, the plasmon modes of an isolated nanoparticle can be found from the eigenvalue equation given in (20). For the collection of N nanoparticles, the surface charge distribution can be expressed as a linear combination of the individual nanoparticle eigenmodes,

$$\sigma(\mathbf{r}) = \sum_{\alpha=1}^{N} \sum_{j} a_{\alpha}^{j} \sigma_{\alpha}^{j}(\mathbf{r}) \tag{22}$$

Using equation (20), we may write

$$\sigma(\mathbf{r}) = \frac{\Lambda}{2\pi} \sum_{\beta,k} \int a_{\beta}^{k} \sigma_{\beta}^{k}(\mathbf{r}_{\beta}) \frac{(\mathbf{r}-\mathbf{r}_{\beta})}{|\mathbf{r}-\mathbf{r}_{\beta}|^{3}} \cdot \hat{\mathbf{n}} dS \tag{23}$$

where σ_{β}^{k} represents the surface charge distribution of kth mode of βth nanoparticle. One can then use the bi-orthogonality between the surface charge distribution, $\sigma(\mathbf{r})$, and the surface dipole distribution, $\tau(\mathbf{r})$, to derive,

$$\begin{aligned} a_{n}^{m} &= \frac{\Lambda}{2\pi} \sum_{\beta,k} a_{\beta}^{k} \left\{ \int \tau_{\alpha}^{m}(\mathbf{r}) \hat{\mathbf{n}} \cdot \int \sigma_{\beta}^{k}(\mathbf{r}_{\beta}) \frac{(\mathbf{r}-\mathbf{r}_{\beta})}{|\mathbf{r}-\mathbf{r}_{\beta}|^{3}} dS_{\beta} dS \right\} \\ &= \frac{\Lambda}{2\pi} \sum_{\beta,k} K_{\alpha\beta}^{mk} a_{\beta}^{k} \end{aligned} \tag{24}$$

The total surface charge distribution in equation (22) must satisfy the boundary integral eigenvalue equation in (20) where the integral is conducted over the surfaces of all nanoparticles. However, the problem now has been transformed to a matrix eigenvalue equation given in (24). Physically, the matrix $\mathbf{K}$ represents the interaction between the surface dipole density of the mth mode of α th particle and the charge distribution of the kth mode of βth particle. Now suppose the arrangement of the nanoparticles is symmetric and certain symmetry operations keep the nanoparticle cluster unchanged. This means the symmetry operators leave the interaction matrix $\mathbf{K}$ invariant. Then, the symmetry operators and matrix $\mathbf{K}$ commute and consequently they share the same eigenvectors. It is now possible to classify all eigenvectors of matrix $\mathbf{K}$ based on how they transform under various symmetry operations. In the language of group theory, the eigenvectors are classified by the irreducible representations of the symmetry group. These eigenvectors possess well-defined transformation characteristics under all symmetry operators that leave the structure invariant and form the symmetry-adapted basis set. The symmetry properties of the plasmon modes proved powerful in analyzing the interaction between various plasmon modes and also between light and plasmons. For example, the incident light interacts with the surface plasmon mode via the induced dipole moment. For a given symmetry group, the dipole operator belongs to a certain irreducible representation. Thus, it naturally follows that only those surface plasmon modes belonging to the same irreducible representation as the dipole operator can interact with light. Other modes are the so-called dark modes that do not interact with incident light and thus cannot be detected by experiments. Furthermore, surface plasmon modes belonging to different irreducible representations do not mix together as their basis functions are mutually orthogonal. These properties provide powerful insights into how complex plasmon modes evolve and interact in complex nanostructures.

2.2 Weakly coupled plasmonic nanostructures

As discussed in the previous section, when the plasmonic nanostructures are only weakly coupled, they may be described well by the effective medium theory. There is a large body of literature on metamaterials which are composed of deep sub-wavelength scale structures and are generally described well by the effective medium theory. The initial development of metamaterials was driven largely by the negative index materials for which excellent reviews are available [58,59]. In this section, we focus on the latest development of plasmonic nanostructures in the weak coupling regime.

2.2.1 Perfect absorber

An interesting application of plasmonic nanostructures is the perfect absorber which absorbs all incident light with minimal reflection and transmission. Such a material would find a wide range of applications in, for example, coatings for optical instruments, sensors, and photothermal or photovoltaic energy conversion devices. An inspection of Fresnel's coefficients immediately reveals that perfect absorption is achieved when the real part of the refractive index is matched with the incident medium and the imaginary part is small. This condition would result in negligibly small reflection and, if the material is thick enough, vanishing transmission as well. This is, in fact, how perfectly absorbing metamaterial was obtained using aligned carbon nanotube arrays [60]. This example illustrates clearly that a perfect absorber does not have to involve a plasmonic nanostructure [60,61]. It is, however, of great technological importance to achieve perfect absorption in thin films and plasmonic nanostructures are ideally suited for that purpose.

One of the simplest examples is a line grating. It is well known that a line grating on a metal surface can couple the free space light into surface plasmons. An ideal grating presents infinitesimal modulation of surface so that the plasmon modes of the original flat surface are well preserved and the grating only provides the extra momentum for coupling. However, it was found that, even in a grating as shallow as 20 nm, additional surface plasmon modes strongly localized at the grating lines exist and interact with the surface plasmon-polariton modes. Lu et al. investigated silver nanograting structures fabricated on the back-electrode of a PCDTBT:PCBM organic photovoltaic device [62]. The system supports two surface plasmon modes when excited by transverse-magnetic (TM, electric field perpendicular to the grating lines) polarized incident light. One is the localized surface plasmon (LSP) mode whose field profile is strongly localized at the grating lines and wavelength insensitive to the incident angle and active layer thickness. The other is the surface plasmon-polariton (SPP) mode strongly interacting with the grating period and thus highly sensitive to incident angle and active layer thickness. Figure 1(a) shows the wavelength of surface plasmon induced absorption peaks as a function of active layer thickness, revealing two branches with distinct dependence on active layer thickness. In general, as the active layer thickness is increased, the refractive index of the dielectric side of the surface is increased, resulting in a red shift in surface plasmon wavelength. The surface plasmon dispersion relation for a dielectric loaded metal surface can be calculated numerically [34]. As shown by the light blue crosses in Figure 1(a), the surface plasmon wavelength exhibits a red shift with increasing active layer thickness and this line coincides well with the SPP branch of the silver nanograting overcoated with PCDTBT:PCBM active layer. This indicates that the 20 nm tall silver nanograting acts as an ideal grating for the SPP modes which retain the characteristics of those in the flat dielectric loaded metal surface. The other branch representing LSP modes is unique to the grating structure and not observed in flat interfaces. Due to the strongly localized nature, it is insensitive to the thickness of the dielectric layer. For thicknesses near 30 nm, the two branches exhibit anti-crossing, indicative of strong mixing between the LSP and SPP modes. The field profiles in Figure 1(b) and (c) show the two contrasting patterns for LSP and SPP modes while those in Figure 1(d) and (e) reveal considerable mixing. Since these modes strongly affect the absorption spectra, the knowledge of the existence of two different modes and their interaction is essential to designing light trapping plasmonic grating structures for photovoltaic devices.

When the grating height is increased, the surface plasmons lose their original characteristics of flat interfaces and the LSP modes become the dominant features. Kravets et al. studied a gold grating with 130 nm height fabricated by depositing gold on a 130 nm thick poly(methyl methacrylate) (PMMA) grating [63]. As shown in Figure 2, this structure exhibited extremely low reflectance and transmittance for TM polarized light, achieving blackbody-like absorption in the visible spectrum with a thickness much smaller than the wavelengths. A theoretical analysis is needed to understand this remarkable behavior. A rigorous theoretical treatment would require full consideration of two sets of gold gratings, one formed in between the PMMA lines and another formed on top of PMMA lines.

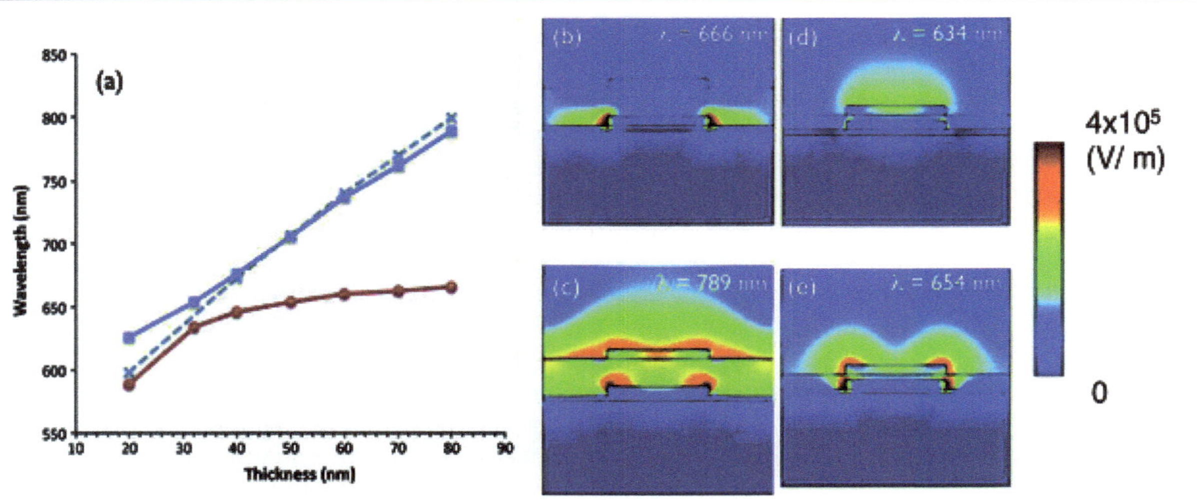

Figure 1 Interaction of surface plasmon. (a) Surface plasmon peak positions as a function of PCDTBT:PCBM active layer thickness exhibiting anti-crossing behavior. Blue and red solid dots represent LSP and SPP peak positions for various PCDTBT:PCBM thicknesses. Blue crosses represent the SPP peak positions calculated from the effective index of the SPP mode supported by the flat silver-PCDTBT:PCBM-air multilayer structure. **(b, c)** Electric field profiles for the surface plasmon modes in the 80 nm thick PCDTBT:PCBM on silver grating and **(d, e)** electric field profile for the surface plasmon modes in the 30 nm thick PCDTBT:PCBM on silver grating. To clarify the structure the bottom grating is the silver nanograting fabricated by nanoimprint lithography. The top surface is also corrugated simply by the result of depositing on a corrugated surface (From Ref. [62]).

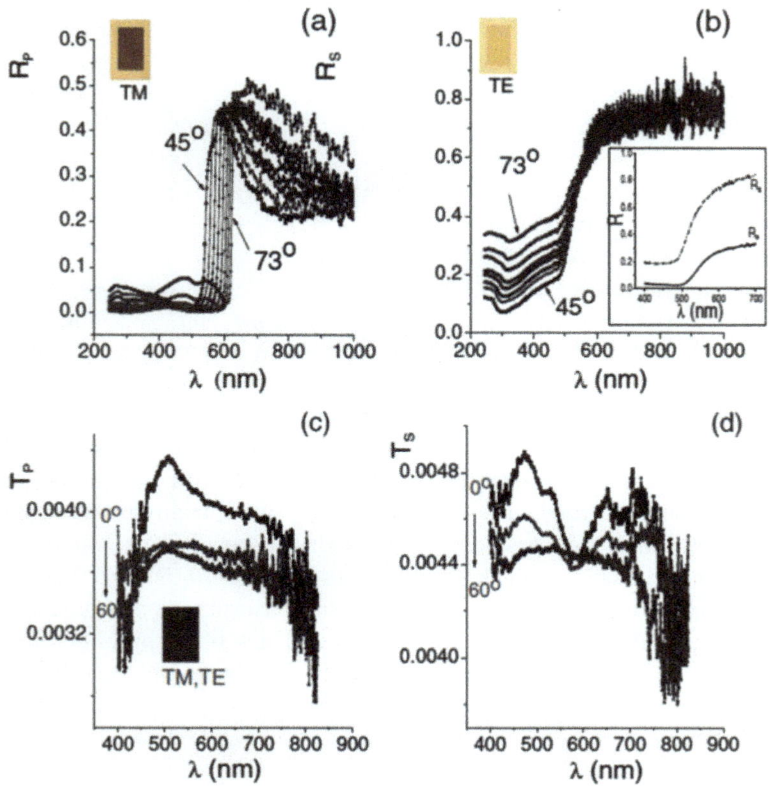

Figure 2 Reflectance and transmittance spectra of the gold nanostripe array with period = 320 nm and linewidth = 150 nm. Insets show the polarization-contrast optical microscopy images. The inset in **(b)** shows reflectance spectra for s- and p-polarizations at normal incidence (From Ref. [63]).

However, since the grating lines are well separated (by more than 100 nm), near-field coupling between adjacent grating lines should be small and therefore it should be possible to use a much simpler effective medium theory. To generalize the Maxwell Garnett results for spherical particles given in equations (14) ~ (16) to non-spherical shapes, we introduce depolarization factor, L. For a general ellipsoidal shape, L is defined as [64]

$$L_i = \frac{a_x a_y a_z}{2}\int_0^\infty \frac{dS}{(s+a_1^2)\sqrt{(s+a_x^2)\left(s+a_y^2\right)(s+a_z^2)}} \text{ where } i = x, y, z \tag{25}$$

where a_i is the semi-axis along the i-direction. For a sphere, $L_x = L_y = L_z = 1/3$. For a cylinder oriented along the x-direction, $L_x = 0$, $L_y = L_z = 1/2$. For a thin film in the yz plane, $L_x = 1$, $L_y = L_z = 0$. We can now write a generalized Maxwell Garnett effective permittivity as follows [65].

$$\varepsilon_{eff}^i = 1 + \frac{f\alpha_i}{1 - fL_i\alpha_i} \text{ where } i = x, y, z \tag{26}$$

where f is the volume fraction as before. Furthermore, the volume-normalized polarizability can be written in terms of the depolarization factor as,

$$\alpha_i = \frac{\varepsilon_s/\varepsilon_m - 1}{1 + L_i(\varepsilon_s/\varepsilon_m - 1)} \text{ where } i = x, y, z \tag{27}$$

where ε_s and ε_m are the permittivity of the ellipsoid and medium, respectively. Combining the two equations above yields

$$\alpha_z = \varepsilon_s/\varepsilon_m - 1 \text{ and } \alpha_y = \alpha_z \frac{2(\varepsilon_s/\varepsilon_m - 1)}{\varepsilon_s/\varepsilon_m - 1} \tag{28}$$

$$\varepsilon_{eff}^x = 1 + f(\varepsilon_s/\varepsilon_m - 1) \text{ and } \varepsilon_{eff}^y = \varepsilon_{eff}^z \frac{\varepsilon_s/\varepsilon_m - 1 + f(\varepsilon_s/\varepsilon_m - 1)}{\varepsilon_s/\varepsilon_m + 1 - f(\varepsilon_s/\varepsilon_m - 1)} \tag{29}$$

for a cylinder oriented along the x-axis. Equation (29) yields the real and imaginary parts of the effective refractive index of the gold grating to be roughly 1.5 and 0.3, respectively, over a broad wavelength range up to 500 nm for TM polarization. These values are close to the optimal complex refractive index value for near-perfect absorption in a thin film. For example, a film with 130 nm thickness on a glass substrate would exhibit

90% absorption at $\lambda = 400$ nm if the refractive index is $1.3 + 0.5i$. A similar approach can be made with nanoparticles instead of nanograting [66]. In this case, the polarizability of a sphere given in equation (15) in Section 2.3 can be used in the Maxwell Garnett effective medium theory. The major difference in the nanoparticle composite medium and the nanograting is the polarization dependence. The inherent isotropy of spherical geometry naturally leads to isotropic, polarization-independent medium in contrast to the strong polarization sensitivity in nanogratings.

A question of both scientific and technological interest is how much absorption is possible in ultrathin films. A simple solution may be found by using total internal reflection. Using the Kretschmann geometry, an ultrathin film absorber can be excited at glancing angle, effectively extending the optical path to infinity and consequently achieving near-perfect absorption. Using this method, 94% absorption in a 4.5 nm thick NbN nanowire array has been demonstrated [67]. It should be noted that this method does not require surface plasmon or any other types of surface resonances, although they could certainly help. While this approach clearly offers a quick solution to achieve perfect absorption, it is not always possible or feasible to use total internal reflection geometry and it would be much more desirable to achieve high absorption at normal incidence or over a broad range of incident angles. To tackle this problem, let us consider a thin film with thickness d and refractive index $m = n + i\kappa$ on a semi-infinite substrate with refractive index n_s and incident medium with refractive index n_i (Figure 3a). The reflection and transmission coefficients for normal incidence can then be written in a general form as follows.

$$r = \frac{r_i - r_s\gamma^2}{1 - r_i r_s \gamma^2} \quad and \quad t = \frac{(1+r_i)(1+r_s)\gamma}{1 - r_i r_s \gamma^2}\exp(-in_s\delta)$$
$$where \quad r_{i,s} = \frac{n_{i,s} - m}{n_{i,s} + m} \quad and \quad \gamma = \exp(im\delta) \tag{30}$$

Also, $\delta = q_0 d$ and q_0 is the vacuum wave number. Generalization to non-normal incidence is straightforwardly done by replacing q_0 with the normal component of incident wave vector. The absorptance is given by where $C = n_s/n_i$ for transverse-electric (TE) incident field and for TM. One can now survey the possible values of m to find the maximum absorption. Figure 3(b ~ e) shows the absorptance at $\lambda_0 = 900$ nm for various film thicknesses of 2, 10, 50 and 150 nm. All cases show a maximum absorptance of approximately 50% along the line $n = \kappa$. Thick films show additional maxima along the horizontal axis due to the Fabry-Perot resonances. In ultrathin films, only those maxima occurring along $n = \kappa$ line are achievable. Normally, for a fixed value of m, the absorption vanishes with vanishing thickness as $A = O(\delta)$. To go beyond this limit, Hägglund et al. sought a refractive index value with dependence where $-1 < \nu < 0$ so that m is large enough to produce substantial absorption while remains small [68]. The optimum value of m_{max} resulting maximum absorption was given as,

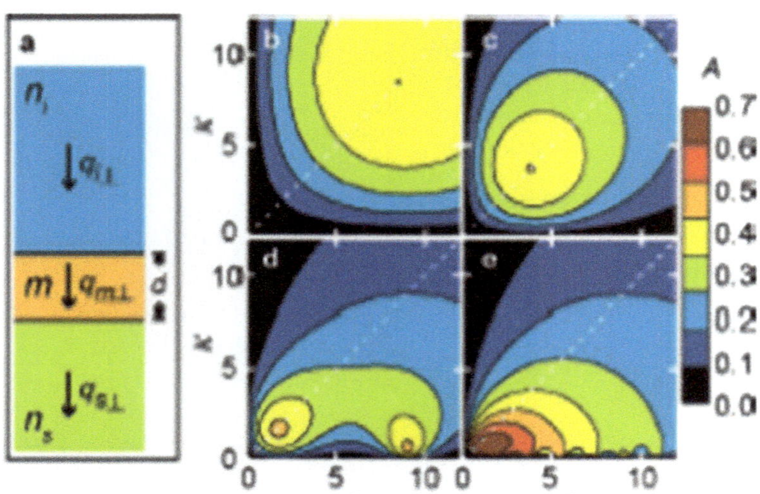

Figure 3 Absorptance of thin film with complex refractive index. (a) Thin homogeneous film with complex refractive index m = n + iκ and thickness d is embedded between two semi-infinite dielectrics of refractive indices n_i and n_s, respectively. In **(b-e)**, absorptance in the film is shown as a function of its optical constants for normal incident light. The wavelength is 900 nm, and the dielectric environment has $n_i = n_s = 1$. The film thicknesses d are 2, 10, 50, and 250 nm in panels **b**, **c**, **d**, and **e**, respectively. The dotted lines mark n = κ about which the shown absorption is nearly symmetric for the two thinnest films **(b, c)**. In these cases, the absorptance reaches a peak value of close to 50% on the symmetry line. For the two thicker films **(d, e)**, the peak height increases and additional maxima appear along the n-axis due to Fabry-Perot modes (From Ref. [68]).

$$m_{\max} = \frac{1+i}{\sqrt{2}}\left(\frac{n_i+n_s}{\delta}\right)^{1/2} + \frac{1+i}{6\sqrt{2}}\frac{(n_i^3+n_s^3)}{(n_i+n_s)^{3/2}}\delta^{1/2} + O(\delta^{3/2}) \quad (31)$$

The maximum absorptance is given as

$$A_{\max} = \frac{n_i}{n_i+n_s}\left[1+\frac{n_s}{6}(3n_s-n_i)\delta^2\right] + O(\delta^4) \quad (32)$$

Equations (31) and (32) show that the maximum absorption occurs at $n = \kappa$ and the maximum absorptance is close to 50%, both of which are shown in Figure 3.

It is obvious that no natural materials would exhibit the optimum refractive index given in equation (31). It is therefore necessary to create a composite medium whose effective refractive index is close to the optimum. It was shown that 50% absorptance could be realized over a broad range of wavelengths using an array of ellipsoidal silver shells with various geometry [68]. This scheme can be extended to a thin film with a back reflector. The reflector automatically eliminates transmission and the problem is reduced to minimizing reflectance. A general system to consider is an ultrathin film with a complex refractive index m and thickness d, separated from a perfect reflector by a spacer layer with refractive index n_s and thickness h. The incident medium index is n_i. The reflection coefficient is then given as

$$r = \frac{r_i - r_d\gamma^2}{1 - r_i r_d\gamma^2} \quad (33)$$

where $r = \dfrac{n_{i,d}-m}{n_{i,d}+m}$, $\gamma = \exp(im\delta)$ and

$$n_s = in_s \cot(\pi\lambda_c/2\lambda)$$

The critical wavelength λ_c is defined as the wavelength satisfying the quarter wavelength condition, $\lambda_c = 4n_s h$ Equation (33) has the same form as equation (30) and thus $\tilde{n}_s$ may be treated as the effective index of the spacer-reflector system. The maximum absorption condition is then found by requiring $r_i = r_d\gamma^2$ in equation (33). Retaining only the leading term the optimum refractive index is given as [69],

$$m^2 = \frac{i(n_i-.n_s)}{\delta} = \frac{\lambda}{2\pi d}\left[in_i + n_s\cot\left(\frac{\pi\lambda_c}{2\lambda}\right)\right] \quad (34)$$

When $\lambda = \lambda_c$, $\tilde{n}_s = 0$ and the optimum refractive index is given as $m = (1+i)\sqrt{n_i/2\delta}$. This represents the critical coupling condition where the required refractive index m is minimum. On the other hand, when the spacer layer thickness tends to zero, n_N diverges and the optimum refractive index cannot be satisfied. Again, the implementation of this concept requires an artificial composite medium made of plasmonic nanostructures. Recently, strong absorption has been experimentally observed using a gold nanoparticle array fabricated by block copolymer lithography and overcoated with a dielectric layer by atomic layer deposition [70]. As shown in Figure 4, the ultrathin composite film with a thickness of ~25 nm exhibited near-perfect absorption at around 600 nm. Vast majority of absorption occurs in the gold nanoparticles, which is not desirable for photovoltaic applications where the absorption must take place in the semiconducting material for current extraction. It is nevertheless noteworthy that the effective absorption coefficient of the gold nanoparticles in this structure was an order of magnitude greater than solid metal. A similar critical coupling leading to perfect absorption has been reported on a gold nanodisk array deposited on a glass substrate [71]. In this case, the critical coupling condition was achieved by tuning the nanodisk density and incident angle of incoming light. When the critical coupling condition is satisfied, reflected light is annihilated by the destructive interference between the light reflected from the substrate and the gold nanodisk array layer.

2.2.2 Invisibility

In addition to enhancing absorption, controlling or suppressing light scattering is also of great interest for many applications. It has been shown that the dominant dipolar contribution to the scattering from a spherical or cylindrical dielectric object could be greatly reduced when coated by a layer of material with negative or very low values of the real part of permittivity [72]. This technique is inherently non-resonant and thus the bandwidth of the scattering cancellation can be quite broad and limited only by the material dispersion of the coating. It was shown that the operational bandwidth could be further extended by use of multiple layers of coatings [73]. Potential applications of this type of cloaking include cloaked sensors [74], cloaked near-field probes [75,76] and reduction of optical forces [77]. For implementation, a natural choice for the coating material is a metal, which exhibits negative permittivity values at frequencies below its plasma frequency. However, in the optical frequency region, most metals tend to exhibit large negative permittivity values. Consequently, an unrealistically thin layer is required to cancel the dipole moment of a dielectric object whose permittivity is typically a small positive number. To circumvent this issue, an effective medium formed by plasmonic nanostructure is needed. Recently, Tamma et al. reported scattering cancellation by a gold nanograting fabricated on the sidewall of a silicon nanorod [78]. As shown in Figure 5(a), the cloaked object was a silicon nanorod fabricated by electron-beam lithography. The gold nanograting was then fabricated on the sidewall by a combination of gold evaporation, focused ion beam milling and silica protective coating deposition by sputtering. The final structure was coated with SU-8 to form a

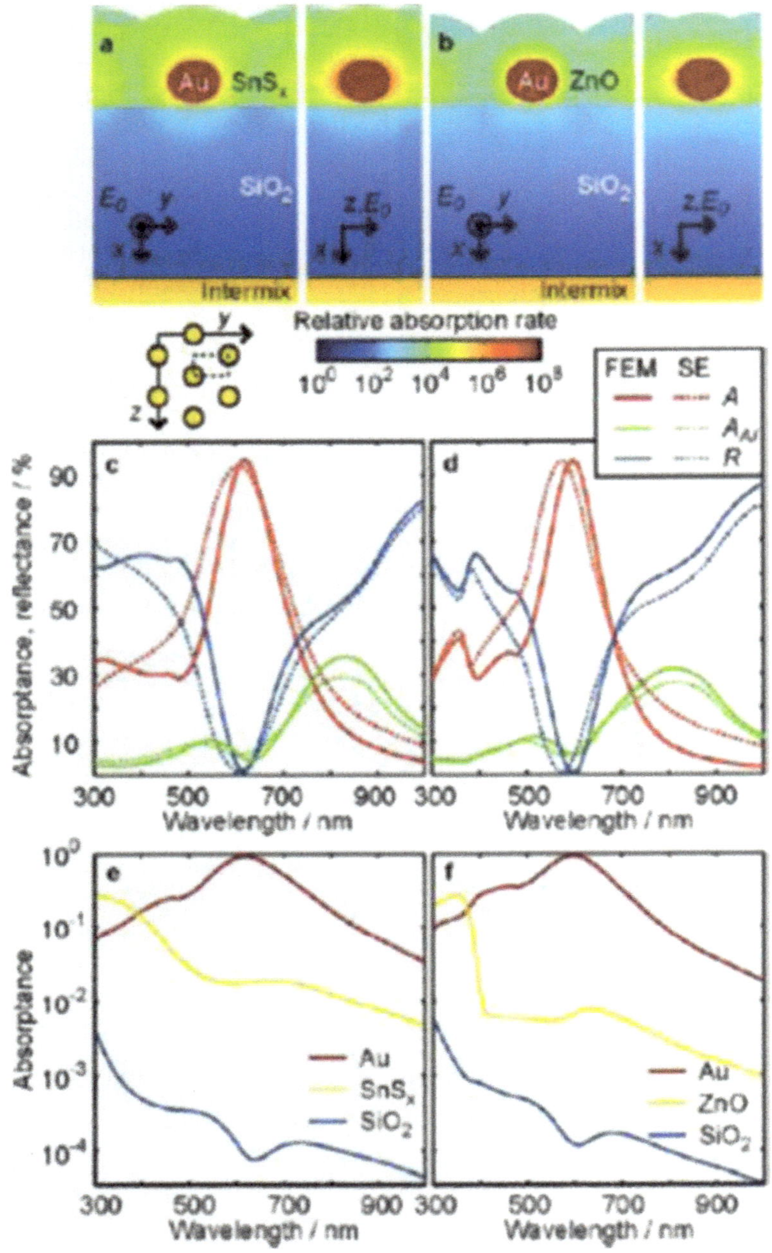

Figure 4 Absorptance of gold nanoparticle embedded thin films. (a) Finite element method (FEM) calculated absorption rates, determined for the sample with the highest absorbing SnS_x coated Au array at its peak wavelength. A plane wave with incident electric field strength E_0 propagates along the normal (x-) direction, with its electric field vector along z. The rectangular unit cell used for the hexagonal lattice is shown below the plot (dashed lines). **(b)** Analogous for ZnO coated sample. **(c, d)** Sample reflectance (R), array absorptance (A), and absorptance in the Al reflector (AAl, including intermix layer), as calculated by FEM and obtained from the spectroscopic ellipsometry (SE) model. **(e)** FEM calculated absorptance in the Au dots of the SnS_x coated sample, in the SnS_x coating, and in the SiO_2 spacer, respectively. **(f)** Analogous for the ZnO coated sample (From Ref. [70]).

waveguide layer and the structure was excited by 1550 nm light coupled into the SU-8 guiding layer. The scattering properties were then investigated by near-field scanning optical microscopy (NSOM). The waveguide geometry used in this study makes it impossible to directly measure the scattered light intensity. Instead, NSOM was used to map the interference fringes produced by the incident light and back-scattered light (Figure 5b). It was clear that the cloaked silicon nanorod scattered much less because the interference fringes were weak and required a long integration time and high gain to obtain reliable data whereas the reference nanorod showed a strong interference pattern. Tamma et al. then used the effective medium theory to relate the effective index of the coating with the

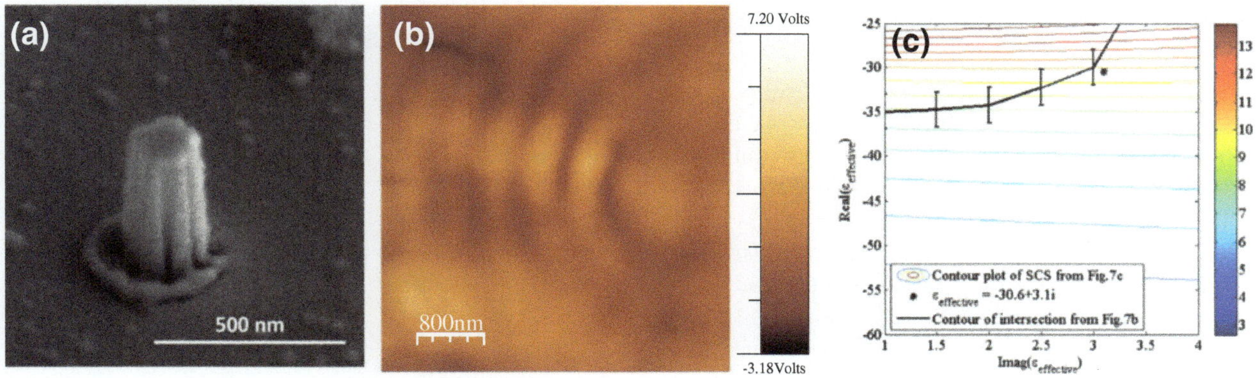

Figure 5 Scattering cancellation by plasmonic nanostructure. (a) Scanning electron micrograph (SEM) of the silicon nanorod with gold nanograting fabricated on the side wall. **(b)** Near-field scanning optical micrograph (NSOM) of scattered light by the cloaked silicon nanorod measured with $\lambda = 1550$ nm. The fringe curvature contains the information on scattering cross section. **(c)** Scattering cross section extracted from the analysis of fringe curvatures. A reduction of 9.8 dB was achieved (From Ref. [78]).

fringe curvature. Here the gold nanograting coating was described as a multilayer structure for which the effective medium theory simple yields [79]

$$\varepsilon_{\|} = f_a\varepsilon_a + f_b\varepsilon_b \quad and \ \varepsilon_{\perp} = (f_a/\varepsilon_a + f_b/\varepsilon_b)^{-1} \tag{35}$$

where the subscripts $||$ and $\perp$ indicates polarization parallel and perpendicular to the interface, respectively. The result of the complete analysis shown in Figure 5(c) indicates that a 9.8 dB reduction in scattering cross section has been achieved. The same approach has been implemented with silver nanoparticles attached on silica spheres [80]. The scattering reduction was however small, ~20%, and the system also exhibited strong absorption.

In concluding the discussion on weakly coupled system that can be described well by the effective medium theory, it is worth noting the robustness of effective medium theory. As discussed in the previous section, the effective medium theory is valid only when the feature size is small compared to the wavelength and the interaction among the nanostructures is weak. When these two conditions are met, the optical properties of the composite structure are determined by the dipole moment of the individual nanostructure without having to include higher order resonances and any higher order interaction. These conditions certainly sound very restrictive but in reality the effective medium theory has been successfully applied to a wide variety of systems which at first glace do not seem to satisfy the conditions for validity. An excellent example is the silicon nanorod cloaked by an array of gold nanowires described above. The individual gold nanowire was 13 nm thick and 20 nm wide and the spacing between two adjacent gold nanowires was 63 nm. This system was modeled as a multilayer composed of 20 nm thick gold and 43 nm thick silica films. Despite this seemingly gross simplification, the effective medium theory provided a remarkably accurate result in good agreement with experiments. It can be argued that the effective medium theory, although oversimplified, has captured the essential part of the light-matter interaction, which in this case concerns the induced surface charge and resultant screening effect. The effective medium theory for multilayers resulting in effective permittivity given in equation [35] considers the induced polarization in each layer. When the incident electric field is perpendicular to the multilayer interfaces, it produces a net interface charge due to the discontinuity in permittivity at the interface. These interface charges then produce a polarization field that counteracts the incident electric field, resulting in a permittivity given as $\varepsilon_{\perp}$ in equation [35]. On the other hand, when the incident electric field is parallel to the interface, no interface charge is formed and thus the effective permittivity is simple given as an arithmetic mean of the constituent permittivities, as given for $\varepsilon_{||}$ in equation [35]. The experiments in the cloaked silicon nanorod were done for TM waveguide mode which has electric field parallel to the gold nanowires. This corresponds to the no interface charge case and the multilayer effective medium theory for no screening case is applicable. If the experiments were done for perpendicular polarization, it would have been important to accurately account for the interface charge density and thus the multilayer effective medium theory could have been inaccurate.

Another case worth mentioning is the nanoparticle array. It is well established that small nanoparticles have dominantly dipole character with negligible contributions by the high order components. Thus, as long as the nanoparticles are separated well enough, effective medium theory should work well. But defining a precise validity limit is difficult because unlike the perturbation theory the effective medium theory does not allow quantitative estimation of errors. For plasmonic nanoparticles, the literature suggests volume fraction of 10 ~ 20% would be the upper limit where the effective medium theory begins to break down [49,81]. In a study of collective magnetic resonance supported by a cluster of plasmonic nanoparticles, Tamma et al. presented a comparison between the rigorous multiple scattering theory and the extended Maxwell Garnett effective medium theory [49]. Figure 6 shows the dispersion curves and effective permittivity for silver nanoparticle arrays with silver volume fractions of 0.1, 0.3 and 0.5. As shown, even at a very high volume fraction of 0.5, the effective medium theory agrees reasonably well with the rigorous multiple scattering theory. The discrepancy is found near the band edge in all cases where the discrepancy remains small for f = 0.1 and becomes significant for higher volume fractions. This discrepancy observed near the band edge captures the nature of approximation in the effective medium theory. Unlike the effective medium theory, the multiple scattering formalism requires the complete knowledge of the nanoparticle positions in the array. For the calculations in Figure 6, the nanoparticles were assumed to be in the face-centered cubic (fcc) structure and the light propagation along the (111) direction was calculated. If the calculations were made for other crystallographic directions, the resulting dispersion curves would be different and the differences would be most pronounced near the band edge where the Bragg resonance condition is met. By construction, the effective medium theory does not require the knowledge of nanoparticle arrangement. This implies the effective medium theory would represent an averaged response of all different crystallographic orientations. It is therefore not surprising to see the major difference between multiple scattering theory and effective medium theory near the band edge. This however does not necessarily represent the deficiency of effective medium theory because most nanocomposite structures do not have a definite crystal structure. A few exceptions are DNA-driven self-assembly of plasmonic nanoparticles that lead to single crystalline fcc and body-centered cubic structures [82,83]. But even in those cases, the achievable structures remain small and multiple crystallographic orientations may be in play. Therefore, although it is advisable to compare with rigorous multiple scattering or other methods whenever possible, the effective medium theory remains highly useful in an extremely wide range of plasmonic nanostructures.

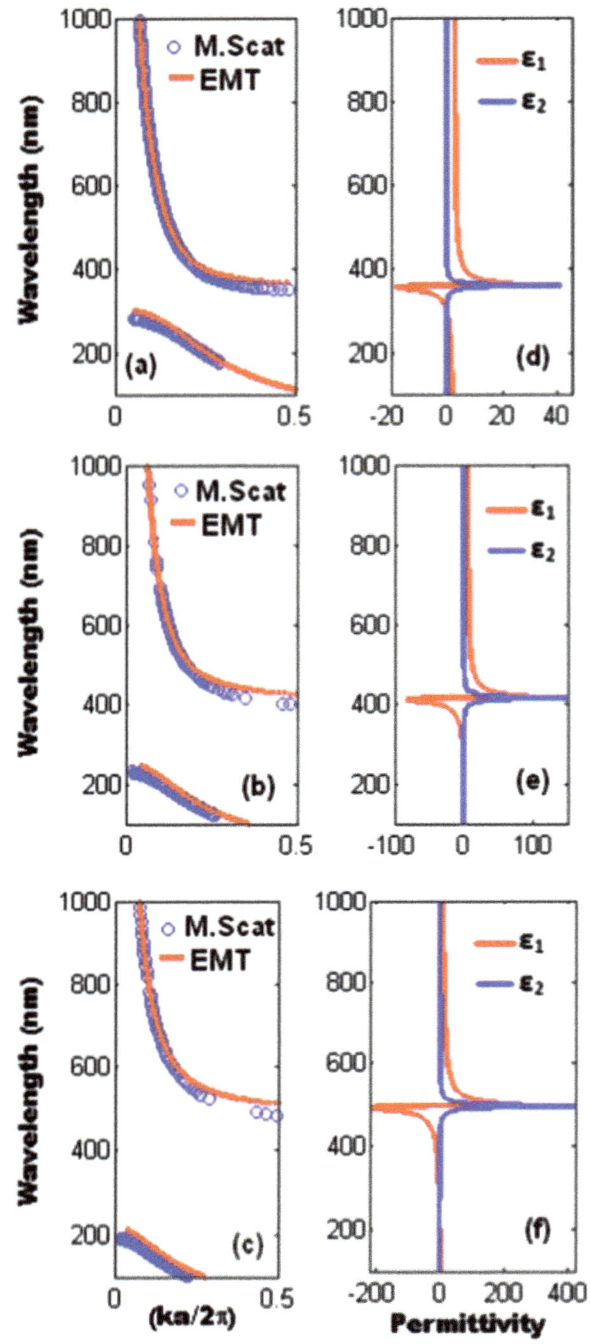

Figure 6 Comparison of multiple scattering theory and effective medium theory. (a), (b), (c) Dispersion curves obtained from multiple scattering (M.Scat) and extended Maxwell-Garnett effective medium theory (EMT) for silver nanoparticle array with volume fractions of 0.1, 0.3 and 0.5, respectively. **(d), (e), (f)** Real and imaginary parts of the effective permittivity calculated by the extended Maxwell-Garnett effective medium for silver volume fractions of 0.1, 0.3 and 0.5, respectively (From Ref. [49]).

2.3 Strongly coupled plasmonic nanostructures

2.3.1 Dimers

Strongly coupled plasmonic nanostructures have been studied extensively during the past decade. For example, the hot spot leading to enhanced Raman scattering is generally attributed to the coupling between two adjacent silver nanoparticles producing strong local field in the gap between the two nanoparticles [84]. A similar phenomenon has been observed in a coupled-dipole nanoantenna composed of two nanorods aligned end-to-end and separated by a small gap [85]. Also, early experimental studies on the simple model system of two closely spaced metal nanoparticles, a system often dubbed as dimer, showed the plasmon resonance exhibits red shift with decreasing gap between the nanoparticles [86-89]. Both the formation of hot spots and the red shift with decreasing gap size are due to the hybridization of individual nanoparticle plasmons. When the gap between the nanoparticles is small enough to allow significant overlap of the plasmon modes of the individual nanoparticles, the strong near-field interaction consequently leads to hybridization of the individual single particle plasmon modes to form a new set of coupled modes delocalized over the entire structure. In a dimer, the single particle plasmon modes couple with each other to form symmetric and anti-symmetric combinations. This is the standard coupled oscillator problem with numerous analogous examples in physics [90]. In the plasmonic dimer, the oscillating entity is the free electron charge in the metal. When the charge density in the two particles oscillates in phase along the axis of the dimer, we have a symmetric mode with strong electric field in the gap. When the two oscillations are 180° out of phase, we have an anti-symmetric mode whose electric field profile has a null in the gap. In molecular analogy, the symmetric mode corresponds to the bonding orbital whereas the anti-symmetric mode is the anti-bonding orbital. Also, the symmetric mode has a lower energy and the anti-symmetric mode has a higher energy, just as the bonding orbital has a lower energy and the anti-bonding orbital has a higher energy. As illustrated in Figure 7 for a nanorod dimer, this hybridization model offers a good explanation for both the hot spot formation and the red shift in plasmon resonance wavelength.

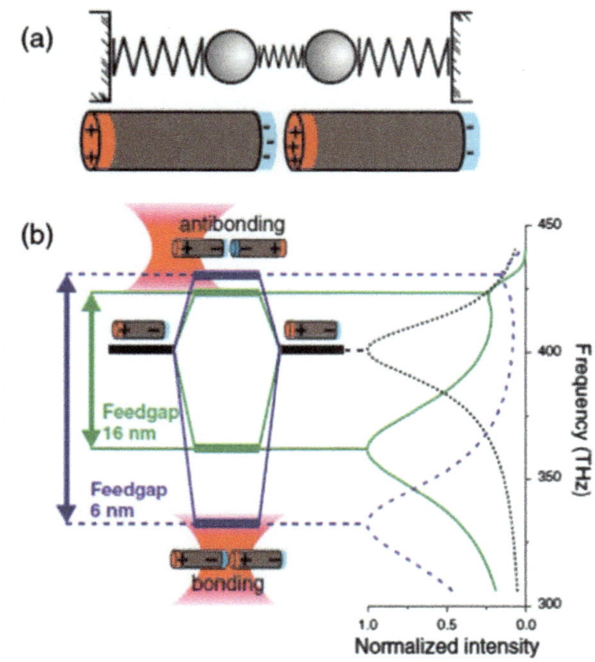

Figure 7 Plasmon coupling in nanorod dimer. (a) Schematic drawings of a coupled-dipole nanoantenna made of two plasmonic nanorods with a small gap and its mechanical analog, a coupled mechanical oscillator. **(b)** Energy-level diagram and simulated near-field intensity spectra for 30 nm high, 50 nm wide and 110 nm long symmetric two-nanorod antennas with 6 nm (blue dashed) and 16 nm (green solid) gap, as well as single-nanorod antennas with the same dimensions (black dotted) (From Ref. [90]).

The hybridized plasmon modes in a dimer can be monitored by directly solving Maxwell's equations numerically using techniques such as finite-difference time-domain (FDTD) [91] or finite-element method (FEM) [92]. Nevertheless, it is always profitable to seek for a theoretical model that could provide deeper insights on the physics governing the hybridization. Also, it is worth mentioning the inability of most brute force numerical simulations to excite anti-symmetric modes. This deficiency stems from the typical simulation geometry in which the nanostructure under investigation is excited by incident light. The simulation then calculates the resultant absorption or scattering which reveals the characteristics of the hybridized plasmon modes. This unfortunately works well only for the symmetric mode, which typically possesses a large dipole moment and thus interacts strongly with light. The anti-symmetric mode, however, represents a state in which the induced dipole moments in the individual nanoparticles align anti-parallel and thus only produce a small net dipole moment. This mode therefore does not interact with light and remain invisible in most experiments and simulations. For this reason, the symmetric mode is often referred to as the bright mode and the anti-symmetric mode as the dark mode. The anti-symmetric mode is nonetheless important as, for example, it can interact with the bright mode to produce Fano resonance as described later. It is therefore useful to have a theoretical model that can describe both the symmetric and anti-symmetric modes equally well. The simplest model employs the electrostatic approximation where the retardation effects are neglected. In this formalism, the polarizability, α, of individual nanoparticles is given by equations (15) and (27) for spheres and ellipsoids, respectively. The induced dipole moment in each particle is then given as $\boldsymbol{\mu} = \alpha \mathbf{E}$. In a dimer, the field, $\mathbf{E}$, contains both the applied field and the field produced by the neighboring particle. Since an individual nanoparticle is treated as a point dipole, we can write

$$\mathbf{E}_1 = \mathbf{E}_{inc} + g\cdot\frac{\boldsymbol{\mu}_2}{\varepsilon_m d^3} \text{ and } \mathbf{E}_2 = \mathbf{E}_{inc} + g\cdot\frac{\boldsymbol{\mu}_1}{\varepsilon_m d^3} \tag{36}$$

Here the subscripts 1 and 2 index the two particles in the dimer, ε_m is the permittivity of the background medium and g is the geometry dependent parameter given as $g = 2$ for end-to-end coupling and $g = -1$ for side-to-side coupling [93,94]. The coupled equation in [36] can be solved to find the polarizability of the bonding mode,

$$\langle\alpha\rangle = \frac{1}{2}\cdot\frac{\alpha_1(1+g\alpha_2/d^3)+\alpha_2(1+g\alpha_1/d^3)}{1-g^2\alpha_1\alpha_2/d^6} \tag{37}$$

The plasmon resonance condition is found by the maximum polarizability and, if the two particles are identical, is given as,

$$\alpha = \frac{d^3}{g} \tag{38}$$

Using the polarizability given in equation (27), we find

$$\frac{\varepsilon_s}{\varepsilon_s} = \frac{1+d^3(1-L_i)/g}{1+d^3/g} \tag{39}$$

Since the permittivity of gold has almost linear dependence on wavelength in the visible region, equation (39) specifies the dependence of the surface plasmon wavelength of a gold dimer on the center-to-center spacing. This simple model correctly predicts the red shift of the symmetric mode for end-to-end aligned dipole pair and the blue shift for side-to-side aligned dipole pair. The opposite behavior originates from the opposite sign of g in equations (37) ~ (39). In plasmonic dimer made of two identical particles, the end-to-end dipole alignment corresponds to the excitation of dimer with incident polarization along the dimer axis while the side-to-side coupling case corresponds to incident polarization perpendicular to the dimer axis. The opposite peak shifts in the two cases have been experimentally observed in the extinction spectra of gold nanodisk pairs fabricated by electron-beam lithography (Figure 8a) [88]. Furthermore, Jain et al. reported the fractional plasmon peak shift, $\Delta\lambda/\lambda_o$, exhibits a universal scaling against the normalized gap width defined as gap/diameter, irrespective of the detailed geometrical parameters, as shown in Figure 8(b) [95]. This universal scaling was also found for dimers of non-circular shapes and nanoshells and also in trimers [96,97]. The origin of the universal scaling behavior may be found in equation (39), which predicts that the fractional peak shift would behave as $(d/D)^{-3}$ or $[(s+D)/D]^{-3}$ where $s = d - D$ is the gap width between the particles and D is the diameter. When the gap width becomes extremely small, the simple dipole-dipole interaction model breaks down and one must take into account the higher order multipole terms, leading to an exponential dependence observed in Figure 8. This omission of multipole interaction terms is one of the key limitations in the dipole-dipole interaction model, which leads to a significant underestimation of plasmon coupling at extremely small gaps. Additionally, the dipole model is essentially based on the electrostatic approximation which ignores any retardation effect. To properly describe the interaction at small gaps including the retardation effects would require the use of numerical modeling techniques such as FDTD and FEM. Before moving on, it is worth mentioning that the universal exponential dependence has led to the derivation of the so-called plasmonic ruler equation, which enables precise determination of distances in biological systems, opening many opportunities for biosensing [95].

Note that the previous discussion of dipole-dipole coupling focused only on the symmetric mode where the two individual dipoles are parallel to each other. As stated earlier, the anti-symmetric mode with anti-parallel dipoles has a small net dipole moment and thus interacts only weakly with light. This is not true when the anti-symmetric mode possesses a large magnetic moment. In this case, the anti-symmetric mode interacts strongly with light through its magnetic field. A dimer of small spherical nanoparticles does not support strong magnetic resonance but a pair of nanorods can. A pair of gold nanorods was shown to support a strong magnetic resonances [98,99] and soon after

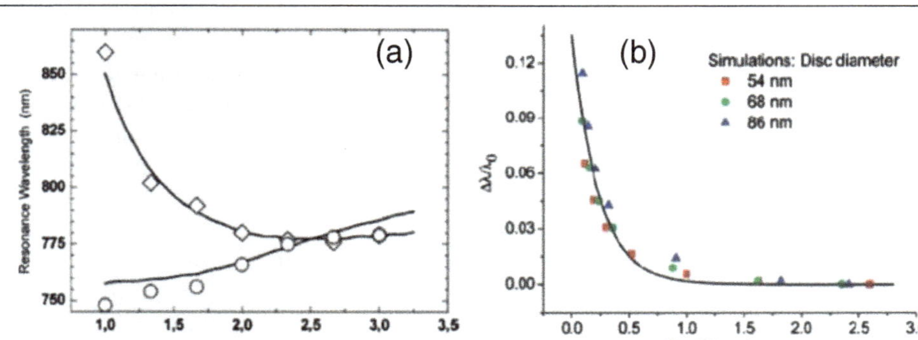

Figure 8 Plasmon coupling in gold nanoparticle dimer. (a) Spectral position of the extinction maximum vs. interparticle distance for seven gold nanodisk pairs fabricated by electron-beam lithography (From [88]). **(b)** Calculated fractional plasmon peak shift vs the ratio of interparticle gap to nanodisk diameter, showing that the scaled data points for the different disc sizes follow a common trend, which can be fit together to the single-exponential decay (solid curves) $y = ae^{-x/\tau}$ with and $\tau = 0.23 \pm 0.03$ (From Ref. [95]).

used to construct negative index material [100]. A similar behavior was also observed in a pair of nanoholes in two closely spaced metal films, which is expected to exhibit the same properties as the nanorod pair through the Babinet principle [101]. The nanohole pair design was then adapted to the fishnet structure which is known to be the most effective design for negative index [102]. Later, nanorod pairs with various different arrangements were investigated [103]. The hybridization scheme for various geometries is shown in Figure 9 together with the surface change densities for various plasmon modes. Here the hybridization scheme was obtained by the plasmon hybridization model and the surface charge density by the boundary integral method, both of which are described in Section 2.4. As shown in Figure 9(a), the single nanorod supports two resonances, longitudinal and transverse modes where the transverse mode has a higher energy due to the smaller transverse dimension. When the two nanorods are placed close together, the near-field interaction leads to hybridization and mode splitting where the symmetric and anti-symmetric combinations form bonding and anti-bonding modes. In the perfectly symmetric arrangements of end-to-end and side-to-side pairing, the anti-symmetric combination results in perfect cancellation of dipole moments. Thus the resultant hybridized mode is dark and does not respond to the incident light. Note that, unlike the nanorod pair used to observe magnetic resonance, which has one nanorod on top of the other so that the induced magnetic moment can align with the incident magnetic field, these nanorod pairs are all fabricated on the same plane and thus no magnetic interaction is expected. The energies of the hybridized modes are easily understood by the electrostatic interaction between electronic charge densities. For example, in the end-to-end pair, the anti-symmetric mode has the same charge concentrated near the gap, resulting in high energy while the symmetric mode has the opposite charge across the gap, leading to lower energy. Thus in this case, the symmetric mode becomes the bonding mode and the anti-symmetric mode forms the anti-bonding mode. For the side-to-side pair, the opposite is the case and the symmetric mode has higher energy than the anti-symmetric mode, making the anti-symmetric mode the bonding mode and the symmetric mode the anti-bonding mode. For other non-symmetric arrangements, the hybridized mode energy depends on the exact geometry. It should be noted that both the plasmon hybridization model and the boundary integral method used to generate Figure 9 take into account only the electrostatic interaction. The energy levels in Figure 9(a) and (b) are thus different from the experimentally measured plasmon peak positions, generally blue shifted from the experimental values due to the neglect of retardation effect.

2.3.2 Heptamers

To go beyond the simple dimer structure to more complex nanostructures, one needs a systematic way to classify various eigenmodes with complex symmetry properties. The group representation theory provides a power means to accomplish that [23]. For a given eigenvalue equation, $\mathbf{A}\mathbf{x}_n = a_n\mathbf{x}_n$, we can define a group of symmetry operators that leaves the operator **A** invariant. Note that the discussion here is generally applicable to any types of eigenvalue equations which could be the Schrödinger's equation in quantum mechanics, Maxwell's equation in optics and electromagnetics, or the fluid dynamics and surface charge

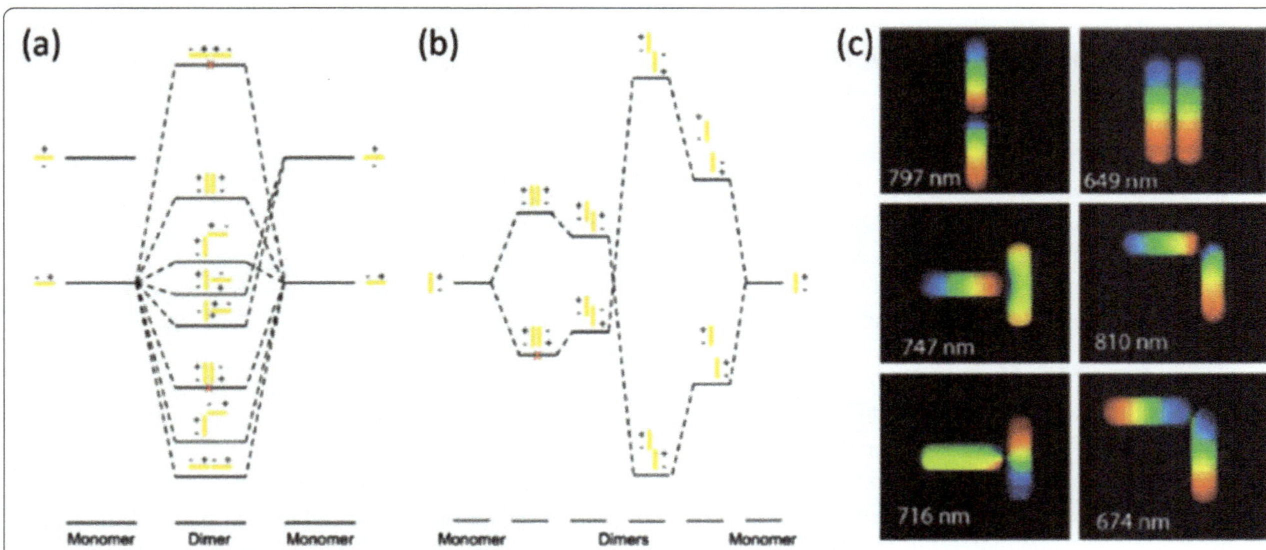

Figure 9 Plasmon hybridization schemes for (a) nanorod dimers in different geometric arrangements and (b) nanorods initially arranged side-to-side and then increasingly longitudinally offset as a function of the center-to-center offset. (c) Surface charge density of interacting gold nanorods placed 1.5 nm apart and with dimensions 78 nm × 24 nm, calculated using the boundary integral method (From Ref. [103]).

density equations used in the plasmon hybridization model and boundary integral formalism, respectively, discussed in Section 2.4. If a symmetry operator, **S**, leaves the operator **A** invariant, it is straightforward to show that the two operators commute with each other and share the same eigenstates. If all symmetry operators that leave the operator **A** invariant are found, then it is possible to classify all eigenstates of operator **A** in terms of how they transform under the various symmetry operators. In the language of group theory, all eigenstates of operator **A** can be indexed by the irreducible representations of the symmetry group. Now let us proceed to the plasmon modes in a trimer using the group representation theory. For a trimer composed of three identical nanoparticles placed at the corners of an equilateral triangle, the symmetry operators that leave the system invariant form the point group D_{3h} and thus the eigenmodes of a trimer can be classified by the irreducible representations of D_{3h}. For simplicity, we treat the plasmon mode of individual nanoparticle as a dipole mode, which is a good approximation for small particles. Then, the individual plasmon mode belongs to the $D^{(l=1)}$ irreducible representation of the full rotation group. For a trimer, we should consider the triple direct product of $D^{(l=1)}$ irreducible representations and their subsequent reduction into a combination of irreducible representations of point group D_{3h}, which yields [104],

$$\Gamma_{trimer} = A_1^{'} + A_2^{'} + 2E' + A_2^{''}E'' \tag{40}$$

The transformation properties of the various irreducible representations of point group D_{3h} can be found in the character table available in the literature [23]. From the character table, one can then construct the projection operators and generate the symmetry-adapted basis functions, as shown in Figure 10, which transform according to the symmetry properties of the irreducible representations. The one-dimensional irreducible representations, $A_1^{'}$, $A_2^{'}$, $A_1^{''}$, $A_2^{''}$, have only one basis function each and the two-dimensional irreducible representations, E', E'', have two which are degenerate. Figure 10 is a powerful illustration of the symmetry properties of the hybridized plasmon modes. First, we notice the planar geometry of trimer naturally separates the in-plane and out-of-plane modes, which belong to the primed and double-primed irreducible representations, respectively. For light incident normally to the plane of the trimer, the out-of-plane modes are orthogonal to the incident electric field oscillation. Therefore, only the in-plane modes would interact with the incident light and the out-of-plane modes would not respond. Furthermore, the two in-plane modes, $A_1^{'}$, $A_2^{'}$, represent perfectly symmetric combinations of the individual dipoles resulting in zero net dipole moment. These modes are therefore dark modes which do not interact with the incident light. The two in-plane E' modes possess net dipole moments and are the ones that will absorb or scatter light. It should be reminded that the actual plasmon modes excited by an incident light are dependent on the precise excitation conditions and would in general not be the E' modes shown in Figure 10. Rather, they may be expressed as a linear combination of the E' modes. A similar analysis has also been carried out for a nanorod trimer [17].

The group theory provides a powerful and effective way to describe more complex structures and extensive studies, both theoretical and experimental, have been conducted on quadrumers [56,104-106], pentamers [20,107], and hexamers [17,18,20]. The structure that received by far the most attention was the heptamer which is constructed by adding a central particle to a hexamer. The main reason for the interest has been the Fano resonance the heptamer supports. In most systems exhibiting resonance, the lineshape function describing the resonance is typically Lorentzian. However, when a discrete state interacts strongly with a continuum, a new resonance characterized by a distinctly asymmetric lineshape function may arise, as first discovered by Fano [108]. Since the origin of Fano resonance is the interference between the discrete and continuum quantum states, it is fundamentally a quantum phenomenon. However, the interference phenomenon is commonly observed in classical optics and Fano-like resonances have been observed in many systems. The oldest example is the Wood's anomaly arising from the interference between the tangentially diffracted wave and the incident wave [109]. It has also been observed in coupled waveguide-cavity systems, photonic crystals and plasmonic nanostructures for which many reviews are available [110-112]. In addition to the rich physics it reveals, Fano resonance also has many technological applications thanks to its high sensitivity to the environmental parameters [19]. Naturally, this led to the demonstration of optical sensing based on Fano resonances [113,114]. Here we review Fano resonances in heptamers in detail.

A heptamer is composed of seven nanoparticles in a geometry with 6-fold rotational symmetry. Six peripheral nanoparticles form a hexamer where the nanoparticles are placed at the corners of a regular hexagon. One central nanoparticle is located at the center of the hexagon. The heptamer has the same point group symmetry as hexamer, D_{6h}. Since the central nanoparticle is not connected to any other nanoparticles by the symmetry operators of point group D_{6h}, the irreducible representations for a heptamer can be found by the union of those for the hexamer formed by the peripheral nanoparticles and the central nanoparticle, $\Gamma_{hep} = \Gamma_{hex} + \Gamma_{cent}$. Similarly to the trimer case, considering only the dipole modes of

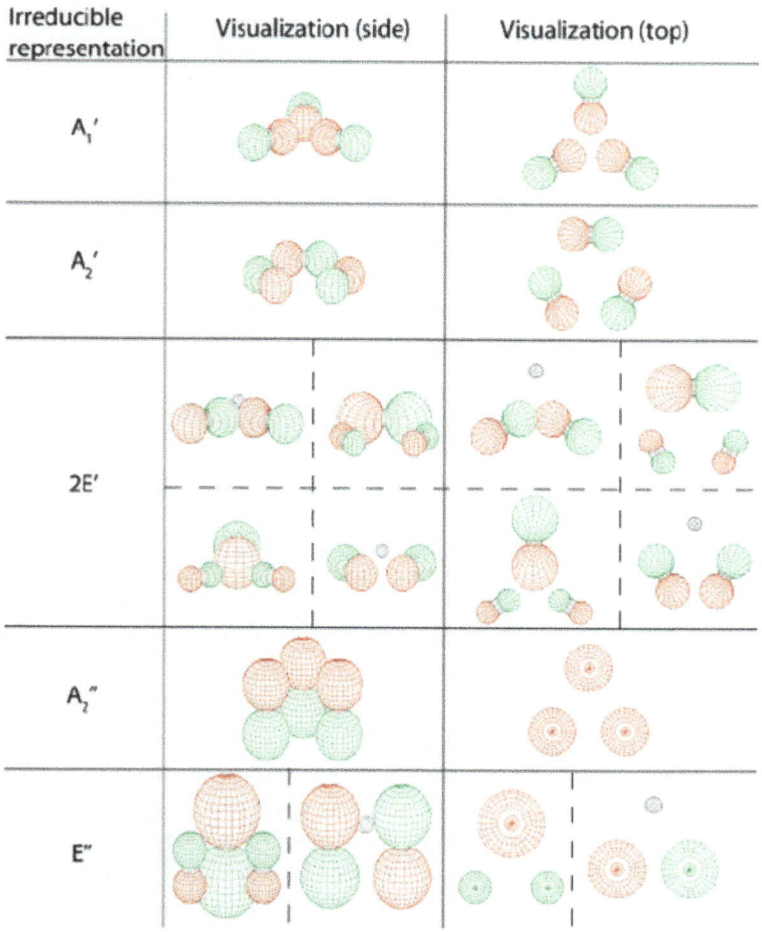

Figure 10 Irreducible representations and symmetric linear combinations of the $l = 1$ dipole modes as predicted by group theory and plasmon hybridization model (From Ref. [104]).

the individual nanoparticles, the irreducible representations of the hexamer, Γ_{hex}, are found by taking the direct product of six $l = 1$ irreducible representations, $\Gamma^{(l=1)}$, of the full rotation group and reducing it into the irreducible representations of the D_{6h} symmetry group. Likewise, the irreducible representation of the central nanoparticle is found by simply reducing $\Gamma^{(l=1)}$. The results are as follows

$$\begin{aligned} \Gamma_{hex} &= A_{1g} + A_{2g} + B_{1u} + B_{2u} + 2E_{2g} + 2E_{1u} \\ \Gamma_{cent} &= E_{1u} \\ \Gamma_{hep} &= A_{1g} + A_{2g} + B_{1u} + B_{2u} + 2E_{2g} + 3E_{1u} \end{aligned} \tag{40}$$

where we only considered the in-plane modes in these reduction schemes. As explained earlier, the out-of-plane dipoles will not interact with the normally incident light. The symmetry-adapted basis functions for these irreducible representations are shown in Figure 11(a) [115]. Since the E irreducible representations are two-dimensional, they have two basis functions each. It is clearly seen that only the E_{1u} irreducible representation exhibits a net dipole moment and will thus interact with incident light. This means all plasmon modes observed in experiments would belong to the E_{1u} mode and the mode profiles can be represented by linear combinations of the basis functions shown in Figure 11(a). To probe the plasmon modes in a heptamer, Mirin et al. calculated the absorption and scattering spectra of a silver nanoparticle heptamer with various sizes. As shown in Figure 11(b) and (c), in the smallest heptamer where the individual nanoparticles have a radius of 10 nm and are separated by 1 nm, three lowest energy peaks are observed in agreement with the plasmon hybridization model which predicts two bright modes at 340 nm and 380 nm and a dark mode at 450 nm. The bright modes are prominent in both the absorption and scattering spectra while the dark mode is weak and only visible in the absorption spectrum. Numerical simulations showed that the bright mode has a field profile indicating that the dipole moment of the central nanoparticle is aligned to that of the hexamer while the dark mode has the two dipoles anti-parallel, resulting in a small net dipole moment. As the particle sizes and gaps are increased proportionally, all modes exhibit red shifts due to the retardation

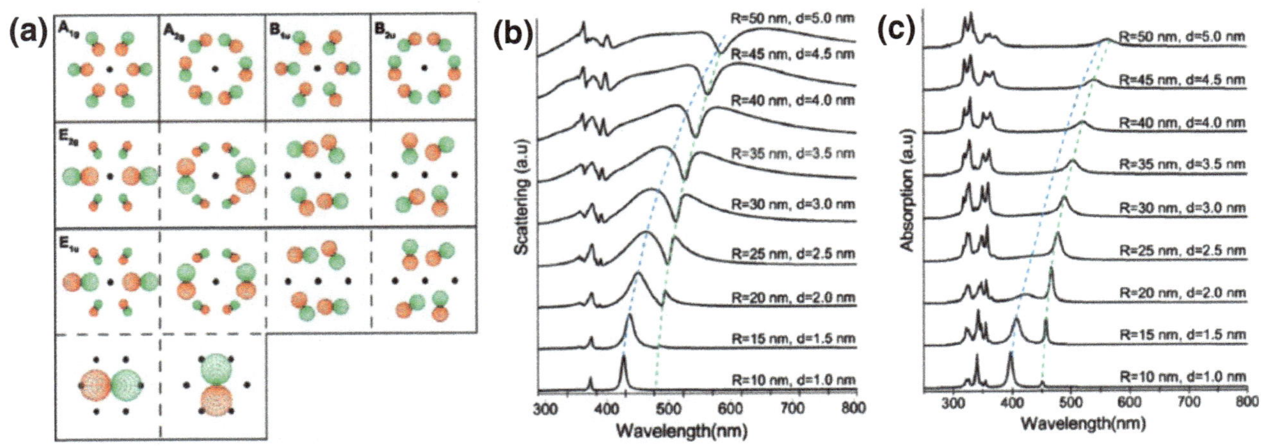

Figure 11 Silver nanoparticle heptamer. (a) Symmetry adapted basis functions of heptamer in-plane modes belonging to the D_{6h} point group. **(b)** Scattering and **(c)** absorption spectra calculated with the finite element method for heptamers with various sizes. The dashed curves show the red-shift of the dark (green) and bright (blue) modes with increasing size (From Ref. [115]).

effect and thus deviate from the results of plasmon hybridization which is valid only in the electrostatic regime. The retardation effect also broadens the peaks, particularly the bright modes, leading to a spectral overlap between the lowest bright mode and the dark mode. This results in the Fano resonance with its distinct asymmetric lineshape clearly visible in the scattering spectra for silver nanoparticle radii of 20 ~ 40 nm. For radius of 50 nm, the energies of the bright and dark modes become almost the same. In the case, the Fano resonance presents itself as an antiresonance with a symmetric dip in the scattering spectrum.

The effect of symmetry lowering on the Fano resonance is of interest to both the fundamental physics and potential applications. Cui et al. fabricated gold heptamers on a flexible poly(dimethylsiloxane) (PDMS) membrane and applied mechanical stress while monitoring the evolution of the Fano resonance [21]. As shown in Figure 12, the Fano dip exhibited a blue shift with increasing mechanical stress for the polarization along the direction of stress while it shifted to lower frequencies for the perpendicular polarization. Also, for the parallel polarization, an additional Fano dip shows up at large mechanical strain values. Figure 12 also shows scattering spectra obtained by FEM simulations which agree very well with the experimental spectra. The same behavior was also confirmed by the generalized multiparticle Mie calculations following the theory described in Section 2.2. To gain insight into the observed behavior, the eigenmodes of the heptamer structure were obtained by solving the boundary integral eigenvalue equations for the charge distribution, as described in Section 2.4. Note that this technique is valid only in the static limit and thus the obtained plasmon energies are not the same as the experimentally observed peak positions. Nevertheless, this study allows for the analysis of the eigenmodes with the group representation theory and thus correlate the symmetry of the system with that of the eigenmodes. As discussed earlier, the heptamer structure has the symmetry of point group D_{6h} and, among the irreducible representations of D_{6h}, E_{1u} is the only one with a net dipole moment and thus optically active. Furthermore, we only consider two lowest energy E_{1u} modes in the unstressed gold heptamer structure because all higher modes are masked by strong absorption by gold. In the first column of Figure 13, we show the charge distribution of the two lowest energy E_{1u} modes in an unstressed gold heptamer structure composed of seven identical gold spheres where the sphere diameter is 150 nm and gap between the spheres is 25 nm. Here the modes shown in Figure 13(a) and (b) belong to the lowest energy E_{1u} mode and (d) and (e) to the second lowest E_{1u} mode. The two-diomensional E_{1u} irreducible representation represetns a doubly degenerate mode with two orthogonal states having net dipole moment in the x and y directions, respectively. Accordingly, the four charge distributions shown in the first column of Figure 13 possess net dipole moment where (a) and (d) are x-dipoles and (b) and (e) are y-dipoles. Depending on the energy and the relative alignment of dipole moment of the center sphere to those of the six satelite spheres, the E_{1u} modes can be classified as dark or bright modes. The lower energy E_{1u} mode shown in Figure 13(a) and (b) is a dark mode where the dipole moment of center particle aligns against the dipole moments of satelite particles, making the total dipole moment small. On the other hand, the higher energy E_{1u} mode shown in Figure 13(d) and (e) is a bright mode where the dipole moments align

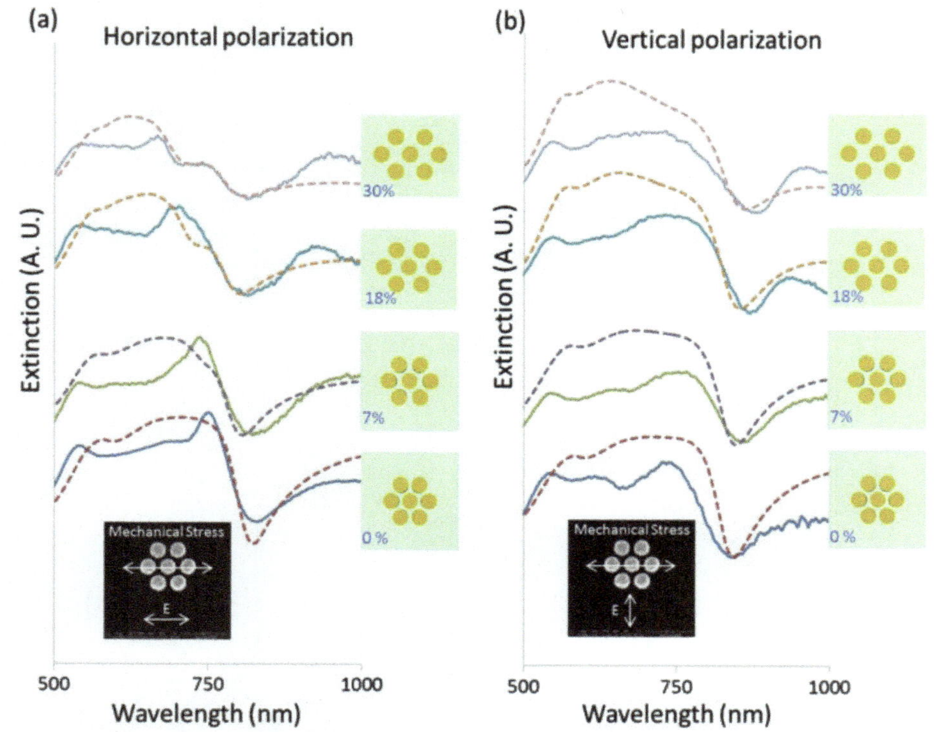

Figure 12 Mechanical tuning of gold nanoparticle heptamer. (a)-(b) show the experimentally measured and simulated extinction spectra for mechanical stress along horizontal direction. Stretching direction and polarizations are shown in the insets. Solid line shows experiment results and dashed line shows simulation results (From Ref. [21]).

together and add. The energies of these two E_{1u} modes were found to be 2.394 eV and 2.459 eV or 517.8 nm and 504.0 nm, respectively. These mode energy values would be accurate only for heptamers made of very small nanoparticles, as the boundary integral method is valid in the static limit only. For larger sizes, retardation effect will shift and broaden the modes. The resultant overlap and interference between the two modes lead to the Fano resonance. The bright mode will broaden much more significantly than the dark mode, resulting in Fano resonance which manifests itself in the form of a dip in the extinction spectrum as observed in Figure 12.

When the heptamer is subject to a uniaxial mechanical stress, the symmetry of the system is lowered to D_{2h}. The doubly degenerate E_{1u} mode splits into two non-degenerate modes belonging to B_{2u} and B_{3u} irreducible representations of the point group D_{2h}. Figure 13 shows the evolution of charge distribution as the mechanical stress is applied along the x direction. It clearly shows the original doubly degenerate modes split into x-dipole (B_{3u}) and y-dipole (B_{2u}) modes. Remarkably, the nature of the modes are preserved. That is, the bright E_{1u} mode splits into bright B_{2u} and B_{3u} modes while the dark E_{1u} mode spawns dark B_{2u} and B_{3u} modes. Also, all modes shift to shorter wavelengths with increasing mechanical strain values. However, the B_{3u} modes which have dipole moment along the direction of mechanical stress shift more than the B_{2u} modes with dipole moment perpendicular to the mechanical stress. This leads to the distinct polarization dependence observed in Figure 12 as the B_{3u} modes interact with x-polarized light and B_{2u} with y-polarized light. Therefore, as the heptamer is stretched along the x direction, x-polairzed light would show resonance features at shorter wavelengths than the y-polarized light. Even when the retardation effects are included and the resonance peaks broaden and shift, this general behavior survives and leads to the experimental observation in Figure 12: the dip in the extinction spectrum due to the Fano resonance blue shifts for polarization parallel to the direction of mechanical stress but red shifts for polarization perpendicular to it. The exact numerical simulations by FEM and generalized multiparticle Mie theory also confirmed this.

In addition to the splitting of E_{1u} modes leading to the poalrization dependence, the scattering spectra for x polarization also show an additional dip at shorter wavelengths. The additional dip is apparent in the spectra for 30% and 45% strain for x polarization but is clearly missing in all spectra for y polarization. The origin of this second dip can be found by the group theoretical analysis. Briefly reiterating, in the unstressed heptamer structure possessing D_{6h} symmetry, only the E_{1u} modes

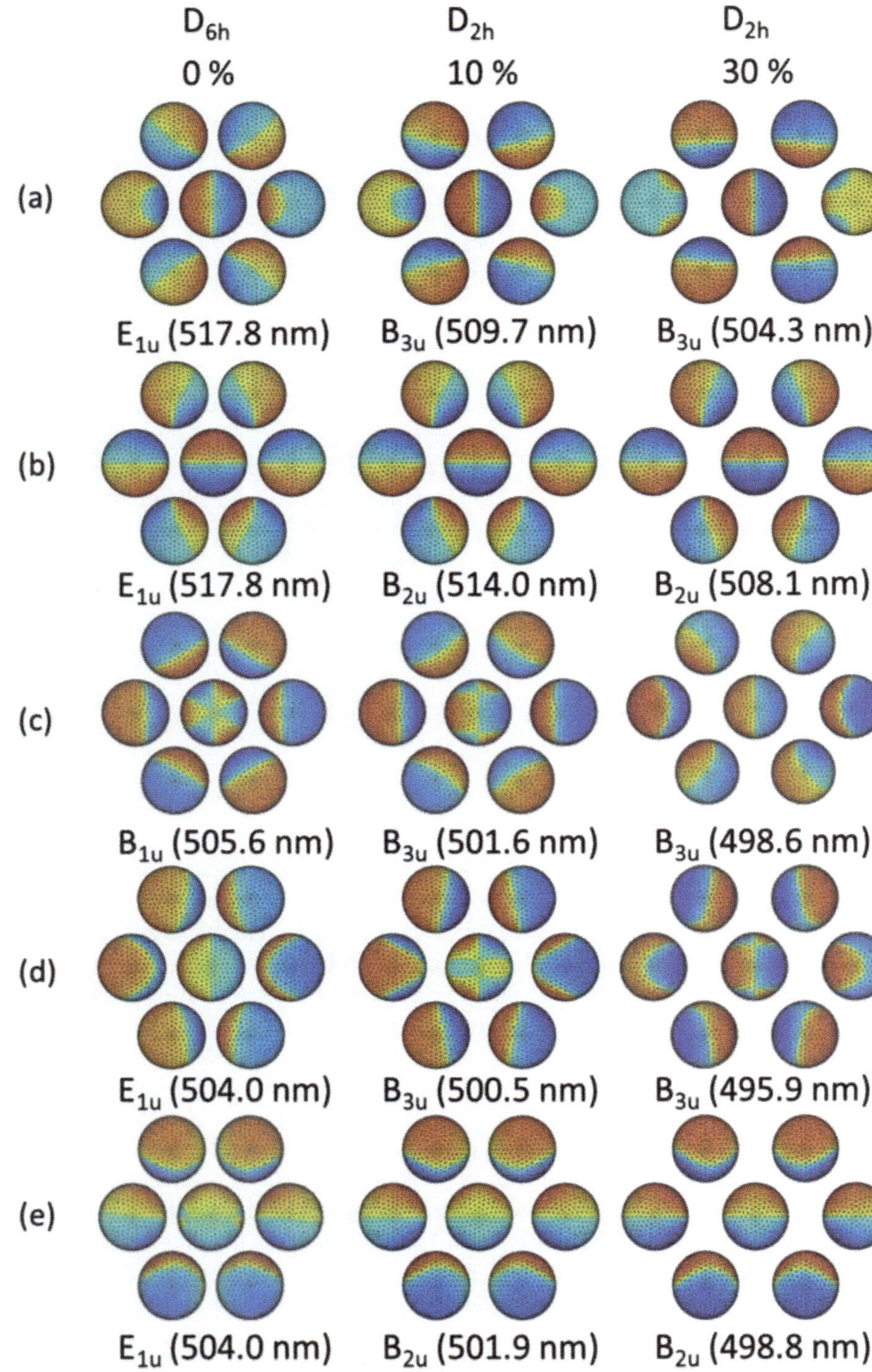

Figure 13 The evolution of charge distribution with uniaxial stress along the x direction. The first column shows the two lowest E_{1u} modes of unstressed heptamer and the B_{1u} mode. Second and third columns show B_{2u} and B_{3u} modes of the heptamer under 10% and 30% strains, respectively. The resonance wavelengths of the heptamers are also shown in paraenthesis (From Ref. [21]).

have non-zero dipole moments and thus are optically active. When the symmetry is lowered to D_{2h} by uniaxial stress, the E_{1u} modes split into B_{2u} and B_{3u} modes which interact with x- and y-polarized light, respectively. Since the unstressed heptamer has two E_{1u} modes in the frequency range of interest, we obtain two B_{2u} and two B_{3u} modes, producing Fano resonance just as in the original unstressed heptamer. However, what is missing in this narrative is that the optically inactive B_{1u} mode in the unstressed heptamer becomes optically active B_{3u} mode under uniaxial stress along the x direction. As shown in Figure 13(c), the charge distribution reveals that this B_{3u} mode is also a dark mode where the dipole moment of the center sphere aligns anti-parallel against those of the satelite spheres, thereby producing a second Fano dip in the scattering spectra. In constrast, there are no other modes of the unstressed heptamer evolving into B_{2u} mode within the frequency range we investigated and thus we do not see any additional dip for the y polarization.

Now we can take a step further and study which dipole components contribute to the formation of Fano resonance. For this, we consider nanorod heptamers composed of a circular center particle surrounded by six nanorods [22]. A circular nanoparticle supports a degenerate dipole mode with two orthogonal dipole moments having the same energy. The anisotropy of nanorod, however, lifts this degeneracy and the transverse mode has a substantially different energy than the longitudinal mode. This allows us to probe how the each dipole components hybridize separately. This design principle is schematically shown in Figure 14(a). The doubly degenerate dipole modes of the circular particle can be arranged into azimuthal dipoles and radial dipoles according to the six-fold rotational symmetry of the heptamer structure. We then replace the circular satellite particles with nanorods arranged azimuthally and radially. By designing the nanorod dimensions so that only the longitudinal modes are in the frequency range of interest, we can construct heptamers with only the azimuthal dipoles and radial dipoles. Subsequent numerical simulations revealed that the azimuthal nanorod heptamer exhibits a strong and well-defined Fano resonance (Figure 14(b)) but the radial nanorod heptamer does not support Fano resonance (Figure 14(c)). We then conducted a series of numerical simulations for a wide range of structural parameters to find that the azimuthal nanorod heptamers always support Fano resonance but the radial nanorod heptamers do not. These results indicate that the Fano resonance observed in circular heptamers are the result of hybridization of azimuthal dipole modes of the satellite particles with the center particle and that the radial dipole modes do not contribute.

The Fano resonance supported by the azimuthal nanorod heptamer was further investigated for their behavior under mechanical stress and symmetry lowering. As before, the mechanical stress was applied along the x (horizontal) direction. Figure 15 shows the simulated extinction and absorption spectra at the induced mechanical strain values of 0%, 13%, 26%, and 42%. For the polarization parallel to the mechanical stress (x direction), the Fano resonance exhibited a moderate blue shift while for the perpendicular polarization it showed a small blue shift. In addition to the spectral shifts with applied mechanical stress, the absorption spectra for the perpendicular polarization showed an additional feature at shorter wavelengths when the heptamer was under mechanical stress. This feature located near 620 nm was particularly noticeable at high strain values of 26% and 42%. The behavior of the spectral features in Figure 15 can be explained by analyzing the eigenmodes obtained by the boundary integral method. The evolution of charge distribution is dictated mainly by the symmetry reduction from D_{6h} to D_{2h} and is similar to that observed in the circular heptamer under mechanical stress described earlier. As was the case in the circular heptamer, the doubly degenerate E_{1u} mode of the azimuthal nanorod heptamer splits into two non-degenerate modes belonging to B_{2u} (y-dipole) and B_{3u} (x-dipole) irreducible representations of the point group D_{2h}. Furthermore, the bright E_{1u} mode was found to split into bright B_{2u} and B_{3u} modes and the dark E_{1u} mode into dark B_{2u} and B_{3u} modes. Both the dark and bright B_{2u} and B_{3u} modes shift to shorter

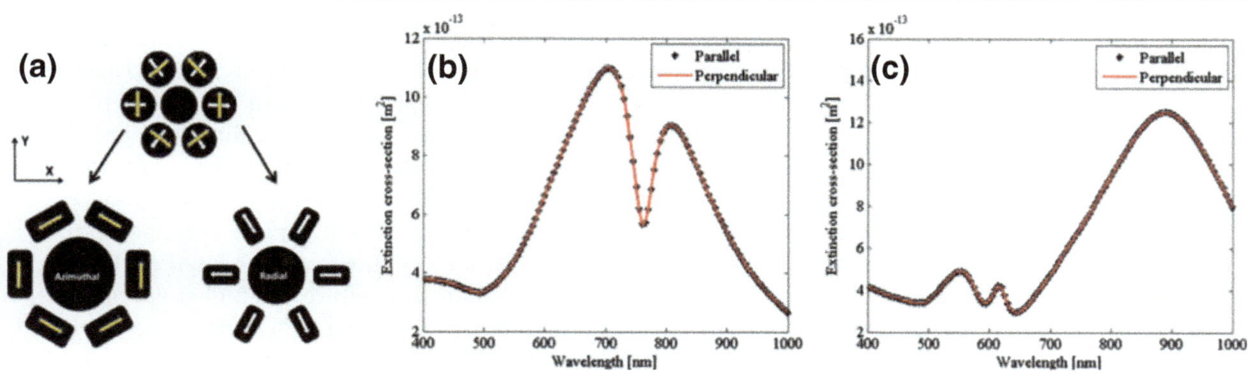

Figure 14 Gold nanorod heptamers. (a) Schematic showing the splitting of doubly degenerate modes in circular heptamer to azimuthal and radial modes in nanorod heptamer. The white colored arrows represent the radial modes and gold colored arrows represent azimuthal modes. Simulated extinction spectra for **(b)** azimuthal and **(c)** radial nanorod heptamers for the same structural parameters: individual nanorods have a length of 95 nm, width of 50 nm and thickness of 40 nm. The center particle has a diameter of 147 nm and gold thickness of 40 nm. The gaps between the center particle and the nanorods are 22 nm (From Ref. [22]).

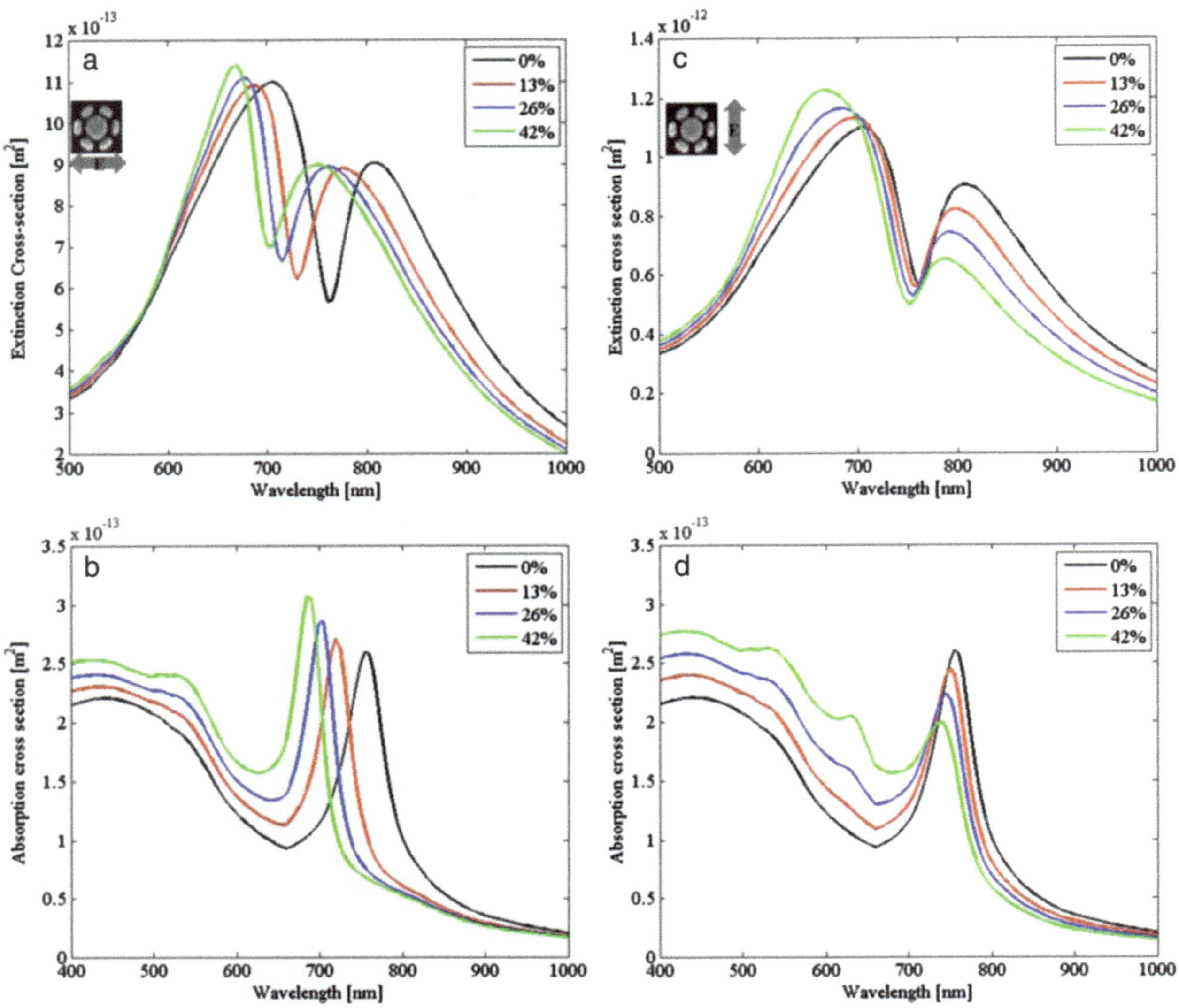

Figure 15 Simulated extinction and absorption spectra for the azimuthal nanorod heptamer for different induced mechanical strain values of 0%, 13%, 26% and 42%. (a) and **(c)** plot the extinction spectra for parallel and perpendicular polarizations (orientations in inset) respectively while **(b)** and **(d)** plot the absorption spectra for parallel and perpendicular polarizations respectively (From Ref. [22]).

wavelengths with increasing mechanical strain values. However, the B_{3u} modes, which have dipole moment along the direction of mechanical stress, shifts more than the B_{2u} modes whose dipole moment is perpendicular to the mechanical stress, leading to the polarization dependence observed in the extinction and absorption spectra in Figure 15. Once again, the retardation effect results in overlap and interference between the dark and bright modes thus causing the Fano resonance to be observed as a dip in the extinction spectrum.

The additional absorption peak observed under stress can be explained by noting that the optically inactive B_{2u} mode in the unstressed heptamer becomes an optically active B_{2u} mode under uniaxial stress along the x direction. Here we have an unfortunate coincidence of having the same label for two very different modes. The B_{2u} irreducible representation in D_{6h} symmetry has no net dipole moment and thus represents an optically inactive mode. However, the B_{2u} irreducible representation in D_{2h} symmetry has a finite dipole moment along the y-direction and thus represents an optically active mode that can interact with y-polarized light. The energy of this B_{2u} mode is higher than the lowest dark B_{2u} mode originating from the dark E_{1u} mode of unstressed heptamer, thereby producing an additional feature at a higher energy. The azimuthal nanorod heptamer differs from the circular heptamer in the behavior of this optically active mode. In the circular heptamer, the optically inactive B_{1u} mode was located between the two lowest energy E_{1u} modes in the unstressed heptamer and became optically active B_{3u} mode under uniaxial stress along the x direction. As a result, in the circular heptamer, the additional mode was observed for x-polarized light. The difference between the circular heptamer and azimuthal nanorod heptamer is readily understood by comparing the eigenmode charge distributions. In the B_{1u} mode observed in the unstressed circular heptamer, the satellite particles have their dipole moments aligned along the radial direction (Figure 13c). In the azimuthal nanorod heptamer, this configuration would have had a much higher energy because the radial dimensions of the nanorods are much smaller than the circular particles used in circular heptamer. On the other hand, the B_{2u} mode in which the satellite particles have their dipole moments aligned along the azimuthal direction has a lower energy because the nanorods have larger dimensions along that direction. For these reasons, we find the B_{2u} mode is located between the two lowest energy E_{1u} modes in the azimuthal nanorod heptamer structure. These behaviors predicted by the theoretical modeling studies have been confirmed by experiments on gold nanorod heptamers fabricated by electron-beam lithography [22].

3 Conclusion and outlook

Plasmonics in mutually interacting nanostructures is an exciting field with rich physics and numerous potential applications. The recent progress clearly demonstrates that plasmonic nanostructures allow us to precisely control the nanoscale optical modes and how they interact with one another. These capabilities make the coupled plasmonic nanostructures an ideal platform to study nanoscale optical phenomena and build optical devices with novel functionalities. Though fundamentally a quantum mechanical entity, surface plasmons can be described well by classical electrodynamics in most cases. The theoretical framework presented in this review provides a firm foundation for designing interesting structures and analyzing their properties. Also, the strong field enhancement near the metal surface and the complex field profiles resulting from the interaction among the nanostructures makes the optical properties highly sensitive to the environment and consequently leads to exotic properties such as perfect absorption, invisibility and Fano resonance.

Some of the latest developments in the coupled plasmonic systems deal specifically with the quantum mechanical nature of the phenomenon. In surface enhanced Raman scattering and many other applications, it is generally advantageous to achieve high local field strength. In the simplest geometry of a dimer, a smaller gap supports higher field. However, when the two nanoparticles in a dimer are nearly touching, the electrons may tunnel through the gap, resulting in a dramatic decrease of local field strength and shift of plasmon energy [26,116]. Furthermore, at such a small length scale, the abrupt interface between two materials is unrealistic and the non-locality of the dielectric function has to be considered [117]. These considerations have led to the development of a quantum-corrected model that could correctly reproduce the fully quantum mechanical calculations [118]. Unveiling the quantum mechanical nature of the gap plasmons in extremely small gaps remains an active research field with hopes of discovering new physics and also with potential for developing novel quantum devices.

Another interesting subject of fundamentally quantum nature is the coupling between plasmon and atomic or molecular exciton. The interaction between plasmon and exciton is of great fundamental interest with strong technological implications in organic electronics and photovoltaics. It has been shown that the surface plasmon can be used to control the relaxation pathways of molecular excitons [119]. Furthermore, in a fashion similar to plasmon hybridization, atomic or molecular exciton can hybridize with plasmon, forming a new quantum mechanically entity sometimes called plexciton [120,121]. The strong coupling between plasmon and exciton has recently been observed in an organic photovoltaic material [122]. The strong interaction could result in dramatic changes in excitation and relaxation dynamics of the molecular excitons, opening new opportunities in controlling and engineering the optical processes in organic materials.

In closing, surface plasmons provide a powerful means to gain access to the near field components that could result in a wide range of unconventional properties. Controlling and manipulating the mutual interactions among the plasmonic nanostructures offer design flexibility and engineering freedom unparalleled by any other photonics technologies. Obtaining high quality materials with low loss in precise nanoscale geometry will continue to be a challenge but the remarkable progress in nanofabrication technologies in the past decades makes it a perfect time to invest in research on the plasmonic nanostructures.

References

1. DK Gramotnev, SI Bozhevolnyi, plasmonics beyond the diffraction limit. Nat. Photon. **4**(2), 83–91 (2010)
2. J-S Lee, MS Han, CA Mirkin, Colorimetric Detection of Mercuric Ion (Hg2+) in Aqueous Media using DNA-Functionalized Gold Nanoparticles. Angew. Chem. **119**(22), 4171–4174 (2007)
3. JJ Storhoff, R Elghanian, RC Mucic, CA Mirkin, RL Letsinger, One-pot colorimetric differentiation of polynucleotides with single base imperfections using gold nanoparticle probes. J. Am. Chem. Soc. **120**(9), 1959–1964 (1998)
4. J Chen, F Saeki, BJ Wiley, H Cang, MJ Cobb, Z-Y Li, L Au, H Zhang, MB Kimmey, Li, Y Xia, Gold Nanocages: Bioconjugation and Their Potential Use as Optical Imaging Contrast Agents. Nano Lett. **5**(3), 473–477 (2005)
5. EB Dickerson, EC Dreaden, X Huang, IH El-Sayed, H Chu, S Pushpanketh, JF McDonald, MA El-Sayed, Gold nanorod assisted near-infrared plasmonic photothermal therapy (PPTT) of squamous cell carcinoma in mice. Cancer Lett. **269**(1), 57–66 (2008)
6. KY Lin, AF Bagley, AY Zhang, DL Karl, SS Yoon, SN Bhatia, Gold nanorod photothermal therapy in a genetically engineered mouse model of soft tissue sarcoma. Nano LIFE **1**(3), 277–287 (2010)
7. S Nie, SR Emory, Probing single molecules and single nanoparticles by surfaceenhanced Raman scattering. Science **275**(5303), 1102–1106 (1997)

8. AM Michaels, M Nirmal, LE Brus, Surface Enhanced Raman Spectroscopy of Individual Rhodamine 6G Molecules on Large Ag Nanocrystals. J. Am. Chem. Soc. **121**(43), 9932–9939 (1999)
9. JR Lakowicz, Radiative decay engineering 5: metal-enhanced fluorescence and plasmon emission. Anal. Biochem. **337**(2), 171–194 (2005)
10. P Andrew, Energy Transfer Across a Metal Film Mediated by Surface Plasmon Polaritons. Science **306**(5698), 1002–1005 (2004)
11. M Lunz, VA Gerard, YK Gun'ko, V Lesnyak, N Gaponik, AS Susha, AL Rogach, AL Bradley, Surface Plasmon Enhanced Energy Transfer between Donor and Acceptor CdTe Nanocrystal Quantum Dot Monolayers. Nano Lett. **11**(8), 3341–3345 (2011)
12. L Novotny, N van Hulst, Antennas for light. Nat. Photon. **5**(2), 83–90 (2011). 54
13. E Prodan, C Radloff, NJ Halas, P Nordlander, A hybridization model for the plasmon response of complex nanostructures. Science **302**(5644), 419–422 (2003)
14. P Nordlander, C Oubre, E Prodan, K Li, MI Stockman, Plasmon hybridization in nanoparticle dimers. Nano Lett. **4**(5), 899–903 (2004)
15. VK Lin, SL Teo, R Marty, A Arbouet, C Girard, E Alarcon-Llado, SH Liu, MY Han, S Tripathy, A Mlayah, Dual wavelength sensing based on interacting gold nanodisk trimers. Nanotechnology **21**(30), 305501 (2010)
16. AJ Mastroianni, SA Claridge, AP Alivisatos, Pyramidal and chiral groupings of gold nanocrystals assembled using DNA scaffolds. J. Am. Chem. Soc. **131**(24), 8455–8459 (2009)
17. DE Gómez, KC Vernon, TJ Davis, Symmetry effects on the optical coupling between plasmonic nanoparticles with applications to hierarchical structures. Phys. Rev. B **81**(7), 075414 (2010)
18. M Hentschel, M Saliba, R Vogelgesang, H Giessen, AP Alivisatos, N Liu, Transition from Isolated to Collective Modes in Plasmonic Oligomers. Nano Lett. **10**(7), 2721–2726 (2010)
19. JB Lassiter, H Sobhani, JA Fan, J Kundu, F Capasso, P Nordlander, NJ Halas, Fano Resonances in Plasmonic Nanoclusters: Geometrical and Chemical Tunability. Nano Lett. **10**(8), 3184–3189 (2010)
20. K Bao, NA Mirin, P Nordlander, Fano resonances in planar silver nanosphere clusters. Appl. Phys. A **100**(2), 333–339 (2010)
21. Y Cui, J Zhou, VA Tamma, W Park, Dynamic Tuning and Symmetry Lowering of Fano Resonance in Plasmonic Nanostructure. ACS Nano **6**(3), 2385–2393 (2012)
22. VA Tamma, Y Cui, J Zhou, W Park, Nanorod orientation dependence of tunable Fano resonance in plasmonic nanorod heptamers. Nanoscale **5**(4), 1592–1602 (2013)
23. M Tinkham, Group Theory and Quantum Mechanics (Dover Publications, Mineola, New York, 2003)
24. F Flores, F Garcia-Moliner, Model-independent theory of surface plasmons. Solid State Commun. **11**(9), 1295–1298 (1972)
25. G Barton, Some surface effects in the hydrodynamic model of metals. Rep. Prog. Phys. **42**(6), 963 (1979)
26. J Zuloaga, E Prodan, P Nordlander, Quantum Description of the Plasmon Resonances of a Nanoparticle Dimer. Nano Lett. **9**(2), 887–891 (2009)
27. Z Jacob, VM Shalaev, Plasmonics Goes Quantum. Science **334**(6055), 463–464 (2011)
28. DJ Bergman, MI Stockman, Surface plasmon amplification by stimulated emission of radiation: quantum generation of coherent surface plasmons in nanosystems. Phys. Rev. Lett. **90**(2), 027402 (2003)
29. A Archambault, F Marquier, J-J Greffet, C Arnold, Quantum theory of spontaneous and stimulated emission of surface plasmons. Phys. Rev. B **82**(3), 035411 (2010)
30. RH Ritchie, HB Eldridge, Optical emission from irradiated foils. I. Phys. Rev. **126**(6), 1935 (1962)
31. J Dionne, L Sweatlock, H Atwater, A Polman, Planar metal plasmon waveguides: frequency-dependent dispersion, propagation, localization, and loss beyond the free electron model. Phys. Rev. B **72**(7), 075405 (2005)
32. J Dionne, L Sweatlock, H Atwater, A Polman, Plasmon slot waveguides: Towards chip-scale propagation with subwavelength-scale localization. Phys. Rev. B **73**(3), 035407 (2006). 55
33. R Zia, M Selker, M Brongersma, Leaky and bound modes of surface plasmon waveguides. Phys. Rev. B **71**(16), 165431 (2005)
34. T Holmgaard, S Bozhevolnyi, Theoretical analysis of dielectric-loaded surface plasmonpolariton waveguides. Phys. Rev. B **75**(24), 245405 (2007)
35. G Mie, Beiträge zur Optik trüber Medien, speziell kolloidaler Metallösungen. Ann. Phys. **330**(3), 377–445 (1908)
36. YL Xu, Electromagnetic scattering by an aggregate of spheres. Appl. Opt. **34**(21), 4573–4588 (1995)
37. Y-L Xu, Electromagnetic scattering by an aggregate of spheres: far field. Appl. Opt. **36**(36), 9496–9508 (1997)
38. Y-L Xu, G BÅ, Comparison between multisphere light-scattering calculations: rigorous solution and discrete-dipole approximation. Astrophys. J. **513**(2), 894 (1999)
39. CF Bohren, DR Huffman, Absorption and Scattering of Light by Small Particles (JWILEY-VCH Verlag GmbH & Co. KGaA, Weinheim, 2008)
40. N Stefanou, V Karathanos, A Modinos, Scattering of electromagnetic waves by periodic structures. J. Phys. Condens. Matter **4**(36), 7389 (1992)
41. X Wang, XG Zhang, Q Yu, B Harmon, Multiple-scattering theory for electromagnetic waves. Phys. Rev. B **47**(8), 4161–4167 (1993)
42. A Moroz, Density-of-states calculations and multiple-scattering theory for photons. Phys. Rev. B **51**(4), 2068 (1995)
43. N Stefanou, A Modinos, Scattering of light from a two-dimensional array of spherical particles on a substrate. J. Phys. Condens. Matter **3**(41), 8135 (1991)
44. N Stefanou, V Yannopapas, A Modinos, Heterostructures of photonic crystals: frequency bands and transmission coefficients. Comput. Phys. Commun. **113**(1), 49–77 (1998)
45. JB Pendry, Photonic Band Structures. J. Mod. Opt. **41**(2), 209–229 (1994)
46. JCM Garnett, Colours in Metal Glasses and in Metallic Films. Philos. Trans. R. Soc. A Math. Phys. Eng. Sci. **203**(359), 385–420 (1904)
47. WT Doyle, Optical properties of a suspension of metal spheres. Phys. Rev. B **39**(14), 9852 (1989)
48. V Yannopapas, Negative refraction in random photonic alloys of polaritonic and plasmonic microspheres. Phys. Rev. B **75**(3), 035112 (2007)
49. VA Tamma, JH Lee, Q Wu, W Park, Visible frequency magnetic activity in silver nanocluster metamaterial. Appl. Opt. **49**(7), A11–A17 (2010)
50. CF Bohren, Applicability of effective-medium theories to problems of scattering and absorption by nonhomogeneous atmospheric particles. J. Atmos. Sci. **43**, 468–475 (1986)
51. S Yoo, Q-H Park, Effective permittivity for resonant plasmonic nanoparticle systems via dressed polarizability. Opt. Express **20**(15), 16480–16489 (2012)
52. TC Choy, Effective Medium Theory (Oxford University Press, New York, 1999)
53. P Berini, Plasmon-polariton waves guided by thin lossy metal films of finite width: Bound modes of symmetric structures. Phys. Rev. B **61**(15), 10484 (2000)
54. P Berini, Plasmon-polariton waves guided by thin lossy metal films of finite width: Bound modes of asymmetric structures. Phys. Rev. B **63**(12), 125417 (2001)
55. H Wang, DW Brandl, P Nordlander, NJ Halas, Plasmonic Nanostructures: Artificial Molecules. Acc. Chem. Res. **40**(1), 53–62 (2007)
56. JA Fan, K Bao, C Wu, J Bao, R Bardhan, NJ Halas, VN Manoharan, G Shvets, P Nordlander, F Capasso, Fano-like Interference in Self-Assembled Plasmonic Quadrumer 56 Clusters. Nano Lett. **10**(11), 4680–4685 (2010)
57. I Mayergoyz, D Fredkin, Z Zhang, Electrostatic (plasmon) resonances in nanoparticles. Phys. Rev. B **72**(15), 155412 (2005)
58. VM Shalaev, Optical negative-index metamaterials. Nat. Photon. **1**(1), 41–48 (2007)
59. W Park, J Kim, Negative-Index Materials: Optics by Design. MRS Bull. **33**(10), 907–914 (2008)
60. Z-P Yang, L Ci, JA Bur, S-Y Lin, PM Ajayan, Experimental Observation of an Extremely Dark Material Made By a Low-Density Nanotube Array. Nano Lett. **8**(2), 446–451 (2008)
61. N Landy, S Sajuyigbe, J Mock, D Smith, W Padilla, Perfect Metamaterial Absorber. Phys. Rev. Lett. **100**(20), 207402 (2008)
62. D Lu, E Rengnath, Y Cui, Z Wang, Y Ding, W Park, Interaction of two plasmon modes in the organic photovoltaic devices with patterned back-electrode. Appl. Phys. Lett. **102**(24), 241114 (2013)
63. V Kravets, F Schedin, A Grigorenko, Plasmonic blackbody: Almost complete absorption of light in nanostructured metallic coatings. Phys. Rev. B **78**(20), 205405 (2008)
64. JD Jackson, Classical Electrodynamics (Wiley, New York, 1998)
65. FJ García-Vidal, JM Pitarke, JB Pendry, Effective medium theory of the optical properties of aligned carbon nanotubes. Phys. Rev. Lett. **78**(22), 4289 (1997)
66. VG Kravets, S Neubeck, AN Grigorenko, AF Kravets, Plasmonic blackbody: Strong absorption of light by metal nanoparticles embedded in a dielectric matrix. Phys. Rev. B **81**(16), 165401 (2010)
67. EFC Driessen, MJA de Dood, The perfect absorber. Appl. Phys. Lett. **94**(17), 171109 (2009)
68. C Hägglund, SP Apell, B Kasemo, Maximized Optical Absorption in Ultrathin Films and Its Application to Plasmon-Based Two-Dimensional Photovoltaics. Nano Lett. **10**(8), 3135–3141 (2010)

69. C Hägglund, SP Apell, Resource efficient plasmon-based 2D-photovoltaics with reflective support. Opt. Express **18**(103), 343–356 (2010)
70. C Hägglund, G Zeltzer, R Ruiz, I Thomann, H-B-R Lee, ML Brongersma, SF Bent, Self-Assembly Based Plasmonic Arrays Tuned by Atomic Layer Deposition for Extreme Visible Light Absorption. Nano Lett. **13**(7), 3352–3357 (2013)
71. M Svedendahl, P Johansson, M Käll, Complete Light Annihilation in an Ultrathin Layer of Gold Nanoparticles. Nano Lett. **13**(7), 3053–3058 (2013)
72. A Alù, N Engheta, Achieving transparency with plasmonic and metamaterial coatings. Phys. Rev. E. **72**(1), 016623 (2005)
73. A Alù, N Engheta, Multifrequency Optical Invisibility Cloak with Layered Plasmonic Shells. Phys. Rev. Lett. **100**(11), 113901 (2008)
74. A Alù, N Engheta, Cloaking a Sensor. Phys. Rev. Lett. **102**(23), 233901 (2009)
75. A Alù, N Engheta, Cloaked Near-Field Scanning Optical Microscope Tip for Noninvasive Near-Field Imaging. Phys. Rev. Lett. **105**(26), 263906 (2010)
76. F Bilotti, S Tricarico, F Pierini, L Vegni, Cloaking apertureless near-field scanning optical microscopy tips. Opt. Lett. **36**(2), 211–213 (2011)
77. S Tricarico, F Bilotti, L Vegni, Reduction of optical forces exerted on nanoparticles covered by scattering cancellation based plasmonic cloaks. Phys. Rev. B **82**(4), 045109 (2010)
78. VA Tamma, Y Cui, W Park, Scattering reduction at near-infrared frequencies using plasmonic nanostructures. Opt. Express **21**(1), 1041–1056 (2013). 57
79. DE Aspnes, Local-field effects and effective-medium theory: A microscopic perspective. Am. J. Phys. **50**, 704 (1982)
80. S Mühlig, A Cunningham, J Dintinger, M Farhat, SB Hasan, T Scharf, T Bürgi, F Lederer, C Rockstuhl, A self-assembled three-dimensional cloak in the visible. Sci. Rep. **3**, 2328 (2013)
81. KL Young, MB Ross, MG Blaber, M Rycenga, Using DNA to Design Plasmonic Metamaterials with Tunable Optical Properties. Adv. Mater. **26**(4), 653–659 (2014)
82. SY Park, AKR Lytton-Jean, B Lee, S Weigand, GC Schatz, CA Mirkin, DNAprogrammable nanoparticle crystallization. Nature **451**(7178), 553–556 (2008)
83. D Nykypanchuk, MM Maye, D van der Lelie, O Gang, DNA-guided crystallization of colloidal nanoparticles. Nature **451**(7178), 549–552 (2008)
84. CE Talley, JB Jackson, C Oubre, NK Grady, CW Hollars, SM Lane, TR Huser, P Nordlander, NJ Halas, Surface-Enhanced Raman Scattering from Individual Au Nanoparticles and Nanoparticle Dimer Substrates. Nano Lett. **5**(8), 1569–1574 (2005)
85. P Muhlschlegel, HJ Eisler, OJF Martin, B Hecht, DW Pohl, Resonant optical antennas. Science **308**(5728), 1607–1609 (2005)
86. T Atay, J-H Song, AV Nurmikko, Strongly Interacting Plasmon Nanoparticle Pairs: From Dipole–Dipole Interaction to Conductively Coupled Regime. Nano Lett. **4**(9), 1627–1631 (2004)
87. KH Su, QH Wei, X Zhang, JJ Mock, DR Smith, S Schultz, Interparticle Coupling Effects on Plasmon Resonances of Nanogold Particles. Nano Lett. **3**(8), 1087–1090 (2003)
88. W Rechberger, A Hohenau, A Leitner, JR Krenn, B Lamprecht, FR Aussenegg, Optical properties of two interacting gold nanoparticles. Opt. Commun. **220**(1), 137–141 (2003)
89. L Gunnarsson, T Rindzevicius, J Prikulis, B Kasemo, M Käll, S Zou, GC Schatz, Confined Plasmons in Nanofabricated Single Silver Particle Pairs: Experimental Observations of Strong Interparticle Interactions. J. Phys. Chem. B **109**(3), 1079–1087 (2005)
90. P Biagioni, J-S Huang, B Hecht, Nanoantennas for visible and infrared radiation. Rep. Prog. Phys. **75**(2), 024402 (2012)
91. A Taflove, SG Johnson, A Oskooi, Advances in FDTD Computational Electrodynamics (Artech House, Norwood, MA, 2013)
92. AC Polycarpou, Introduction to the Finite Element Method in Electromagnetics (Morgan & Claypool Publishers, 2006).
93. U Kreibig, M Vollmer, Optical Properties of Metal Clusters (Springer Verlag, Berlin, 2010)
94. M Gluodenis, CA Foss, The Effect of Mutual Orientation on the Spectra of Metal Nanoparticle Rod–Rod and Rod–Sphere Pairs. J. Phys. Chem. B **106**(37), 9484–9489 (2002)
95. PK Jain, W Huang, MA El-Sayed, On the Universal Scaling Behavior of the Distance Decay of Plasmon Coupling in Metal Nanoparticle Pairs: A Plasmon Ruler Equation. Nano Lett. **7**(7), 2080–2088 (2007)
96. PK Jain, MA El-Sayed, Universal Scaling of Plasmon Coupling in Metal Nanostructures: Extension from Particle Pairs to Nanoshells. Nano Lett. **7**(9), 2854–2858 (2007)
97. PK Jain, MA El-Sayed, Surface Plasmon Coupling and Its Universal Size Scaling in Metal Nanostructures of Complex Geometry: Elongated Particle Pairs and Nanosphere 58 Trimers. J. Phys. Chem. C **112**(13), 4954–4960 (2008)
98. VA Podolskiy, AK Sarychev, VM Shalaev, Plasmon modes in metal nanowires and left-handed materials. J. Nonlinear Optic. Phys. Mat. **11**(1), 65–74 (2002)
99. VA Podolskiy, AK Sarychev, VM Shalaev, Plasmon modes and negative refraction in metal nanowire composites. Opt. Express **11**(7), 735–745 (2003)
100. VM Shalaev, W Cai, UK Chettiar, HK Yuan, AK Sarychev, VP Drachev, AV Kildishev, Negative index of refraction in optical metamaterials. Opt. Lett. **30**(24), 3356–3358 (2005)
101. S Zhang, W Fan, N Panoiu, K Malloy, R Osgood, S Brueck, Experimental Demonstration of Near-Infrared Negative-Index Metamaterials. Phys. Rev. Lett. **95**(13), 137404 (2005)
102. G Dolling, C Enkrich, M Wegener, CM Soukoulis, S Linden, Simultaneous negative phase and group velocity of light in a metamaterial. Science **312**(5775), 892–894 (2006)
103. AM Funston, C Novo, TJ Davis, P Mulvaney, Plasmon Coupling of Gold Nanorods at Short Distances and in Different Geometries. Nano Lett. **9**(4), 1651–1658 (2009)
104. DW Brandl, NA Mirin, P Nordlander, Plasmon Modes of Nanosphere Trimers and Quadrumers. J. Phys. Chem. B **110**(25), 12302–12310 (2006)
105. Z Zhang, A Weber-Bargioni, SW Wu, S Dhuey, S Cabrini, PJ Schuck, Manipulating nanoscale light fields with the asymmetric bowtie nano-colorsorter. Nano Lett. **9**(12), 4505–4509 (2009)
106. M Rahmani, T Tahmasebi, Y Lin, B Lukiyanchuk, TYF Liew, MH Hong, Influence of plasmon destructive interferences on optical properties of gold planar quadrumers. Nanotechnology **22**(24), 245204 (2011)
107. SJ Barrow, X Wei, JS Baldauf, AM Funston, P Mulvaney, The surface plasmon modes of self-assembled gold nanocrystals. Nat. Commun. **3**, 1275 (2012)
108. U. Fano, "On the absorption spectrum of noble gases at the arc spectrum limit," *arXiv.org*, vol. cond-mat.other. 08-Feb-2005 http://arxiv.org/abs/cond-mat/0502210.
109. A Hessel, AA Oliner, A new theory of Wood's anomalies on optical gratings. Appl. Opt. **4**(10), 1275–1297 (1965)
110. B luk'yanchuk, NI Zheludev, SA Maier, NJ Halas, P Nordlander, H Giessen, CT Chong, The Fano resonance in plasmonic nanostructures and metamaterials. Nat. Mater. **9**(9), 707–715 (2010)
111. A Miroshnichenko, S Flach, Y Kivshar, Fano resonances in nanoscale structures. Rev. Mod. Phys. **82**(3), 2257–2298 (2010)
112. Y Francescato, V Giannini, SA Maier, Plasmonic Systems Unveiled by Fano Resonances. ACS Nano **6**(2), 1830–1838 (2012)
113. N Liu, T Weiss, M Mesch, L Langguth, U Eigenthaler, M Hirscher, C Sönnichsen, H Giessen, Planar Metamaterial Analogue of Electromagnetically Induced Transparency for Plasmonic Sensing. Nano Lett. **10**(4), 1103–1107 (2010)
114. C Wu, AB Khanikaev, R Adato, N Arju, AA Yanik, H Altug, G Shvets, Fanoresonant asymmetric metamaterials for ultrasensitive spectroscopy and identification of molecular monolayers. Nat. Mater. **11**(1), 69–75 (2012)
115. NA Mirin, K Bao, P Nordlander, Fano Resonances in Plasmonic Nanoparticle Aggregates †. J. Phys. Chem. A **113**(16), 4028–4034 (2009)
116. J Zuloaga, E Prodan, P Nordlander, Quantum Plasmonics: Optical Properties and Tunability of Metallic Nanorods. ACS Nano **4**(9), 5269–5276 (2010)
117. FJ García de Abajo, Nonlocal Effects in the Plasmons of Strongly Interacting Nanoparticles, Dimers, and Waveguides. J. Phys. Chem. C **112**(46), 17983–59–17987 (2008)
118. R Esteban, AG Borisov, P Nordlander, J Aizpurua, Bridging quantum and classical plasmonics with a quantum-corrected model. Nat. Commun. **3**, 825 (2012)
119. G Wiederrecht, J Hall, A Bouhelier, Control of Molecular Energy Redistribution Pathways via Surface Plasmon Gating. Phys. Rev. Lett. **98**(8), 083001 (2007)
120. NT Fofang, NK Grady, Z Fan, AO Govorov, NJ Halas, Plexciton Dynamics: Exciton–Plasmon Coupling in a J-Aggregate–Au Nanoshell Complex Provides a Mechanism for Nonlinearity. Nano Lett. **11**(4), 1556–1560 (2011)
121. A Manjavacas, FG de Abajo, P Nordlander, Quantum plexcitonics: strongly interacting plasmons and excitons. Nano Lett. **11**(6), 2318–2323 (2011)
122. N Azarova, AJ Ferguson J, van de Lagemaat, E Rengnath, W Park, JC Johnson, Coupling between a Molecular Charge-Transfer Exciton and Surface Plasmons in a Nanostructured Metal Grating, J. Phys. Chem. Lett 2658–2663 (2013).

2

Highly enhanced optical properties of indocyanine green/perfluorocarbon nanoemulsions for efficient lymph node mapping using near-infrared and magnetic resonance imaging

Pan Kee Bae, Juyeon Jung and Bong Hyun Chung*

Abstract

The near-infrared (NIR) fluorescence probe has better tissue penetration and lower autofluorescence. Indocyanine green (ICG) is an NIR organic dye for extensive biological application, and it has been clinically approved for human medical imaging and diagnosis. However, application of this dye is limited by its numerous disadvantageous properties in aqueous solution, including its concentration-dependent aggregation, poor aqueous stability in vitro, and low quantum yield. Its use in molecular imaging probes is limited because it loses fluorescence after binding to nonspecific plasma proteins, leading to rapid elimination from the body with a half-life of 2 – 4 min. In this study, the multifunctional perfluorocarbon (PFC)/ICG nanoemulsions were investigated with the aim of overcoming these limitations. The PFC/ICG nanoemulsions as a new type of delivery vehicle for contrast agents have both NIR optical imaging and ^{19}F-MR imaging moieties. These nanoemulsions exhibited less aggregation, increased fluorescence intensity, long-term stability, and physicochemical stability against external light and temperature compared to free aqueous ICG. Also, the PFC/ICG bimodal nanoemulsions allow excellent detection of lymph nodes in vivo through NIR optical imaging and ^{19}F-MR imaging. This result showed the suitability of the proposed nanoemulsions for non-invasive lymph node mapping as they enable long-time detection of lymph nodes.

Keywords: NIR optical imaging; Indocyanine green; bimodal imaging; ^{19}F-MR imaging; Lymph node mapping

1 Background

Medical imaging has experienced explosive growth over the past few decades and now plays a central role in non-invasive imaging technologies for biomedical research. Multimodal imaging will allow clinicians not only to see where a tumor is located in the body but also to visualize the expression and activity of specific molecules and biological processes that influence tumor behaviour and/or its response to therapy. The multimodal imaging probes with optical imaging dyes and magnetic resonance (MR) imaging contrast agents have been exploited for targeted molecular imaging, disease diagnosis, and in vivo animal studies [1,2]. Targeted magnetic resonance imaging (MRI) has emerged as a promising diagnostic approach that offers a high-resolution depiction of pathological anatomy and the ability to detect associated disease biomarkers [3,4]. Fluorophores have long been used as luminescence probes in various biological and biomedical applications [5–7].

Perfluorocarbons (PFCs) have been widely employed as ^{19}F-MR agents for imaging modalities due to their biological and chemical inertness. PFCs have been used clinically as blood substitutes because of their high gas dissolving capacity for oxygen and carbon-dioxide as well as their chemical and metabolic stability [8]. The ^{19}F in the PFC nanoparticles have low background biological abundance and provide excellent signal sensitivity compared with ^{1}H [9–11]. The encapsulation or conjugation of a wide variety of contrast agents onto PFCs for multimodal imaging and therapeutics, in combination with antibodies or other targeting ligands, causes them to accumulate in specific sites, holding great potential for medical applications. The PFC nanoemulsions have been widely

* Correspondence: chungbh@kribb.re.kr
Bionanotechnology Research Center, Korea Research Institute of Bioscience and Biotechnology, Daejeon 305-806, South Korea

explored for various applications, such as ultrasound imaging as a vehicle for targeted delivery of contranst agents and drug delivery [12], partial liquid lung ventilation [13], and blood substitutes [8,14].

Indocyanine green (ICG) is a water-soluble tricarbocyanine dye with substantial absorption and fluorescence in the near-infrared region (NIR) [15,16]. The United States Food and Drug Administration (FDA) [17] has approved ICG for use in diagnostic applications in clinical settings including its use as an optical contrast agent in the imaging of cardiac and hepatic vascular systems [18,19], the retinal and choroidal vasculature [20], and lymphatic systems [21–23]. ICG has also been investigated in laser-mediated therapeutic applications. This dye can transform from the absorbed NIR light energy into free-oxygen species and heat, which further expands its therapeutic application to photothermal and photodynamic therapies [24–26]. Moreover, ICG, a NIR fluorescence probe, has several advantages over visible optical probes for in vivo imaging applications, including improved deep-tissue penetration, lower absorption and scattering by blood and tissue components, and minimal autofluorescence [18,27]. However, ICG has several intrinsic limitations in optical imaging applications: 1) ICG is rapidly cleared from the circulatory system (half-life of 2-4 min) [28,29]; 2) ICG forms aggregates depending on its concentration and interacts non-covalently with various proteins such as lipoproteins, plasma proteins, and human serum albumin via physical mechanisms due to its amphiphilic properties [30,31]; 3) in aqueous solutions, ICG is unstable as the compound undergoes physicochemical transformations such as thermal degradation and photodegradation [28,32]. Such changes result in discoloration, decreased light absorption, decreased fluorescence, and a shift in the wavelength of maximum absorption.

Recently, researchers have proposed nanomaterial-based ICG probes to overcome the high degradation rate and the short plasma half-life of ICG [33–35]. Polymeric nanoparticles and inorganic nanoparticles containing ICG could increase the stability of ICG and improve its physicochemical stability [34,35]. Also, nanoparticles with a polymer core and lipid shell provided great targeting capability [36]. Liposomes have been widely used as fluorescence probes for in vivo applications. In this study, we synthesized multifunctional PFC/ICG nanoemulsions as a new type of delivery vehicles for ICG to overcome the aforementioned limitations. The PFC/ICG nanoemulsions have both NIR optical imaging and ^{19}F-MR imaging moieties. We demonstrated the stability and fluorescence intensity of PFC/ICG nanoemulsions in vivo and in vitro and their physicochemical stability against exterior light and temperature. Also, we used multi-modal PFC/ICG nanoemulsions to indentify sentinel lymph nodes.

2 Methods

2.1 Materials

Perfluoro-15-crown ether (PFCE) was obtained from SynQuest Laboratories Inc. (Alachua, FL). L-a-phosphatidylcholine (Egg PC), and 1,2-distearoyl-*sn*-glycero-3-phosphoethanolamine-*N*-[methoxy(polyethylene glycol)2000] (DSPE-mPEG$_{2000}$) were purchased from Avanti Polar Lipids Inc. (Alabaster, AL). Indocyanine green, and cholesterol were obtained from Sigma-Aldrich Co. (St. Louis, MO).

2.2 Preparation of PFC/ICG nanoemulsions

To synthesize the PFC/ICG nanoemulsions, PFCE liquids were emulsified in an aqueous solution using a lipid mixture. The lipid compositions of the PFC/ICG nanoemulsions were PC/cholesterol/DSPE-mPEG$_{2000}$ in a molar ratio of 70:20:10, respectively. The lipid mixture was reacted for 1 h at room temperature by the addition of 2 mg ICG, evaporated with a rotary evaporator to ensure the production of a thin lipid film, and dried in a vacuum oven (25°C) for 24 h. The lipid film was rehydrated with phosphate-buffered saline (PBS), and the resulting solution was sonicated in a bath sonicator followed by five cycles of freezing and thawing. The rehydrated lipid mixture (2% w/v) and PFCE solution (20% v/v) were mixed for 4 min using a homogenizer, followed by microfluidisation [37]. A M-110S microfluidiser (Microfluidics Inc., Newton, MA) operating at a liquid pressure of approximately 20,000 psi was used for nanoemulsion preparations. The PFC/ICG nanoemulsions were stored at 4°C.

2.3 Characterization of PFC/ICG nanoemulsions

To evaluate the characteristics of the PFC/ICG nanoemulsions, a JEOL FE-TEM (transmission electron microscope) was utilized, and the TEM images were captured at 200 kV using a device from Tecnai. The PFC/ICG nanoemulsions were drop-cast onto carbon-coated TEM grids preliminarily stained with 2% uranyl acetate, and the solution was dried in a vacuum oven.

The emission and absorption spectra were obtained on a Perkin-Elmer LS-55 and a Beckman Coulter UV–VIS spectrophotometer (DU 800). The size of the PFC/ICG nanoemulsions was analyzed via dynamic light scattering using an electrophoretic light scattering photometer (ELS-Z, Otsuka Electronics, Osaka, Japan). The NIR fluorescence images of the PFC/ICG nanoemulsions were obtained using the IVIS Lumina imaging system (Caliper Life Science, MA) with an ICG filter set.

2.4 ICG and PFC loading efficiency

The ICG loading efficiency was analyzed using a previously reported method [29]. The quantity of ICG loaded into the PFC/ICG nanoemulsions was determined from the free ICG that was not incorporated into the PFC/ICG nanoemulsions. A 1-mL sample of the PFC/ICG nanoemulsion was centrifuged, and the supernatant was removed and stored in a centrifuge tube; the PFC/ICG nanoemulsion was dispersed in a PBS solution. The centrifugation was repeated, and the collected supernatants were combined. The ICG concentration was quantified via UV–Vis spectroscopy. The quantity of ICG inside the PFC/ICG nanoemulsions was also measured to verify the accuracy of the method. Selected PFC/ICG nanoemulsion samples were treated with an HNO_3 solution to induce capsule disassembly and to release the ICG into the solution. For all tested samples, the quantity of ICG released and the unencapsulated ICG equaled the quantity of the ICG precursor, indicating that the mass balance was conserved. The loading efficiency was calculated as the mass of ICG incorporated by the PFC/ICG nanoemulsions divided by the total ICG mass added to the nanoemulsion aggregate suspension.

To evaluate the PFC loading efficiency in PFC/ICG nanoemulsions, 19 F-MR imaging was performed on serial dilutions (0 – 0.4 ml) of PFCE liquids using a 4.7 T Bruker scanner (Biospec, Rheinstetten, Germany). The 19 F MR signal intensity was determined from the PFC signals originating from the PFCE liquids within a region of interest (ROI). We generated a calibration curve from the serial dilutions of the PFCE liquids and calculated the PFC loading efficiency in PFC/ICG nanoemulssions. The loading efficiency of PFC onto the PFC/ICG nanoemulsions was approximately 75.5 ± 3.2%.

2.5 Physicochemical stability of PFC/ICG nanoemulsions

The ICG and PFC/ICG nanoemulsions were diluted with distilled water to a final concentration of 1 μg/ml and were loaded onto a 12-well plate. The samples were irradiated with 760 nm NIR light from an LED for a predetermined time of 10, 20, 30, 60, or 120 min at room temperature. The effect of the light exposure on the degradation of the PFC/ICG nanoemulsions was determined with visible light at room temperature. The fluorescence intensity was measured for up to 6 days. After incubation, the remaining fluorescence of each sample was measured using a spectrofluorometer with excitation and emission wavelengths of 760 nm and 820 nm, respectively. For the quantitative analysis, we normalized the fluorescence signal intensity. This processing normalized the signal data points to the range [0, 1].

$$Y' = (Y - Y_{\min})/(Y_{\max} - Y_{\min})$$

Y denotes the y values of input curve, and Y' is the normalized curve.

2.6 Cell culture

The HeLa (human cervical cancer cells) and Raw264.7 (Murine macrophage cells) cell lines were obtained from the American Type Culture Collection (Rockville, MD). These cell lines were grown and maintained in Dulbecco's modified Eagle's medium (DMEM; Gibco BRL, Grand Island, NY) supplemented with 10% heat-inactivated fetal bovine serum (FBS), 50 IU/ml penicillin, and 50 μg/ml streptomycin. The cultures were maintained at 37°C/5% CO_2 in tissue culture plates. The DC2.4 cells, previously characterized as an immature murine dendritic cell line, were obtained from Dr. Kenneth L. Rock (Dana-Farber Cancer Institute, Boston, MA) [38]. This cell line was grown and maintained in DMEM supplemented with 10% heat-inactivated FBS, 50 IU/ml penicillin, and 50 μg/ml streptomycin.

2.7 Cell fluorescence imaging

To determine the intracellular delivery capacity of PFC/ICG nanoemulsions, the HeLa, Raw264.7, and DC2.4 cells were incubated with 10 μl/ml PFC/ICG nanoemulsions in μ-slide 8-well microscopy chamber at a density of 1×10^4 cells per well for 6 h at 37°C. The culture medium was then carefully aspirated, and the cells were washed three times. The labeled cells were fixed with 4% paraformaldehyde and stained with DAPI. The NIR fluorescence images were obtained on a Deltavision RT deconvolution microscope (Applied Precision Technologies, Issaquah, WA) using a filter set (excitation: 775/50, emission: 845/55; Omega Optical, Brattleboro, VT).

2.8 In vitro ^{19}F-MR and NIR fluorescence imaging

HeLa, Raw, or DC2.4 cells (1×10^6) were seeded on each well of a 6-well plate and grown for 24 h. The cells were then incubated with a medium containing 10 μl/ml PFC/ICG nanoemulsions. After 6 h, the medium was removed, and the cells were washed three times with PBS. The cell pellets were suspended with a 2% solution of low-melting agarose. The cells were collected in 0.2-mL tubes, and the MR and NIR fluorescence signals were measured. All ^{19}F-MR imaging of the PFC/ICG nanoemulsions was performed with a 4.7 T Bruker scanner using a double-tuned $^1H/^{19}F$ quadrature birdcage RF resonator. The ^{19}F-MR image was captured with a FLASH sequence (128×128 matrix; 30×30 mm^2 FOV; 50 ms TR; 2.6 ms TE; 10 mm slice thickness; 256 NEX). The NIR fluorescence images were obtained using the IVIS Lumina imaging system (Caliper Life Science, MA) with an ICG filter set.

2.9 Cell cytotoxicity assays

The cell cytotoxicity was assessed using a modified 3-(4, 5-dimethylthiazol-2-yl)-2, 5-diphenyltetrazolium bromide (MTT) assay. Raw, HeLa, or DC2.4 cells were seeded in a 96-well plate (Corning Costar, Cambridge, MA) at 1×10^4 cells/well. After incubation for 24 or 48 h, several different concentrations of the prepared PFC/ICG nanoemulsions

(0.38 ug/ul of ICG, 0.3 ul/ul of PFC) were poured into the wells. After incubation for a predetermined time, the residual nanoemulsions were removed, and a 2.5 mg/ml MTT solution was added to each well. The wells were then incubated in a humidified CO_2 incubator at 37°C for 2 h. An acidified isopropanol/10% Triton X-100 solution (100 µl) was then added, and the plates were shaken to dissolve the formazan products. The absorbance was measured using a microplate reader at 570 nm. The cell survival rate in the control wells without the PFC/ICG nanoemulsions was considered 100% cell survival. The cytotoxic concentration (CC_{80}) was defined as the concentration of the compound that reduced the absorbance of the control samples by 80%.

2.10 In vivo tracking of PFC/ICG nanoemulsions using NIR fluorescence and ^{19}F-MR imaging

Female hairless mice, 5–6 weeks of age, were purchased from SLC, Inc. (Japan). The mice were maintained at the KRIBB animal facility under pathogen-free conditions. All animal care and experimental procedures were approved by the Animal Care Committees of the KRIBB.

For the in vivo NIR fluorescence and ^{19}F-MR imaging of the sentinel lymph nodes, hairless mice were injected with 20 µl (25 µM of ICG, 15 ul/ml of PFC) of the PFC/ICG nanoemulsions (n = 5) or free ICG solutions (n = 5) in the footpad of the foreleg. Prior to the fluorescence imaging experiments, the mice were anesthetized with 200 µl of a 2.5% avertin solution (2, 2, 2-tribromoethanol-tert amyl alcohol, Sigma) throughout the experiments. After a predetermined time, the fluorescence intensity was quantitatively analyzed using the IVIS Lumina imaging system. Thereafter, the ^{19}F-MR images of the mice were obtained with a 4.7 T Bruker scanner using a double-tuned $^{1}H/^{19}F$ Birdcage coil design (inner diameter: 35 mm; length: 78 mm). After acquiring the morphological ^{1}H images, the resonator was tuned to ^{19}F. For the ^{19}F-MR image, the mouse was imaged with a gradient echo sequence (128 × 128 matrix; 3 cm FOV; 56.0 ms TR; 2.6 ms TE; 20 mm slice thickness; 60° flip angle; 256 NEX; 30 min total scan time).

2.11 Statistical analysis

The statistical evaluations of the experiments were performed by ANOVA analysis followed by a Newman-Keuls multiple comparison test.

3 Results and discussion

3.1 Characteristics of PFC/ICG nanoemulsions

To develop suitable MR imaging probes with improved sensitivity for noninvasive in vivo imaging at the cellular

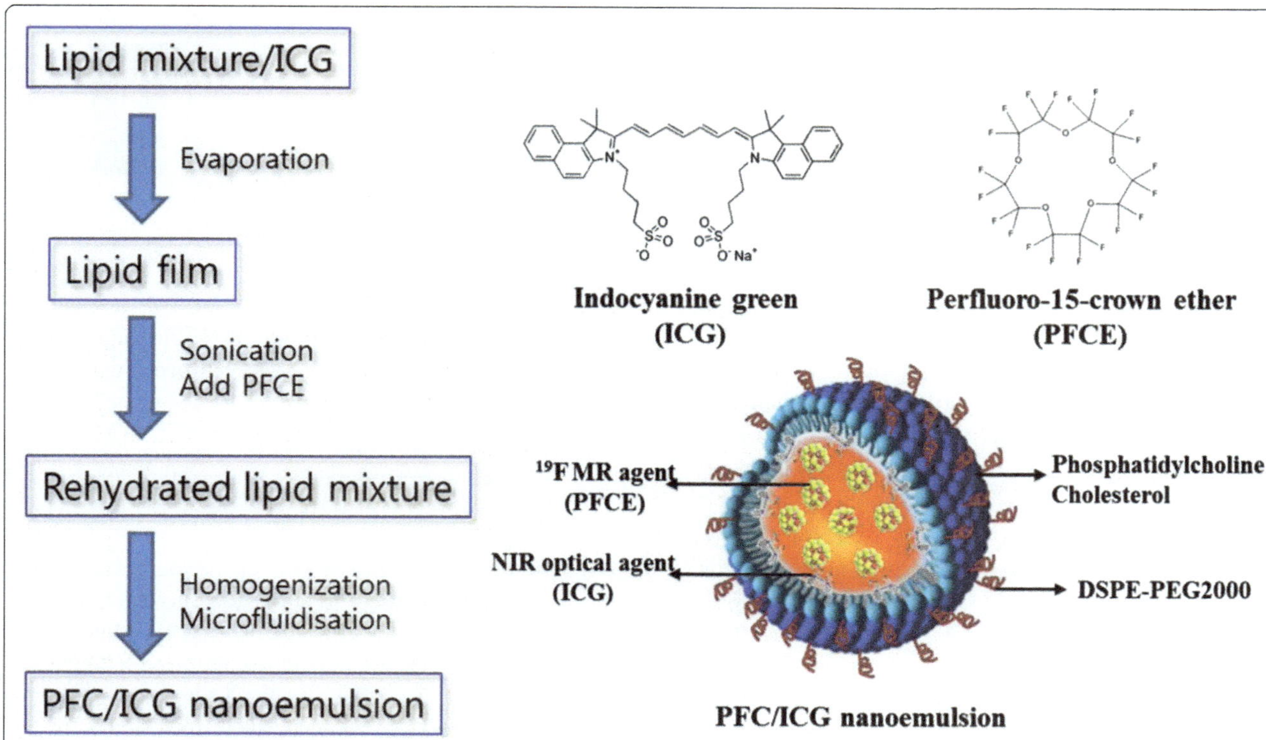

Scheme 1 Schematic illustration of the PFC/ICG nanoemulsions having both ^{19}F-MR and NIR optical imaging capabilities.

and molecular levels, we synthesised multifunctional PFC-based nanoemulsion containing ICG as a NIR organic dye, which provides simultaneous ^{19}F-based MR imaging and NIR optical imaging. ICG has been approved by the FDA for human medical and diagnosis [17]. Also, PFCs were developed for use as a blood substitute [8], and no toxicity, carcinogenicity, or mutagenic effects of PFCs have been reported for pure fluorocarbons [39]. Our synthesis strategy for preparing the multifunctional imaging probe is illustrated in Scheme 1. The bimodal PFC/ICG nanoemulsions were prepared via thin film hydration followed by microfluidisation, as described in the Methods section. The emission spectra of the PFC/ICG nanoemulsion obtained using a fluorescence spectrometer peaked at approximately 825 nm, with a full-width-at-half-maximum (FWHM) of 30 - 35 nm (Figure 1B). Compared with the ICG in water, the absorption and emission spectra of the PFC/ICG nanoemulsions were bathochromically shifted by approximately 20 - 30 nm (Figure 1). These shifts of the PFC/ICG nanoemulsions correspond to the changes in the physicochemical environment and are consistent with a specific interaction of the ICG molecules [16,30]. The red shift toward longer wavelengths resulted in a dramatic decrease of in vivo background signal during detection, leading to an improved signal-to-noise ratio in vivo [22]. The interaction of the ICG molecules within the PFC/ICG nanoemulsions leads to a chemical stabilization of the ICG molecules and can reduce the formation of aggregates [32]. The ^{19}F-based MR spectroscopy data for the PFC/ICG nanoemulsions exhibited a singlet peak (at approximately 36.3 ppm) (Figure 1C). This peak was chosen as the excitation frequency for the PFC/ICG nanoemulsion ^{19}F-MR imaging. The average size and zeta potential of PFC/ICG nanoemulsions were measured using dynamic light scattering analysis (Table 1). The mean diameter of the PFC/ICG nanoemulsions was 119.1 ± 25.1 nm with a polydispersity index of 0.07, and the zeta potential analysis revealed a surface charge of -15.3 ± 0.58 mV (Table 1). The loading efficiency of ICG onto the PFC/ICG nanoemulsions was approximately 95.1 ± 2.2% (Table 1). For comparison, ICG liposomes without a PFC solution had a loading efficiency of 20 ± 3.5%. Encapsulation within the PFC/ICG nanoemulsions resulted in an approximately 4.7-fold increase. The ICG was almost completely encapsulated within the PFC-based nanoemulsion. The PFC nanoparticles core material in PFC/ICG nanoemulsions is surrounded by a lipid monolayer, because these molecules have the property of being both lipophobic and hydrophobic. The ICG molecules can be carried in the PFC-based nanoemulsions for NIR optical imaging. The ICG molecules within these nanoemulsions could successfully lead to the chemical stabilization of the ICG molecules in aqueous solution because of the lipophobic/hydrophobic properties of the PFC molecules.

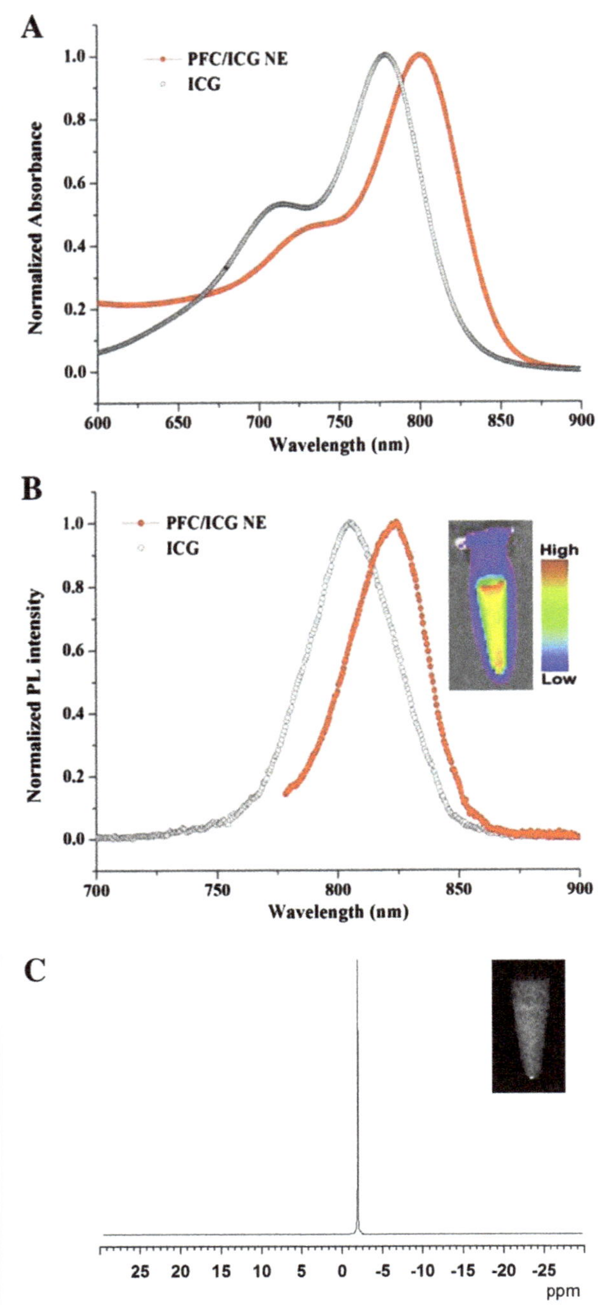

Figure 1 Characteristics of the PFC/ICG nanoemulsions.
The excitation **(A)** and emission **(B)** spectra of the nanoemulsions. The emission spectra were measured at λex = 760 nm. **(C)** ^{19}F magnetic resonance spectra and imaging of the PFC/ICG nanoemulsion.

3.2 Physicochemical stability of PFC/ICG nanoemulsions

To investigate the influence of the ICG concentration on fluorescence in aqueous solutions, the fluorescence intensity of PFC/ICG nanoemulsions was measured using ICG concentrations ranging from 0.4 to 50 μM and comparing them to that of free ICG dissolved in water (Figure 2A). The fluorescence intensity increased with increasing ICG concentration up to a maximum of 6.25 μM. Further

Table 1 Summary of the physicochemical properties and cytotoxicity activity of the PFC/ICG nanoemulsions

Sample	Emission peak (nm)	Diameter (nm)	Polydispersity	Loading efficiency (%)	CC_{80} (μl/ml)[a]		
					DC2.4	HeLa	Raw264.7
PFC/ICG nanoemulsions	825	119.1 ± 25.1	0.07	95.1 ± 2.2	> 100	> 100	> 100
ICG liposome	825	63.4 ± 15.6	0.09	20 ± 3.5	> 100	> 100	> 100

[a] cytotoxicity activity.

increase in the ICG concentration causes a gradual decrease in the fluorescence intensity. The PFC/ICG nanoemulsions emitted a 2.3-fold higher fluorescence intensity than the free-ICG solution. This is due to the increased aggregation and self-quenching of free-ICG molecules at high concentration [29]. However, The ICG in PFC/ICG nanomulsions maintains the chemical stabilization and reduces the ICG molecular aggregates. The influence of the external environment on the ICG degradation in the PFC/ICG nanoemulsions was also investigated. The thermal stability of the PFC/ICG nanoemulsions in the dark was determined at 4, 25, and 37°C over 3 months (Figure 2B). Little decrease was noted in the fluorescence intensity of the PFC/ICG nanoemulsions stored at 4°C over the 4-month period (Additional file 1: Figure S2). However, increasing in temperature significantly enhanced the rate of ICG degradation in the PFC/ICG nanoemulsions stored at 25 and 37°C. The fluorescence intensities observed in the PFC/ICG nanoemulsions at 25 and 37°C were approximately 74.6% and 58.9%, respectively. The fluorescence intensity of free-ICG solutions stored at 4, 25, and 37°C disappeared completely within 10 days. To examine whether the encapsulation of the ICG within the PFC/ICG nanoemulsions could enhance the photostability of the ICG fluorescence against external light, the PFC/ICG nanoemulsions were exposed to a NIR LED

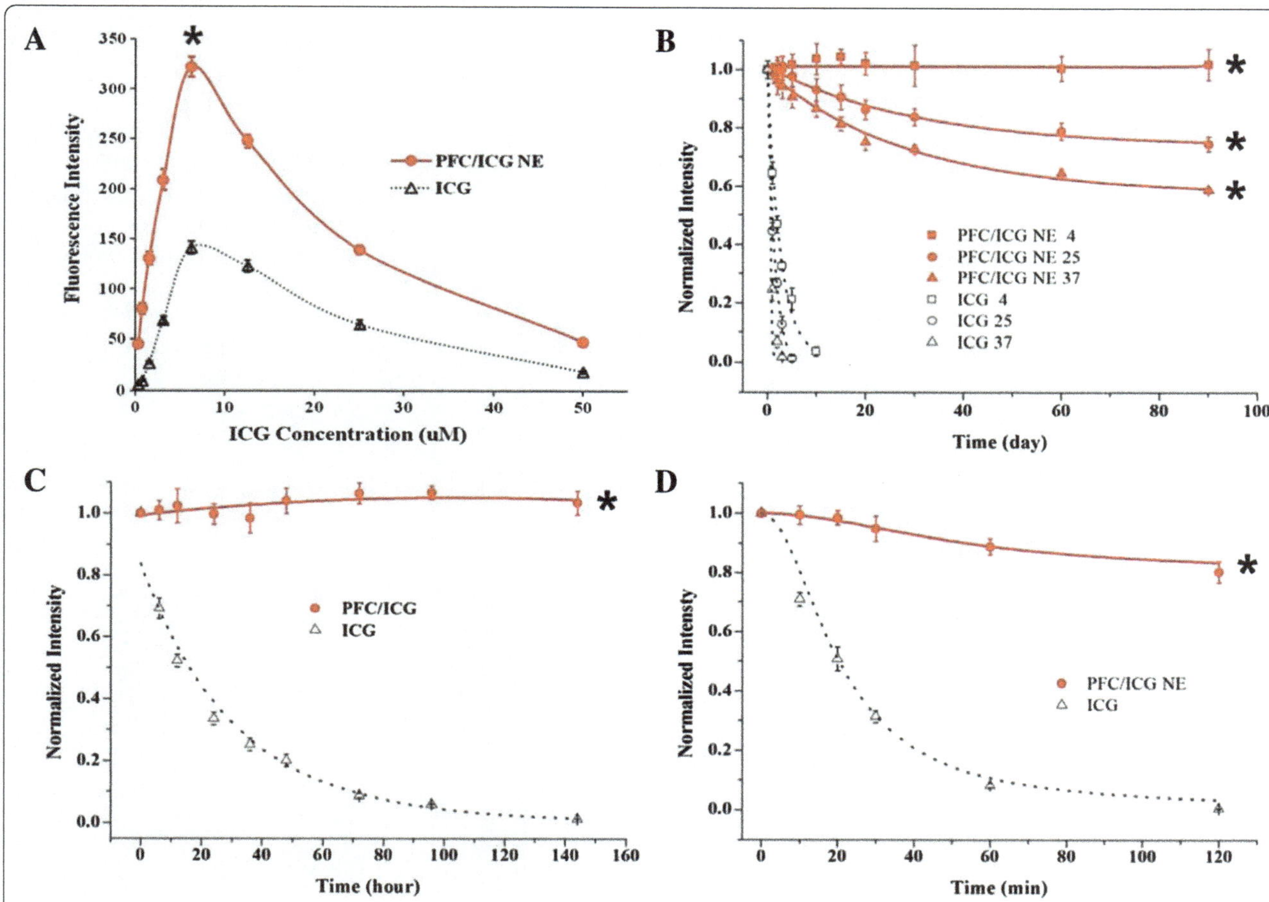

Figure 2 Physicochemical properties of PFC/ICG nanoemulsions. (A) Plot of fluorescence intensity versus the ICG concentration in the PFC/ICG nanoemulsions or aqueous free ICG. Influence of temperature **(B)**, visible light exposure **(C)**, and NIR lamp **(D)** on the ICG degradation in the PFC/ICG nanoemulsions or ICG solution. The fluorescence intensity of each sample was measured using a spectrofluorometer with excitation and emission wavelengths of 760 nm and 820 nm, respectively. (mean ± SD, *n* = 6). *P < 0.001 compared with the free ICG aqueous.

light at 760 nm or visible light for a predetermined period, and the fluorescence intensity was analyzed. The PFC/ICG nanoemulsions exposed to the visible ambient light maintained their fluorescence intensity over the 6-day period. However, after 1 day of exposure to visible ambient light, the fluorescence intensity of the free-ICG solution was reduced to less than 50% of its initial value and was 3 times less than that of the PFC/ICG nanoemulsions. The exposure to visible light did not affect the fluorescence of the PFC/ICG nanoemulsions. Figure 2D demonstrates the influence of exposure to NIR light on the fluorescence intensity of the PFC/ICG nanoemulsions. After 1 hour of exposure to NIR light, the fluorescence emitted by the PFC/ICG nanoemulsions exhibited an approximately 10.9-fold increase over that of the free-ICG solutions. At the endpoint, the fluorescence of the PFC/ICG nanoemulsions decreased slightly approximately 20%, but the fluorescence of the free-ICG solution completely disappeared. This result suggests that encapsulation of the ICG within PFC/ICG nanoemulsions could successfully protect the ICG fluorescence signals against external environments. The degradation of the ICG with exposure to NIR or visible light is due to the production of photo-excited ICG molecules [40]. The ICG molecules in the PFC/ICG nanoemulsions could efficiently prevent them from aggregating, thus decreasing the fluorescence self-quenching. Therefore, these nanoemulsions could significantly increase the stability and the fluorescence intensity of the ICG and improve its physicochemical stability against external light and temperature changes.

3.3 In vitro fluorescence and ^{19}F-MR imaging

NIR fluorescence and ^{19}F-MR imaging were performed on PFC/ICG nanoemulsions containing ICG and PFC using IVIS Lumina imaging system (Caliper Life Science, MA) and 4.7 T MR scanner (Biospec, Rheinstetten, Germany), respectively. Figure 3A and B shows the ^{19}F-MR and NIR fluorescence images of three cell lines (Raw264.7, HeLa, and DC2.4) loaded with the PFC/ICG nanoemulsions. The three cell lines were incubated with the PFC/ICG nanoemulsions for 6 h at 37°C, and the NIR fluorescence and ^{19}F-MR images were obtained using IVIS Lumina imaging system and 4.7 T MR scanner, respectively. The NIR fluorescence signals and ^{19}F-MR signals were observed for the three pelleted cell lines. We performed the regions of interest (ROI) analysis to confirm the ^{19}F-MR and NIR fluorescence signals and observed the variations in intensity among the cells (Figure 3C and 3D). These data indicated that the PFC/ICG nanoemulsions could efficiently monitor the intracellular uptake in two phagocytic cell lines (Raw264.7 and DC2.4) suggesting that the PFC/ICG nanoemulsions accumulated in the lymph nodes because they were phagocytosed by phagocytic cells. To evaluate the cell cytotoxicity of the PFC/ICG nanoemulsion, three cell lines were exposed to various concentrations of the PFC/ICG nanoemulsion ranging from 0 to 100 μl/ml for 24 h or 48 h and were evaluated using an MTT assay (Additional file 1: Figure S3). The data revealed that the labeling of the three cell lines with the PFC/ICG nanoemulsion was nontoxic over a broad concentration range,

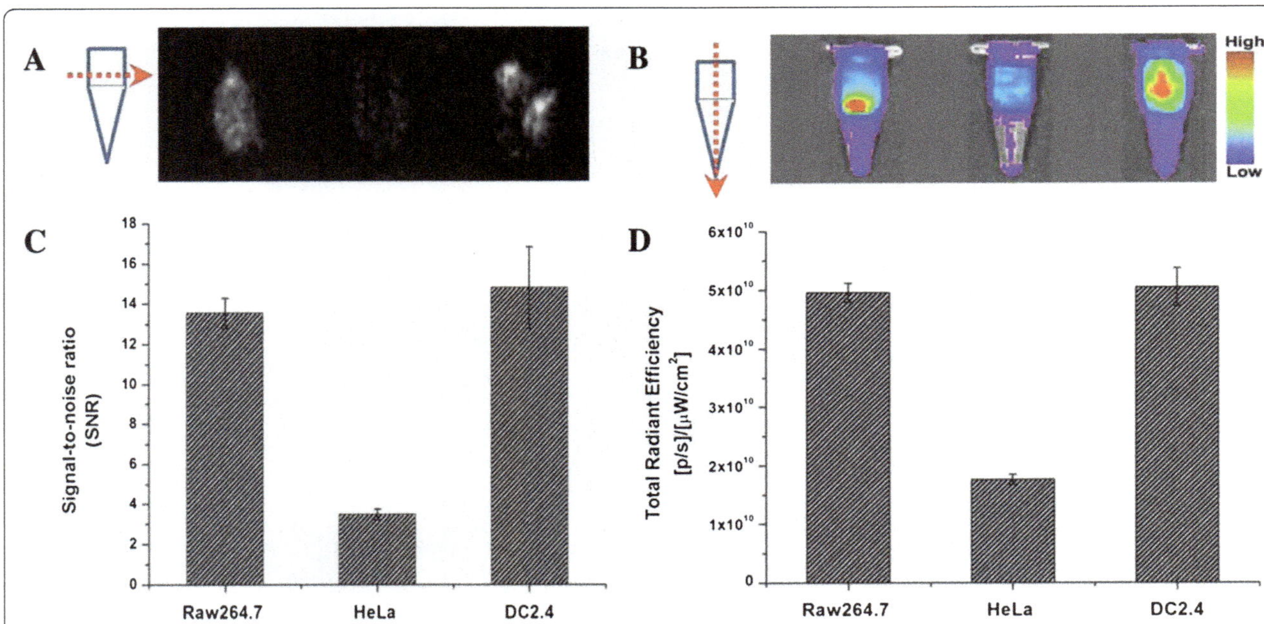

Figure 3 In vitro ^{19}F-MR and NIR fluorescence imaging. ^{19}F-MR images **(A)** and NIR fluorescence images **(B)** of three cell lines (Raw264.7, HeLa, and DC2.4) after incubation with the PFC/ICG nanoemulsions. **(C)** Signal-to-noise ratio for in vitro ^{19}F-MR images in Figure 3A. **(D)** The quantitative analysis of fluorescence intensity in Figure 3B.

with a CC_{80} of >100 μl/ml (Table 1), indicating that these nanoemulsions are suitable for in vivo animal studies, clinical applications, and biological applications. To determine the intracellular localization of the PFC/ICG nanoemulsions, three cell lines were incubated with the PFC/ICG nanoemulsions for 6 h at 37°C and then stained with a DAPI solution to observe the nuclei. The NIR signals inside the cell membrane exhibited the intracellular uptake of the PFC/ICG nanoemulsions (Figure 4).

3.4 Lymph node mapping

The use of PFC/ICG nanoemulsion-based bimodal imaging contrast agents was investigated for lymph-node mapping using an IVIS Lumina imaging system and a 4.7 T MRI scanner. For the NIR fluorescence imaging and the in vivo ^{19}F-MR imaging of sentinel lymph nodes, the PFC/ICG nanoemulsions (20 μl of 25 μM) were injected into the footpad of the foreleg of hairless mice. As shown in Figure 5, the ICG NIR signal and the PFC ^{19}F-MR signal were observed in the lymph nodes (red circles). Ex vivo NIR images were obtained of the lymph nodes dissected from the PFC/ICG nanoemulsions-injected mice (Figure 5D), confirming the in vivo imaging results. We performed the ROI analysis to confirm the NIR fluorescence signals (Additional file 1: Figure S4). The NIR fluorescence signal of the PFC/ICG nanoemulsions in the lymph nodes was significantly prolonged compared with that of the free-ICG solution. The PFC/ICG nanoemulsions presented significant NIR fluorescence signals even at 72 h post-injection. In contrast, no detectable signal was recorded in the lymph nodes from the free-ICG solution at 6 h post-injection. The relatively short in vivo fluorescence of the free-ICG is attributed to the fluorescence quenching of the free-ICG in physiological environments and its rapid aggregation and clearance from the body [16,30,41]. Because ICG is amphiphilic properties, it can bind to lipoprotein and plasmatic proteins. This binding leads to its removal by hepatic parenchymal cells and its secreted into the bile [41]. However, the ICG molecules encapsulated inside the PFC/ICG nanoemulsions were protected from the physicochemical environment, suggesting that the lymph node mapping could be tracked effectively using PFC/ICG nanoemulsions as bimodal imaging contrast agents.

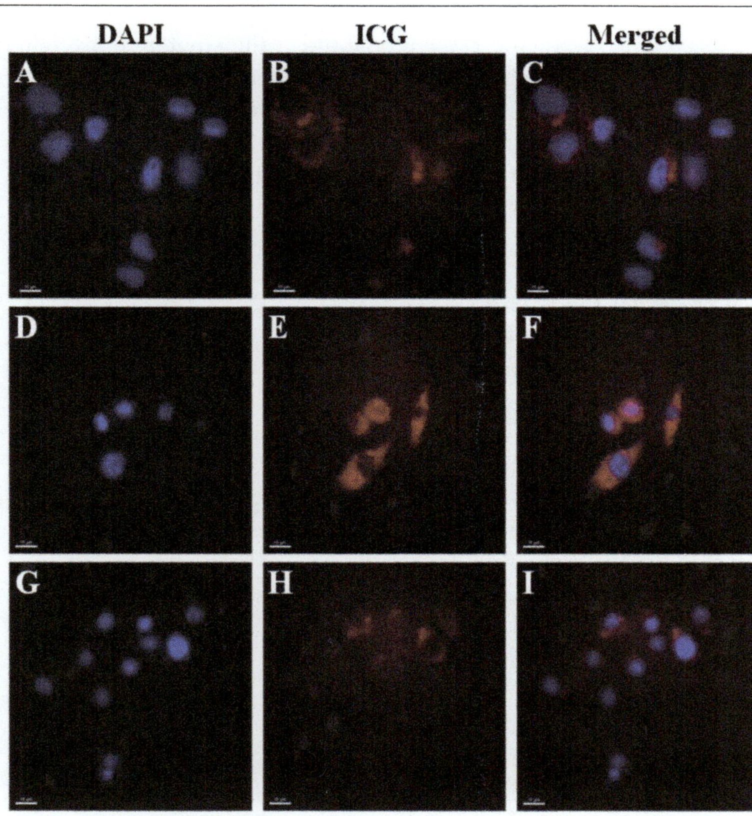

Figure 4 Cell fluorescence imaging with the PFC/ICG nanoemulsions. Fluorescence images of HeLa **(A to C)**, DC2.4 **(D to F)**, and Raw264.7 **(G to I)** cells after incubation with the PFC/ICG nanoemulsions. The PFC/ICG nanoemulsions are red, and the DAPI-stained nuclei are blue. Bar, 15 μm.

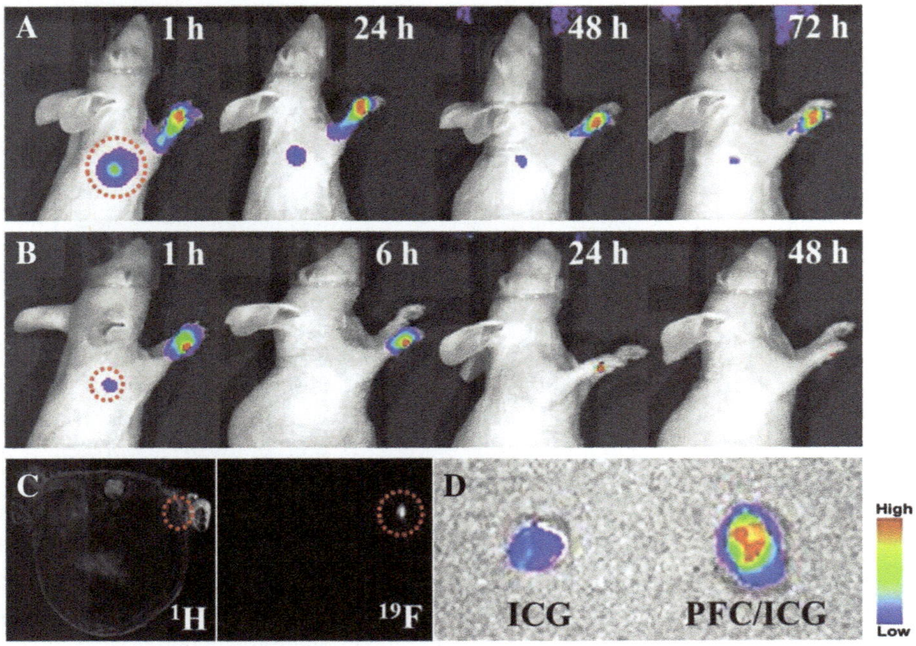

Figure 5 In vivo NIR image and ^{19}F-MR image of the sentinel lymph nodes. NIR fluorescence images at the indicated time following intradermal injection of PFC/ICG nanoemulsions **(A)** and free-ICG solution **(B)** into the right paw of a mouse. **(C)** ^{1}H-MR (left) and ^{19}F-MR (right) image following intradermal injection of PFC/ICG nanoemulsions into right paw of a mouse. **(D)** Ex vivo NIR fluorescence images of dissected lymph node at 1 h post-injection.

4 Conclusion

In summary, we encapsulated ICG molecules inside the PFC/ICG nanoemulsions as a novel bimodal imaging probe to allow for simultaneous ^{19}F-MR imaging and NIR optical imaging. The conjugation chemistry of ICG molecules is difficult due to their amphiphilicity and few functional groups. However, ICG molecules were encapsulated through a simple method to improve their properties. The ICG molecules protected the PFC/ICG nanoemulsions from aggregation and thus decreased the fluorescence self-quench. Therefore, these nanoemulsions could significantly increase the in vivo and in vitro stability and fluorescence intensity of ICG and improve its physicochemical stability against external light and temperature. Also, we used multi-modal PFC/ICG nanoemulsions to indentify the sentinel lymph nodes. Lymph nodes were detected by NIR optical imaging and ^{19}F-MR imaging. For sentinel lymph node biopsy, the incision procedure of sentinel nodes should be performed within short time (~30 min) because of the easy diffusion of free ICG and the decrease of fluorescence signal. The accuracy of sentinel lymph node biopsy depends upon the detection of sentinel nodes with high sensitivity and long-lasting vital dye. This result showed the suitability of the proposed nanoemulsions for noninvasive lymph node mapping as they enable long-time detection of lymph nodes. In the future, molecular probes in combination with various imaging modalities will provide more effective image-guided therapeutic tools for diagnostics, prognostics, and the treatment of diseases in diverse clinical settings.

Additional file

Additional file 1: Figure S1. TEM image of PFC/ICG nanoemulsion. **Figure S2.** Fluorescence intensity of the PFC/ICG nanoemulsions or free ICG at varying incubation times. Samples were excited at 760 nm. The emission spectra of the PFC/ICG nanoemulsions and free ICG solutions were collected at 825 nm or 805 nm, respectively. (mean ± SD, $n = 6$). **Figure S3.** In vitro cytotoxicity of the PFC/ICG nanoemulsions in HeLa, DC2.4, and Raw264.7 cells. The cells were incubated with the PFC/ICG nanoemulsion for 24 and 48 h at 37°C and the viability of the cells were evaluated with increasing concentrations of the nanoemulsions ranging from 0.41 to 100 μL mL^{-1} using an MTT assay (mean ± SD, $n = 6$). **Figure S4.** The quantitative analysis of the fluorescence intensity after intradermal injection of either the PFC/ICG nanoemulsions or free-ICG solution into the foodpad of the foreleg. A) foot, B) Lymph node.

Competing interests

The authors declare that they have no competing interests.

Authors' contributions

PK Bae carried out the animal studies and nanoemulsion synthesis and drafted the manuscript. JY Jung participated in the design of the study and performed the statistical analysis. BH Chung conceived of the study, and participated in its design and coordination. All authors read and approved the final manuscript.

Authors' information

Dr. Pan Kee Bae is currently a postdoctoral research associate at BioNanotechnology Research Center in KRIBB. He obtained his B.S. from ChungNam University in 2000 and Ph.D. from department of Biology, Yonsei

University in 2006. His recent research is focused on the design of nanoparticles for bio-imaging against cancers and infectious disease.
Dr. Juyeon Jung is a senior research scientist in the BioNanotechnology Research Center in KRIBB. She obtained her B.S. degrees in Biochemistry from Michigan State University in 1998 and her Ph.D. from State University of New York at Stony Brook in 2004. After a postdoctoral study at Weill Medical College of Cornell University, she joined Elusys Therapeutics, Inc. in 2007 and then moved to KRIBB in 2009. Her current research focuses on the development of diagnostic/therapeutic antibodies and biointerfacing technologies for efficient biomolecules immobilization on solid surfaces.
Prof. Dr. Bong Hyun Chung received his PhD degree at Korea Advanced Institute of Science and Technology (KAIST) in 1987. He did his postdoctoral work at Caltech from 1990 to 1991. He is currently a director of the BioNanotechnology Research Center in KRIBB. He is also a professor of Nanobio Major at the University of Science and Technology (UST). His main research interests include (a) bio-nano interfacing science, (b) biochip/biosensor, (c) nanomedicine.

Acknowledgements
This research was supported by the Pioneer Research Center Program through the National Research Foundation of Korea funded by the Ministry of Science, ICT & Future Planning (Grant No.2011-0001952). This work was also supported by the KRIBB Initiative Research Program, Republic of Korea.

References

1. JR McCarthy, FA Jaffer, R Weissleder, Small **2**, 983 (2006)
2. MJ Pittet, FK Swirski, F Reynolds, L Josephson, R Weissleder, Nat. Protoc. **1**, 73 (2006)
3. ME Juweid, BD Cheson, N. Engl. J. Med. **354**, 496 (2006)
4. R Weissleder, Nat. Rev. Cancer. **2**, 11 (2002)
5. B Dubertret, P Skourides, DJ Norris, V Noireaux, AH Brivanlou, A Libchaber, Science **298**, 1759 (2002)
6. Y Li, YT Cu, D Luo, Nat. Biotechnol. **23**, 885 (2005)
7. C Wu, J Hong, X Guo, C Huang, J Lai, J Zheng, J Chen, X Mu, Y Zhao, Chem. Commun. 750 (2008)
8. JG Riess, Artif. Cells Blood Substit. Immobil. Biotechnol. **34**, 567 (2006)
9. ET Ahrens, R Flores, H Xu, PA Morel, Nat. Biotechnol. **23**, 983 (2005)
10. M Higuchi, N Iwata, Y Matsuba, K Sato, K Sasamoto, TC Saido, Nat. Neurosci. **8**, 527 (2005)
11. GM Lanza, PM Winter, AM Neubauer, SD Caruthers, FD Hockett, SA Wickline, Curr. Top. Dev. Bio. **70**, 57 (2005)
12. TD Tran, SD Caruthers, M Hughes, JN Marsh, T Cyrus, PM Winter, AM Neubauer, SA Wickline, GM Lanza, Int. J. Nanomedicine **2**, 515 (2007)
13. MA Croce, TC Fabian, JH Patton Jr, SM Melton, M Moore, LL Trenthem, J. Trauma. **45**, 273 (1998)
14. AJ McGoron, R Pratt, J Zhang, Y Shiferaw, S Thomas, R Millard, Artif. Cells Blood Substit. Immobil. Biotechnol. **22**, 1243 (1994)
15. RC Benson, HA Kues, Phys. Med. Biol. **23**, 159 (1978)
16. ML Landsman, G Kwant, GA Mook, WG Zijlstra, J. Appl. Physiol. **40**, 575 (1976)
17. J Malicka, I Gryczynski, CD Geddes, JR Lakowicz, J. Biomed. Opt. **8**, 472 (2003)
18. JV Frangioni, Curr. Opin. Chem. Biol. **7**, 626 (2003)
19. JS Slakter, LA Yannuzzi, DR Guyer, JA Sorenson, DA Orlock, Curr. Opin. Ophthalmol. **6**, 25 (1995)
20. VL Dzurinko, AS Gurwood, JR Price, Optometry **75**, 743 (2004)
21. T Kitai, T Inomoto, M Miwa, T Shikayama, Breast Cancer **12**, 211 (2005)
22. ST Proulx, P Luciani, S Derzsi, M Rinderknecht, V Mumprecht, JC Leroux, M Detmar, Cancer Res. **70**, 7053 (2010)
23. JC Rasmussen, IC Tan, MV Marshall, CE Fife, EM Sevick-Muraca, Curr. Opin. Biotechnol. **20**, 74 (2009)
24. BM Barth, EI Altinoğlu, SS Shanmugavelandy, JM Kaiser, D Crespo-Gonzalez, NA DiVittore, C McGovern, TM Goff, NR Keasey, JH Adair, TP Loughran Jr, DF Claxton, M Kester, ACS. Nano **5**, 5325 (2011)
25. S Fickweiler, RM Szeimies, W Bäumler, P Steinbach, S Karrer, AE Goetz, C Abels, F Hofstadter, M Landthaler, J. Photochem. Photobiol. B **38**, 178 (1997)
26. VV Tuchin, EA Genina, AN Bashkatov, GV Simonenko, OD Odoevskaya, GB Altshuler, Lasers Surg. Med. **33**, 296 (2003)
27. KE Adams, S Ke, S Kwon, F Liang, Z Fan, Y Lu, K Hirschi, ME Mawad, MA Barry, EM Sevick-Muraca, J. Biomed. Opt. **12**, 024017 (2007)
28. B Hollins, B Noe, JM Henderson, Clin. Chem. **33**, 765 (1987)
29. V Saxena, M Sadoqi, J Shao, J. Pharm. Sci. **92**, 2090 (2003)
30. S Mordon, JM Devoisselle, S Soulie-Begu, T Desmettre, Microvasc. Res. **55**, 146 (1998)
31. G Patonay, J Salon, J Sowell, L Strekowski, Moleucles **9**, 40 (2004)
32. J Gathje, RR Steuer, KR Nicholes, J. Appl. Physiol. **29**, 181 (1970)
33. VB Rodriguez, SM Henry, AS Hoffman, PS Stayton, X Li, SH Pun, J. Biomed. Opt. **13**, 014025 (2008)
34. V Saxena, M Sadoqi, J Shao, J. Photochem. Photobiol. B **74**, 29 (2004)
35. J Yu, MA Yaseen, B Anvari, MS Wong, Chem. Mater. **19**, 1277 (2007)
36. JM Chan, L Zhang, KP Yuet, G Liao, JW Rhee, R Langer, OC Farokhzad, Biomaterials **30**, 1627 (2009)
37. YT Lim, YW Noh, JN Kwon, BH Chung, Chem. Commun. **45**, 6952 (2009)
38. Z Shen, G Reznikoff, G Dranoff, KL Rock, J. Immunol. **158**, 2723 (1997)
39. E Longstaff, Ann. N. Y. Acad. Sci. **534**, 283 (1998)
40. W Holzer, M Mauerer, A Penzkofer, RM Szeimies, C Abels, MM Landthaler, W Baumler, J. Photochem. Photobiol. B **47**, 155 (1998)
41. T Desmettre, JM Devoisselle, S Mordon, Surv. Ophthalmol. **45**, 15 (2000)

3

Carbon nanotubes and graphene towards soft electronics

Sang Hoon Chae[1,2] and Young Hee Lee[1,2*]

Abstract

Although silicon technology has been the main driving force for miniaturizing device dimensions to improve cost and performance, the current application of Si to soft electronics (flexible and stretchable electronics) is limited due to material rigidity. As a result, various prospective materials have been proposed to overcome the rigidity of conventional Si technology. In particular, nano-carbon materials such as carbon nanotubes (CNTs) and graphene are promising due to outstanding elastic properties as well as an excellent combination of electronic, optoelectronic, and thermal properties compared to conventional rigid silicon. The uniqueness of these nano-carbon materials has opened new possibilities for soft electronics, which is another technological trend in the market. This review covers the recent progress of soft electronics research based on CNTs and graphene. We discuss the strategies for soft electronics with nano-carbon materials and their preparation methods (growth and transfer techniques) to devices as well as the electrical characteristics of transparent conducting films (transparency and sheet resistance) and device performances in field effect transistor (FET) (structure, carrier type, on/off ratio, and mobility). In addition to discussing state of the art performance metrics, we also attempt to clarify trade-off issues and methods to control the trade-off on/off versus mobility). We further demonstrate accomplishments of the CNT network in flexible integrated circuits on plastic substrates that have attractive characteristics. A future research direction is also proposed to overcome current technological obstacles necessary to realize commercially feasible soft electronics.

Keywords: Carbon nanotube; Graphene; Nano-carbon; Soft electronics; Flexible; Stretchable; Transparent conducting film; Thin film transistor

1 Introduction

Since the invention of the transistor, the semiconductor industry has affected nearly every aspect of our daily life [1,2]. One main stream technological trend of the silicon industry is scaling down the device sizes. For instance, the gate length has been reduced down to ~20 nm under current optical lithography technique, and the count of transistors in a commercially available CPU numbers more than 5 billion [3]. In spite of the tremendous progress of miniaturized silicon technology, further development to soft electronics is still limited by the rigidity of the materials themselves. Electronic devices on flexible and stretchable substrates, defined as soft electronics, are contrasted to traditional rigid chips using conventional silicon and metals. The strategies for developing soft electronics are driven by the investigation of new materials which are bendable, twistable, flexible and stretchable. Toward the basic requirement of replacing traditional rigid silicon electronics by new materials, structure engineering, such as structures in "wavy" layouts and the open mesh geometry have also been investigated to achieve stretchability [4–6].

Figure 1 shows the development of materials for achieving soft electronics from traditional rigid chips. Amorphous silicon (a-Si), low temperature polycrystalline silicon (p-Si), semiconducting metal oxides, nanowires, and organic semiconductors are promising candidates for flexible electronics from a materials perspective, but several challenges must be overcome prior to their practical use. a-Si is low-cost and is applicable for large-area displays, but suffers from poor mobility and flexibility [7]. Low temperature p-Si has the advantage of relatively high mobility but has low uniformity and processability [8]. Metal oxides are costly due to the shortage of rare earth elements and display poor environmental stability. Polymers have

* Correspondence: leeyoung@skku.edu
[1]Center for Integrated Nanostructure Physics (CINAP), Institute for Basic Science (IBS), Suwon 440-746, Republic of Korea
[2]Department of Energy Science, Department of Physics, Sungkyunkwan University (SKKU), Suwon 440-746, Republic of Korea

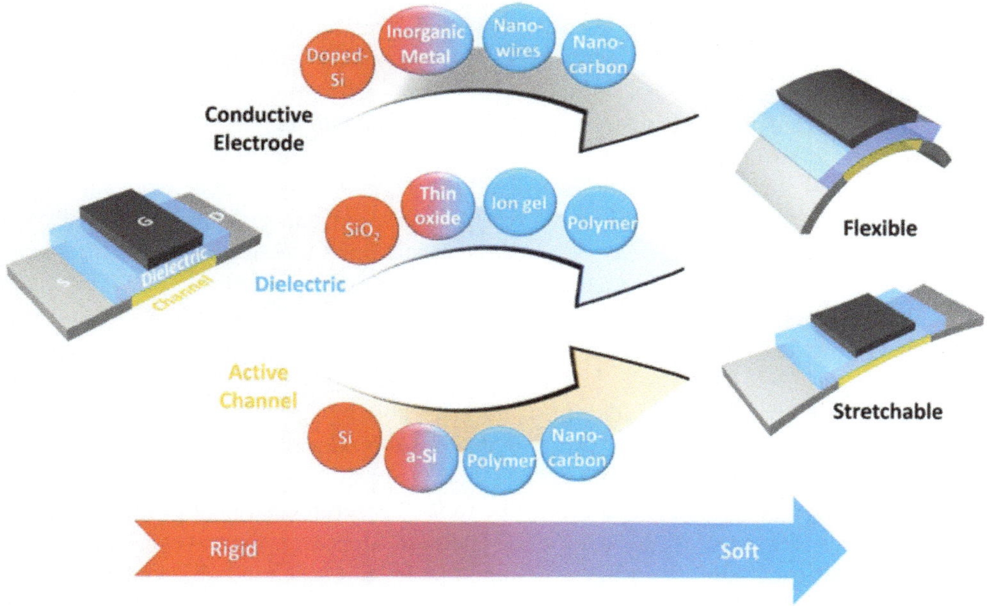

Figure 1 Classification of materials from rigid to soft. Conventional Si-based materials need to be replaced by new materials to realize soft (flexible/stretchable) electronics. With good electrical and mechanical properties, materials such as a-Si, organic polymer, nanowires, and nano-carbon materials are good candidates for next-generation soft applications.

substantial bendability, but have poor mobility and chemical stability.

Nano-carbons such as one-dimensional carbon nanotubes (CNTs) and two-dimensional graphene layers have been widely studied to open a new technology platform based on flexible electronics requiring high transmittance, bendability, and high mobility [9–12]. Figure 2 shows various types of carbon-based materials - fullerene, CNT, graphene, graphite, graphene oxide (GO), and diamond.

The extraordinary electrical, physical, and chemical properties of CNTs and graphene have been attractive since their discoveries. Both materials exhibit outstanding carrier mobility, which is attractive for applications to

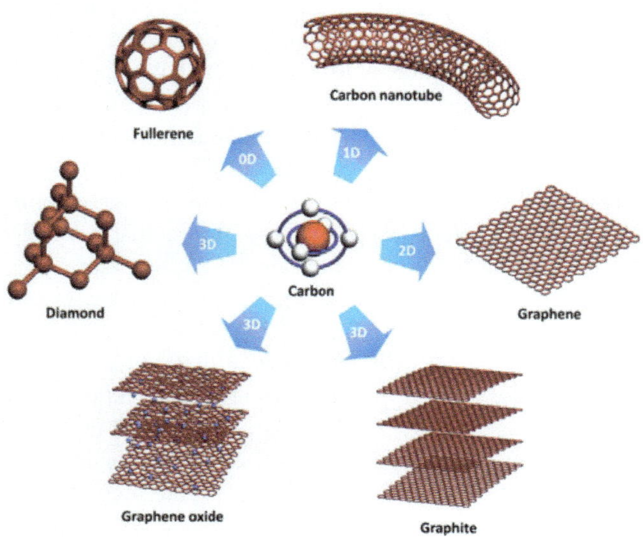

Figure 2 Carbon-based nanomaterials. Nano-carbon materials including 0D fullerene, 1D CNT, 2D graphene, 3D graphite, 3D graphene oxide, and 3D diamond are demonstrated.

electronic devices. The carrier mobility in semiconducting single-walled carbon nanotubes (SWCNTs) has been shown to be as high as ~80,000 $cm^2 V^{-1} s^{-1}$ [13], while the mobility of exfoliated graphene ranges from ~100,000 $cm^2 V^{-1} s^{-1}$ [14] on insulating substrates to 230,000 $cm^2 V^{-1} s^{-1}$ in suspended structures [15]. These ultra-high mobility values suggest that these materials have the potential to outperform established materials for next-generation high-speed electronics. The electric current capacity for both CNTs and graphene are reported above 10^9 A cm^{-2} [16,17]. At room temperature, CNTs exhibit a thermal conductivity up to 3,500 W m^{-1} K^{-1} [18], and graphene has a value of 5,300 W m^{-1} K^{-1} [19] with a high transmittance of nearly 97% [20]. In addition to high flexibility and stretchability, both materials also have superb mechanical strength (Young's modulus of 1.0 TPa and tensile strength of 130 GPa) [21]. For these reasons, CNTs and graphene are regarded as the most promising materials to realize next-generation electronics.

The purpose of this article is to summarize the recent progresses of both CNTs and graphene in soft electronics, and furthermore, to provide guidance for future nano-carbon research by clarifying feasible approaches which will most likely lead to soft applications. We first discuss several successful attempts to synthesize CNTs and graphene. Variations in transfer techniques for both materials are discussed thoroughly. For the use of CNTs and graphene for transparent conducting films (TCFs), the characteristics of TCFs using both nano-carbon materials are compared in depth, together with ITO. Furthermore, various types of field-effect devices using different forms of CNT FETs such as single CNT FET, random network CNT FET, aligned CNT FET, and different forms of graphene FETs such as single layer graphene (SLG), bilayer graphene (BLG), and graphene nanoribbon (GNR) are compared. Moreover, the specific FET device performances related to material preparation and fabrication techniques are also discussed. Finally, the logic level, flexibility, and stretchability of devices with a combination of graphene and CNTs along with their utilizations in logic circuits are further discussed. The systematic deep analyses of the device properties of graphene and CNTs highlight excellent opportunities for future flexible electronics. We conclude with a brief perspective on the research directions of soft electronics in future.

2 Review

2.1 Material preparations

The preparation techniques for CNTs and graphene are the most important fundamental research areas providing realistic applications. From the discovery of CNTs and graphene, diverse work has been done to improve the quality of the materials (crystallinity and uniformity) and to control other parameters (chirality, density, and doping levels) and morphology (length, area, dimension, and thickness). This section describes some of the most successful methods for synthesis of nano-carbon materials.

2.1.1 Carbon nanotubes

The CNT synthesis techniques aim to provide control over the tube density, spatial distribution, length, and orientation. Controlling the tube diameter and ratio of semiconducting to metallic SWCNTs have been a critical issue for electrical applications [22,23]. The conventional growth methods for large-scale CNTs include arc discharge, laser ablation, and chemical vapor deposition (CVD) [24–28]. While CNTs grown by arc discharge and laser ablation usually have fewer structural defects than those produced by CVD techniques, the CVD method is intrinsically scalable for realistic applications due to its low setup cost, high production yield, and ease of scale-up. Moreover, long average tube lengths can be obtained from CVD method, which lead to generally better electrical properties in CNTs. The challenge to control alignment and geometry of SWCNTs is mitigated by the CVD method as well. As a one-dimensional material, the as-grown CNTs have various geometries, as shown in Figure 3.

Individual CNTs are horizontally grown on the substrate by CVD, as shown in Figure 3a. Horizontally aligned SWCNTs can be grown using stable and laminar gas flow, which can be determined by the Reynolds number, which depends on volumetric flow rate, viscosity of gases, and the hydraulic diameter of the quartz tube [29,30]. Both the buoyancy effect induced by gas temperature and gas flow stability play a dominant role in preparing batch-scale SWCNT arrays [31]. In Figure 3b shows scanning electron microscopy (SEM) images of an aligned SWCNT film grown from Fe catalyst patterned into narrow stripes oriented perpendicular to the growth direction on quartz [32]. The CVD process on ST-cut quartz wafers using patterned stripes of Fe catalyst leads to the highest levels of alignment and density of CNTs. Linear alignment of individual SWCNTs was achieved with an average diameter of ~1 nm, and a density approaching ~10 SWCNT/μm. Figure 3c shows that vertically stacked CNT films can stand on a SiO_2 substrate. The CVD growth was carried out on various catalysts, including Fe nano-particles and metal thin films (Fe, Al/Fe, Al_2O_3/Co) on Si wafers, quartz, and metal foils to synthesize CNT forest [33,34]. Depending on the collection time, the thickness of CNT films can be changed from micrometers to a few centimeters [35]. Highly-stacked nanotube structures were successfully fabricated on wafer-scale substrates with different thicknesses, which are robust for numerous applications as a conducting film [36,37]. Efficient field emission has been demonstrated where the screening of the field emission current is determined by the ratio of

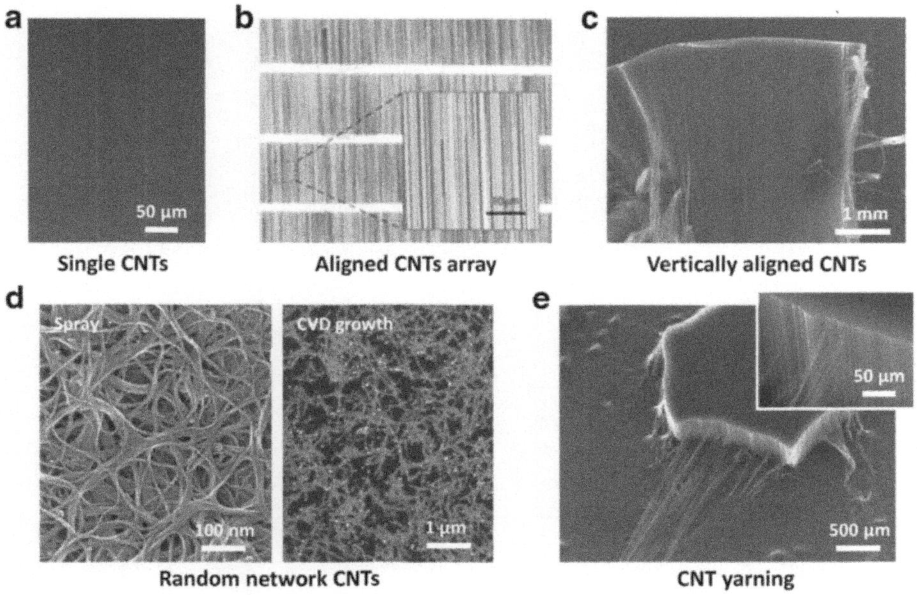

Figure 3 Various methods of CNT film preparation. **a**, CVD-grown aligned individual SWCNTs on SiO_2 substrate using stable and laminar gas flow. **b**, Aligned array of CNTs on ST-cut quartz with narrow strip pattern of Fe catalyst. Reproduced with permission [32]. Copyright 2007, Nature Publishing Group. **c**, Array of vertically aligned MWCNTs on Fe/Al/SiO_2 substrate. **d**, Random network SWCNTs prepared by spray of CNT solution (left) and CVD-grown on SiO_2 substrate (right). **e**, Yarning of vertically aligned MWCNT film.

the interlayer spacing to CNT length [38,39]. Figure 3d shows SEM images of random network geometry CNTs. The network geometry can be achieved easily by printing SWCNTs from a solution suspension [40,41]. Solution methods such as spray, filtering, dip-coating, and ink-jet printing have been commonly used for random network type CNT films [42–46]. One serious drawback of the solution approach is the bundling of individual CNTs. This degrades the performance of transparent conducting films (sheet resistance vs transmittance) and transistors (on/off ratio vs mobility) [47]. Random network CNT films prepared directly from CVD or arc discharge can also produce CNT networks and improve the device performance [48,49]. The bundling of CNTs can be avoided and rather clean CNTs can be retained through the CVD method without worrying about the addition of additives that are used in solution approach [41]. By controlling the concentration of catalysts of Fe/Co/Mo, the density of CNTs can be modified, due to increased surface area, pore volume, and catalytic activity [50]. Nevertheless, realizing large-area with good uniformity is still challenging with the CVD method. Owing to their strength, toughness, capabilities of mechanical energy damping, and resistance to knot-induced failure, yarns made from vertically aligned films of MWCNTs are promising multifunctional materials [51–53]. Figure 3e shows an example of the yarning process for a vertically aligned MWCNT film. A beneficial feature of these yarns is the diameter, which can be as little as 2% of the diameter of a human hair, making them ideal as an artificial muscle actuator or artificial muscle, and for storing energy as part of a fiber supercapacitor or battery. MWCNT fibers could also replace rigid metal wires in electronic textiles, such as in heated blankets, where the rigidity of the metal wires can be uncomfortable. Replacing wires with conducting fibers can also provide radio or microwave absorption, electrostatic discharge protection, other types of textile heating, or for simple wiring applications such as headphones where flexibility is important [37,54].

2.1.2 Graphene

Since graphene was first electrically isolated from graphite using a mechanical exfoliation method, many efforts have been studied to synthesize thin graphene films such as the CVD method, reduction of graphene oxide (GO), epitaxial growth on SiC, and chemical molecular assembly method.

As shown in Figure 4a, the mechanical exfoliation technique offers high quality but small flakes of graphene. Tape was used as the micromechanical cleavage layer to detach graphene samples from graphite. The exfoliation method was followed by the identification and selection of monolayers by using an optical microscopy, scanning electron microscopy (SEM), and atomic force microscopy (AFM) [55,56]. However, the practical use of such a graphene for electronics applications is limited by the tiny size of the exfoliated graphene films, despite

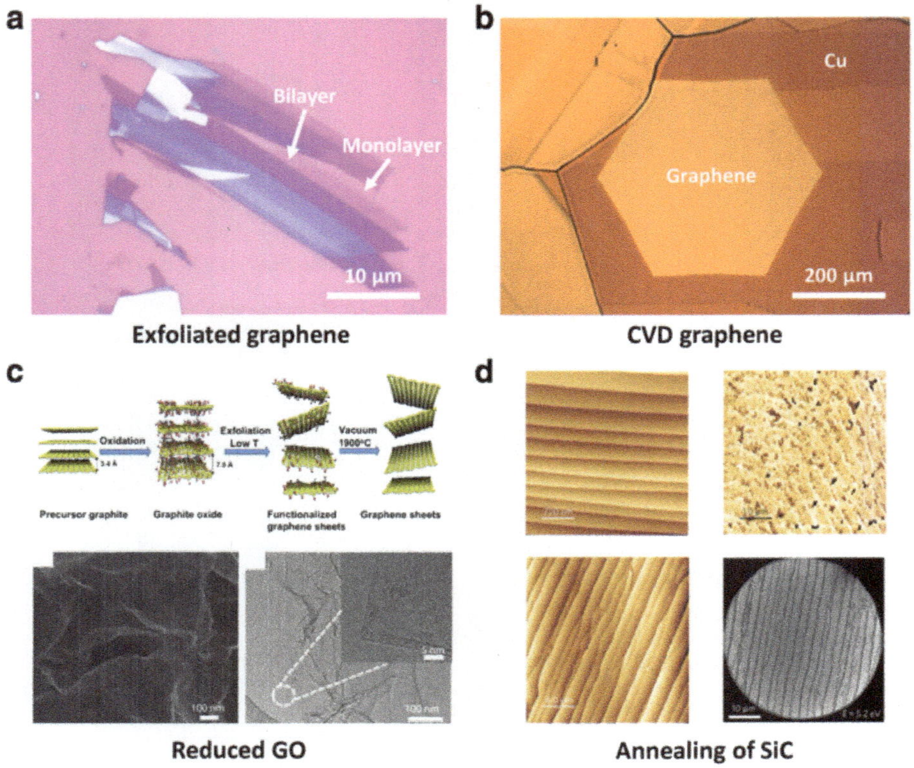

Figure 4 Various methods of graphene synthesis. a, Exfoliated graphene (monolayer, bilayer, and other thick layer) obtained by taping from graphite. **b**, Graphene flake is grown on Cu foil by CVD. **c**, Schematic procedure to generate high quality graphene powder obtained from reduced graphite oxide and the electron diffraction pattern. Adapted with permission [73]. **d**, Images of monolayer graphene on 6H–SiC(0001) for explaining epitaxial growth of graphene. Reproduced with permission [74]. Copyright 2009, Nature Publishing Group.

their high crystallinity. The preparation of graphene using the CVD method has been reported for the feasible use of graphene [57–59]. Figure 4b shows that the graphene flake was grown on Cu foil under an atmospheric CVD system. The CVD approach is attractive because it allows fabrication over large-area, and expansion of the applicability of graphene to flexible or stretchable devices. Although quality and size of graphene keep improving, field effect mobilities of devices using CVD graphene exhibit still lower values compared to those of devices with exfoliated or epitaxial graphene. Yet, the presence of defects such as point defects, grain boundaries, and wrinkles is unavoidable in the CVD process [60]. Grain boundaries and defects reduce the conductivity of the film and therefore it is highly desired to remove them during growth. Observations and controlling such defects are key research topics in the CVD method. Atomic rearrangement at graphene grain boundaries has been observed using transmission electron microscopy (TEM) and scanning tunneling microscopy (STM). Recent works use optical microscopy to observe the grain boundaries realized by selectively oxidizing the underlying copper foil through graphene grain boundaries functionalized with –O and –OH radicals generated by ultraviolet irradiation [61] and sodium chloride solution [62]. Graphene can be also prepared by a liquid-phase exfoliation or reduction of GO, which has advantages in quantity, yield and cost [63–67]. Large quantities of GO can be prepared by the traditional Brodie and Hummer method, although these methods can be slightly modified to improve the quality of GO [68–71]. Several reducing agents have been used to achieve reduced GO [72]. Although these methods are advantageous for mass production, the complete removal of epoxy and hydroxyl groups and defect generation are an unsolved problem at the present time, unlike the high quality pristine graphene. A simple thermal exfoliation followed by high temperature annealing up to 1500°C in vacuum provides a route of obtaining better quality graphene powder (Figure 4c) [73]. This graphene powder method is challenging but certainly advantageous for conducting film and electrode applications. The fabrication of graphene using the epitaxial growth of graphene directly on rigid insulating silicon carbide (SiC) wafers has been also reported (Figure 4d) [74]. A carbon-included material like SiC is used as a substrate for graphene preparation with high temperature annealing (around 1,500°C) [75]. Graphene obtained with epitaxial growth is highly crystalline, thus is intensely studied to fabricate transistors that operate at high frequencies [76,77]. Wafer-scale graphene can be produced by epitaxial growth on SiC, but those graphenes are not suitable for practical purposes because it is hard to detach graphene from the SiC substrate. Although a solid source molecular beam epitaxy method was also reported to fabricate graphene directly on Si(111), the high cost of molecular beam epitaxy will likely prevent the method from being commercially viable [78].

2.1.3 Transfer methods

Most CVD approaches for synthesizing CNTs and graphene require high temperatures which prevent direct growth of nano-carbon materials on plastic and other soft target substrates. CNTs and graphene located on a catalytic substrate need to be transferred onto a target substrate. Transferring graphene from the metal substrates onto desired substrates without degrading the quality of graphene is the critical step to use CVD-grown graphene for most practical applications.

Wet etching processes are commonly used to detach as-grown materials from the mother substrates using chemical solutions. $FeCl_3$ or $(NH_4)_2S_2O_8$ are often used for removing Cu, and NaOH or KOH for sapphire [79,80]. The most popular binder to hold graphene during wet etching is poly(methyl methacrylate) (PMMA), but this process unavoidably damages and contaminates the graphene layer with residuals, and is not desirable for scale-up fabrication. The dry printing (or stamping) technique uses polydimethylsiloxane (PDMS) stamp to transfer SWCNTs and graphene films from the growth substrates such as SiO_2/Si and metal films, still has problems with mechanical damage [81]. The roll to roll (R2R) lamination process can produce a large-area graphene film on flexible substrates [82,83]. The R2R transfer technique uses a thermal release layer as a temporary support and enables the continuous production of graphene film on 44 inch-scale flexible substrates. The synthesized graphene with Cu foil was laminated with the assistance of an adhesive layer, poly(ethylene co-vinyl acetate, EVA) with vinyl acetate (VA) as a supporting layer, to plastic film, followed by Cu etching, as shown in Figure 5a [82]. The transferred graphene film has appropriate uniformity with a resistance deviation of less than 10%. However, the graphene surface is still contaminated by organic adhesive from the thermal release tape using this transfer approach, which may fairly degrade the electrical properties of the film. Undesired mechanical defects also can be caused by this R2R transfer on graphene film. A bubbling method for transferring graphene films to target substrates is nondestructive not only to graphene but also to the mother-substrate (Figure 5b) [84]. The PMMA/graphene/Pt(or Cu, Ni) was dipped into NaOH solution and was used as the cathode with a constant current supply. At the negatively charged cathode, H_2 gas is produced by a water reduction reaction, and the PMMA/graphene layer detaches from Pt substrate due to the H_2 bubbles at the interface between the PMMA/graphene and Pt substrate. Damage of the mother-substrate is reduced considerably, and the substrate can be used repeatedly for the next CVD growth. In addition, the transferred graphene is free of metal particles, which are commonly found in graphene transferred by the metal etching process. Figure 5c explains the "clean-lifting transfer (CLT)" method, which uses electrostatic forces to transfer graphene onto target substrates, and doesn't use a PMMA adhesive layer [85]. An electrostatic generator (SIMCO, 18 kV) was placed at a distance of one inch away from the substrate, then the discharge process occurred via the electrostatic generator, followed by a pressing process to enable more uniform attachment between graphene and substrate. After the Cu foil was etched, the remaining graphene film on the target substrate was rinsed with deionized water to remove the residual etchant. The methods described so far are a rather simple transfer process that does not take account of positioning. There is an interesting transfer method for aligning 2D flakes to a desired location. In order to fabricate stacked graphene on BN devices, a few-micro-size flakes of graphene and BN should be positioned at a desired location (Figure 5d) [86]. Graphene was exfoliated separately onto a polymer stack consisting of a water-soluble polyvinyl alcohol (PVA) and a PMMA layer. When dipped into water, PVA was dissolved and the graphene/PMMA layer was detached from substrate and was floated on the surface of water bath. The PMMA membrane was securely adhered to a holder, which has a tiny hole to identify the top flake onto the PMMA layer during the aligned transfer process. The holder was clamped on the arm of a micro-positioner and then mounted on an optical microscope. The graphene was precisely aligned to the target BN flake by using the microscope to locate the position and the two (PMMA/graphene and BN) brought into contact. The demand for stacked layered structures has been growing [87–90]. A better strategy for transfer in a large-area without damages and residues on graphene is required for profound study.

2.2 Carbon-based elements

Common electronic devices require conducting, semiconducting, and insulating materials. For conducting elements, several conducting polymers such as polyacetylene, polypyrrole, polythiophene, polyaniline, and poly(3,4-ethylenedioxythiophene) poly(styrenesulfonate) (PEDOT:PSS) have been investigated for future applications to replace conventional rigid conducting and semiconducting materials [91,92]. However, these polymers have a relatively

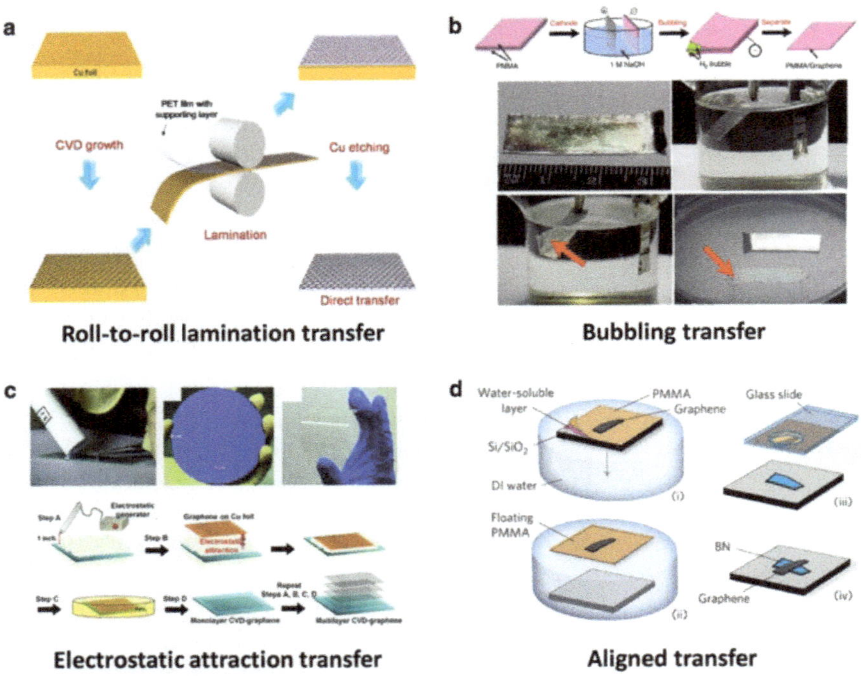

Figure 5 Various transfer methods of graphene. a, Schematic demonstration of Roll-to-Roll lamination transfer using a thermal release layer. Adapted with permission [82]. **b**, Schematic and photography images of bubbling process. The PMMA/graphene/Pt was dipped into NaOH solution with a constant current supply. Reproduced with permission [84]. Copyright 2012, Nature Publishing Group. **c**, "Clean-lifting transfer (CLT)" method, which uses electrostatic forces to transfer graphene. Adapted with permission [85]. **d**, Aligned transfer for placing graphene and BN to a desired location. Reproduced with permission [86]. Copyright 2010, Nature Publishing Group.

low electrical conductance and poor stability, compared with metal electrodes [93]. a-Si, p-Si, semiconducting metal oxides, nanowires, and organic semiconductors are promising candidates for the active channel, but several challenges - including rigidity and electrical performance issues - must be overcome prior to practical uses. CNTs and graphene electrodes can be an alternative not only to conducting electrodes but also to a semiconducting channel.

2.2.1 Conducting electrodes

Electrical conducting materials would have potential for consumer applications, such as soft displays, energy generators, and human bio-devices. In such applications, metal oxides such as IZO and ITO are the most widely used materials [94–96]. However, they have several limitations: i) They are costly and a predicted shortage of indium is a concern, and ii) fracture strain less than 1% limits the mechanical ability of flexible devices. Nano-carbon materials can overcome many of these limitations and open a new technology platform due to their outstanding electronic, optoelectronic, thermal, and mechanical properties. Here, we describe nano-carbon materials as conductive electrodes and the development of TCF using CNTs and graphene, where the aim is to replace ITO for certain applications.

During the past few years, much effort has been given in synthesizing CNT films as a conducting element [44,97–99]. Such CNT films have many applications including flexible and stretchable transparent loudspeakers [100], electrodes for LEDs, [101] lithium-ion batteries [102], and touch panels [103]. Figure 6a shows a practical touch panel assembled by directly yarning vertically aligned CNTs. Although the idea of utilizing CNT films as conducting materials is simple, controlling density, average tube length, tube diameter and mixture of metallic and semiconducting CNTs is still challenging. Even with optimized growth conditions, one serious drawback is the relatively high sheet resistance compared to that of conventional ITO [104]. Highly flexible, transparent, and conducting SWCNT films are one of the recent emerging technologies [105–107]. The pristine SWCNT TCF have a reported 360 Ω/sq sheet resistance at transmittance of 90% [43]. This sheet resistance could be dramatically improved by chemical doping treatments. Once such method using nitric acid removes the remaining surfactant from the CNT network and can lower the sheet resistance to a 150 Ω/sq at transmittance of 90% [108]. Further doping with Au^{3+} ions has also been shown to reduce sheet resistance to 110 Ω/sq at a transmittance of 90% [109,110]. While not surpassing the electrical performance of ITO, these films have

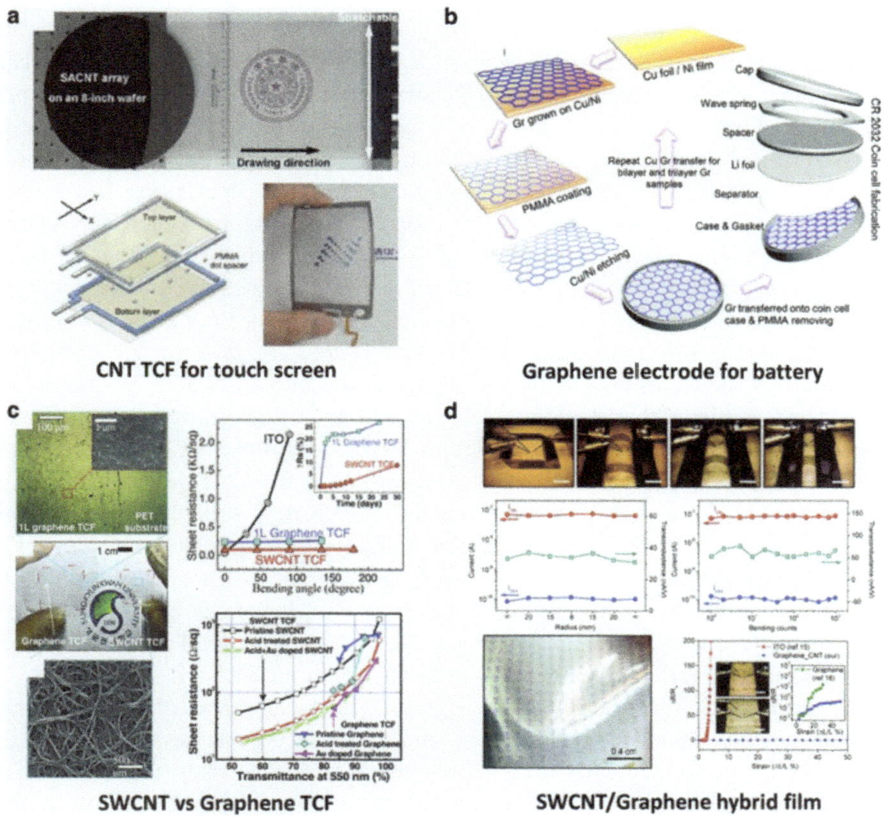

Figure 6 CNTs and graphene as conducting electrodes. **a**, Touch screen using yarned CNT film from vertically aligned CNTs. Adapted with permission [103]. **b**, Li-ion battery using CVD graphene as an electrode. Reproduced with permission [111]. Copyright 2012, American Chemical Society. **c**, Comparison of the properties (bending angle vs sheet resistance, and transmittance vs sheet resistance) of CNT- and graphene-based TCF with ITO film. Adapted with permission [47]. **d**, Mechanical advantage of SWCNT/graphene hybrid electrode. Reproduced with permission [123]. Copyright 2011, American Chemical Society.

the advantage of better mechanical stability and are fabricated from a more ubiquitous chemical element, carbon. Figure 6b shows an example that graphene can be used as electrodes to study Li ion diffusion through graphite in lithium-ion batteries [111]. Together with CNTs, graphene is attractive as a conducting film [112,113], due to a large theoretically-predicted conductivity and good chemical stability. In particular, a scalable CVD process to produce large sheets of graphene with high transmittance and robust adhesion to plastic polymers opens the possibility of using graphene in numerous applications in soft electronics. Still the improvement of sheet resistance of the film is an important issue for conducting films. Similar to CNT films, the chemical doping approach has been widely studied for conductivity improvement in graphene films [114–116]. A new approach of layer-by-layer (LbL) doping to improve the conductivity of transparent graphene films has been proposed [117]. Each layer was transferred to a polyethylene terephthalate (PET) substrate followed by $AuCl_3$ doping. This approach demonstrates not only improvement of sheet resistance and uniformity but also better environmental stability compared to topmost layer doping. The optimized LbL-doped four-layer graphene shows a sheet resistance of 54 Ω/sq and a transmittance of 85% (at 550 nm) with a robust bending stability. The performance of the graphene conducting films need to be further tuned and improved to meet different requirements of practical flexible products [118,119]. Both CNTs and graphene TCFs have a remarkable spectral response in the UV region, compared to the poor response of ITO films, as shown in Figure 6c [47]. While ITO shows a rapid increase in the sheet resistance due to cracking of the film as the bending angle increases, SWCNTs and graphene films show almost no significant change in the sheet resistance. One drawback of the CNT TCF film is that the performance strongly relies on the dispersion of CNTs in solution. In graphene case, the bottleneck process is the transfer process, which often involves wrinkles and crack formation. Compared to a two-dimensional graphene film, the SWCNT/graphene hybrid electrode is interesting due to its enhanced mechanical properties

[120–122]. The SWCNT/graphene hybrid electrode showed a 36% resistance change at a 50% strain, as shown in Figure 6d [123]. The resistance change is remarkably smaller than found in ITO electrodes (i.e., 2000% at 5% strain) and even in a few layers of graphene (i.e., 200% at 30% strain). This superb stretching performance results from the use of graphene and SWCNT network. A continuous and robust contact can be formed between the SWCNT network and the graphene electrode even with graphene layer cracks under strain. This one- and two-dimensional material combination could well provide CNTs and graphene as an appropriate soft and transparent electrode. Table 1 summarizes the transmittance and sheet resistance of various films. It seems that doping is very necessary to reduce sheet resistance. It is also noted that the CNT/graphene hybrid may improve the sheet resistance. This will be a future research direction.

2.2.2 Active channel - CNT and graphene FETs

Miniaturization is the most important issue not only to increase device integration density but also to improve FET performance for complicated operations. Semiconducting Si technology has given great contributions to society, but now faces scaling which involves heat and power consumption issues due to the fundamental limitations of Si. Atomic-thick nano-carbon materials might satisfy the scaling issue and give great benefits with combination of electrical/mechanical/optical advantages. As an active channel component, SWCNTs and graphene have been studied for fabricating FETs and p – n junctions to demonstrate their potential to outperform established materials for next-generation electronics [125–128]. Here, we discuss extensively the advantages and challenges of such nano-carbon materials for the use of FETs and furthermore their adaptability to silicon technology.

Figure 7 shows that various kinds of FETs using nano-carbon materials-based active channel. Diverse geometries of FETs based on semiconducting SWCNTs have been the subject of intensive research [129–131]. An individual SWCNT FET shows favorable device characteristics such as large on-off ratio ($>10^5$), at room-temperature operation [132–134]. With single CNT studies, it has also been demonstrated that the saturated on-current level can be simply determined from the work function difference between the CNT and metal or Schottky barrier height formed at the junction, as shown in Figure 7a [135]. For fabricating this transistor, e-beam lithography is used to pattern the electrodes to desired positions, but has limitations for realistic multi-array transistors. An alternative easy fabrication method without e-beam lithography is required for large-scale integration for practical electronic device applications. Although isolated SWCNTs are not relevant to future applications at their current stage, numerous works show that the aligned arrays of SWCNTs or random networks can serve as an active channel component. Figure 7b shows FETs with aligned arrays of SWCNTs. The use of dense aligned arrays of linear SWCNTs was used as an effective semiconducting channel suitable for integration into transistors and other classes of electronic devices [32]. The tubes were parallel to one another to better than 0.1 degree. The average CNT density can be as high as 10 SWCNT/μm, and the film provides good device-level performance characteristics with mobility of ~1,000 $cm^2\ V^{-1}\ s^{-1}$ [136,137]. Figure 7c shows an array of FETs with random network SWCNTs that were synthesized on a catalyst (0.01 M of ferrocene) array by using a plasma-enhanced chemical vapor deposition (PECVD) method at low temperature (450°C) [138]. SWCNTs network was placed between the source and drain electrodes and played a role of active channel path. This random network type morphology has the potential applicability from CNT thin film transistors (TFTs) to large-scale flexible electronics due to its good uniformity and processability over a large-area, which is alternative to conventional organic or other classes of semiconductors for integrated circuitry applications [126,139]. However, the gate modulation is degraded due to the inclusion of some metallic CNTs in the channel. Strategies to reduce metallic CNTs in the channel will be discussed in the next Section 3.2.1. Figure 7d shows an example of graphene channel FETs on a flexible plastic substrate [140]. In graphene, the charge carriers in the two-dimensional (2D) channel can change from electrons to holes subject to electrostatic gate with a minimum

Table 1 Performance comparisons for TCFs based on graphene and carbon nanotubes

Material	Preparation method	Transmittance (% at 300 Ω/sq sheet resistance)	Sheet resistance (Ω/sq at 90% transmittance)	Flexibility	Stretchability
Random network CNTs [108]	Spray & $AuCl_3$ doping	95.7	110	O	O
Yarning CNTs [103]	Laser trimming & Metal deposition	91	208	O	O
CVD Graphene [117]	Layer-by-layer doping	97	108	O	O
CNT-Graphene hybrid [123]	Solid-phase layer-stacking	70	735	O	O
Metal-Graphene hybrid [124]	Metal grid & Graphene transfer	-	20	O	O
ITO [104]	Sputtering	91	80	Poor	-

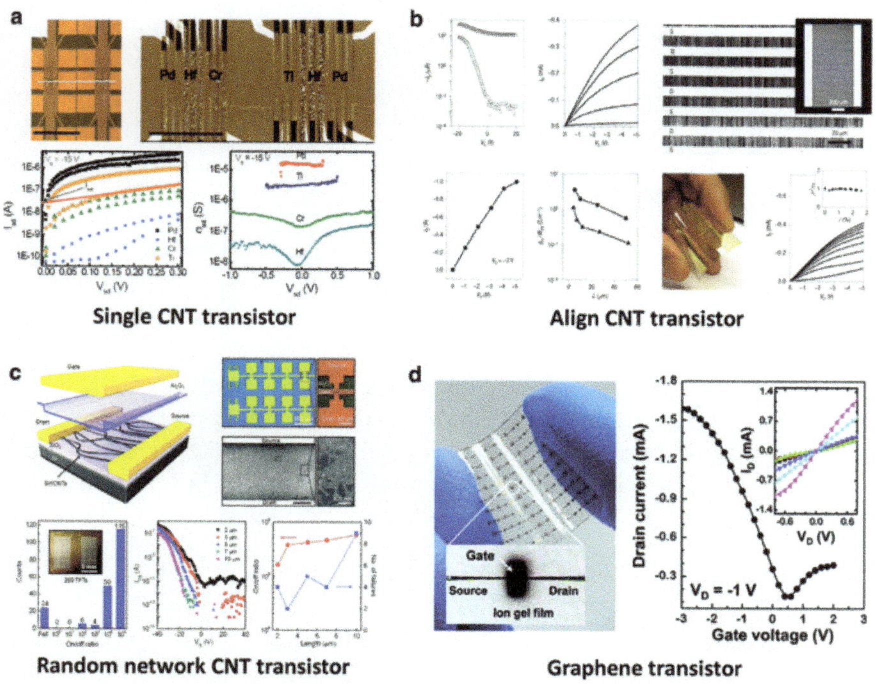

Figure 7 Morphologies and characteristics of CNTs and graphene FETs. a, Single CNT transistors with different metal electrodes (Pd, Hf, Cr, and Ti). Reproduced with permission [135]. Copyright 2011, American Chemical Society. **b**, Electrical performance, SEM images, and optical microscopy images of flexible TFTs using aligned CNTs array. Reproduced with permission [32]. Copyright 2007, Nature Publishing Group. **c**, Array of FETs with random network SWCNTs. Reproduced with permission [138]. Copyright 2009, American Chemical Society. **d**, Flexible graphene transistor with ion gel dielectric. Reproduced with permission [140]. Copyright 2010, American Chemical Society.

density (or Dirac) point characterizing the transition [127,141–143]. The experimental graphene FETs have extremely large mobility compared to SWCNT FETs, while on/off ratio is as low as ~10 due to zero band gap. Despite low on/off ratio, high transconductances and current saturation are achieved, making graphene devices suited for analogue applications [144].

2.2.2.1 Performance control – on/off ratio control
One of the key issues in high-performance TFTs is high on/off ratio for efficient switching behavior. In the case of a CNT channel, the as-grown CNT network usually contains both semiconducting and metallic CNTs [145]. These metallic CNT paths reduce the on/off ratio of the transistor [146]. Since controlling the ratio of semiconducting to metallic CNTs leads to a trade-off between on/off ratio and charge carrier mobility of a transistor, engineering the proper parameter is important in terms of the type of applications. In the case of zero band gap graphene, opening the band gap is a big challenge in the way of achieving a higher on/off ratio in transistors [127]. Here, we introduce several strategies for increasing the on/off ratio of a transistor. In CNTs, electrical thinning and selective channel cutting, and separation approaches are described below. BLG and nanoribbon approaches will be discussed for increasing the on/off ratio in graphene transistors.

One method to obtain high on/off ratio involves electrical thinning of the thick MWCNTs and CNT bundles, as shown in Figure 8a [147]. The electrical thinning process involves sweeping the drain voltage from 0 V to negative values while holding the gate voltage at a just above the threshold. Multiple sweeps with increasing voltage eventually eliminate metallic CNT channels or thin nanotubes (or bundles) to increase on/off ratio [32,148]. After this procedure the off-state current in the devices is reduced to values consistent with semiconducting CNTs alone. A striping technique was used to cut metallic CNT paths [123,149]. Figure 8b shows the schematic image and SEM image of a region of the random network SWCNT channel. By inserting the cutting line perpendicular to the channel length direction, the metallic CNTs can be terminated and the on/off ratio increases. The critically important role of the cutting width in determining the electrical characteristics can be quantified. For cutting widths of 5 mm, the etched lines increase the on/off ratio by up to four orders of magnitude, while reducing the transconductance by only 40%. It is now possible to obtain uniform CNT thin films with only semiconducting behavior by the techniques of

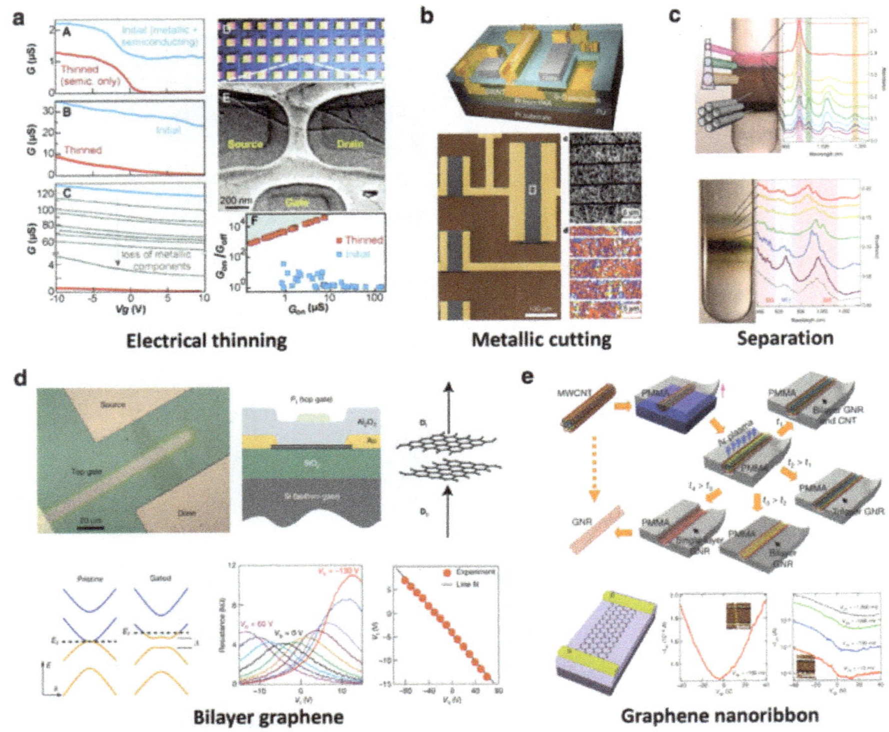

Figure 8 Various methods of improving on/off ratio of FET based on CNTs and graphene. a, Thinning of MWCNTs and CNT bundles by applying bias. Reproduced with permission [147]. Copyright 2001, American Association for the Advancement of Science. **b**, Schematic and SEM image of a region of the random network SWCNT channel. A striping technique was used to cut metallic CNT paths. Reproduced with permission [149]. Copyright 2008, Nature Publishing Group. **c**, Separation of semiconducting CNTs and metallic CNTs by density-gradient method. Reproduced with permission [153]. Copyright 2006, Nature Publishing Group. **d**, BLG transistor with top and bottom gate to open band gap. Applying perpendicular field from bottom gate, band gap of the BLG can opened up to 250 meV. Reproduced with permission [157]. Copyright 2009, Nature Publishing Group. **e**, Graphene nanoribbons with a width below 10 nm were obtained by upzipping CNTs. By narrowing the width of graphene to a few nanometers, a quantum confinement effect of carriers happens to open the band gap. Reproduced with permission [162]. Copyright 2009, Nature Publishing Group.

semiconducting/metallic CNT separation in solution [150–152]. The purification processes produce separated CNTs in solution of the same chirality, diameter, length and semiconducting/metallic type. A self-sorting method to achieve a chirality separated CNT thin film by controlling surface chemistry and a further large-scale demonstration was reported in Figure 8c [153]. The representative techniques are density gradient ultracentrifugation (DGU) and gel chromatography, which can produce >99% semiconducting CNTs and continue to improve. Despite the quite low productivity, yield, and high process cost, this DGU technique appears to be the most promising method to prepare semiconducting CNT materials [153]. The gel chromatography separation method, much simpler than DGU method, is based on the strength of the structure-dependent interaction of CNTs with an allyl dextran-based gel [152]. TFTs based on such separated CNTs also provide high on/off ratio. BLG has a unique dispersion relationship whereby application of a strong transverse electric field breaks electron–hole inversion symmetry [154–156]. Experimentally, it has been reported that an optical bandgap of ~ 250 meV is possible. The effective electrical gap is smaller than the reported optical gap, typically due to the presence of disorder and sample imhomogeneities. Even so, large improvements in on/off ratios and the existence of an insulating state at charge neutrality have been observed (Figure 8d) [157]. In these dual-gate BLG transistors, on/off ratios of ~ 100 and ~2000 at room temperature and 20 K have been reported, respectively [158]. BLG is disadvantageous compared to graphene monolayer since acoustic-phonon scattering is increased strongly, optical-phonon scattering is reduced, and a parabolic band dispersion near the band edge reduces carrier mobility compared with monolayer graphene [159]. Moreover, the band structure of BLG can be modified, with a larger bandgap possible by applying a combination of strain (along z axis) and an electrical field. However, this approach is unfeasible with current technology. A new strategy demonstrated that benzyl viologen (BV) as an electron-donating group and bis(trifluoromethanesulfonyl)imide (TFSI) as an electron-withdrawing group are conjugated on the top and bottom sides of bilayer graphene to open the band gap [160,161]. This compensation doping induces a high local electric field in the bilayer, but has the limitation of weak field-effect due to a large disorder potential. The graphene nanoribbon (GNR) strategy is to ideally introduce a quantum confinement effect of carriers to open the band gap by narrowing the width of graphene to a nanometer scale. In reality, this strategy is limited by fabrication procedures. Instead of confinement-induced gap, this leads to a coulomb blockade effect that is strongly enhanced for dimensions below 20 nm. Graphene nanoribbons with a width below 10 nm can be obtained by upzipping CNTs (Figure 8e) [162] and by solution-phase stripping from bulk graphite [163]. The GNR transistors exhibited an on-off ratio of $\sim 10^7$ at room temperature [162–165]. Similar to the GNR method, graphene with a nanomesh structure can open up a band gap and shows an on/off ratio of $>10^2$ in a large sheet of graphene [166,167]. However, these GNR and graphene nanomesh transistors have poor on-state conductivity and cannot be used for high-speed devices unless a new method is found due to reduce scattering at the edges. The band gap of graphene can be modulated by chemical and physical doping processes. Band gaps of boron- and nitrogen-doped graphene transistors showed an on/off ratio of >100 [168,169]. It also has been reported that by patterned adsorption of atomic H onto the graphene surface, surface absorption can induce a band gap in graphene of at least 450 meV around the Fermi level [170]. Yet, again the degradation of mobility due to sp^3 hybridization with atomic H makes this approach impractical.

2.2.2.2 Performance Control – Polarity Control Although CNTs and graphene intrinsically have an ambipolar transport property, both show p-type behavior under ambient conditions due to contacts, doping by oxidizing acids, or doping by the adsorption of atmospheric oxygen molecules and/or moisture. It is important to control the carrier type of nano-carbon transistors for applying "complementary metal-oxide-semiconductor (CMOS) technology" because high noise immunity and low static power consumption are critical issues in the modern semiconductor industry. Therefore, it is desired to control the major carrier types of CNTs and graphene FETs by chemical and/or nonchemical doping methods. Here, we introduce several polarity control methods to modify the majority carriers in CNT- and graphene-based transistors such as chemical doping, oxygen doping, electrostatic doping, trap charge-induced doping, and metal work function engineering.

In order to have n-type conversion and p-type enhancement behavior in CNTs under ambient conditions, various chemical doping strategies have been investigated [171–179]. The choice of chemical dopant is complicated by the fact that the redox potential of CNTs is strongly diameter-dependent, as shown in Figure 9a [108]. The values in parentheses indicate the chiral index of the SWCNTs and the reduction potentials of dopants (BV, NADH, DDQ, $NOBF_4$, and $AuCl_3$) are also indicated as dotted lines. As shown in Figure 9a, the Au^{3+} ion has the large reduction potential of 1.50 V, which acts as p-type doping in CNTs. BV has an oxidation potential of −1.1 V, which implies that BV can act as an n-type dopants. BV donates electrons to the empty conduction band of semiconducting CNTs [180]. The right panel of Figure 9a shows an example of n-type CNT transistor by precisely positioning BV with inkjet printing on CNTs channel region [181]. Using β-nicotinamide adenine dinucleotide (reduced dipotassium salt, NADH), a type conversion in CNTs is also demonstrated distinctly [182]. A reduction potential of tetrafluorotetracyano-p-quinodimethane (F4TCNQ) in the range of 0.1 V to 0.2 V makes it an electron extractor and p-type dopant [183]. For graphene, it has been demonstrated that the work function of CVD graphene can be modulated up to 1.1 eV with BV doping [184]. Similarly, other work showed GO doping with Au allowed control of the work function [185]. For BLG, surface chemical doping in BLG can be utilized to induce a vertical displacement field. Interestingly, tunable Dirac points can be rationally controlled by the amount of BV doping, providing complementary inverter circuits [186]. Figure 9b shows a simple way to control polarity by just annealing the p-CNT FET in vacuum, converting it to an n-CNT FET [187]. One of the reasons for having p-character in CNT FETs is due to the interaction with O_2 physisorbs on the CNT surface [148,188,189]. Originally a p-type CNT FET was converted to n-type after annealing process for removing O_2 molecules [187]. It has been shown that the type conversion of CNT FETs could be possible by electrostatic doping using a charge-trap layer between the gate electrode and CNT channel [190,191]. Figure 9c shows the transfer characteristics of p-type and n-type CNT FETs converted using an Au floating gate. At high negative gate bias range, positive charges are trapped in the trap layer, and the threshold voltage is shifted in the negative bias direction. Therefore, the FETs show n-type characteristics in relatively small gate voltage sweep range. On the contrary, when high positive gate bias is initially applied, which traps the negative charges, the FETs show p-type characteristics in a relatively small gate voltage sweep range. Figure 9d shows the electrical performance of an initially p-type characteristic as it is gradually changed to n-type via

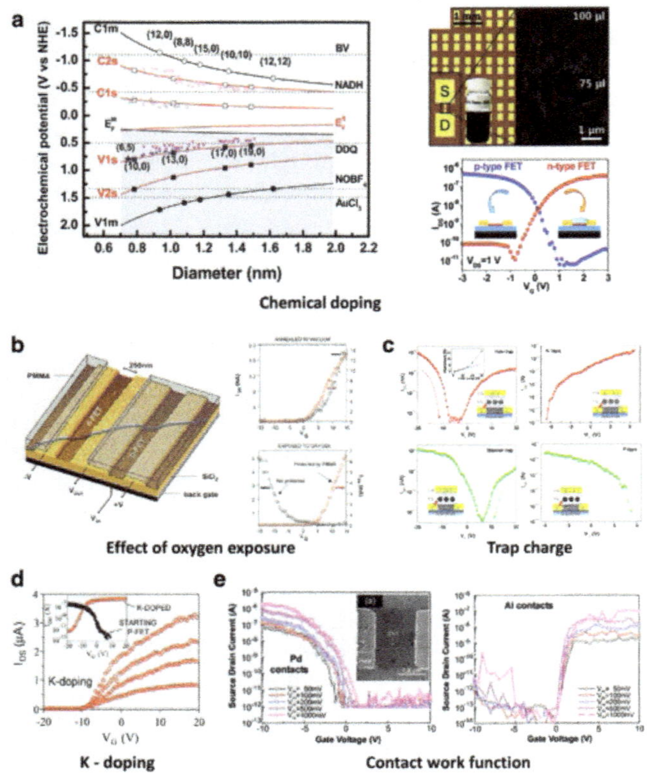

Figure 9 Various methods of polarity control of FETs based on CNTs and graphene. a, Redox potential of nanotubes as a function of the diameter (left). This Reproduced with permission [108]. Copyright 2010, Royal Society of Chemistry. Array of n-type CNT transistor by precisely positioning an air-stable BV. Reproduced with permission [181]. Copyright 2011, American Chemical Society. **b**, Effect of oxygen on p-doping. I-V curves of originally p-type CNT FET, with the nanotube capped with PMMA, have been converted to n-type. Reproduced with permission [187]. Copyright 2001, American Chemical Society. **c**, The type conversion of CNT FETs by trap layer-induced electrostatic doping. Adapted with permission [190]. **d**, I-V characteristic of an initially p-type characteristic in SWCNT FET, gradually changed to n-type caused by increasing amounts of K. Reproduced with permission [187]. Copyright 2001, American Chemical Society. **e**, Polarity control by metal (Pd and Al) work function. Reproduced with permission [193]. Copyright 2005, American Institute of Physics.

increasing amounts of K on the nanotube [187,192]. Potassium ions have a high oxidation potential of –0.7 V and act as an electron donor (n-type dopants) for CNTs. Logic circuits and pn junctions were fabricated by covering half of a CNT FET with PMMA and K-doping the exposed regions. The electrical polarity of SWCNT FETs can be affected by the work function of the contact metal, especially by the contact barrier control for the injection of carriers [148,175,193]. Figure 9e shows the transfer characteristics of CNT FETs using different metal contact electrodes such as Pd and Al [193]. The transfer characteristics show the presence of a p-type on-state but no n-branch in the case of high-work function metals such as Pd and Ti, ambipolar behavior in the case of Mg, and n-type only behavior in the case of Ca electrodes [194]. By varying the work function, the band alignment for a Mg-contacted device has efficient hole and electron injection, resulting in ambipolar characteristics. Conversely, due to work function and surface dipole formation, CNTs contacted by Ca electrodes have a suppressed p-type branch due to large energy barrier for holes. Although this method works to control the injection of carriers in single devices, the use of different metal electrodes in high-density devices is commercially unreasonable and resulting devices still have highly variable contact properties.

Numerous efforts have been made to get higher on/off ratios and better control of carrier type in nano-carbon transistors. In order to understand advantages and disadvantages for CNT and graphene FETs, a side-by-side comparison is required. Table 2 shows the comparison for FET performance of CNTs and graphene devices. CNT FET and graphene devices exhibit output performances in a different manner. Moreover, the performances are distinct in different types of FET devices consisting of different forms of CNTs (single CNT, aligned CNT network, random CNT network) and graphene (CVD graphene, exfoliated BLG, GNR) with different gate structures. Nevertheless, a clear trade-off behavior between on/off ratio and mobility for each device was shown.

Table 2 Device performance of various CNTs and graphene FETs

Channel	Preparation method	Transistor structure	Gate dielectric	Gate length (μm)	Carrier type	On/Off ratio	Mobility (cm^2/Vs)
Single CNT [32]	CVD on quartz	Back gate	SiO_2	5	p-type	10^5	$636^{(C)}$
Aligned CNTs [32]	Electrical breakdown	Back gate	HfO_2	12	p-type	$2 \rightarrow 10^4$	$570^{(C)} \rightarrow 200^{(C)}$
Random network CNTs [149]	Channel cutting	Top gate	HfO_2	100	p-type	$10 \rightarrow 10^4$	$200^{(C)} \rightarrow 80^{(C)}$
Random network CNTs [153]	97% separated CNTs	Back gate	SiO_2	20	p-type	10^4	$20^{(p)}$
Random network CNTs [181]	Viologen doped CNTs	Back gate	HfO_2	9	p → n-type	10^3	$2^{(p)}$
Exfoliated graphene [141]	Monolayer graphene	Back gate	SiO_2	4	Ambipolar	10	$10{,}000^{(p)}$
CVD grown graphene [195]	Monolayer graphene	Back gate	SiO_2	5	Ambipolar	10	$1{,}100^{(p)}$
Exfoliated graphene [158]	Bilayer graphene	Dual gate	SiO_2 (Back) HfO_2 (Top)	1.6	Ambipolar	5 → 100	-
Graphene nanoribbon [162]	16 →6 nm nanoribbon	Back gate	SiO_2	0.25	Ambipolar → p-type	1.5 → 100	-

p: Parallel plate Model, c: Cylindrical Model, h: Hole Mobility, e: Electron Mobility.

2.3 Flexible electronics

2.3.1 Integrated logic circuits

Next-generation military and industrial radio-frequency (RF) surveillance systems will benefit from flexibility and stretchability of circuits for increased resilience. A realistic short- and medium-term goal for carbon electronics is utilizing the combination of electrical, mechanical, and optical properties of CNTs and graphene thin films to replace organic semiconductors and a-Si in these flexible/stretchable systems [196–204]. In this section, we introduce recent progress for integrating high-quality circuits on plastic substrates.

Figure 10a shows an integrated circuit fabricated with monolayer graphene as the electrodes and a SWCNT network for the channel [123]. Using this layout, transparent logic circuit arrays (inverters, NOR gates, and NAND gates) using SWCNT-channel/graphene-electrodes transistors were fabricated with a high yield of 80%. The authors connected two p-type transistors to create a PMOS inverter with gain of approximately 1.4, with an operating voltage range of 0–5 V. PMOS NOR and NAND logic gates were similarly constructed using three SWCNT/graphene transistors. The graphene electrode and the SWCNT network channel are desirable not only for flexible and stretchable electronics, but also for use with invisible electronics due to the high transparency of atomically thin materials. Figure 10b shows a flexible four-bit row decoder circuit using SWCNT as the channel and metal electrodes [149]. A binary-encoded input of four data bits is successfully decoded using this decoder circuit. Due to the high mobility of the SWCNT thin films, even with critical dimensions (100 μm) these decoder circuits can successfully operate in the kHz region. With such large channel lengths, cheap and scalable patterning methods such as screen printing are possible. More complex device structures are also easily possible such as master–slave delay flip-flops and 21-stage ring oscillators which were fabricated on PEN substrates [205]. Figure 10c demonstrates flexible complementary graphene inverters prepared on a plastic substrate by connecting two graphene transistors with a coplanar gate configuration. Fabrication was achieved using only two materials: graphene and an ion gel gate dielectric [206]. Unlike conventional solid state dielectrics, the operation of ion-gel gated transistors is based on the formation of a high capacitance electric double layer (EDL) under an electric field. The graphene inverter operates uniquely with two identical ambipolar transistors, unlike complementary inverters based on separate n- and p-channel transistors. Also in contrast to typical CMOS inverters, the output voltage did not saturate to zero or the supply voltage (V_{DD}) due to the zero band gap of graphene [206]. With an estimated maximum voltage gain of 2.6, the technology is sufficient to drive subsequent components in logic circuits. Graphene-based frequency doublers and modulators on rigid substrates have been reported to demonstrate the feasible usage of graphene in analogue electronics [207–211]. Figure 10d shows a flexible all-graphene modulator circuit for quaternary digital modulations, which can encode two bits of information per symbol [212]. A couple of transistors are required for these two quaternary modulations.

2.3.2 Other Flexible Applications

Applications ranging from flexible solar cells, displays, e-papers, wearable and biomedical skin-like devices open up new opportunities in the field of electronics. In this section, we describe applications of several flexible devices possible with carbon electronics jsuch as sensors, LEDs, RF devices, stimulators, and memory devices.

As an example of further applications of flexible devices, Figure 11a demonstrates an active-matrix backplane for an artificial electronic skin (e-skin) device, capable of spatial touch mapping [213]. The SWCNT TFTs are used for a mechanically flexible backplane with polyimide as a

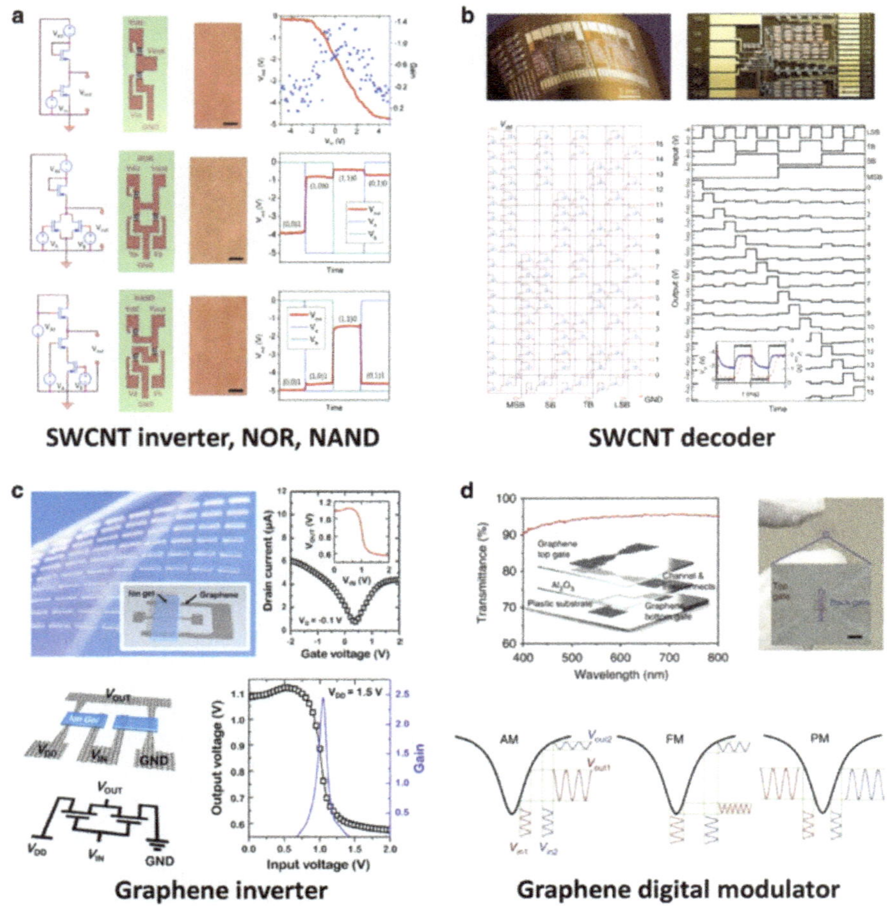

Figure 10 Flexible logic circuits using CNTs and graphene. a, Transparent and flexible logic circuits (inverter, NAND, and NOR) using graphene as electrodes and random network CNTs as the channel. Reproduced with permission [123]. Copyright 2011, American Chemical Society. **b**, Flexible four-bit row decoder circuit using SWCNT channel and metal electrodes. Reproduced with permission [149]. Copyright 2008, Nature Publishing Group. **c**, Flexible complementary graphene inverters prepared on plastic substrate with ion-gel gate dielectric. Reproduced with permission [206]. Copyright 2012, American Chemical Society. **d**, Transparent and flexible all-graphene digital modulator for quaternary digital modulations. Reproduced with permission [212]. Copyright 2012, Nature Publishing Group.

support substrate. The polyimide film substrate was utilized as a honeycomb mesh structure to make the substrate more robust against strain. Each pixel of pressure sensor is actively controlled by a SWCNT TFT. The sensor sensitivity shows ~ 30 $\mu SkPa^{-1}$, which is three times larger than previous NW-based sensors [214]. Figure 11b shows the flexible active-matrix design with SWCNTs as the channel material. In these devices, high current drive is needed to actively switch OLEDs [215]. Each pixel is controlled by a SWCNT TFT that acts as a switch for an active-matrix of OLED and pressure sensor. Alternating current electroluminescence devices on flexible PET substrates were also demonstrated based on monolayer graphene electrodes [216]. Graphene seems to be an ideal material for high-speed systems owing to its extremely high carrier mobility. Despite poor switching behavior of graphene transistors limits their usage in digital/logic applications, they are still promising in the analogue/RF applications due to their atomic-thick layout that allows for shorter scaling of channel length. The combination of high speed and flexibility is a big challenge for flexible graphene RF devices [217–221]. RF devices using graphene have achieved cut-off frequencies between 100–300 GHz. Figure 11c shows the flexible solution-based graphene transistors at GHz frequencies with a current gain cut-off frequency of 2.2 GHz and a power gain cut-off frequency of 550 MHz [217]. Noninvasive probing and manipulation of biological tissue is another field where graphene is useful. Figure 11d reports a nonvascular surgical method to increase cerebral blood volume using a flexible, transparent, and biocompatible graphene electrical field stimulator [222]. The flexible graphene stimulator was placed onto the cortical brain without tissue damage or unnecessary neuronal activation. A noncontact electric field was

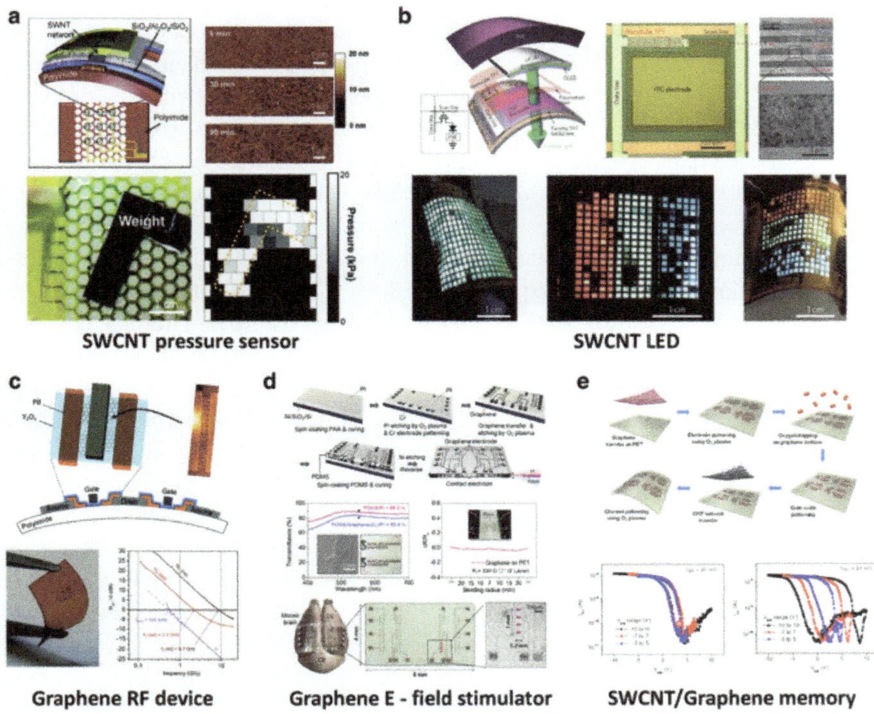

Figure 11 Various flexible applications using CNTs and graphene. a, Active-matrix of SWCNT TFTs for a pressure sensor device. Reproduced with permission [213]. Copyright 2011, American Chemical Society. **b**, Flexible active-matrix design using SWCNTs as the channel material of the TFTs in OLEDs. Reproduced with permission [215]. Copyright 2013, Nature Publishing Group. **c**, Flexible RF device using solution-based graphene. Graphene is an ideal material for high-speed communication systems owing to its uniquely high carrier mobility. Reproduced with permission [217]. Copyright 2012, American Chemical Society. **d**, Flexible, transparent, and noncytotoxic graphene electric field stimulator. Reproduced with permission [222]. Copyright 2013, American Chemical Society. **e**, Transparent and flexible memory devices using SWCNT channel and graphene electrodes. The oxygen-decorated graphene electrode revealed an initially large hysteresis in SWCNT/graphene TFT. Adapted with permission [223].

applied at a specific local blood vessel to detect effective cerebral blood volume increases in mouse brains using *in vivo* optical recordings of signal imaging. In Figure 11e, transparent and flexible memory devices were fabricated using graphene electrodes and SWCNT channel [223]. The original electrical characteristics of the FET using graphene electrode without ozone treatment show small hysteresis. When the graphene gate was treated under an ozone generator, oxygen atoms and graphene have bonding as C-O-C, C = O, and C-OH, which acted as charge trap sites. The FET with oxygen-decorated graphene electrode exhibits large hysteresis. This hysteresis-controllable FET can act as memory device, and showed no degradation of transmittance after oxygen decoration. This result is noticeable, compared to Au and Al nanoparticle trap layers that provided an 11.4% and 25% decrease in transmittance, respectively [224]. Flexible organic resistive memory devices with multilayer graphene electrodes were also reported [225]. Memory devices using a graphene oxide film were also fabricated on flexible substrates with reliable memory performance in terms of retention and endurance [226].

2.4 Stretchable electronics

Stretchability is a key parameter in the development of wearable devices that can be embedded into clothes and garments or even attached directly to the skin, where high levels of strain will be encountered. Possible applications include the human-friendly devices for detecting human motions, monitoring health system, and healing. In addition to flexibility, all these stretchable applications demand tolerance of large levels of strain (> > 1%) without fracture or significant degradation in electronic properties. The mainstream strategy to realize improved stretchability focuses on the development of stretchable materials including organic polymers, networks of 1-D wires, and nano-carbons [227–231]. Owing to the difficulties in developing new stretchable materials, geometrical engineering of the structures also needs to be addressed [6]. For example, ultrathin buckled geometries and pre-strained geometrically wavy materials offer stretchability with applied strain [232–235]. These devices can be integrated into larger systems containing conventional rigid materials. In this section, we introduce developed classes of material-based stretchable devices that use CNTs and graphene thin films on elastomer substrates.

2.4.1 Stretchable conducting films

Loading a SWCNT random network onto an elastomeric substrate simply affords a stretchable conducting film with the ability to accommodate strains greater than 20% [236–238]. The left panel of Figure 12a shows transparent, conducting spray-deposited films of SWCNTs that can be stretched by applying strain along each axis [239]. This stretchable SWCNT film accommodates the stretchability by up to 150% with conductivities as high as 2,200 S cm^{-1} at the strain of 150%. This property can be utilized to construct strain sensors, with performance comparable to conventional metal-strain gauges. Using a nonlinear buckling process as shown in the right panel of Figure 12a, ribbon arrays of CNT films can be modified into a "wavy" layout [231]. With a pre-strain (100%) method, the wavy CNT ribbon can accommodate large stretching with the 4.1% resistance increases when the wavy CNT ribbon is stretched to the pre-strain stage. Applied strains lead to a reversible deformation of these buckled patterns which change the electrical properties. Together with the good optical and electrical properties, graphene films have excellent mechanical properties applicable to stretchable electrodes. One such example consists of few-layer CVD grown graphene films transferred onto elastic substrates, as shown in the left panel of Figure 12b [240]. The transferred film on an unstrained substrate recovers its original resistance after stretching by ~6%. In this work, the authors also transferred the film to pre-strained (12%) substrates to enhance the electromechanical stabilities. Both longitudinal and transverse resistances (R_y and R_x) were stable up to ~11% stretching with only one order of magnitude change at ~25% stretching. 3D-graphene macroscopic structures formed with a foam-like network of graphene was also developed using template-directed CVD (right panel of Figure 12b) [241]. The composites fabricated by this approach are a monolithic 3D-graphene network, in which electrical and mechanical properties were improved by using continuous CVD grown graphene building blocks. The results of graphene composites show stretchability over 50% with resistance changes stable after the fifth cycle of stretch-release.

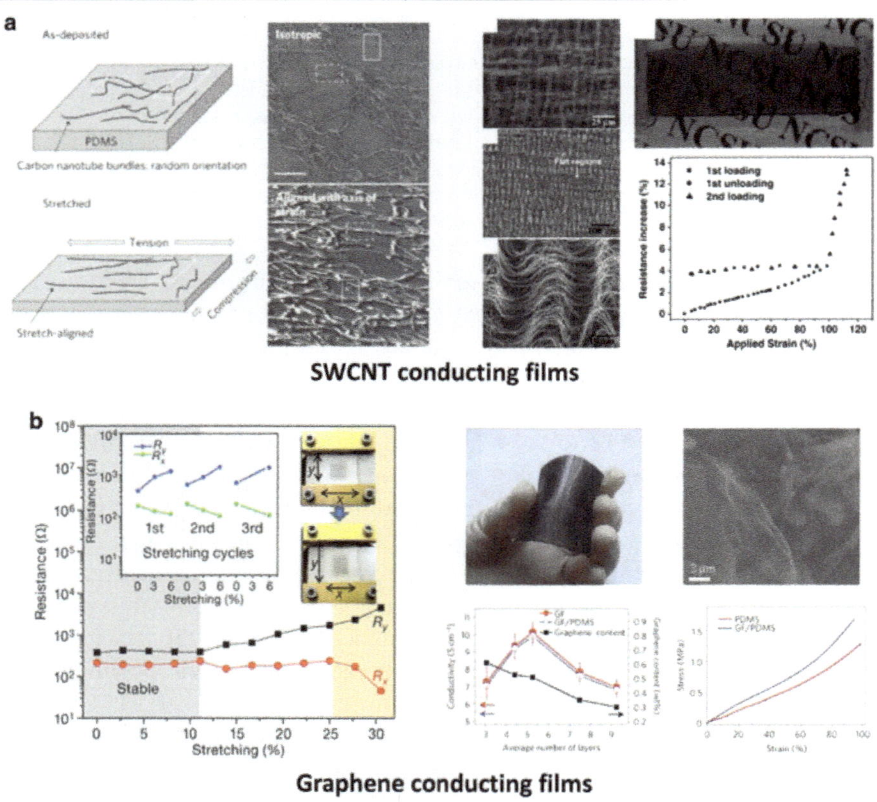

Figure 12 Stretchable conducting films using CNTs and graphene. a, Transparent, conducting spray-deposited films of SWCNTs that can be rendered stretchable by applying strain along each axis (left). Reproduced with permission [239]. Copyright 2011, Nature Publishing Group. Wavy ribbons of CNTs are embedded in elastomeric substrates to fabricate stretchable conductors (right). Adapted with permission [231]. **b**, Stretchable conducting films using few-layer CVD grown graphene (left). 3D-graphene macroscopic structure with a foam-like network graphene (right). Reproduced with permission [240,241]. Copyright 2009 and 2011, Nature Publishing Group.

2.4.2 Stretchable applications

Extreme difficulties are associated with the development of complete sets of stretchable electronics because all elements of the system need to be stretched out together. For instance, currently available carbon-based devices such as TFTs usually exhibit limited flexibility and stretchability owing to the use of fragile oxide dielectrics such as Al_2O_3 and SiO_2. Polymer dielectrics have modest electrical performance despite their excellent bendability [242]. In this section, we introduce several strategies to fabricate stretchable devices using CNTs and graphene.

Reproduced with permission [246] Copyright 2011, Nature Publishing Group.

Figure 13a shows transparent and stretchable integrated circuits composed of CNTs and polymer dielectric [243]. The active channel and electrodes were all fabricated from CNTs (semiconducting and metallic), with PMMA dielectric layer and a plastic substrates. Although these were fabricated on plastic substrate, thermo-pressure was used for forming dorm-shape biaxial strain. The devices exhibit biaxial stretchability of up to 18% and the level of logic circuits include inverters, ring oscillators, NOR, NAND, XOR gates, and static random access memory (SRAM) cells. In Figure 13b, a graphene FET array on a stretchable rubber substrate with ion-gel dielectric is introduced [244]. Such all-graphene devices (graphene composes both the channel and electrodes) exhibit hole and electron mobilities of ~1188 and ~422 $cm^2V^{-1}\ s^{-1}$, respectively with stable operation up to 5% stretching. Although the stretchability of transistors is moderate, impressively the electrical properties were invariant even after 1000 cycles. Figure 13c shows a new approach for preparing a wrinkled gate dielectric using a transfer method to maximize the performance of the oxide without compromising the ability to stretch and bend [245]. A 50 nm aluminum oxide (Al_2O_3) layer was deposited onto rough Cu foil using atomic layer deposition. After coating with PMMA, Cu foil was chemically etched, and the Al_2O_3 layer was then transferred as dielectric layer. This transferred Al_2O_3 layer was wrinkled with a "wavy" structure, which was robust under high tensile strain. The resulting TFTs exhibited device-acceptable electrical

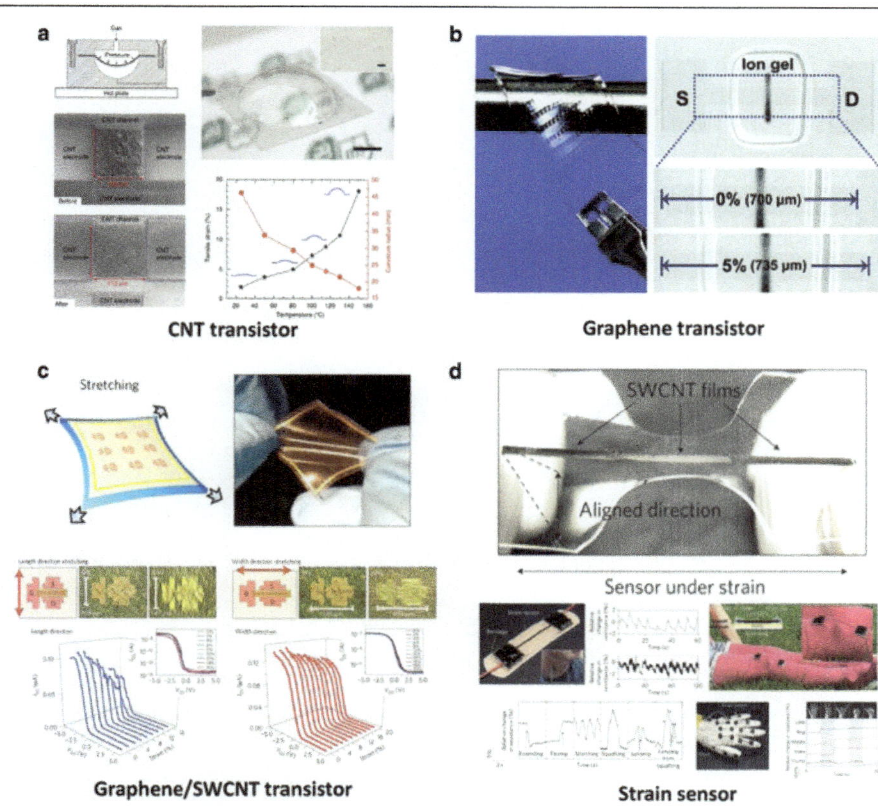

Figure 13 Stretchable applications using CNTs and graphene. a, Transparent and soft integrated circuits with random network SWCNT channel and PMMA dielectric layer on the PEN substrate. Reproduced with permission [243]. Copyright 2013, Nature Publishing Group. **b**, Graphene FET array on a stretchable rubber substrate with ion-gel dielectric. Reproduced with permission [244]. Copyright 2011, American Chemical Society. **c**, Stretchable and transparent TFTs combining SWCNTs/graphene with a geometrically wrinkled Al_2O_3 dielectric layer. Reproduced with permission [245]. Copyright 2013, Nature Publishing Group. **d**, Wearable and stretchable strain sensors fabricated from thin films of aligned SWCNTs. Reproduced with permission [246]. Copyright 2011, Nature Publishing Group.

performance with small gate leakage current due to the build-in air gap between wrinkled Al_2O_3 and graphene gate. The devices were stretched along the length direction (16% strain) and along the width direction (20% strain), as shown in Figure 13c. The devices were stretched and released up to a maximum of 1,000 times without deterioration. Figure 13d shows a class of wearable and stretchable devices fabricated from thin films of aligned SWCNTs [246]. When stretched, the films fracture into gaps and islands with tube bundles bridging the gaps. This mechanism allows the films to act as strain sensors with capabilities extending up to 280% strain, which is 50 times more than conventional metal strain gauges, with high durability (10,000 cycles at 150% strain), and fast response (delay time of 14 ms). When the CNT sensors were assembled on stockings, bandages and gloves to fabricate devices, the devices were able to detect human movement, typing, breathing and speech, each unique applications useful for developing human-friendly and bio-integrated devices [239]. Figure 14 shows a summary of the flexible/stretchable device layouts and circuit levels of devices using nano-carbon, followed by the demonstrations of electrical, optical and mechanical properties.

3 Conclusions

3.1 Summary and prospects

We have reviewed the current status of CNTs and graphene in diverse applications of soft electronics from material preparation to performance in logic circuits. Low-dimensional carbon materials exhibit superb electronic properties and promising performance and are attractive for future electronics. Methods for synthesizing one-dimensional CNT and two-dimensional graphene films, as

	Electrode-Channel	Logic level	Gate dielectric	Mobility (cm^2/Vs)	On/Off Ratio	Operation Voltage (V)	Transmittance (% at 550nm)	Flexibility	Stretchability (Over 1% strain)
a 	Au-CNT[149]	Inverter, NOR, NANAD, Decoder	HfO_2	80 (C)	10^4	2	-	O	-
b 	Au-CNT[205]	Inverter, Oscillator, NOR, NAND, Flip-flop	Al_2O_3	35 (p) 634 (C)	10^6	5	-	O	-
c 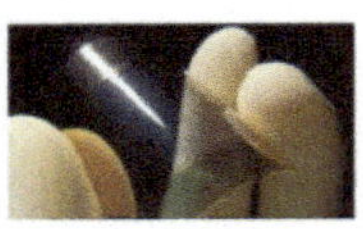	Graphene-CNT[123]	Inverter, NOR, NANAD	Al_2O_3	80 (p)	10^4	5	83.6	O	-
d 	Graphene-Graphene[244]	Transistor	Ion gel	1,131±96 (h) 362±45 (e)	4	2	83	O	5%
e 	CNT-CNT[243]	Inverter, Oscillator, XOR, SRAM	PMMA	1,027 (C)	10^5	10	85	O	18%
f 	Graphene-CNT[245]	Transistor	Wrinkled Al_2O_3	40.2 (p) 624.8 (C)	10^6	1	78	O	20%

p : Parallel plate Model, c : Cylindrical Model, h : Hole Mobility, e : Electron Mobility

Figure 14 Performance comparisons between graphene and carbon nanotube logic circuits in terms of their logic level, device characteristics, flexibility, and stretchability. a, Integrated circuits (inverter, NOR, NAND, and Decoder) with random network SWCNT channel on flexible plastic substrates. Since this device uses metal electrodes, there is no transmittance data. Reproduced with permission [149]. Copyright 2008, Nature Publishing Group. **b**, Flexible integrated circuits (inverter, oscillator, NOR, NAND, and Flip-flop) with random network SWCNT channel on the PEN substrate. Reproduced with permission [205]. Copyright 2011, Nature Publishing Group. **c**, Transparent and flexible logic circuits (inverter, NOR, and NAND) using graphene as electrodes and random network SWCNTs as an active channel. Reproduced with permission [123]. Copyright 2011, American Chemical Society. **d**, Graphene FET array on a stretchable rubber substrate with ion-gel dielectric. Reproduced with permission [244]. Copyright 2011, American Chemical Society. **e**, Transparent and soft integrated circuits (inverter, oscillator, XOR, and SRAM) with random network SWCNT channel and PMMA dielectric layer on the PEN substrate. Reproduced with permission [243]. Copyright 2013, Nature Publishing Group. **f**, Stretchable and transparent TFTs combining SWCNT/graphene with a geometrically wrinkled Al_2O_3 dielectric layer. Reproduced with permission [245]. Copyright 2013, Nature Publishing Group.

well as procedures for device fabrication on soft substrates have been discussed here. Both CNTs and graphene exemplify TCF properties including a high operational flexibility and stretchability that are not accessible with transparent ITO electrodes. Likewise, field effect mobilities of carbon-based transistors have reached levels unfeasible by organic semiconductors/a-Si. CNT FETs, whether composed of a single CNT, aligned CNTs, or random network CNTs, show high on/off ratio and mobility. Graphene FETs provide extremely high mobility but poor on/off ratio due to zero band gap. Engineering for on/off ratio increase and carrier polarity control were summarized. For applications in active electronics, SWCNT and graphene transistors can be assembled on a variety of substrates including flexible plastic and stretchable elastomers. Various complex integrated circuits based on nano-carbon materials have been demonstrated in the literature, as well. Each of these topics requires significant future exploration in order to realize commercialized applications of the immense potential of nano-carbon in next-generation electronics.

In spite of recent progress demonstrating the unique advantages of CNTs and graphene, the possible applications, social influence, addressable markets, and related economic issues will eventually decide the success of these nano-carbon materials. Both have unique and superb properties which open the possibility for soft electronics. Nevertheless, applications are limited by a different set of factors. Assemblies of CNTs are practical compared to the use of individual CNTs, but require the positioning of the CNTs in a specific direction, with desired density, and of desired metallicity/chirality. Methods to achieve this control are a current hot topic, but adoption of a particular method will require a high yield for industrial utilization even in niche applications. Conversely, graphene can be prepared in a large-area format. Yet, the transfer to a desired substrate may provoke damage in the graphene layer and degrade device performance. Therefore, developing a smart way of assembling CNTs to maximize the device performance and robust method of transfer of large-area graphene are two key ingredients that are unsolved but required for application. On a systems level, future electronics including biomedical applications with biocompatibility will require further research. For instance, CNTs and graphene combine synergistically, showing better flexibility and stretchability with no degradation of electrical performance when engineered to maximize potential. Additionally, combining both stretchable materials and stretchable geometries can allow for extremely stretchable systems. Aside from the engineering challenges of applying nano-carbon to soft electronics, CNTs and graphene are outstanding materials for demonstrating a number of basic science concepts in the fields of quantum electrodynamics, quantum optics, and quantum chemistry. Controlled synthesis and application of monolayer materials also allows exploration into a new class of vertical tunneling devices. Aside from carbon, other classes of graphene-like 2D materials such as transition-metal dichalcogenide (TMD) materials and boron nitride (BN), might also be promising in the field of soft electronics when a band gap or other electrical/mechanical properties are required. These related engineering opportunities in areas with the broad range of influential research topics provides strong motivation for continued efforts in human-friendly soft electronics.

Competing interests

The authors declare no competing financial interests.

Authors' contributions

SHC and YHL contributed to this work in the manuscript preparation. Both authors read and approved the final manuscript.

Authors' information

Sang Hoon Chae: Sang Hoon Chae is a researcher of the Center for Integrated Nanostructure Physics (CINAP), Institute of Basic Science (IBS) in Korea. He is a Ph.D. candidate within the Department of Energy Science at Sungkyunkwan University. He received his B.Sc. in the Department of Semiconductor Systems Engineering from Sungkyunkwan University, Korea, in 2010. His research interests include flexible and stretchable electronics using carbon nanotubes, graphene, and other 2D materials.

Young Hee Lee: Young Hee Lee is a Director of the Center for Integrated Nanostructure Physics (CINAP), Institute of Basic Science (IBS) in Korea. He is a professor in the Department of Energy Science and Physics at Sungkyunkwan University, Korea. He received his B.Sc. in physics from Chonbuk National University, Korea and his Ph.D. in physics from Kent State University, USA. His research focus include fundamental studies of nanomaterials and their applications to electronic and optical devices, energy harvesting, and nanobiomedical areas: transparent, flexible, and stretchable transistors, supercapacitors, nanobatteries, hydrogen storage, neuroscience, and cancer therapy using noncontact electrical stimulators.

Acknowledgements

This work was supported by the Institute for Basic Science (IBS) and in part by BK-Plus through Ministry of Education, Korea. We thank David Perello (CINAP) and Alex Patterson (MIT) for valuable discussions during the preparation of this manuscript.

References

1. HJ Leamy, JH Wernick, Semiconductor silicon: the extraordinary made ordinary. MRS Bull. **22**, 47–55 (1997)
2. HR Huff, An electronics division retrospective 1952–2002 and future opportunities in the twenty-first century. J. Electrochem. Soc. **149**, S35–S58 (2002). doi:10.1149/1.1471893
3. TP Morgan, *Intel Xeon Phi Battles GPUs, Defends x86 in Supercomputers*, 2012. http://www.theregister.co.uk/2012/11/12/intel_xeon_phi_coprocessor_launch/
4. JA Rogers, T Someya, Y Huang, Materials and mechanics for stretchable electronics. Science **327**, 1603–1607 (2010). doi:10.1126/science.1182383
5. T Someya, Y Kato, T Sekitani, S Iba, Y Noguchi, Y Murase, H Kawaguchi, T Sakurai, Conformable, flexible, large-area networks of pressure and thermal sensors with organic transistor active matrixes. Proc. Natl. Acad. Sci. U. S. A. **102**, 12321–12325 (2005). doi:10.1073/pnas.0502392102
6. D-H Kim, JA Rogers, Stretchable electronics: materials strategies and devices. Adv. Mater. **20**, 4887–4892 (2008). doi:10.1002/adma.200801788
7. A Shah, J Meier, A Buechel, U Kroll, J Steinhauser, F Meillaud, H Schade, D Dominé, Towards very low-cost mass production of thin-film silicon photovoltaic (PV) solar modules on glass. Thin Solid Films **502**, 292–299 (2006). doi:10.1016/j.tsf.2005.07.299

8. MD Jacunski, MS Shur, M Hack, Threshold voltage, field effect mobility, and gate-to-channel capacitance in polysilicon TFT's. IEEE Trans. Electron. Devices. **43**, 1433–1440 (1996). doi:10.1109/16.535329
9. DH Oh, YH Lee, Stability and cap formation mechanism of single-walled carbon nanotubes. Phys. Rev. B **58**, 7407–7411 (1998). doi:10.1103/PhysRevB.58.7407
10. E Artukovic, M Kaempgen, DS Hecht, S Roth, G GrUner, Transparent and flexible carbon nanotube transistors. Nano Lett. **5**, 757–760 (2005). doi:10.1021/nl050254o
11. L Hu, W Yuan, P Brochu, G Gruner, Q Pei, Highly stretchable, conductive, and transparent nanotube thin films. Appl. Phys. Lett. **94**, 161108 (2009). doi:10.1063/1.3114463
12. Q Liu, Z Liu, X Zhang, L Yang, N Zhang, G Pan, S Yin, Y Chen, J Wei, Polymer photovoltaic cells based on solution-processable graphene and P3HT. Adv. Funct. Mater. **19**, 894–904 (2009). doi:10.1002/adfm.200800954
13. T Durkop, SA Getty, E Cobas, MS Fuhrer, Extraordinary mobility in semiconducting carbon nanotubes. Nano Lett. **4**, 35–39 (2004). doi:10.1021/nl034841q
14. AS Mayorov, RV Gorbachev, SV Morozov, L Britnell, R Jalil, LA Ponomarenko, P Blake, KS Novoselov, K Watanabe, T Taniguchi, AK Geim, Micrometer-scale ballistic transport in encapsulated graphene at room temperature. Nano letters. **11**, 2396–2399 (2011). doi:10.1021/nl200758b
15. KI Bolotin, KJ Sikes, Z Jiang, M Klima, G Fudenberg, J Hone, P Kim, HL Stormer, Ultrahigh electron mobility in suspended graphene. Solid State Commun. **146**, 351–355 (2008). doi:10.1016/j.ssc.2008.02.024
16. A Javey, P Qi, Q Wang, H Dai, Ten- to 50-nm-long quasi-ballistic carbon nanotube devices obtained without complex lithography. Proc. Natl. Acad. Sci. U. S. A. **101**, 13408–13410 (2004). doi:10.1073/pnas.0404450101
17. J Yu, G Liu, AV Sumant, V Goyal, AA Balandin, Graphene-on-diamond devices with increased current-carrying capacity: carbon sp2-on-sp3 technology. Nano letters. **12**, 1603–1608 (2012). doi:10.1021/nl204545q
18. E Pop, D Mann, Q Wang, KE Goodson, HJ Dai, Thermal conductance of an individual single-wall carbon nanotube above room temperature. Nano Lett. **6**, 96–100 (2006). doi:10.1021/nl052145f
19. AA Balandin, S Ghosh, W Bao, I Calizo, D Teweldebrhan, F Miao, CN Lau, Superior thermal conductivity of single-layer graphene. Nano Lett. **8**, 902–907 (2008). doi:10.1021/nl0731872
20. RR Nair, P Blake, AN Grigorenko, KS Novoselov, TJ Booth, T Stauber, NMR Peres, AK Geim, Fine structure constant defines visual transparency of graphene. Science **320**, 1308–1308 (2008). doi:10.1126/science.1156965
21. C Lee, X Wei, JW Kysar, J Hone, Measurement of the elastic properties and intrinsic strength of monolayer graphene. Science **321**, 385–388 (2008). doi:10.1126/science.1157996
22. JU Park, MA Meitl, SH Hur, ML Usrey, MS Strano, PJ Kenis, JA Rogers, In situ deposition and patterning of single-walled carbon nanotubes by laminar flow and controlled flocculation in microfluidic channels. Angew. Chem. **45**, 581–585 (2006). doi:10.1002/anie.200501799
23. MA Meitl, YX Zhou, A Gaur, S Jeon, ML Usrey, MS Strano, JA Rogers, Solution casting and transfer printing single-walled carbon nanotube films. Nano Lett. **4**, 1643–1647 (2004). doi:10.1021/nl0491935
24. JM Moon, KH An, YH Lee, YS Park, DJ Bae, GS Park, High-yield purification process of singlewalled carbon nanotubes. J. Phys. Chem. B **105**, 5677–5681 (2001). doi:10.1021/jp0102365
25. R Vajtai, BQ Wei, PM Ajayan, Controlled growth of carbon nanotubes. Philos. Trans. R. Soc. A-Math. Phys. Eng. Sci. **362**, 2143–2160 (2004). doi:10.1098/rsta.2004.1433
26. B Kitiyanan, WE Alvarez, JH Harwell, DE Resasco, Controlled production of single-wall carbon nanotubes by catalytic decomposition of CO on bimetallic Co-Mo catalysts. Chem. Phys. Lett. **317**, 497–503 (2000). doi:10.1016/s0009-2614(99)01379-2
27. A Thess, R Lee, P Nikolaev, HJ Dai, P Petit, J Robert, CH Xu, YH Lee, SG Kim, AG Rinzler, DT Colbert, GE Scuseria, D Tomanek, JE Fischer, RE Smalley, Crystalline ropes of metallic carbon nanotubes. Science **273**, 483–487 (1996). doi:10.1126/science.273.5274.483
28. MJ Bronikowski, PA Willis, DT Colbert, KA Smith, RE Smalley, Gas-phase production of carbon single-walled nanotubes from carbon monoxide via the HiPco process: A parametric study. J. Vac. Sci. Technol. A: Vac, Surf. Films. **19**, 1800 (2001). doi:10.1116/1.1380721
29. Z Jin, H Chu, J Wang, J Hong, W Tan, Y Li, Ultralow feeding gas flow guiding growth of large-scale horizontally aligned single-walled carbon nanotube arrays. Nano Lett. **7**, 2073–2079 (2007). doi:10.1021/nl070980m
30. W Zhou, Z Han, J Wang, Y Zhang, Z Jin, X Sun, Y Zhang, C Yan, Y Li, Copper catalyzing growth of single-walled carbon nanotubes on substrates. Nano Lett. **6**, 2987–2990 (2006). doi:10.1021/nl061871v
31. BH Hong, JY Lee, T Beetz, YM Zhu, P Kim, KS Kim, Quasi-continuous growth of ultralong carbon nanotube arrays. J. Am. Chem. Soc. **127**, 15336–15337 (2005). doi:10.1021/ja054454d
32. SJ Kang, C Kocabas, T Ozel, M Shim, N Pimparkar, MA Alam, SV Rotkin, JA Rogers, High-performance electronics using dense, perfectly aligned arrays of single-walled carbon nanotubes. Nature nanotechnology. **2**, 230–236 (2007). doi:10.1038/nnano.2007.77
33. I-H Lee, J-W Im, U-J Kim, E-J Bae, K-K Kim, E-H Lee, Y-H Lee, S-H Hong, Y-S Min, Low temperature growth of single-walled carbon nanotube forest. Bull. Kor. Chem. Soc. **31**, 2819–2822 (2010). doi:10.5012/bkcs.2010.31.10.2819
34. K Hata, DN Futaba, K Mizuno, T Namai, M Yumura, S Iijima, Water-assisted highly efficient synthesis of impurity-free single-walled carbon nanotubes. Science **306**, 1362–1364 (2004). doi:10.1126/science.1104962
35. ZF Ren, Synthesis of large arrays of well-aligned carbon nanotubes on glass. Science **282**, 1105–1107 (1998). doi:10.1126/science.282.5391.1105
36. UJ Kim, IH Lee, JJ Bae, S Lee, GH Han, SJ Chae, F Guenes, JH Choi, CW Baik, SI Kim, JM Kim, YH Lee, Graphene/Carbon nanotube hybrid-based transparent 2D optical array. Adv. Mater. **23**, 3809–+ (2011). doi:10.1002/adma.201101622
37. D Wang, P Song, C Liu, W Wu, S Fan, Highly oriented carbon nanotube papers made of aligned carbon nanotubes. Nanotechnology **19**, 075609 (2008). doi:10.1088/0957-4484/19/7/075609
38. SC Lim, K Lee, IH Lee, YH Lee, Field emission and application of carbon nanotubes. Nano **2**, 69–89 (2007)
39. SC Lim, HK Choi, HJ Jeong, YI Song, GY Kim, KT Jung, YH Lee, A strategy for forming robust adhesion with the substrate in a carbon-nanotube field-emission array. Carbon **44**, 2809–2815 (2006). doi:10.1016/j.carbon.2006.03.030
40. L Hu, DS Hecht, G Gruner, Percolation in transparent and conducting carbon nanotube networks. Nano Lett. **4**, 2513–2517 (2004). doi:10.1021/nl048435y
41. A Moisala, AG Nasibulin, EI Kauppinen, The role of metal nanoparticles in the catalytic production of single-walled carbon nanotubes - a review. J. Phys-Condense Matter. **15**, S3011–S3035 (2003). doi:10.1088/0953-8984/15/42/003
42. RC Tenent, TM Barnes, JD Bergeson, AJ Ferguson, B To, LM Gedvilas, MJ Heben, JL Blackburn, Ultrasmooth, large-area, high-uniformity, conductive transparent single-walled-carbon-nanotube films for photovoltaics produced by ultrasonic spraying. Adv. Mater. **21**, 3210–3216 (2009). doi:10.1002/adma.200803551
43. H-Z Geng, KK Kim, KP So, YS Lee, Y Chang, YH Lee, Effect of acid treatment on carbon nanotube-based flexible transparent conducting films. J. Am. Chem. Soc. **129**, 7758–+ (2007). doi:10.1021/ja0722224
44. Z Wu, Z Chen, X Du, JM Logan, J Sippel, M Nikolou, K Kamaras, JR Reynolds, DB Tanner, AF Hebard, AG Rinzler, Transparent, conductive carbon nanotube films. Science **305**, 1273–1276 (2004). doi:10.1126/science.1101243
45. W Jong Yu, S Yol Jeong, K Kang Kim, B Ram Kang, D Jae Bae, M Lee, S Hong, S Prabhu Gaunkar, D Pribat, D Perello, M Yun, J-Y Choi, Y Hee Lee, Bias-induced doping engineering with ionic adsorbates on single-walled carbon nanotube thin film transistors. New J. Phys. **10**, 113013 (2008). doi:10.1088/1367-2630/10/11/113013
46. K Kordas, T Mustonen, G Toth, H Jantunen, M Lajunen, C Soldano, S Talapatra, S Kar, R Vajtai, PM Ajayan, Inkjet printing of electrically conductive patterns of carbon nanotubes. Small **2**, 1021–1025 (2006). doi:10.1002/smll.200600061
47. C Biswas, YH Lee, Graphene versus carbon nanotubes in electronic devices. Adv. Funct. Mater. **21**, 3806–3826 (2011). doi:10.1002/adfm.201101241
48. YX Zhou, A Gaur, SH Hur, C Kocabas, MA Meitl, M Shim, JA Rogers, p-channel, n-channel thin film transistors and p-n diodes based on single wall carbon nanotube networks. Nano Lett. **4**, 2031–2035 (2004). doi:10.1021/nl048905o
49. M Glerup, J Steinmetz, D Samaille, O Stéphan, S Enouz, A Loiseau, S Roth, P Bernier, Synthesis of N-doped SWNT using the arc-discharge procedure. Chem. Phys. Lett. **387**, 193–197 (2004). doi:10.1016/j.cplett.2004.02.005
50. G Zhang, D Mann, L Zhang, A Javey, Y Li, E Yenilmez, Q Wang, JP McVittie, Y Nishi, J Gibbons, H Dai, Ultra-high-yield growth of vertical single-walled carbon nanotubes: hidden roles of hydrogen and oxygen. Proc. Natl. Acad. Sci. U. S. A. **102**, 16141–16145 (2005). doi:10.1073/pnas.0507064102

51. M Zhang, KR Atkinson, RH Baughman, Multifunctional carbon nanotube yarns by downsizing an ancient technology. Science **306**, 1358–1361 (2004). doi:10.1126/science.1104276
52. K Liu, Y Sun, R Zhou, H Zhu, J Wang, L Liu, S Fan, K Jiang, Carbon nanotube yarns with high tensile strength made by a twisting and shrinking method. Nanotechnology **21**, 045708 (2010). doi:10.1088/0957-4484/21/4/045708
53. X Zhang, K Jiang, C Feng, P Liu, L Zhang, J Kong, T Zhang, Q Li, S Fan, Spinning and processing continuous yarns from 4-inch wafer scale super-aligned carbon nanotube arrays. Adv. Mater. **18**, 1505–1510 (2006). doi:10.1002/adma.200502528
54. K Jiang, J Wang, Q Li, L Liu, C Liu, S Fan, Superaligned carbon nanotube arrays, films, and yarns: a road to applications. Adv. Mater. **23**, 1154–1161 (2011). doi:10.1002/adma.201003989
55. KS Novoselov, AK Geim, SV Morozov, D Jiang, Y Zhang, SV Dubonos, IV Grigorieva, AA Firsov, Electric field effect in atomically thin carbon films. Science **306**, 666–669 (2004). doi:10.1126/science.1102896
56. KS Novoselov, AK Geim, SV Morozov, D Jiang, MI Katsnelson, IV Grigorieva, SV Dubonos, AA Firsov, Two-dimensional gas of massless Dirac fermions in graphene. Nature **438**, 197–200 (2005). doi:10.1038/nature04233
57. SJ Chae, F Güneş, KK Kim, ES Kim, GH Han, SM Kim, H-J Shin, S-M Yoon, J-Y Choi, MH Park, CW Yang, D Pribat, YH Lee, Synthesis of Large-Area Graphene Layers on Poly-Nickel Substrate by Chemical Vapor Deposition: Wrinkle Formation. Adv. Mater. **21**, 2328–2333 (2009). doi:10.1002/adma.200803016
58. A Reina, X Jia, J Ho, D Nezich, H Son, V Bulovic, MS Dresselhaus, J Kong, Large area, few-layer graphene films on arbitrary substrates by chemical vapor deposition. Nano Lett. **9**, 30–35 (2009). doi:10.1021/nl801827v
59. GH Han, SJ Chae, ES Kim, F Guenes, IH Lee, SW Lee, SY Lee, SC Lim, HK Jeong, MS Jeong, YH Lee, Laser thinning for monolayer graphene formation: heat sink and interference effect. Acs Nano. **5**, 263–268 (2011). doi:10.1021/nn1026438
60. GH Han, F Gunes, JJ Bae, ES Kim, SJ Chae, HJ Shin, JY Choi, D Pribat, YH Lee, Influence of copper morphology in forming nucleation seeds for graphene growth. Nano letters. **11**, 4144–4148 (2011). doi:10.1021/nl201980p
61. DL Duong, GH Han, SM Lee, F Gunes, ES Kim, ST Kim, H Kim, QH Ta, KP So, SJ Yoon, SJ Chae, YW Jo, MH Park, SH Chae, SC Lim, JY Choi, YH Lee, Probing graphene grain boundaries with optical microscopy. Nature **490**, 235–239 (2012). doi:10.1038/nature11562
62. TH Ly, DL Duong, QH Ta, F Yao, QA Vu, HY Jeong, SH Chae, YH Lee, Nondestructive characterization of graphene defects. Adv. Funct. Mater. **23**, 5183–5189 (2013). doi:10.1002/adfm.201300493
63. Y Hernandez, V Nicolosi, M Lotya, FM Blighe, Z Sun, S De, IT McGovern, B Holland, M Byrne, YK Gun'Ko, JJ Boland, P Niraj, G Duesberg, S Krishnamurthy, R Goodhue, J Hutchison, V Scardaci, AC Ferrari, JN Coleman, High-yield production of graphene by liquid-phase exfoliation of graphite. Nature nanotechnology. **3**, 563–568 (2008). doi:10.1038/nnano.2008.215
64. X Wang, L Zhi, K Muellen, Transparent, conductive graphene electrodes for dye-sensitized solar cells. Nano Lett. **8**, 323–327 (2008). doi:10.1021/nl072838r
65. C Mattevi, G Eda, S Agnoli, S Miller, KA Mkhoyan, O Celik, D Mastrogiovanni, G Granozzi, E Garfunkel, M Chhowalla, Evolution of electrical, chemical, and structural properties of transparent and conducting chemically derived graphene thin films. Adv. Funct. Mater. **19**, 2577–2583 (2009). doi:10.1002/adfm.200900166
66. G Eda, G Fanchini, M Chhowalla, Large-area ultrathin films of reduced graphene oxide as a transparent and flexible electronic material. Nat. Nanotechnol. **3**, 270–274 (2008). doi:10.1038/nnano.2008.83
67. M Jin, H-K Jeong, T-H Kim, KP So, Y Cui, WJ Yu, EJ Ra, YH Lee, Synthesis and systematic characterization of functionalized graphene sheets generated by thermal exfoliation at low temperature. J. Phys. D. Appl. Phys. **43**, 275402 (2010). doi:10.1088/0022-3727/43/27/275402
68. H-K Jeong, M Jin, EJ Ra, KY Sheem, GH Han, S Arepalli, YH Lee, Enhanced electric double layer capacitance of graphite oxide intercalated by poly (sodium 4-styrensulfonate) with high cycle stability. Acs Nano. **4**, 1162–1166 (2010). doi:10.1021/nn901790f
69. H-K Jeong, L Colakerol, MH Jin, P-A Glans, KE Smith, YH Lee, Unoccupied electronic states in graphite oxides. Chem. Phys. Lett. **460**, 499–502 (2008). doi:10.1016/j.cplett.2008.06.042
70. H-K Jeong, YP Lee, MH Jin, ES Kim, JJ Bae, YH Lee, Thermal stability of graphite oxide. Chem. Phys. Lett. **470**, 255–258 (2009). doi:10.1016/j.cplett.2009.01.050
71. H-K Jeong, MH Jin, KH An, YH Lee, Structural stability and variable dielectric constant in poly sodium 4-styrensulfonate intercalated graphite oxide. J. Phys. Chem. C **113**, 13060–13064 (2009). doi:10.1021/jp9026282
72. H-J Shin, KK Kim, A Benayad, S-M Yoon, HK Park, I-S Jung, MH Jin, H-K Jeong, JM Kim, J-Y Choi, YH Lee, Efficient reduction of graphite oxide by sodium borohydride and its effect on electrical conductance. Adv. Funct. Mater. **19**, 1987–1992 (2009). doi:10.1002/adfm.200900167
73. M Jin, TH Kim, SC Lim, DL Duong, HJ Shin, YW Jo, HK Jeong, J Chang, S Xie, YH Lee, Facile physical route to highly crystalline graphene. Adv. Funct. Mater. **21**, 3496–3501 (2011). doi:10.1002/adfm.201101037
74. KV Emtsev, A Bostwick, K Horn, J Jobst, GL Kellogg, L Ley, JL McChesney, T Ohta, SA Reshanov, J Rohrl, E Rotenberg, AK Schmid, D Waldmann, HB Weber, T Seyller, Towards wafer-size graphene layers by atmospheric pressure graphitization of silicon carbide. Nat. Mat. **8**, 203–207 (2009). doi:10.1038/nmat2382
75. A Ouerghi, MG Silly, M Marangolo, C Mathieu, M Eddrief, M Picher, F Sirotti, S El Moussaoui, R Belkhou, Large-area and high-quality epitaxial graphene on off-axis SiC wafers. Acs Nano. **6**, 6075–6082 (2012). doi:10.1021/nn301152p
76. JS Moon, D Curtis, M Hu, D Wong, C McGuire, PM Campbell, G Jernigan, JL Tedesco, B VanMil, R Myers-Ward, C Eddy Jr, DK Gaskill, RF Epitaxial-Graphene, Field-effect transistors on Si-Face 6H-SiC substrates. IEEE Electr. Device L. **30**, 650–652 (2009). doi:10.1109/led.2009.2020699
77. Y Wu, YM Lin, AA Bol, KA Jenkins, F Xia, DB Farmer, Y Zhu, P Avouris, High-frequency, scaled graphene transistors on diamond-like carbon. Nature **472**, 74–78 (2011). doi:10.1038/nature09979
78. J Hackley, D Ali, J DiPasquale, JD Demaree, CJK Richardson, Graphitic carbon growth on Si(111) using solid source molecular beam epitaxy. Appl. Phys. Lett. **95**, 133114 (2009). doi:10.1063/1.3242029
79. YH Lee, L Yu, H Wang, W Fang, X Ling, Y Shi, CT Lin, JK Huang, MT Chang, CS Chang, M Dresselhaus, T Palacios, LJ Li, J Kong, Synthesis and transfer of single-layer transition metal disulfides on diverse surfaces. Nano letters. **13**, 1852–1857 (2013). doi:10.1021/nl400687n
80. X Li, Y Zhu, W Cai, M Borysiak, B Han, D Chen, RD Piner, L Colombo, RS Ruoff, Transfer of large-area graphene films for high-performance transparent conductive electrodes. Nano Lett. **9**, 4359–4363 (2009). doi:10.1021/nl902623y
81. S-H Hur, OO Park, JA Rogers, Extreme bendability of single-walled carbon nanotube networks transferred from high-temperature growth substrates to plastic and their use in thin-film transistors. Appl. Phys. Lett. **86**, 243502 (2005). doi:10.1063/1.1947380
82. GH Han, H-J Shin, ES Kim, SJ Chae, J-Y Choi, YH Lee, Poly(Ethylene Co-Vinyl Acetate)-Assisted one-step transfer of ultra-large graphene. Nano **06**, 59–65 (2011). doi:10.1142/s1793292011002342
83. S Bae, H Kim, Y Lee, X Xu, JS Park, Y Zheng, J Balakrishnan, T Lei, HR Kim, YI Song, YJ Kim, KS Kim, B Ozyilmaz, JH Ahn, BH Hong, S Iijima, Roll-to-roll production of 30-inch graphene films for transparent electrodes. Nat. Nanotechnol. **5**, 574–578 (2010). doi:10.1038/nnano.2010.132
84. L Gao, W Ren, H Xu, L Jin, Z Wang, T Ma, LP Ma, Z Zhang, Q Fu, LM Peng, X Bao, HM Cheng, Repeated growth and bubbling transfer of graphene with millimetre-size single-crystal grains using platinum. Nat. Commun. **3**, 699 (2012). doi:10.1038/ncomms1702
85. DY Wang, IS Huang, PH Ho, SS Li, YC Yeh, DW Wang, WL Chen, YY Lee, YM Chang, CC Chen, CT Liang, CW Chen, Clean-lifting transfer of large-area residual-free graphene films. Adv. Mater. **25**, 4521–4526 (2013). doi:10.1002/adma.201301152
86. CR Dean, AF Young, I Meric, C Lee, L Wang, S Sorgenfrei, K Watanabe, T Taniguchi, P Kim, KL Shepard, J Hone, Boron nitride substrates for high-quality graphene electronics. Nat. Nanotechnol. **5**, 722–726 (2010). doi:10.1038/nnano.2010.172
87. L Wang, I Meric, PY Huang, Q Gao, Y Gao, H Tran, T Taniguchi, K Watanabe, LM Campos, DA Muller, J Guo, P Kim, J Hone, KL Shepard, CR Dean, One-dimensional electrical contact to a two-dimensional material. Science **342**, 614–617 (2013). doi:10.1126/science.1244358
88. L Wang, Z Chen, CR Dean, T Taniguchi, K Watanabe, LE Brus, J Hone, Negligible environmental sensitivity of graphene in a hexagonal Boron Nitride/Graphene/h-BN Sandwich structure. Acs Nano. **6**, 9314–9319 (2012). doi:10.1021/nn304004s
89. T Taychatanapat, K Watanabe, T Taniguchi, P Jarillo-Herrero, Electrically tunable transverse magnetic focusing in graphene. Nat. Phys. **9**, 225–229 (2013). doi:10.1038/nphys2549

90. B Hunt, JD Sanchez-Yamagishi, AF Young, M Yankowitz, BJ LeRoy, K Watanabe, T Taniguchi, P Moon, M Koshino, P Jarillo-Herrero, RC Ashoori, Massive Dirac fermions and Hofstadter butterfly in a van der Waals Heterostructure. Science **340**, 1427–1430 (2013). doi:10.1126/science.1237240
91. PS Rao, DN Sathyanarayana, Synthesis of electrically conducting copolymers of o-/m-toluidines and o-/m-amino benzoic acid in an organic peroxide system and their characterization. Synth. Met. **138**, 519–527 (2003). doi:10.1016/s0379-6779(02)01253-5
92. XS Wang, HP Tang, XD Li, X Hua, Investigations on the mechanical properties of conducting polymer coating-substrate structures and their influencing factors. Int. J. Mol. Sci. **10**, 5257–5284 (2009). doi:10.3390/ijms10125257
93. YL Wu, YN Li, BS Ong, Printed silver ohmic contacts for high-mobility organic thin-film transistors. J. Am. Chem. Soc. **128**, 4202–4203 (2006). doi:10.1021/ja058725w
94. E Fortunato, A Pimentel, A Gonçalves, A Marques, R Martins, High mobility amorphous/nanocrystalline indium zinc oxide deposited at room temperature. Thin Solid Films **502**, 104–107 (2006). doi:10.1016/j.tsf.2005.07.311
95. B Yaglioglu, H-Y Yeom, DC Paine, Crystallization of amorphous In[sub 2]O[sub 3]–10 wt % ZnO thin films annealed in air. Appl. Phys. Lett. **86**, 261908 (2005). doi:10.1063/1.1977209
96. T Minami, T Yamamoto, Y Toda, T Miyata, Transparent conducting zinc-co-doped ITO films prepared by magnetron sputtering. Thin Solid Films **373**, 189–194 (2000). doi:10.1016/s0040-6090(00)01132-9
97. F Yao, SC Lim, WJ Yu, IH Lee, F Guenes, HR Hwang, SB Yang, KP So, GH Han, AC Lee YH, Response to gas exposure in vertically aligned multiwalled carbon nanotube electrode. J. Phys. Chem. C **114**, 3659–3663 (2010). doi:10.1021/jp910678m
98. A Ghosh, YH Lee, Carbon-based electrochemical capacitors. Chem Sus Chem **5**, 480–499 (2012). doi:10.1002/cssc.201100645
99. HJ Jeong, HK Choi, GY Kim, YI Song, Y Tong, SC Lim, YH Lee, Fabrication of efficient field emitters with thin multiwalled carbon nanotubes using spray method. Carbon **44**, 2689–2693 (2006). doi:10.1016/j.carbon.2006.04.009
100. L Xiao, Z Chen, C Feng, L Liu, Z-Q Bai, Y Wang, L Qian, Y Zhang, Q Li, K Jiang, S Fan, Flexible, stretchable, transparent carbon nanotube thin film loudspeakers. Nano Lett. **8**, 4539–4545 (2008). doi:10.1021/nl802750z
101. D Zhang, K Ryu, X Liu, E Polikarpov, J Ly, ME Tompson, C Zhou, Transparent, conductive, and flexible carbon nanotube films and their application in organic light-emitting diodes. Nano Lett. **6**, 1880–1886 (2006). doi:10.1021/nl0608543
102. J Chen, Y Liu, AI Minett, C Lynam, J Wang, GG Wallace, Flexible, aligned carbon nanotube/conducting polymer electrodes for a lithium-ion battery. Chem. Mater. **19**, 3595–3597 (2007). doi:10.1021/cm070991g
103. C Feng, K Liu, J-S Wu, L Liu, J-S Cheng, Y Zhang, Y Sun, Q Li, S Fan, K Jiang, Flexible, Stretchable, Transparent conducting films made from superaligned carbon nanotubes. Adv. Funct. Mater. **20**, 885–891 (2010). doi:10.1002/adfm.200901960
104. J-Y Lee, ST Connor, Y Cui, P Peumans, Solution-processed metal nanowire mesh transparent electrodes. Nano Lett. **8**, 689–692 (2008). doi:10.1021/nl073296g
105. H-Z Geng, DS Lee, KK Kim, GH Han, HK Park, YH Lee, Absorption spectroscopy of surfactant-dispersed carbon nanotube film: modulation of electronic structures. Chem. Phys. Lett. **455**, 275–278 (2008). doi:10.1016/j.cplett.2008.02.102
106. H-Z Geng, KK Kim, K Lee, GY Kim, HK Choi, DS Lee, KH An, YH Lee, Y Chang, YS Lee, B Kim, YJ Lee, Dependence of material quality on performance of flexible transparent conducting films with single-walled carbon nanotubes. Nano **2**, 157–167 (2007). doi:10.1142/s1793292007000532
107. M Razeghi, H-Z Geng, KK Kim, YH Lee, D Pribat, YH Lee, Recent progress in carbon nanotube-based flexible transparent conducting film. Proc. of SPIE. **7037**, 70370A-1-70370A-14 (2008). doi:10.1117/12.796143
108. KK Kim, S-M Yoon, HK Park, H-J Shin, SM Kim, JJ Bae, Y Cui, JM Kim, J-Y Choi, YH Lee, Doping strategy of carbon nanotubes with redox chemistry. New J. Chem. **34**, 2183–2188 (2010). doi:10.1039/c0nj00138d
109. SM Kim, KK Kim, YW Jo, MH Park, SJ Chae, DL Duong, CW Yang, J Kong, YH Lee, Role of anions in the AuCl3-Doping of carbon nanotubes. Acs Nano. **5**, 1236–1242 (2011). doi:10.1021/nn1028532
110. KK Kim, JJ Bae, HK Park, SM Kim, H-Z Geng, KA Park, H-J Shin, S-M Yoon, A Benayad, J-Y Choi, YH Lee, Fermi level engineering of single-walled carbon nanotubes by AuCl3 doping. J. Am. Chem. Soc. **130**, 12757–12761 (2008). doi:10.1021/ja8038689
111. F Yao, F Gunes, HQ Ta, SM Lee, SJ Chae, KY Sheem, CS Cojocaru, SS Xie, YH Lee, Diffusion mechanism of lithium ion through basal plane of layered graphene. J. Am. Chem. Soc. **134**, 8646–8654 (2012). doi:10.1021/ja301586m
112. B Kumar, KY Lee, H-K Park, SJ Chae, YH Lee, S-W Kim, Controlled growth of semiconducting nanowire, nanowall, and hybrid nanostructures on graphene for piezoelectric nanogenerators. Acs Nano. **5**, 4197–4204 (2011). doi:10.1021/nn200942s
113. D Choi, MY Choi, WM Choi, HJ Shin, HK Park, JS Seo, J Park, SM Yoon, SJ Chae, YH Lee, SW Kim, JY Choi, SY Lee, JM Kim, Fully rollable transparent nanogenerators based on graphene electrodes. Adv. Mater. **22**, 2187–2192 (2010). doi:10.1002/adma.200903815
114. KK Kim, A Reina, Y Shi, H Park, LJ Li, YH Lee, J Kong, Enhancing the conductivity of transparent graphene films via doping. Nanotechnology **21**, 285205 (2010). doi:10.1088/0957-4484/21/28/285205
115. H-J Shin, WM Choi, D Choi, GH Han, S-M Yoon, H-K Park, S-W Kim, YW Jin, SY Lee, JM Kim, J-Y Choi, YH Lee, Control of electronic structure of graphene by various dopants and their effects on a nanogenerator. J. Am. Chem. Soc. **132**, 15603–15609 (2010). doi:10.1021/ja105140e
116. F Guenes, GH Han, KK Kim, ES Kim, SJ Chae, MH Park, H-K Jeong, SC Lim, YH Lee, Large-area graphene-based flexible transparent conducting films. Nano **4**, 83–90 (2009). doi:10.1142/s1793292009001538
117. F Gunes, H-J Shin, C Biswas, GH Han, ES Kim, SJ Chae, J-Y Choi, YH Lee, Layer-by-layer doping of few-layer graphene film. Acs Nano. **4**, 4595–4600 (2010). doi:10.1021/nn1008808
118. WH Lee, J Park, SH Sim, SB Jo, KS Kim, BH Hong, K Cho, Transparent flexible organic transistors based on monolayer graphene electrodes on plastic. Adv. Mater. **23**, 1752–1756 (2011). doi:10.1002/adma.201004099
119. Y Chen, Y Xu, K Zhao, X Wan, J Deng, W Yan, Towards flexible all-carbon electronics: flexible organic field-effect transistors and inverter circuits using solution-processed all-graphene source/drain/gate electrodes. Nano Res. **3**, 714–721 (2010). doi:10.1007/s12274-010-0035-3
120. C Li, Z Li, H Zhu, K Wang, J Wei, X Li, P Sun, H Zhang, D Wu, Graphene Nano-"patches" on a carbon nanotube network for highly transparent/conductive thin film applications. J. Phys. Chem. C **114**, 14008–14012 (2010). doi:10.1021/jp1041487
121. T-K Hong, DW Lee, HJ Choi, HS Shin, B-S Kim, Transparent, flexible conducting hybrid multi layer thin films of multiwalled carbon nanotubes with graphene nanosheets. Acs Nano. **4**, 3861–3868 (2010). doi:10.1021/nn100897g
122. VC Tung, L-M Chen, MJ Allen, JK Wassei, K Nelson, RB Kaner, Y Yang, Low-temperature solution processing of graphene-carbon nanotube hybrid materials for high-performance transparent conductors. Nano Lett. **9**, 1949–1955 (2009). doi:10.1021/nl9001525
123. WJ Yu, SY Lee, SH Chae, D Perello, GH Han, M Yun, YH Lee, Small hysteresis nanocarbon-based integrated circuits on flexible and transparent plastic substrate. Nano letters. **11**, 1344–1350 (2011). doi:10.1021/nl104488z
124. Y Zhu, Z Sun, Z Yan, Z Jin, JM Tour, Rational design of hybrid graphene films for high-performance transparent electrodes. Acs Nano. **5**, 6472–6479 (2011). doi:10.1021/nn201696g
125. Q Cao, SJ Han, GS Tulevski, Y Zhu, DD Lu, W Haensch, Arrays of single-walled carbon nanotubes with full surface coverage for high-performance electronics. Nat. Nanotechnol. **8**, 180–186 (2013). doi:10.1038/nnano.2012.257
126. VK Sangwan, RP Ortiz, JMP Alaboson, JD Emery, MJ Bedzyk, LJ Lauhon, TJ Marks, MC Hersam, Fundamental performance limits of carbon nanotube thin-film transistors achieved using hybrid molecular dielectrics. Acs Nano. **6**, 7480–7488 (2012). doi:10.1021/nn302768h
127. I Meric, MY Han, AF Young, B Ozyilmaz, P Kim, KL Shepard, Current saturation in zero-bandgap, top-gated graphene field-effect transistors. Nat. Nanotechnol. **3**, 654–659 (2008). doi:10.1038/nnano.2008.268
128. C Biswas, SY Lee, L Thuc Hue, A Ghosh, D Quoc Nguyen, YH Lee, Chemically doped random network carbon nanotube p-n junction diode for rectifier. Acs Nano. **5**, 9817–9823 (2011). doi:10.1021/nn203391h
129. A Javey, H Kim, M Brink, Q Wang, A Ural, J Guo, P McIntyre, P McEuen, M Lundstrom, H Dai, High-k dielectrics for advanced carbon-nanotube transistors and logic gates. Nat. Mater. **1**, 241–246 (2002). doi:10.1038/nmat769
130. A Javey, J Guo, DB Farmer, Q Wang, E Yenilmez, RG Gordon, M Lundstrom, HJ Dai, Self-aligned ballistic molecular transistors and electrically parallel nanotube arrays. Nano Lett. **4**, 1319–1322 (2004). doi:10.1021/nl049222b

131. C Klinke, J Chen, A Afzali, P Avouris, Charge transfer induced polarity switching in carbon nanotube transistors. Nano Lett. **5**, 555–558 (2005). doi:10.1021/nl048055c
132. SJ Tans, ARM Verschueren, C Dekker, Room-temperature transistor based on a single carbon nanotube. Nature **393**, 49–52 (1998)
133. A Javey, J Guo, Q Wang, M Lundstrom, HJ Dai, Ballistic carbon nanotube field-effect transistors. Nature **424**, 654–657 (2003). doi:10.1038/nature01797
134. A Bachtold, P Hadley, T Nakanishi, C Dekker, Logic circuits with carbon nanotube transistors. Science **294**, 1317–1320 (2001). doi:10.1126/science.1065824
135. DJ Perello, SC Lim, SJ Chae, I Lee, MJ Kim, YH Lee, M Yun, Thermionic field emission transport in carbon nanotube transistors. Acs Nano. **5**, 1756–1760 (2011). doi:10.1021/nn102343k
136. C Kocabas, SJ Kang, T Ozel, M Shim, JA Rogers, Improved synthesis of aligned arrays of single-walled carbon nanotubes and their implementation in thin film type transistors. J. Phys. Chem. C **111**, 17879–17886 (2007). doi:10.1021/jp071387w
137. L Ding, D Yuan, J Liu, Growth of high-density parallel arrays of long single-walled carbon nanotubes on quartz substrates. J. Am. Chem. Soc. **130**, 5428–+ (2008). doi:10.1021/ja8006947
138. WJ Yu, UJ Kim, BR Kang, IH Lee, E-H Lee, YH Lee, Adaptive logic circuits with doping-free ambipolar carbon nanotube transistors. Nano Lett. **9**, 1401–1405 (2009). doi:10.1021/nl803066v
139. SW Lee, SY Lee, SC Lim, K Y-d, J-S Yoon, K Uh, YH Lee, Positive gate bias stress instability of carbon nanotube thin film transistors. Appl. Phys. Lett. **101**, (2012). doi:10.1063/1.4740084
140. BJ Kim, H Jang, SK Lee, BH Hong, JH Ahn, JH Cho, High-performance flexible graphene field effect transistors with ion gel gate dielectrics. Nano letters. **10**, 3464–3466 (2010). doi:10.1021/nl101559n
141. AK Geim, KS Novoselov, The rise of graphene. Nat. Mater. **6**, 183–191 (2007). doi:10.1038/nmat1849
142. J Lee, TJ Ha, KN Parrish, SF Chowdhury, L Tao, A Dodabalapur, D Akinwande, High-performance current saturating graphene field-effect transistor with hexagonal boron nitride dielectric on flexible polymeric substrates. IEEE Electron. Dev. L. **34**, 172–174 (2013). doi:10.1109/led.2012.2233707
143. C Yan, JH Cho, JH Ahn, Graphene-based flexible and stretchable thin film transistors. Nanoscale. **4**, 4870–4882 (2012). doi:10.1039/c2nr30994g
144. F Schwierz, ELECTRONICS Industry-compatible graphene transistors. Nature **472**, 41–42 (2011)
145. L Chico, VH Crespi, LX Benedict, SG Louie, ML Cohen, Pure carbon nanoscale devices: nanotube heterojunctions. Phys. Rev. Lett. **76**, 971–974 (1996). doi:10.1103/PhysRevLett.76.971
146. HJ Dai, EW Wong, CM Lieber, Probing electrical transport in nanomaterials: conductivity of individual carbon nanotubes. Science **272**, 523–526 (1996). doi:10.1126/science.272.5261.523
147. PG Collins, MS Arnold, P Avouris, Engineering carbon nanotubes and nanotube circuits using electrical breakdown. Science **292**, 706–709 (2001). doi:10.1126/science.1058782
148. P Avouris, Carbon nanotube electronics. Chem. Phys. **281**, 429–445 (2002). doi:10.1016/s0301-0104(02)00376-2
149. Q Cao, HS Kim, N Pimparkar, JP Kulkarni, C Wang, M Shim, K Roy, MA Alam, JA Rogers, Medium-scale carbon nanotube thin-film integrated circuits on flexible plastic substrates. Nature **454**, 495–500 (2008). doi:10.1038/nature07110
150. AA Green, MC Hersam, Processing and properties of highly enriched double-wall carbon nanotubes. Nat. Nanotechnol. **4**, 64–70 (2009). doi:10.1038/nnano.2008.364
151. S Ghosh, SM Bachilo, RB Weisman, Advanced sorting of single-walled carbon nanotubes by nonlinear density-gradient ultracentrifugation. Nat. Nanotechnol. **5**, 443–450 (2010). doi:10.1038/nnano.2010.68
152. H Liu, D Nishide, T Tanaka, H Kataura, Large-scale single-chirality separation of single-wall carbon nanotubes by simple gel chromatography. Nat. Commun. **2**, 309 (2011). doi:10.1038/ncomms1313
153. MS Arnold, AA Green, JF Hulvat, SI Stupp, MC Hersam, Sorting carbon nanotubes by electronic structure using density differentiation. Nat. Nanotechnol. **1**, 60–65 (2006). doi:10.1038/nnano.2006.52
154. JB Oostinga, HB Heersche, X Liu, AF Morpurgo, LM Vandersypen, Gate-induced insulating state in bilayer graphene devices. Nat. Mater. **7**, 151–157 (2008). doi:10.1038/nmat2082
155. T Ohta, A Bostwick, T Seyller, K Horn, E Rotenberg, Controlling the electronic structure of bilayer graphene. Science **313**, 951–954 (2006). doi:10.1126/science.1130681
156. J Velasco Jr, L Jing, W Bao, Y Lee, P Kratz, V Aji, M Bockrath, CN Lau, C Varma, R Stillwell, D Smirnov, F Zhang, J Jung, AH MacDonald, Transport spectroscopy of symmetry-broken insulating states in bilayer graphene. Nat. Nanotechnol. **7**, 156–160 (2012). doi:10.1038/nnano.2011.251
157. Y Zhang, TT Tang, C Girit, Z Hao, MC Martin, A Zettl, MF Crommie, YR Shen, F Wang, Direct observation of a widely tunable bandgap in bilayer graphene. Nature **459**, 820–823 (2009). doi:10.1038/nature08105
158. F Xia, DB Farmer, YM Lin, P Avouris, Graphene field-effect transistors with high on/off current ratio and large transport band gap at room temperature. Nano Lett. **10**, 715–718 (2010). doi:10.1021/nl9039636
159. X Li, KM Borysenko, MB Nardelli, KW Kim, Electron transport properties of bilayer graphene. Phys. Rev. B **84**, (2011). doi:10.1103/PhysRevB.84.195453
160. J Park, SB Jo, YJ Yu, Y Kim, JW Yang, WH Lee, HH Kim, BH Hong, P Kim, K Cho, KS Kim, Single-Gate Bandgap Opening of Bilayer Graphene by Dual Molecular Doping. Advanced materials. (2011). doi:10.1002/adma.201103411
161. DL Duong, SM Lee, SH Chae, QH Ta, SY Lee, GH Han, JJ Bae, YH Lee, Band-gap engineering in chemically conjugated bilayer graphene: Ab initio calculations. Phys. Rev. B **85**, (2012). doi:10.1103/PhysRevB.85.205413
162. L Jiao, L Zhang, X Wang, G Diankov, H Dai, Narrow graphene nanoribbons from carbon nanotubes. Nature **458**, 877–880 (2009). doi:10.1038/nature07919
163. X Li, X Wang, L Zhang, S Lee, H Dai, Chemically derived, ultrasmooth graphene nanoribbon semiconductors. Science **319**, 1229–1232 (2008). doi:10.1126/science.1150878
164. M Han, B Özyilmaz, Y Zhang, P Kim, Energy band-gap engineering of graphene nanoribbons. Phys. Rev. Lett. **98**, (2007). doi:10.1103/PhysRevLett.98.206805
165. X Wang, Y Ouyang, X Li, H Wang, J Guo, H Dai, Room-temperature all-semiconducting Sub-10-nm graphene nanoribbon field-effect transistors. Phys. Rev. Lett. **100**, (2008). doi:10.1103/PhysRevLett.100.206803
166. J Bai, X Zhong, S Jiang, Y Huang, X Duan, Graphene nanomesh. Nat. Nanotechnol. **5**, 190–194 (2010). doi:10.1038/nnano.2010.8
167. M Kim, NS Safron, E Han, MS Arnold, P Gopalan, Fabrication and characterization of large-area, semiconducting nanoperforated graphene materials. Nano letters. **10**, 1125–1131 (2010). doi:10.1021/nl9032318
168. Y-B Tang, L-C Yin, Y Yang, X-H Bo, Y-L Cao, H-E Wang, W-J Zhang, I Bello, S-T Lee, H-M Cheng, C-S Lee, Tunable band gaps and p-Type transport properties of boron-doped graphenes by controllable ion doping using reactive microwave plasma. Acs Nano. **6**, 1970–1978 (2012). doi:10.1021/nn3005262
169. D Wei, Y Liu, Y Wang, H Zhang, L Huang, G Yu, Synthesis of N-Doped graphene by chemical vapor deposition and its electrical properties. Nano Lett. **9**, 1752–1758 (2009). doi:10.1021/nl803279t
170. R Balog, B Jorgensen, L Nilsson, M Andersen, E Rienks, M Bianchi, M Fanetti, E Laegsgaard, A Baraldi, S Lizzit, Z Sljivancanin, F Besenbacher, B Hammer, TG Pedersen, P Hofmann, L Hornekaer, Bandgap opening in graphene induced by patterned hydrogen adsorption. Nat. Mater. **9**, 315–319 (2010). doi:10.1038/nmat2710
171. J Chen, C Klinke, A Afzali, P Avouris, Self-aligned carbon nanotube transistors with charge transfer doping. Appl. Phys. Lett. **86**, 123108 (2005). doi:10.1063/1.1888054
172. MS Strano, CA Dyke, ML Usrey, PW Barone, MJ Allen, H Shan, C Kittrell, RH Hauge, JM Tour, RE Smalley, Electronic structure control of single-walled carbon nanotube functionalization. Science **301**, 1519–1522 (2003). doi:10.1126/science.1087691
173. DL Duong, IH Lee, KK Kim, J Kong, SM Lee, YH Lee, Carbon nanotube doping mechanism in a salt solution and hygroscopic effect: density functional theory. Acs Nano. **4**, 5430–5436 (2010). doi:10.1021/nn1011489
174. S-M Yoon, UJ Kim, A Benayad, IH Lee, H Son, H-J Shin, WM Choi, YH Lee, YW Jin, E-H Lee, SY Lee, J-Y Choi, JM Kim, Thermal conversion of electronic and electrical properties of AuCl3-Doped single-walled carbon nanotubes. Acs Nano. **5**, 1353–1359 (2011). doi:10.1021/nn103055u
175. S Heinze, J Tersoff, R Martel, V Derycke, J Appenzeller, P Avouris, Carbon nanotubes as schottky barrier transistors. Phys. Rev. Lett. **89**, (2002). doi:10.1103/PhysRevLett.89.106801

176. T Takenobu, T Kanbara, N Akima, T Takahashi, M Shiraishi, K Tsukagoshi, H Kataura, Y Aoyagi, Y Iwasa, Control of carrier density by a solution method in carbon-nanotube devices. Adv. Mater. **17**, 2430–2434 (2005). doi:10.1002/adma.200500759
177. V Derycke, R Martel, J Appenzeller, P Avouris, Controlling doping and carrier injection in carbon nanotube transistors. Appl. Phys. Lett. **80**, 2773 (2002). doi:10.1063/1.1467702
178. C Zhou, Modulated chemical doping of individual carbon nanotubes. Science **290**, 1552–1555 (2000). doi:10.1126/science.290.5496.1552
179. IH Lee, UJ Kim, H Bin Son, S-M Yoon, F Yao, WJ Yu, DL Duong, J-Y Choi, JM Kim, EH Lee, YH Lee, Hygroscopic effects on AuCl3-Doped carbon nanotubes. J. Phys. Chem. C **114**, 11618–11622 (2010). doi:10.1021/jp1036662
180. SM Kim, JH Jang, KK Kim, HK Park, JJ Bae, WJ Yu, IH Lee, G Kim, DD Loc, UJ Kim, E-H Lee, H-J Shin, J-Y Choi, YH Lee, Reduction-controlled viologen in bisolvent as an environmentally stable n-Type dopant for carbon nanotubes. J. Am. Chem. Soc. **131**, 327–331 (2009). doi:10.1021/ja807480g
181. SY Lee, SW Lee, SM Kim, WJ Yu, YW Jo, YH Lee, Scalable complementary logic gates with chemically doped semiconducting carbon nanotube transistors. Acs Nano **5**, 2369–2375 (2011). doi:10.1021/nn200270e
182. BR Kang, WJ Yu, KK Kim, HK Park, SM Kim, Y Park, G Kim, H-J Shin, UJ Kim, E-H Lee, J-Y Choi, YH Lee, Restorable type conversion of carbon nanotube transistor using pyrolytically controlled antioxidizing photosynthesis coenzyme. Adv. Funct. Mater. **19**, 2553–2559 (2009). doi:10.1002/adfm.200801712
183. SM Kim, YW Jo, KK Kim, DL Duong, H-J Shin, JH Han, J-Y Choi, J Kong, YH Lee, Transparent organic P-Dopant in carbon nanotubes: Bis(trifluoromethanesulfonyl)imide. Acs Nano. **4**, 6998–7004 (2010). doi:10.1021/nn102175h
184. HK Jeong, K-j Kim, SM Kim, YH Lee, Modification of the electronic structures of graphene by viologen. Chem. Phys. Lett. **498**, 168–171 (2010). doi:10.1016/j.cplett.2010.08.065
185. A Benayad, H-J Shin, HK Park, S-M Yoon, KK Kim, MH Jin, H-K Jeong, JC Lee, J-Y Choi, YH Lee, Controlling work function of reduced graphite oxide with Au-ion concentration. Chem. Phys. Lett. **475**, 91–95 (2009). doi:10.1016/j.cplett.2009.05.023
186. WJ Yu, L Liao, SH Chae, YH Lee, X Duan, Toward tunable band gap and tunable dirac point in bilayer graphene with molecular doping. Nano letters. **11**, 4759–4763 (2011). doi:10.1021/nl2025739
187. V Derycke, R Martel, J Appenzeller, P Avouris, Carbon nanotube inter- and intramolecular logic gates. Nano Lett. **1**, 453–456 (2001). doi:10.1021/nl015606f
188. GU Sumanasekera, CKW Adu, S Fang, PC Eklund, Effects of gas adsorption and collisions on electrical transport in single-walled carbon nanotubes. Phys. Rev. Lett. **85**, 1096–1099 (2000). doi:10.1103/PhysRevLett.85.1096
189. PG Collins, Extreme oxygen sensitivity of electronic properties of carbon nanotubes. Science **287**, 1801–1804 (2000). doi:10.1126/science.287.5459.1801
190. WJ Yu, BR Kang, IH Lee, YS Min, YH Lee, Majority carrier type conversion with floating gates in carbon nanotube transistors. Adv. Mater. **21**, 4821–4824 (2009). doi:10.1002/adma.200900911
191. WJ Yu, YH Lee, Strategy for carrier control in carbon nanotube transistors. ChemSusChem **4**, 890–904 (2011). doi:10.1002/cssc.201000412
192. M Bockrath, J Hone, A Zettl, PL McEuen, AG Rinzler, RE Smalley, Chemical doping of individual semiconducting carbon-nanotube ropes. Phys. Rev. B **61**, 10606–10608 (2000)
193. MH Yang, KBK Teo, WI Milne, DG Hasko, Carbon nanotube Schottky diode and directionally dependent field-effect transistor using asymmetrical contacts. Appl. Phys. Lett. **87**, 253116 (2005). doi:10.1063/1.2149991
194. Y Nosho, Y Ohno, S Kishimoto, T Mizutani, Relation between conduction property and work function of contact metal in carbon nanotube field-effect transistors. Nanotechnology **17**, 3412–3415 (2006). doi:10.1088/0957-4484/17/14/011
195. Y Lee, S Bae, H Jang, S Jang, SE Zhu, SH Sim, YI Song, BH Hong, JH Ahn, Wafer-scale synthesis and transfer of graphene films. Nano letters. **10**, 490–493 (2010). doi:10.1021/nl903272n
196. Q Cao, JA Rogers, Ultrathin films of single-walled carbon nanotubes for electronics and sensors: a review of fundamental and applied aspects. Adv. Mater. **21**, 29–53 (2009). doi:10.1002/adma.200801995
197. S Jang, H Jang, Y Lee, D Suh, S Baik, BH Hong, JH Ahn, Flexible, transparent single-walled carbon nanotube transistors with graphene electrodes. Nanotechnology **21**, 425201 (2010). doi:10.1088/0957-4484/21/42/425201
198. JH Chen, M Ishigami, C Jang, DR Hines, MS Fuhrer, ED Williams, Printed Graphene Circuits. Adv. Mater. **19**, 3623–3627 (2007). doi:10.1002/adma.200701059
199. JU Park, S Nam, MS Lee, CM Lieber, Synthesis of monolithic graphene-graphite integrated electronics. Nat. Mat. **11**, 120–125 (2012). doi:10.1038/nmat3169
200. DM Sun, C Liu, WC Ren, HM Cheng, A review of carbon nanotube- and graphene-based flexible thin-film transistors. Small **9**, 1188–1205 (2013). doi:10.1002/smll.201203154
201. C-C Lu, Y-C Lin, C-H Yeh, J-C Huang, P-W Chiu, High mobility flexible graphene field-effect transistors with self-healing gate dielectrics. Acs Nano. **6**, 4469–4474 (2012). doi:10.1021/nn301199j
202. T Kim, H Kim, SW Kwon, Y Kim, WK Park, DH Yoon, AR Jang, HS Shin, KS Suh, WS Yang, Large-scale graphene micropatterns via self-assembly-mediated process for flexible device application. Nano letters. **12**, 743–748 (2012). doi:10.1021/nl203691d
203. SJ Park, OS Kwon, SH Lee, HS Song, TH Park, J Jang, Ultrasensitive flexible graphene based field-effect transistor (FET)-type bioelectronic nose. Nano letters. **12**, 5082–5090 (2012). doi:10.1021/nl301714x
204. Q Cao, SH Hur, ZT Zhu, YG Sun, CJ Wang, MA Meitl, M Shim, JA Rogers, Highly Bendable, transparent thin-film transistors that use carbon-nanotube-based conductors and semiconductors with elastomeric dielectrics. Adv. Mater. **18**, 304–309 (2006). doi:10.1002/adma.200501740
205. DM Sun, MY Timmermans, Y Tian, AG Nasibulin, EI Kauppinen, S Kishimoto, T Mizutani, Y Ohno, Flexible high-performance carbon nanotube integrated circuits. Nat. Nanotechnol. **6**, 156–161 (2011). doi:10.1038/nnano.2011.1
206. BJ Kim, S-K Lee, MS Kang, J-H Ahn, JH Cho, Coplanar-gate transparent graphene transistors and inverters on plastic. Acs Nano. **6**, 8646–8651 (2012). doi:10.1021/nn3020486
207. H Wang, D Nezich, J Kong, T Palacios, Graphene frequency multipliers. IEEE Electron. Dev. L. **30**, 547–549 (2009). doi:10.1109/led.2009.2016443
208. X Yang, G Liu, AA Balandin, K Mohanram, Triple-mode single-transistor graphene amplifier and its applications. Acs Nano. **4**, 5532–5538 (2010). doi:10.1021/nn1021583
209. Z Wang, Z Zhang, H Xu, L Ding, S Wang, L-M Peng, A high-performance top-gate graphene field-effect transistor based frequency doubler. Appl. Phys. Lett. **96**, 173104 (2010). doi:10.1063/1.3413959
210. H Wang, A Hsu, J Wu, J Kong, T Palacios, RF Graphene-Based Ambipolar, Mixers. IEEE Electron. Dev. Lett. **31**, 906–908 (2010). doi:10.1109/led.2010.2052017
211. A Hsu, H Wang, KK Kim, J Kong, T Palacios, High frequency performance of graphene transistors grown by chemical vapor deposition for mixed signal applications. Jpn. J. Appl. Phys. **50**, 070114 (2011). doi:10.1143/jjap.50.070114
212. S Lee, K Lee, CH Liu, GS Kulkarni, Z Zhong, Flexible and transparent all-graphene circuits for quaternary digital modulations. Nat. Commun. **3**, 1018 (2012). doi:10.1038/ncomms2021
213. T Takahashi, K Takei, AG Gillies, RS Fearing, A Javey, Carbon nanotube active-matrix backplanes for conformal electronics and sensors. Nano letters. **11**, 5408–5413 (2011). doi:10.1021/nl203117h
214. JA Rogers, Z Bao, K Baldwin, A Dodabalapur, B Crone, VR Raju, V Kuck, H Katz, K Amundson, J Ewing, P Drzaic, Paper-like electronic displays: large-area rubber-stamped plastic sheets of electronics and microencapsulated electrophoretic inks. Proc. Natl. Acad. Sci. U. S. A. **98**, 4835–4840 (2001). doi:10.1073/pnas.091588098
215. C Wang, D Hwang, Z Yu, K Takei, J Park, T Chen, B Ma, A Javey, User-interactive electronic skin for instantaneous pressure visualization. Nat Mater. **12**, 899–904 (2013). doi:10.1038/nmat3711
216. Z-g Wang, Y-f Chen, P-j Li, X Hao, J-b Liu, R Huang, Y-r Li, Flexible Graphene-Based Electroluminescent Devices. Acs Nano. **5**, 7149–7154 (2011). doi:10.1021/nn2018649
217. C Sire, F Ardiaca, S Lepilliet, JW Seo, MC Hersam, G Dambrine, H Happy, V Derycke, Flexible gigahertz transistors derived from solution-based single-layer graphene. Nano letters. **12**, 1184–1188 (2012). doi:10.1021/nl203316r
218. OM Nayfeh, Graphene transistors on mechanically flexible polyimide incorporating atomic-layer-deposited gate dielectric. IEEE Electron. Dev. L. **32**, 1349–1351 (2011). doi:10.1109/led.2011.2163489
219. J Lee, T-J Ha, H Li, KN Parrish, M Holt, A Dodabalapur, RS Ruoff, D Akinwande, 25 GHz embedded-gate graphene transistors with high-K dielectrics on extremely flexible plastic sheets. Acs Nano. **7**, 7744–7750 (2013)

220. N Petrone, I Meric, J Hone, KL Shepard, Graphene field-effect transistors with gigahertz-frequency power gain on flexible substrates. Nano letters. **13**, 121–125 (2013). doi:10.1021/nl303666m
221. L Liao, YC Lin, M Bao, R Cheng, J Bai, Y Liu, Y Qu, KL Wang, Y Huang, X Duan, High-speed graphene transistors with a self-aligned nanowire gate. Nature **467**, 305–308 (2010). doi:10.1038/nature09405
222. C Heo, SY Lee, A Jo, S Jung, M Suh, YH Lee, Flexible, transparent, and noncytotoxic graphene electric field stimulator for effective cerebral blood volume enhancement. Acs Nano. **7**, 4869–4878 (2013). doi:10.1021/nn305884w
223. WJ Yu, SH Chae, SY Lee, DL Duong, YH Lee, Ultra-transparent, flexible single-walled carbon nanotube non-volatile memory device with an oxygen-decorated graphene electrode. Adv. Mater. **23**, 1889–1893 (2011). doi:10.1002/adma.201004444
224. B Park, K Cho, S Kim, S Kim, Transparent nano-floating gate memory on glass. Nanotechnology **21**, 335201 (2010). doi:10.1088/0957-4484/21/33/335201
225. Y Ji, S Lee, B Cho, S Song, T Lee, Flexible organic memory devices with multilayer graphene electrodes. Acs Nano. **5**, 5995–6000 (2011). doi:10.1021/nn201770s
226. HY Jeong, JY Kim, JW Kim, JO Hwang, JE Kim, JY Lee, TH Yoon, BJ Cho, SO Kim, RS Ruoff, SY Choi, Graphene oxide thin films for flexible nonvolatile memory applications. Nano letters. **10**, 4381–4386 (2010). doi:10.1021/nl101902k
227. TS Hansen, K West, O Hassager, NB Larsen, Highly stretchable and conductive polymer material made from Poly(3,4-ethylenedioxythiophene) and polyurethane elastomers. Adv. Funct. Mater. **17**, 3069–3073 (2007). doi:10.1002/adfm.200601243
228. F Xu, W Lu, Y Zhu, Controlled 3D buckling of silicon nanowires for stretchable electronics. Acs Nano. **5**, 672–678 (2011). doi:10.1021/nn103189z
229. P Lee, J Lee, H Lee, J Yeo, S Hong, KH Nam, D Lee, SS Lee, SH Ko, Highly stretchable and highly conductive metal electrode by very long metal nanowire percolation network. Adv. Mater. **24**, 3326–3332 (2012). doi:10.1002/adma.201200359
230. C Yu, C Masarapu, J Rong, B Wei, H Jiang, Stretchable supercapacitors based on buckled single-walled carbon-nanotube macrofilms. Adv. Mater. **21**, 4793–4797 (2009). doi:10.1002/adma.200901775
231. F Xu, X Wang, Y Zhu, Y Zhu, Wavy ribbons of carbon nanotubes for stretchable conductors. Adv. Funct. Mater. **22**, 1279–1283 (2012). doi:10.1002/adfm.201102032
232. J Xiao, A Carlson, ZJ Liu, Y Huang, H Jiang, JA Rogers, Stretchable and compressible thin films of stiff materials on compliant wavy substrates. Appl. Phys. Lett. **93**, 013109 (2008). doi:10.1063/1.2955829
233. J Jeong, S Kim, J Cho, Y Hong, Stable stretchable silver electrode directly deposited on wavy elastomeric substrate. IEEE Electron. Dev. Lett. **30**, 1284–1286 (2009). doi:10.1109/led.2009.2033723
234. Y Sun, V Kumar, I Adesida, JA Rogers, Buckled and wavy ribbons of gaas for high-performance electronics on elastomeric substrates. Adv. Mater. **18**, 2857–2862 (2006). doi:10.1002/adma.200600646
235. DY Khang, HQ Jiang, Y Huang, JA Rogers, A stretchable form of single-crystal silicon for high-performance electronics on rubber substrates. Science **311**, 208–212 (2006). doi:10.1126/science.1121401
236. C Hierold, A Jungen, C Stampfer, T Helbling, Nano electromechanical sensors based on carbon nanotubes. Sensors Actuators A Phys. **136**, 51–61 (2007). doi:10.1016/j.sna.2007.02.007
237. TW Tombler, CW Zhou, L Alexseyev, J Kong, HJ Dai, L Lei, CS Jayanthi, MJ Tang, SY Wu, Reversible electromechanical characteristics of carbon nanotubes under local-probe manipulation. Nature **405**, 769–772 (2000)
238. H Maune, M Bockrath, Elastomeric carbon nanotube circuits for local strain sensing. Appl. Phys. Lett. **89**, 173131 (2006). doi:10.1063/1.2358821
239. DJ Lipomi, M Vosgueritchian, BC Tee, SL Hellstrom, JA Lee, CH Fox, Z Bao, Skin-like pressure and strain sensors based on transparent elastic films of carbon nanotubes. Nat. Nanotechnol. **6**, 788–792 (2011). doi:10.1038/nnano.2011.184
240. KS Kim, Y Zhao, H Jang, SY Lee, JM Kim, KS Kim, JH Ahn, P Kim, JY Choi, BH Hong, Large-scale pattern growth of graphene films for stretchable transparent electrodes. Nature **457**, 706–710 (2009). doi:10.1038/nature07719
241. Z Chen, W Ren, L Gao, B Liu, S Pei, HM Cheng, Three-dimensional flexible and conductive interconnected graphene networks grown by chemical vapour deposition. Nat. Mater. **10**, 424–428 (2011). doi:10.1038/nmat3001
242. ME Roberts, N Queralto, SCB Mannsfeld, BN Reinecke, W Knoll, Z Bao, Cross-linked polymer gate dielectric films for low-voltage organic transistors. Chem. Mater. **21**, 2292–2299 (2009). doi:10.1021/cm900637p
243. DM Sun, MY Timmermans, A Kaskela, AG Nasibulin, S Kishimoto, T Mizutani, EI Kauppinen, Y Ohno, Mouldable all-carbon integrated circuits. Nat. Commun. **4**, 2302 (2013). doi:10.1038/ncomms3302
244. SK Lee, BJ Kim, H Jang, SC Yoon, C Lee, BH Hong, JA Rogers, JH Cho, JH Ahn, Stretchable graphene transistors with printed dielectrics and gate electrodes. Nano letters. **11**, 4642–4646 (2011). doi:10.1021/nl202134z
245. SH Chae, WJ Yu, JJ Bae, DL Duong, D Perello, HY Jeong, QH Ta, TH Ly, QA Vu, M Yun, X Duan, YH Lee, Transferred wrinkled Al2O3 for highly stretchable and transparent graphene-carbon nanotube transistors. Nat. Mater. **12**, 403–409 (2013). doi:10.1038/nmat3572
246. T Yamada, Y Hayamizu, Y Yamamoto, Y Yomogida, A Izadi-Najafabadi, DN Futaba, K Hata, A stretchable carbon nanotube strain sensor for human-motion detection. Nat. Nanotechnol. **6**, 296–301 (2011). doi:10.1038/nnano.2011.36

Integration of monolithic porous polymer with droplet-based microfluidics on a chip for nano/ picoliter volume sample analysis

Jin-young Kim[1*], Soo-Ik Chang[2], Andrew J deMello[3] and Danny O'Hare[4]

Abstract

In this paper, a porous polymer nanostructure has been integrated with droplet-based microfluidics in a single planar format. Monolithic porous polymer (MPP) was formed selectively within a microfluidic channel. The resulting analyte bands were sequentially comartmentalised into droplets. This device reduces band broadening and the effects of post-column dead volume by the combination of the two techniques. Moreover it offers the precise control of nano/picoliter volume samples.

1 Background

Microfabricated systems have emerged as promising tools for chemical and biological analysis due to the speed, throughput and control of minute samples that they make possible [1-8]. In recent years, there has been considerable interest in droplet-based microfluidics because flow segmentation enables compartmentalisation of reagent volumes from fL to μL within a continuous and immiscible fluid, as well as the generation of monodisperse droplets and the precise control of them [9-11].

MPP potentially offers the advantages of simple control of permeability and surface areas as well as easy preparation within a micro-fluidic channel for various applications of lab-on-a-chip (LOC) [12-15]. Various trials have been conducted to combine separation and droplet functions on a chip using beads-packed channels or capillary electrophoresis (CE) separation [16-20]. However, integration of MPP with droplet microfluidics has not been reported yet. Integration of MPP with droplet-based microfluidics on a single chip will provide various benefits. For example, MPP can be used as a nanostructure for a column or membrane to separate and filter analytes in the microfluidic channel. Subsequent compartmentalisation by droplets following MPP dramatically reduces Taylor dispersion of the effluent and minimizes dead volume effects; in addition, it allows further analysis downstream or offline with the encapsulated analytes. Furthermore, it is easier (in comparison to the widely-used beads) to form a porous nanostructure in a channel using selective UV polymerization and to control properties such as permeability or porosity by adjusting the composition of the MPP solution.

Our device consists of two parts: an MPP-filled channel and a droplet-generation zone. Thermoset polyester (TPE) was chosen as a device material because of its excellent mechanical properties, allowing it to withstand pressures of up to 18 Mpa [21]. It makes possible a broad range of device applications including membrane, separation column and high-frequency droplet generation; these require operational stability under high pressure, which is not feasible when using typical materials, such as polydimethylsiloxane (PDMS) for the microfluidic device.

* Correspondence: jinyoung.kim@bsse.ethz.ch
[1]Department of Biosystems Science and Engineering, Bioegineering Laboratory, ETH Zurich, CH-4058 Basel, Switzerland
Full list of author information is available at the end of the article

2 Methods

2.1 Materials

TPE (CFS Fibreglass, UK) was prepared by mixing it with its polymerization catalyst, methyl ethyl ketone peroxide (MEKP), and polyethylene terephthalate, PET, (Daedong Polymer, South Korea) was used as substrate. The mixture of ethylene diacrylate (EDA, monomer) 0.485 g, methyl methacrylate (MMA, cross-linker) 0.485 g and benzophenone (BP, photo-initiator) 0.03 g was prepared for the grafting layer. Butyl methacrylate (BuMA, monomer) 0.6 g, ethylene dimethacrylate (EDMA, cross-linker) 0.4 g, 1-dodecanol (progen) 1.5 g and 2,2'-dimethoxy-2-

phenylacetophenone (DMPAP, photo-initiator) 0.01 g were mixed for the monolithic porous polymer.

For droplet experiments, a 10% (v/v) mixture of fluorocarbon oil, FC-70 (3 M Fluorinert, USA), and 1H, 1H, 2H, 2H-perfluorooctanol (PFO, Sigma-Aldrich) was used for the continuous phase. The mobile phase was a mixture of 26 mM phosphate buffer (pH 7) and methanol (HiPerSolve for HPLC, BDH Prolabo) in a ratio of 5: 95 (v/v)

Fluorescein isothiocyanate, FITC, (Sigma-Aldrich, USA) and Alexa Fluor® 488, AF 488, (Invitrogen, USA) were diluted to 0.02 μM and 50 μM concentration in the mobile phase as the analytes.

2.2 Fabrication procedure

The fabrication process of the MPP-integrated droplet device is shown schematically in Figure 1. It consists of two major steps: forming the TPE microfluidic channel followed by the creation of a monolithic porous polymer. The former was made with an SU-8 master mould, fabricated using standard photolithography. PDMS was poured onto the SU-8 mould, cured for 4 h at 65°C, and then peeled off. The embossed microfluidic channel pattern on the PDMS was surrounded with 4-mm-thick walls in order to define the final outside dimensions of the device. TPE resin was mixed with the MEKP catalyst in a ratio of 100: 1 (w/w) and degassed and decanted into the PDMS mould. It was partially cured for 10 min at 60°C then cooled down to room temperature for 5 min before being separated from the PDMS mould. In the meantime, the PET substrate was cleaned by sonification in isopropyl alcohol (IPA) and was treated by O_2 plasma at 70 mW for 12 sec in order to obtain a robust bonding. While the fully cured TPE is hard material, the semi-cured TPE is and has gel-like property. It was carefully removed and bonded to the PET substrate. To connect the microchannels with the syringe, PEEK unions (Phenomenex, USA) were placed on the inlets and outlets. The assembly was then put into a vacuum desiccator to remove residual gas. Finally, the TPE device was cured at 76°C for 1 h. This two-stsge process was adopted because semi-cured TPE substrate can be easily removed from the PDMS mould but adheres and potentially damages the SU-8 master. In addition, the flexibility of the PDMS working mould

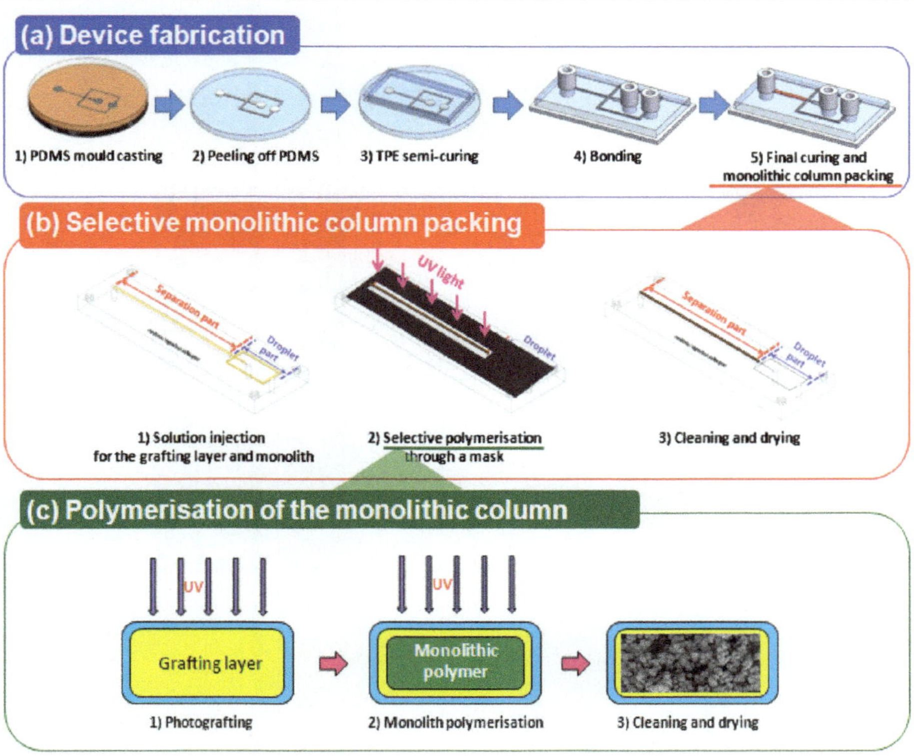

Figure 1 Schematic of the hybrid monolithic LC-droplet device fabrication, (a) overall procedures - 1) PDMS mould casting from SU-8 master mould, 2) peeling of the PDMS, 3) TPE pouring and semi-curing, 4) TPE bonding to the substrate and attaching the interconnects, 5) full-curing and monolithic column packing, (b) selective packing of the monolithic column within a microchannel – 1) solutions injection, 2) selective UV exposure through a mask, 3) cleaning and drying and (c) polymerisation of the monolithic column – 1-1) injection of the grafting layer solution, 1-2) UV polymerisation through a mask, 1-3) cleaning and drying, 2-1) injection of the monolithic polymer solution, 2-2) UV polymerisation through a mask, 2-3) cleaning and drying.

greatly facilitates the disassembly of the substrate and mould.

Next, as shown in Figure 1(b) and (c), the separation zone of the channel was selectively packed with poly(methyl acrylate) MPP. The TPE channel was selectively exposed to a UV light through a film mask during the polymerisation of the grafting layer and MPP. N_2 gas was blown for 10 min into the grafting layer mixture (See section 2.1.) before filling the channel in order to remove oxygen and avoid expansion and subsequent heat-induced voids during polymerisation. Then it was radiated to UV light (broadband 290-385 nm, 12.22 mW/cm2) for 10 mins. The channel was flushed with 10 volumes of cleaning solvent (methanol : DI water = 1:1 (v/v)) to remove the unreacted polymer and was then dried at 40C. The monolithic polymer mixture was prepared and irradiated in a similar fashion as the grafting layer. Subsequently, it was cleaned by flowing the solvent at 10ul/min for 1 h to remove the remaining porogenic solvent and photo-initiators. It was then left to dry at 40°C overnight. UV irradiation was conducted from below since PET has much higher transmittance in the UV range than TPE does.

2.3 Characterisation

The mobile phase was introduced into the MPP-filled channel at 10 µL/min by HP 1050 HPLC pump (Agilent, USA). 5 µL of the mixture of two dyes in the mobile phase was injected; then the effluent was segmented by the oil injected at 100 µL/min using precision syringe pumps (Harvard Apparatus, USA). Droplets containing the dyes were recorded by a high speed camera (Phantom, USA) and detected by laser-induced fluorescence (LIF) detection with a beam from 488 nm Ar^+ diode laser (Omnichrome, Melles Griot, Cambridge, UK).

3 Results and discussion

The fabricated MPP-integrated droplet device is illustrated in Figure 2. The poly(methyl acrylate) MPP was a selectively-formed well in the separation part only and the flow-focusing channel for droplet generation flanked the MPP within a single layer. Figure 2(a) shows SEM images of the poly(methyl acrylate) MPP within TPE channels. It consists of a number of globes that are cross-linked to each other and the globular structure includes numerous mesopores (2-50 nm) between the globes. Furthermore micropores (< 2 nm) were observed

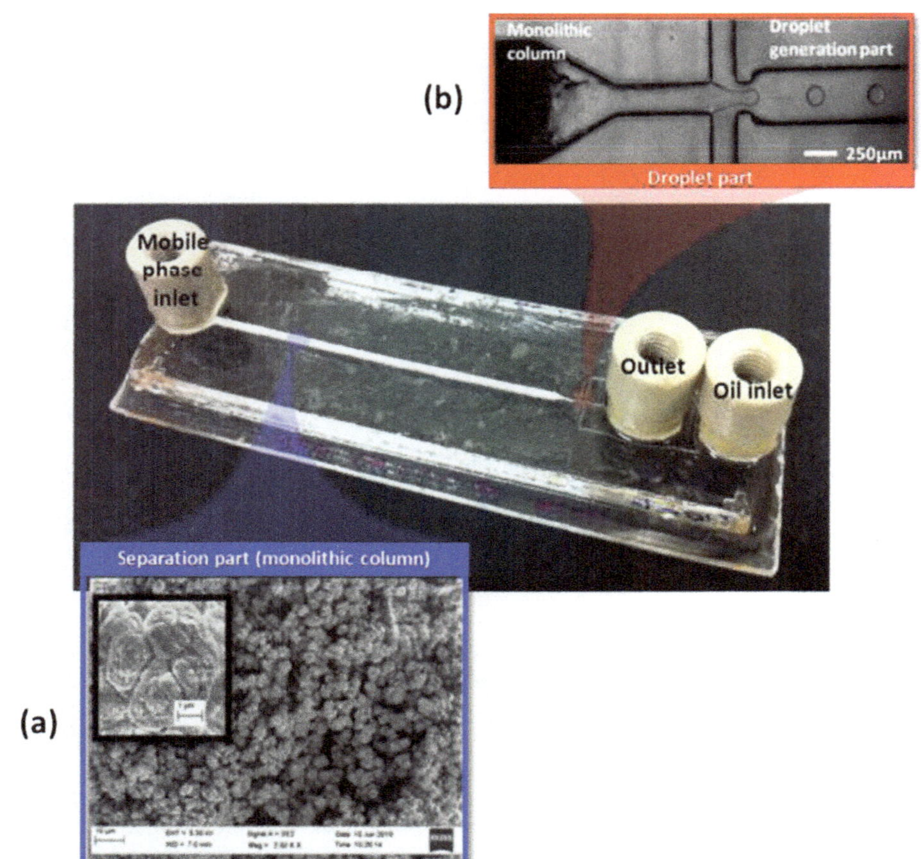

Figure 2 The MPP-integrated droplet device (a) an image of the monolithic column in 1 mm (w) × 0.25 mm (H) × 50 mm (L) separation channel and (b) the 250 µm wide droplet channel.

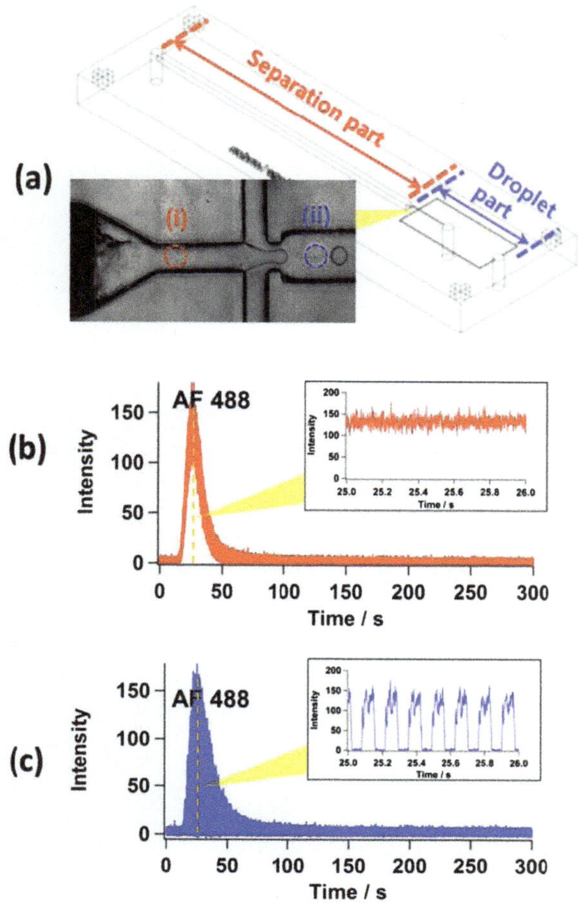

Figure 3 LIF detection of AF 488 and FITC, (a) schematic of the movement of fluids in the channels and the two laser detection points, (b) the chromatograms of the separated dyes from point (i) and (c) compartmentalised dyes from point (ii).

on the surface of the globes. These pores significantly contribute to increase the surface area which results in more reactions or interaction sites. Each globe is approximately 2 μm in diameter and the monolith is bonded securely to the channel wall by the grafting layer.

Two dyes, AF 488 and FITC, were injected and fluorescence was measured at two different points: before (point i) and after (point ii) the droplet generation to assess the efficacy of compartmentalisation in preserving concentration gradients, as shown in Figure 3(a). Figure 3(b) and (c) describe fluorescent peaks from the two measurement points. While the continuous signal was observed at point i, the segmented signals associated with the compartmentalised analyte by droplets are evident at point ii. No dislocation or clogging of MPP in the channel was observed during the run for over 1 hr. Moreover, since concentrations in the eluted bands are approximately Gaussian at the MPP outflow, variation in concentration between droplets can be controlled [20,22]. However separation between the dyes was poor. Only AF 488 was observed without the FITC peak, even though the measurement continued for up to 1 hr. Therefore the separation conditions for the dyes in the fabricated MPP column such as mobile phase composition and flow rate must be optimized.

To investigate diffusional band broadening, the chromatograms in (b) and (c) were quantitatively compared by extracting the maximum fluorescence of each droplet.

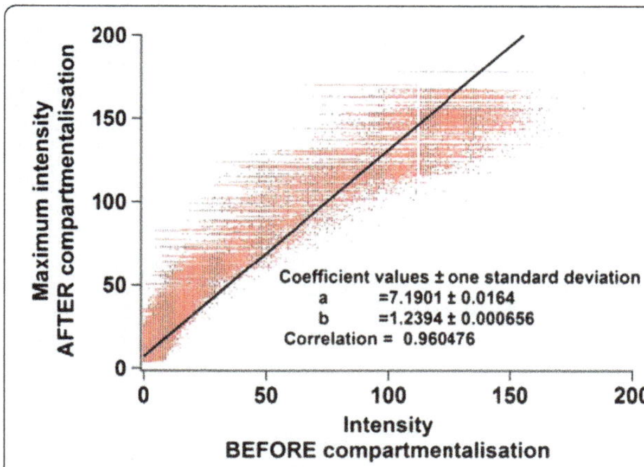

Figure 4 Linear best fit between the chromatograms before (Figure 3(b)) and after (Figure 3(c)) compartmentalisation.

A digital maximum filter with a 100-point window size (Igor Pro 6, USA) (slightly longer than the maximum interval between the droplets) was applied. As shown in Figure 4, the two chromatograms were strongly correlated with a slope of 1.2394 ± 0.000656 and a correlation coefficient of 0.96, which means there is no significant additional band broadening between due to compartmentalisation of the analyte and that droplet flow can act as an effective fraction collector

4 Conclusions

MPP has been integrated with droplet microfluidic channel on a monolithic chip. Straightforward fabrication of monolithic porous nanostructure within a microfluidic channel has been demonstrated. Band broadening and dead volume issues have been decreased as well by sequential compartmentalisation of analytes at the MPP outflow, encapsulating them in defined droplet volumes. This confirms the potential of MPP for separation or filtering function in a microfluidic chip and the ability of droplets to act as a fraction collector for the handling of nano/picoliter volume samples.

Competing interests

The authors declare that they have no competing interests.

Authors' contributions

JY Kim carried out all experiments and drafted the manuscript. All authors participated in data analysis and the manuscript revision. All authors read and approved the final manuscript.

Acknowledgements

This work was supported by the Global Research Laboratory Program of the National Research Foundation of Korea (NRF) grant (Grant Number K20904000004-13A0500-00410) and Global Frontier Project (Grant Number H-GUARD_2013M3A6B2078957) of BioNano Health-Guard Research Center, which are funded by the Ministry of Science, ICT & Future Planning (MSIP) of Korea.

Author details

[1]Department of Biosystems Science and Engineering, Bioegineering Laboratory, ETH Zurich, CH-4058 Basel, Switzerland. [2]Department of Biochemistry, Chungbuk National University, Cheongju, Chungbuk 361-763, Korea. [3]Institute of Chemical and Bioengineering, Department of Chemistry and Applied Bioscience, ETH Zurich, CH-8093 Zurich, Switzerland. [4]Department of Bioengineering, Imperial College London, London SW7 2AZ, UK.

References

1. GM Whitesides, Nature **442**, 368–373 (2006)
2. AJ deMello, Nature **442**, 394–402 (2006)
3. J Hong, M Choi, AJ DeMello, JB Edel, BioChip J. **3**, 203–207 (2009)
4. A Arora, G Simone, GB Salieb-Beugelaar, JT Kim, A Manz, Anal. Chem. **82**, 4830–4847 (2010)
5. GB Salieb-Beugelaar, G Simone, A Arora, A Philippi, A Manz, Anal. Chem. **82**, 4848–4864 (2010)
6. C Lim, J Hong, BG Chung, AJ deMello, J Choo, Analyst **135**, 837–844 (2010)
7. XCI Solvas, A deMello, Chem. Commun. **47**, 1936–1942 (2011)
8. SW Cho, DK Kang, J Choo, AJ deMello, SI Chang, BMB Rep. **44**, 705–712 (2011)
9. H Song, DL Chen, RF Ismagilov, Angew. Chem. Int. Ed. **45**, 7336–7356 (2006)
10. Y Schaerli, F Hollfelder, Mol. BioSyst. **5**, 1392–1404 (2009)
11. J Hong, M Choi, JB Edel, AJ deMello, Lab. Chip. **10**, 2702–2709 (2010)
12. K Miyabe, G Guiochon, J. Sep. Sci. **27**, 853–873 (2004)
13. S Eeltink, WMC Decrop, GP Rozing, PJ Schoenmakers, WT Kok, J. Sep. Sci. **27**, 1431–1440 (2004)
14. H Zou, X Huang, M Ye, Q Luo, J. Chromatogr. A **954**, 5–32 (2002)
15. R Wu, L Hu, F Wang, M Ye, H Zou, J. Chromatogr. A **1184**, 369–392 (2008)
16. JS Edgar, G Milne, YQ Zhao, CP Pabbati, DSW Lim, DT Chiu, Angew. Chem. Int. Ed. **48**, 2719–2722 (2009)
17. XZ Niu, B Zhang, RT Marszalek, O Ces, JB Edel, DR Klug, AJ Demello, Chem. Commun. (41), 6159–6161 (2009)
18. AB Theberge, G Whyte, WTS Huck, Anal. Chem. **82**, 3449–3453 (2010)
19. S Hamed, B Shay, AS Basu, *In proceeding of 2011 Annual International Conference of the IEEE Engineering in Medicine and Biology Society (Embc)*, 2011, pp. 8396–8399
20. JY Kim, SW Cho, DK Kang, JB Edel, SI Chang, AJ deMello, D O'Hare, Chem. Commun. **48**, 9144–9146 (2012)
21. JY Kim, AJ deMello, SI Chang, J Hong, D O'Hare, Lab. Chip. **11**, 4108–4112 (2011)
22. OJ Miller, A El Harrak, T Mangeat, JC Baret, L Frenz, B El Debs, E Mayot, ML Samuels, EK Rooney, P Dieu, M Galvan, DR Link, AD Griffiths, Proc. Natl. Acad. Sci. U. S. A. **109**, 378–383 (2012)

5

A silicon nanowire photodetector using Au plasmonic nanoantennas

Sang-Won Jee[1,3], Keya Zhou[1,4], Dong-Wook Kim[2*] and Jung-Ho Lee[1*]

Abstract

We suggest a silicon nanowire (SiNW) photodetector side-contacted with hemispherical Au nanoantennas (NAs) that can remarkably amplify the intensity of the near-infrared optical field. The plasmonic NA suppresses visible-range guided mode excitation in SiNW, which enables our photodetector to possess a near-infrared-selective response and overcome the inherent poor optical absorption of Si. The NA can form Schottky contacts with SiNWs for enhancing carrier collection. The vertical NW array also has the geometrically beneficial effects of wide tolerance in light polarization with reduced material consumption.

Keywords: Si nanowire; Photodetector; Plasmonics; Subwavelength structure; Optical absorption; Plasmonic nanoantennas

1 Background

Si near-infrared (NIR) photodetectors are compatible with complementary metal oxide semiconductor (CMOS) technology [1], which can be fabricated with a production cost even lower than that of devices using other materials such as Ge [2] or InGaAs [3]. The indirect bandgap of Si (~1.12 eV), however, leads to poor optical absorption in the NIR region (750 nm < λ < 1100 nm), which hinders the development of Si photodetectors for optoelectronic on-chip integration. Semiconducting nanowires (NWs) might enable us to overcome the limitations of conventional bulk and thin film devices because of their unique geometrical advantages. NWs are known to possess an vastly larger optical cross-section than geometrical cross-section, due to the guided mode [4] and Fabry-Pérot resonances [5,6], as well as anti-reflection effects [7,8]. As a vertical platform of an NW array, spectral dependence of the optical response can be further modified via controlling its period, filling ratio, and symmetric arrangement [9-11]. Moreover, the radial *p-n* junctions offer an efficient carrier collection with the aid of orthogonal collection of charge carriers and incident photons [11-13]. Thus, there have been intensive studies to realize low cost and high efficiency NW-based optoelectronic devices [4-16].

Surface plasmon (SP) provides a new means to amplify the intensity of a local electromagnetic (EM) field, while tuning the optical responses throughout the visible-to-NIR spectral range [14-21]. It is known that the resonance energy of localized surface plasmon (LSP) is highly dependent on the geometry and dielectric functions of the metal nanoparticles (NPs) and supporting materials [17-19]. Thus, metallic NPs have been used as light harvesting nanoantennas (NAs), allowing efficient conversion of propagating light into a nanoscale confined, strong optical field [20-22]. Especially, Au NA shows a high radiation efficiency at long wavelengths (λ ≥ 600 nm) due to a large negative real part of dielectric function ε(ω), while Ohmic absorption losses are kept low with a small imaginary value of ε(ω) [20]. This guides us to simultaneously utilize the Si NWs and Au NA for the development of a new type of photodetector with enhanced NIR responsivity. Recently, there has been increasing research interest in optical and electrical characterizations of Si-based photodetectors with metallic NP [14-16]. Interesting works focused on either single NW devices [15,16] or sophisticated top-down fabrication [22] can offer useful insights to understand the fundamental mechanism; however, cost-efficient and simple fabrication methods are highly desirable for practical applications.

* Correspondence: dwkim@ewha.ac.kr; jungho@hanyang.ac.kr
[2]Department of Physics, Ewha Womans University, Seoul 120-750, South Korea
[1]Department of Chemical Engineering, Hanyang University, Ansan 426-791, South Korea
Full list of author information is available at the end of the article

Here, we propose SiNW-array-based plasmonic photodetectors with Au hemispherical NPs (hNPs) as optical NA, of which photoresponsivity in the NIR range is remarkably enhanced due to intriguing SP resonance effects at the hNP/Si interface. Excitation of the guided mode resonance in the SiNW is suppressed, since incoming light is dissipated and scattered by the hNPs at wavelengths (≤600 nm). The Au hNPs also form Schottky contacts with Si to promote efficient collection of photoexcited carriers. This approach provides an opportunity for low-cost fabrication of a highly sensitive SiNW NIR photodetector.

2 Methods

Vertical SiNW arrays (5 Ω-cm, 6 μm long) were fabricated on 15-μm-thick Si(n/n$^+$)-on-Insulator (SOI) wafers by metal-assisted electroless etching, in which the space filling ratio (FR) of SiNWs was ~0.28. Prior to the SiNW preparation, 50-nm-thick Au films were thermally evaporated on the wafers using a shadow mask to form top electrodes of the photodetectors. Subsequently, Ag nanoparticles were uniformly dispersed on the Si top surface while dipping in a mixed solution of de-ionized water, HF (4.8 M), and $AgNO_3$ (0.01 M) at 20°C for 1 min. Chemical etching was followed in a mixed solution of deionized water, HF (4.8 M), and H_2O_2 (0.5 M) to form a vertically aligned SiNW array with a diameter of 60 ± 33 nm. After etching, Ag nanoparticles that remained at the bottom of SiNWs were removed by nitric acid solution. Au was thermally evaporated onto the SiNWs with a very low deposition rate (0.1 Å/sec) so that the deposited Au agglomerated to form hNPs. Most hNPs were found within 1 μm from the top-ends of the SiNWs due to a directional tendency of thermal evaporation. Figure 1(a) and (b) show the scanning electron microscopy (SEM) and transmission electron microscopy (TEM) images of bare SiNW (BS) and Au NP-coated SiNW (AS) arrays, respectively. The histogram shows that the average diameter of AS is 20 ± 11 nm as shown in Figure 1(c). For Ohmic contacts, the backsides of SOI wafers were patterned using standard photolithography and etching of the oxide layer. The exposed backside n^+ regions of the wafers were attached to Au pads using Ag paste. Spectral photoresponses were measured using monochromatic light generated by a 300 W Xenon lamp. Photovoltaic measurements were carried out at one – sun illumination (AM 1.5G, 100 mW cm^{-2}) using a solar simulator (Peccel PEC-L11). The structural characterization of the NWs was carried out using field emission SEM (Hitach s-4800) and TEM (FEI Tecnai F30 Super-Twin). Nanoscale analyses using monochromated electron energy loss spectra (EELS) were carried out to estimate the SP energies for the Au NPs coated on SiNW. SiNWs suspended in isopropanol were dropped onto a carbon-coated copper grid for TEM and EELS observation.

3 Results and discussion

3-1 Optical Characteristics

To characterize the optical absorption properties of BS and AS arrays, a polydimethylsiloxane (PDMS) solution was allowed to infiltrate the wafer, and viscous PDMS

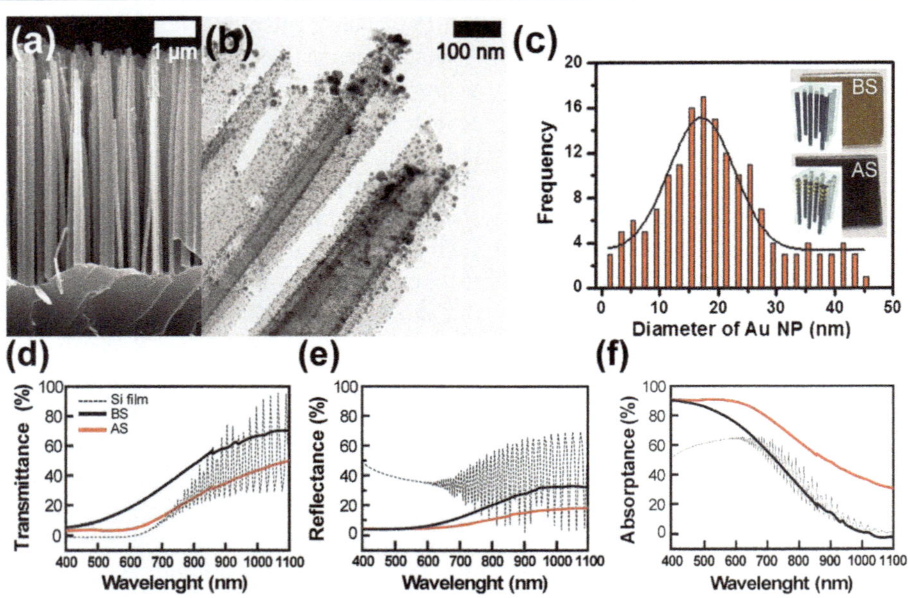

Figure 1 Morphology and optical features of bare SiNWs. (a) SEM image of bare SiNWs (BS, 6 μm-long) on a Si substrate. **(b)** TEM image of a top-end region (<1 μm) of Au NP-coated SiNWs (AS). **(c)** Histogram of the Au NP diameters in the AS arrays. (Insets show the illustration and optical photographs of BS and AS embedded in PDMS). The experimental optical spectra of BS (black) and AS (red) embedded in PDMS: **(d)** transmittance, **(e)** reflectance, and **(f)** absorptance. Dotted lines shown in panels d-f correspond to calculation results of a 6-μm-thick Si film.

was cured in a vacuum oven at 50°C for 3 hrs to fill the SiNW array. Then, flexible PDMS layers embedded with SiNW matrix were mechanically peeled from the Si wafers. The entire regions of the NWs were immersed inside PDMS, but the bottoms of them were exposed to air after peeling from the Si wafer. The optical reflectance and transmittance of NW embedded in PDMS were characterized using a UV–VIS/NIR spectrophotometer (Perkin-Elmer Lambda 750) equipped with an integrating sphere (60 mm) for collecting the diffused light. The optical transmittance, $T(\lambda)$, and reflectance, $R(\lambda)$, spectra of BS and AS were measured in the wavelength range from 400 to 1100 nm (Figure 1(d) and (e)) and compared with those of a 6-μm-thick film, which was obtained from calculations using the transfer matrix method (TMM). The optical absorptance, $A(\lambda) = 100 - T(\lambda) - R(\lambda)$, was estimated as shown in Figure 1(f). $A(\lambda)$ values of AS increased in the visible range ($500 < \lambda < 750$ nm) due to depression of $T(\lambda)$, which made the samples darker as clearly shown in the optical photographs (inset of Figure 1(c)). Also noteworthy is that the absorptance of AS in the NIR range ($\lambda \geq 750$ nm) is much larger than those of BS and 6-μm-thick film (Figure 1(f)). Since both $T(\lambda)$ and $R(\lambda)$ of AS decreased in the NIR range, the enhanced absorptance is resulting from the absorption by interaction of Au NPs and SiNWs. Regarding $A(\lambda)$ of AS, it should be clearly understood how large portion of light is absorbed in the SiNWs rather than being dissipated as Ohmic loss inside the NP. This is a critical concern for realizing high photoresponsivity of the plasmonic photodetectors [15-17].

The near-field interaction of the Au NPs with the SiNW was investigated using monochromated EELS, as shown in Figure 2. This technique could measure the LSP resonance energy in a single Au NP by subnanometer-sized e-beam at low energy range (<5 eV). The energy loss peaks lower than 1.3 eV (~950 nm) were not distinguishable because of a large full-width at half-maximum (FWHM: 0.25 – 0.35 eV) of the zero loss peak from the e-beam. The EELS spectra were obtained at four different positions ($P_1 - P_4$) in the TEM image of Figure 2(a), in which the P_1 and P_4 correspond to the cores of a hemispherical NP (hNP; 20 nm) and a spherical NP (sNP: 5 nm), respectively. P_2 and P_3 denote the hNP/SiNW interface and a SiNW, respectively. Except for P_3, all other positions show SP-related energy loss peaks: some portion of the incident electron energy is transferred to the LSP energy. For 20 nm-sized hNP, the single peak feature is observed at 2.4 eV (~520 nm) for P_1 and at 1.6 eV (~800 nm) for P_2, respectively (Figure 2(b)). The redshifted LSP resonance peak is observed at P_2 due to interaction between the oscillating free electrons in the hNPs and the charge carriers in the underlying Si [18,19]. Knight *et al.* found that the substrate with a larger permittivity would also give rise to a larger image charge with stronger interactions [19]. P_4 at the small NP, which has a spherical shape (as shown in Figure 2 (a)), has two peaks. The stronger peak at 3.0 eV and the weaker one at 2.5 eV are originated from the interband absorption and the LSP resonance, respectively. The interband transition in small-sized NPs (<10 nm) can have clear, intense feature contrary to the 20 nm-sized hNP, since electrons emerge as discrete energy levels in the NPs by their quantum size effect [23]. The LSP-related energy loss peak can be weakened for such small NPs because their dimensions are appreciably smaller (<10 nm) than a mean free path of oscillating electrons. The absorption feature at wavelengths below 600 nm reflects energy dissipation in NPs [23], not promoting absorption in the SiNWs. Thus, the broadband absorption enhancement in AS at NIR range ($\lambda > 750$ nm, Figure 1(f)) is mainly attributed to the LSP resonance at the hNP (>10 nm)/SiNW interfaces.

3-2 Simulation results

To understand the importance of the Au NP shape in the near-field interaction with the SiNW, spherical and hemispherical nanoparticle (sNP and hNP) with the same diameter of 20 nm were compared on semi-infinite

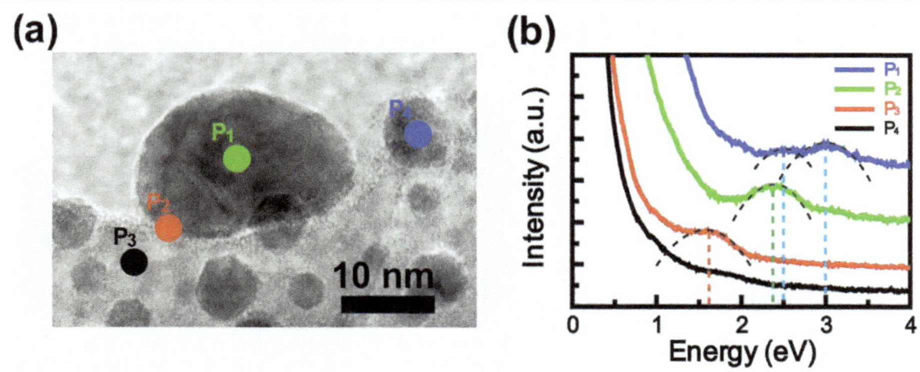

Figure 2 TEM and EELS characteristics of a Au NP-coated-SiNW. (a) TEM image of a Au NP-coated-SiNW. **(b)** EELS spectra extracted from the green (P_1), red (P_2), black (P_3), and blue (P_4) region in **(a)**.

Si substrate, as shown in Figure 3. Three dimensional (3D) simulations of an electromagnetic (EM) field were conducted using COMSOL multiphysics (ver. 4.2) based on a finite element method (FEM).

The total power absorbed by the single Au NP, $P_{abs,Au}(\omega)$, was calculated by integration of photon flux over a closed surface surrounding the NP;

$$P_{abs,Au}(\omega) = \int \frac{1}{2}\omega |E(\omega, x, y, z)|^2 \varepsilon_{Au}(\omega) dV \qquad (1)$$

in which ω, ε_{Au}, and $|E(\omega)|$ are the frequency of the light, the imaginary part of the Au permittivity, and the magnitude of the electric field, respectively. The absorption efficiency (Q_{abs}) of a single Au NP is defined as $Q_{abs}(\omega) = P_{abs,Au}(\omega)/[I_{in}(\omega)A_{geom}]$, where $I_{in}(\omega)$ and A_{geom} are the incident light intensity and the geometrical cross section, respectively. The Au sNP shows a single LSP resonance peak in the visible range: $\lambda = 500$ nm in air and 510 nm on Si (Figure 3(a)). The sNP contributes only slightly to the light absorption into a substrate even at the LSP resonance ($\lambda = {\sim}500$ nm) because the strong LSP-induced near-field occurs far from the Si surface as shown in Figure 3(b). The hNP in air also shows a single peak at $\lambda = 520$ nm (Figure 3(c)). In contrast, the hNP on Si reveals two peaks at $\lambda = {\sim}750$ nm and ~920 nm. At these wavelengths, substantial enhancement in a local field is observed at the hNP/Si interfaces (Figure 3(e)). As a comparison, similar field enhancement cannot be observed in the sNP/Si interface, as shown in Figure 3(d) ($\lambda = 920$ nm). All these results suggest that the NIR absorption peaks ($\lambda \geq 750$ nm) in AS should be caused by LSP-mediated field enhancement at hNP/SiNW interfaces. The slight variations in the peak positions between the experiments (Figure 2(b)) and the simulations (Figure 3(a)) can be originated from the consideration of a specific size of the hNP (i.e., 20 nm) and a simplified Si geometry (i.e., planar wafer) given in the simulation.

Figure 4(a) shows the absorption enhancement spectra of AS arrays with different NP size (diameter: 20, 30, and 40 nm), obtained by the FEM simulations. The selected NP sizes cover the statistical distribution of the NPs (Figure 1(c)). Small-sized NPs (<10 nm) were disregarded since their contribution to the absorption in SiNW is negligible as aforementioned. The unit cells, representing periodic BS and AS arrays, were modelled in the simulation. Monochromatic plane waves with polarization parallel to the x-axis were incident from the top of the NW array along the z axis. Boundary conditions of perfect magnetic conductors (PMCs) were used for $y = 0$ and $y = p/2$ planes (p: period of the NW array), and perfect electric conductor (PEC) conditions were used for the other two planes, $x = 0$ and $x = p/2$. Perfect matched layer (PML) boundary conditions were used at the top and bottom faces of the unit cell. As a result, the electric field could be mirrored to the whole periodic space. Geometric parameters of the NW arrays were chosen to have the representative values of SiNW arrays used in the experimental studies: diameter (d) = 60 nm, $p = 100$ nm, and a space filling ratio ($FR = \pi d^2/4p^2 = 0.28$). A length of NW was fixed 1 μm because the most of NPs were found within 1 μm from the top-ends of the SiNW array. For the sake of simplicity, an ordered arrangement of the Au hNPs was assumed in the calculations: four hNPs were orthogonally placed around a SiNW at the same height and nine pairs of the hNP were positioned along the axial direction of 1-μm-long SiNWs with a pitch of 100 nm.

The theoretical absorption enhancement in SiNW is defined as the absorbed power of SiNW in AS divided by that of BS. The absorbed power in the SiNW ($P_{abs,SiNW}$) can be evaluated through the integration of the photon

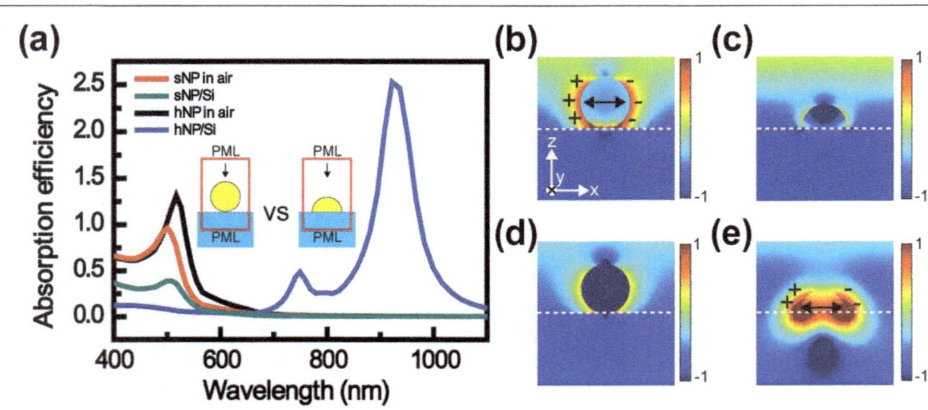

Figure 3 Absorption efficiency and electric field intensity for single Au sNP and hNP. (a) Absorption efficiency spectra of single Au sNP and hNP (20 nm) in air and on a planar Si substrate, respectively. A perfectly-matched-layer (PML) boundary condition was used for the simulation volume surface. Incident light is a plane wave propagating along the −z axis and linearly polarized along the x axis. Cross-sectional distributions of the electric-field intensity at a NP/Si interface are shown for wavelengths of **(b, c)** 500 nm and **(d, e)** 920 nm.

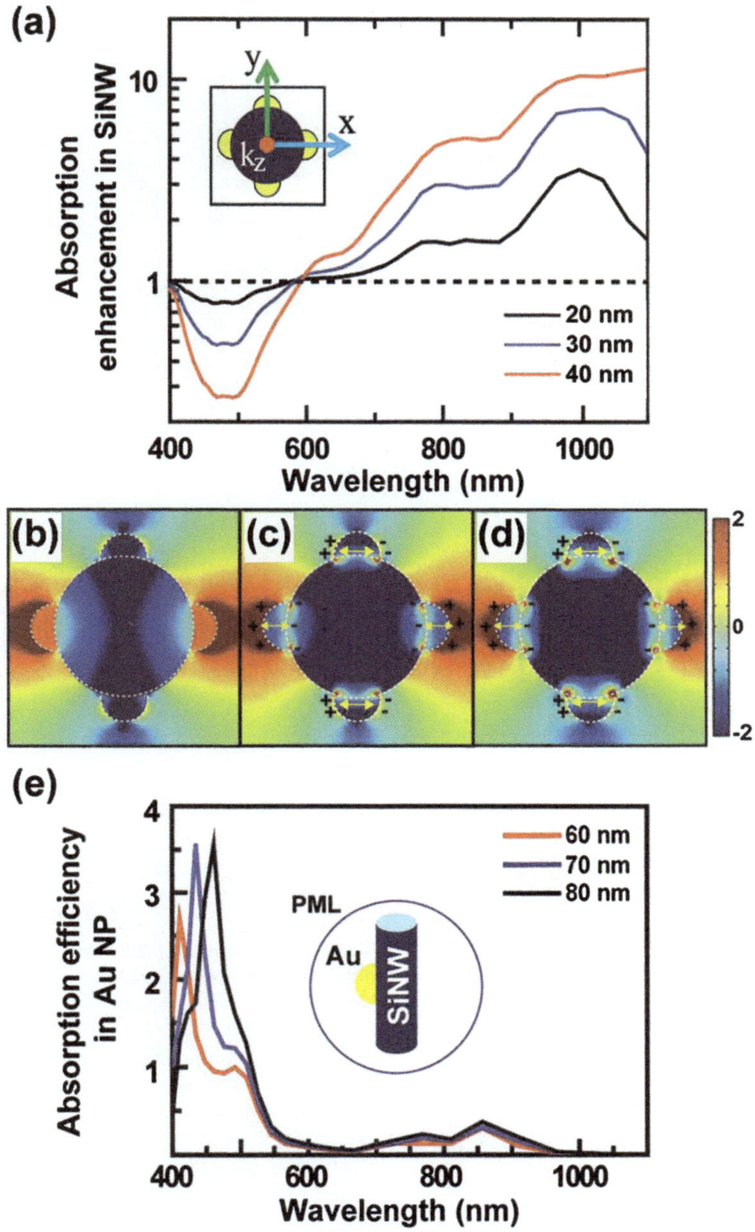

Figure 4 Absorption enhancement spectra for SiNW array coated with NPs. (a) Absorption enhancement spectra (simulation) of the SiNW arrays coated with NPs (20 nm for black, 30 nm for blue, and 40 nm for red). Inset shows a cross-sectional top-view of a unit cell for simulation in the x-y plane. k_z represents a wave vector of the incident light and the electric-field is polarized along the x axis. **(b-d)** Distributions of the electric-field intensity in a unit cell of the array are shown for wavelengths of 490, 800, and 1000 nm in the x-y plane, respectively. Yellow arrows inside Au hNPs represent dipolar oscillations. **(e)** Absorption efficiency (Q_{abs}) spectra of Au NP for a single Au NP-coated SiNW (A spherical PML: blue circle in inset).

flux within the volume of the SiNW according to a following equation,

$$P_{\mathrm{abs,SiNW}}(\omega) = \int_{\mathrm{SiNW}} \frac{1}{2}\omega |E(\omega, x, y, z)|^2 \varepsilon_{\mathrm{Si}}(\omega) dV \qquad (2)$$

in which $\varepsilon_{\mathrm{Si}}(\omega)$ is the imaginary part of the Si permittivity. The absorption enhancement exhibits two broad peaks at ~800 and ~1000 nm, corresponding to the LSP resonance wavelengths. Upon increasing the NP size from 20 nm to 40 nm, the enhancement at $\lambda = 1000$ nm is also increased from 2 to 10 due to the stronger LSP – substrate interaction. Local hot spots along the perimeters of the hNP/SiNW interface are observed in the simulated field distribution (Figure 4(c) and (d)) so that each hNP, as an optical NA, concentrates the incoming light into a strongly localized field. Very interestingly, field enhancement is observed not only around the NPs placed along the polarization direction (x axis), but also those placed orthogonally (y axis) to the polarization direction. This clearly demonstrates a beneficial role of the NW geometry for enhanced near-field coupling that minimizes the directional limitation of light polarization. In addition, we could also understand the reason why the simulation results assuming the periodic arrangement of NPs show good agreement with the experimental data obtained from randomly coated NPs on SiNWs. The NP-induced field enhancement is as large as 63 at the wavelength of ~1000 nm. The NIR absorption is enhanced with increasing NP size and the peak is further red-shifted due to the enlarged contact area of the NP/SiNW (Figure 4(a)).

The absorption of the NP-coated SiNWs is smaller than that of the bare SiNWs at $\lambda < 600$ nm (Figure 4(a)); it has a minimum of $\lambda = 490$ nm, corresponding to the guided mode resonance of SiNW ($d = 60$ nm). The SiNWs show strong absorption at their resonance modes. The resonant wavelength is determined by the NW diameter, d: for example, $\lambda = 550$–750 nm for $d = 90$–140 nm [4]. At $\lambda = 490$ nm (Figure 4(b)), a large electric field is seen inside the bulk of hNPs, indicating significant absorption by the Au hNP. Strong radiation from the hNPs is also observed along the polarization direction of the incident light (x axis). These features show that much of incoming light is absorbed in the Au NPs and reradiated into air rather than absorbed in the SiNWs. In order to examine the influence of the guided mode resonance, the $Q_{abs,\ Au}$ for a single Au NP attached on a SiNW, with $d = 60$, 70, and 80 nm, was compared. A spherical perfectly matched layer (PML: blue circle in inset) with a thickness of 30 nm and a radius of 200 nm was introduced to absorb the light scattered by the Au NP and SiNW. As shown in Figure 4(e), the $Q_{abs}(\omega)$ of the Au hNP (diameter: 20 nm) has two peaks; a larger one at $\lambda < 600$ nm and a smaller one at $\lambda > 800$ nm are related to the NW-guided-mode resonance and LSP-mediated resonance, respectively. The peak position for the former is well matched with the guided mode resonance wavelengths, as reported [4]. The peaks for the latter are nearly invariant because the NP size is fixed. Note that the Au hNPs perturb the optical confinement of the resonance modes in the SiNWs; moreover, suppress visible range ($\lambda < 600$ nm) optical absorption enhancement in the SiNWs as shown in Figure 4(a).

3-3 Electrical characteristics

Spectral responsivities for the BS and AS devices are measured, as shown in Figure 5a. A schematic diagram of our device is illustrated in the inset of Figure 5(a). The responsivity (R) can be given by the following equation;

$$R = \frac{I_{light} - I_{dark}}{P_{inc}} \quad (3)$$

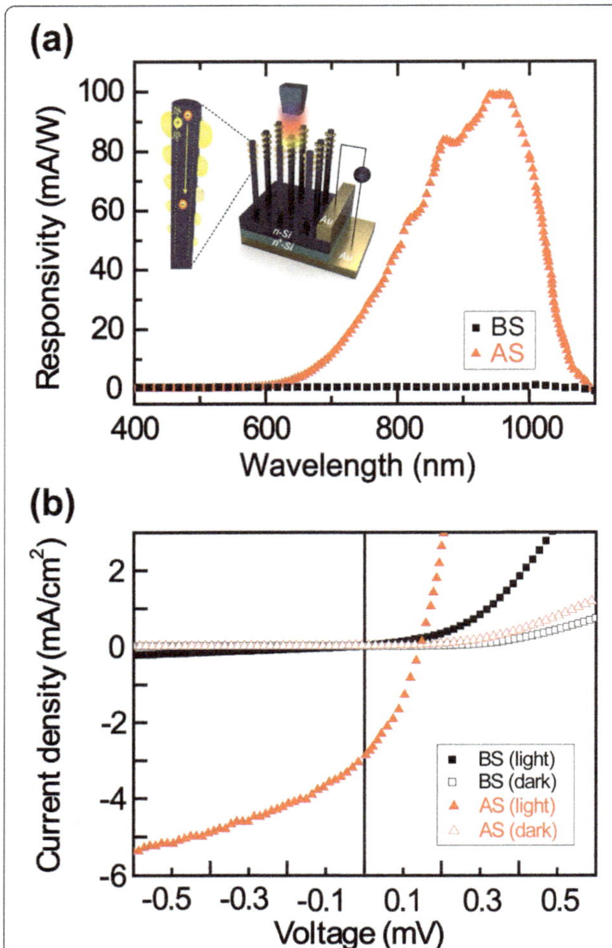

Figure 5 Responsivity spectra and J-V plots for SiNWs with and without Au NPs. (a) Responsivity (experimental) spectra for BS (black), AS (red) photodetectors. (Inset is schematic illustration of our photodetector consisting of Au hNP-coated SiNWs.) **(b)** The *J-V* curves of the SiNW photodetectors without (BS for black) and with Au NPs (AS for red). Open and filled symbols indicate *J-V* curves in dark and under illumination at AM 1.5G, respectively.

where I_{light}, I_{dark}, and P_{inc} are the current under illumination, the dark current, and the incident illumination power, respectively. R of the BS device is nearly zero over the entire wavelength range, whereas R of the AS device is very large at wide wavelength range. R approaches zero as the photon energy becomes smaller than the bandgap of Si (i.e., $\lambda = \sim 1{,}100$ nm). R of the AS device has a broad peak in the NIR region, where the absorption enhancement appears from the experiments (Figure 1(f)) and simulation (Figure 4(a)).

Figure 5(b) shows the current density-voltage (J-V) characteristics of the BS and AS devices under dark and light illumination at AM 1.5G. The J-V curves under both conditions exhibit a rectifying behavior since the Au top-electrodes (finger bar) form Schottky contacts with n-Si substrate. The short circuit current density (J_{sc}) and the open circuit voltages (V_{oc}) for the BS device are only 0.2 mA/cm^2 and nearly zero, respectively. J_{sc} and V_{oc} of the AS device are estimated to be 2.9 mA/cm^2 and 150 mV, respectively. In the BS device, the photogenerated electron–hole pairs (EHPs) suffer from severe recombination in Si before reaching the electrodes. On the other hand, the Au NPs can also act as a Schottky junctions in contact to a SiNW surface for spatially separating the photogenerated EHPs into the orthogonal direction to the NW axis, driven by the built-in-field in space charge region (SCR) (inset in Figure 5(a)). As the excited electrons reach a neutral region of *n*-type Si bulk, a charge state of a substrate tends to be more negative, which results in further enhancement of *n*-type conductance. This explains why the larger J_{sc} and V_{oc} were observed from the NPs-coated SiNW devices compared to those of BS.

4 Conclusions

We have demonstrated that the vertical SiNW array coated with Au hemispherical nanoparticles (hNPs) exhibit remarkable performance as a near-infrared (NIR) photodetector, compared with the bare SiNW array. Experimental and simulation results clearly show that the Au hNPs (>10 nm) act as nanoantennas, then induce strong NIR absorption. hNPs suppress the guided-mode-induced absorption in SiNW for reducing a visible range photocurrent. This allows the hNP-coated SiNW array to possess the unique NIR-selective response. The Schottky contact formation at the hNP/SiNW interface supports the photo-excited carrier separation and improves the responsivity. Our devices offer an alternative route for conventional NIR photodetectors based on a microfabrication method, without the need of any cost-increasing procedures.

Competing interests

The authors declare that they have no competing interest.

Authors' contributions

SWJ carried out experimental work. SWJ, JHL, and DWK analysed the data to draw physical insights. KZ carried out the calculations and plotted simulation results. SWJ, KZ, DWK, JHL wrote the article and approved the final manuscript. All authors read and approved the final manuscript.

Acknowledgements

This work was supported by the National Research Foundation of Korea (NRF) grant funded by the Korea government (MSIP) (No. 2011–0028604). This work was also supported by the Human Resources Development program (No. 20124030200130) of the Korea Institute of Energy Technology Evaluation and Planning (KETEP) grant funded by the Korea government Ministry of Trade, Industry and Energy.

Author details

[1]Department of Chemical Engineering, Hanyang University, Ansan 426-791, South Korea. [2]Department of Physics, Ewha Womans University, Seoul 120-750, South Korea. [3]Department of Mechanical Engineering, Texas A&M University, College Station, TX, USA. [4]Department of Physics, Harbin Institute of Technology, Harbin, China.

References

1. M Csutak, JD Schaub, WE Wu, JC Campbell IEEE Photon. Tech. Lett. **14**, 516 (2002)
2. J Michel, J Liu, LC Kimerling, Nat. Photon. **4**, 527 (2010)
3. AM Joshi, FJ Effenberger, M Grieco, G Feng, W Zhong, J Ott. Proc. SPIE **2290**, 430 (1994)
4. K Seo, M Wober, P Steinvurzel, E Schonbrun, Y Dan, T Ellenbogen, K Crozier Nano Lett. **11**, 1851 (2011)
5. T Kempa, J Cahoon, S-K Kim, R Day, D Bell, H-G Park, C Lieber. Proc. Natl. Acad. Sci. U. S. A. **109**, 1407 (2012)
6. BC Sturmberg, KB Dossou, LC Botten, AA Asatryan, CG Poulton, CM De Sterke, RC McPhedran. Opt. Express **19**, A1067 (2011)
7. J Zhu, Z Yu, G Burkhard, C-M Hsu, S Connor, Y Xu, Q Wang, M McGehee, S Fan, Y Cui. Nano Lett. **9**, 279 (2009)
8. J-Y Jung, Z Guo, S-W Jee, H-D Um, K-T Park, J-H Lee. Opt. Express **18**, A286 (2010)
9. L Hu, G Chen. Nano Lett. **7**, 3249 (2007)
10. Q Du, C Kam, H Demir, H Yu, X Sun. Opt. Lett. **36**, 1884 (2011)
11. M Kelzenberg, S Boettcher, J Petykiewicz, D Turner-Evans, M Putnam, E Warren, J Spurgeon, R Briggs, N Lewis, H Atwater. Nat. Mater. **9**, 239 (2010)
12. KQ Peng, ST Lee. Adv. Mater. **23**, 198 (2011)
13. B Tian, X Zheng, TJ Kempa, Y Fang, N Yu, G Yu, J Huang, CM Lieber. Nature **449**, 885 (2007)
14. K Zhou, S-W Jee, Z Guo, S Liu, J-H Lee. Appl. Opt. **50**, G63 (2011)
15. S Brittman, H Gao, EC Garnett, P Yang. Nano Lett. **11**, 5189 (2011)
16. JK Hyun, LJ Lauhon. Nano Lett. **11**, 2731 (2011)
17. VE Ferry, JN Munday, HA Atwater. Adv. Mater. **22**, 4794 (2010)
18. KR Catchpole, A Polman. Appl. Phys. Lett. **93**, 191113 (2008)
19. MW Knight, Y Wu, JB Lassiter, P Nordlander, NJ Halas. Nano Lett. **9**, 2188 (2009)
20. KB Crozier, A Sundaramurthy, GS Kino, CF Quate. J. Appl. Phys. **94**, 4632 (2003)
21. V Giannini, AI Fernandez-Dominguez, SC Heck, SA Maier. Chem. Rev. **111**, 3888 (2011)
22. MW Knight, H Sobhani, P Nordlander, NJ Halas. Science **332**, 702 (2011)
23. B Balamurugana, T Maruyama. Appl. Phys. Lett. **87**(143), 105 (2005)

6

Nanoscale investigation of surface potential distribution of $Cu_2ZnSn(S,Se)_4$ thin films grown with additional NaF layers

Gee Yeong Kim[1], Juran Kim[1], William Jo[1*], Dae-Ho Son[2], Dae-Hwan Kim[2] and Jin-Kyu Kang[2]

Abstract

CZTS precursors [SLG/Mo (300 nm)/ZnS (460 nm)/SnS (480 nm)/Cu (240 nm)] were deposited by RF/DC sputtering, and then NaF layers (0, 15, and 30 nm) were grown by electron beam evaporation. The precursors were annealed in a furnace with Se metals at 590°C for 20 minutes. The final composition of the CZTSSe thin-films was of Cu/(Zn + Sn) ~ 0.88 and Zn/Sn ~ 1.05, with a metal S/Se ratio estimated at ~0.05. The CZTSSe thin-films have different NaF layer thicknesses in the range from 0 to 30 nm, achieving a ~3% conversion efficiency, and the CZTSSe thin-films contain ~3% of Na. Kelvin probe force microscopy was used to identify the local potential difference that varied according to the thickness of the NaF layer on the CZTSSe thin-films. The potential values at the grain boundaries were observed to increase as the NaF thickness increased. Moreover, the ratio of the positively charged GBs in the CZTSSe thin-films with an NaF layer was higher than that of pure CZTSSe thin-films. A positively charged potential was observed around the grain boundaries of the CZTSSe thin-films, which is a beneficial characteristic that can improve the performance of a device.

Keywords: $Cu_2ZnSn(S,Se)_4$; Na; Grain boundary; Surface potential; Kelvin probe force microscopy

1 Background

$Cu_2ZnSn(S,Se)_4$ (CZTSSe) quaternary compound semiconductors are promising materials with characteristics that are desirable for thin-film solar cells. First of all, CZTSSe consists of Zn and Sn, which are abundant, low-cost materials relative to the In and Ga required for $Cu(In,Ga)Se_2$ (CIGS), and therefore CZTSSe has a composition that is promising for commercialization [1,2]. Second, CZTSSe has a direct band gap between 1.0 and 1.7 eV which can be tuned by adjusting the S and Se composition, with the optimum value for solar cells at 1.4 eV [3]. Third, CZTSSe has a high absorption coefficient of approximately 10^5 cm^{-1}, which is similar to that of CIGS [4], and thus the photo-generated carrier can efficiently be absorbed just a few micrometer. However, CZTSSe thin-film solar cells have a conversion efficiency of 12.6%, which is lower than the 20.8% of CIGS solar cells. Also, there are some limitations in producing CZTSSe thin-film solar cells where, for example, the optimal growth conditions are still not well known. In order to reduce gap in the efficiency with CIGS, there are many issues related to CZTSSe compounds that must be solved.

One issue is the effect that Na has on the material. In previous studies, the introduction of Na to CIGS solar cells enhanced the open-circuit voltage (V_{OC}) and the fill factor (FF) [5]. The Na from the soda lime glass diffused out to the CIGS layer, resulting in an improvement in the crystallinity while prohibiting the inter-diffusion of In and Ga. Furthermore, Na increased the hole concentration of the CIGS absorber layer, and it also enhanced the *p*-type conductivity [6,7]. Rudmann *et al.* reported that Na was located at the grain boundaries (GBs) in polycrystalline CIGS and not in the bulk of single CIS [5,8]. Na incorporates into the In and Ga lattice sites as an acceptor, and the In or Ga atoms on the Cu sites [$(In, Ga)_{Cu}$ defects] were replaced with Na on the Cu sites (Na_{Cu}). Thus, Na can reduce the amount of In-Cu antisites (In_{Cu} donors), enhancing the crystallinity [9-11]. In other studies, Na was observed to enhance the efficiency of the solar cells by changing the electronic properties of the cells, such as the electrical passivation, and by improving the atomic motilities [12]. In addition, the Na

* Correspondence: wmjo@ewha.ac.kr
[1]Department of Physics, Ewha Womans University, Seoul, Korea
Full list of author information is available at the end of the article

removed the defect states in the CIGS thin-films, and therefore many acceptor and donor states, as well as defects, around the GB could be altered.

The GB states in CIGS have an increased concentration of free carriers as a result of the passivation by Na. This results in the occurrence of acceptor-like states or in the elimination of donor-like states at the GBs. The effective electron diffusion length of CIGS can then affect the V_{OC} [12]. Yan *et al.* confirmed that Na segregates across the GBs in CIS thin-films grown on SLG. Kelvin prove force microscopy (KPFM) measurements revealed that the GBs in CIS films exhibited a positively charged potential (~200 meV) [13], but on the other hand, the surface of the CIS thin-films grown on borosilicate glass revealed a small potential value lower than 30 meV. Therefore, Na provides a beneficial effect to CIGS thin-films by forming an electrical potential at the GBs due the segregation of Na across GB. Consequently, the positively charged potential at the GBs due to the Na segregation could repel holes and attract electrons, suppressing recombination [13]. The grain boundaries (GBs) of the CIGS act as a factor that improves the performance of the cells, contrary to single crystal thin-film solar cells where GBs act as defects and recombination centers [14,15].

We have already published how the dominant positively charged GBs and the negatively charged IGs on CZTSSe thin-films assist in the efficient carrier collection and carriers separation near the GBs [16,17]. Also, Sutter-Fella *et al.* reported that Na in CZTSSe thin-films formed liquid Na_2Se_x phases, enhancing the grain growth and thereby improving the electronic properties at an optimal thickness of the NaF layer (10 nm) [6]. They suggested that Na accumulated around the GBs, thus improving the carrier collection in the CZTSSe thin-film solar cells. However, a high amount of Na was not sufficiently incorporated into CZTSSe absorber layer. Details on the effects of Na on the GBs of CZTSSe are still lacking. Thus, in this study we used KPFM to carefully investigate the potential distribution around the GBs with respect to the variation of the NaF layer on CZTSSe thin-films.

2 Method

The metal precursors of the CZTSSe layers were grown via RF/DC sputtering, and the staking order was SLG/Mo/ZnS/SnS/Cu. The thickness of each stacked layer was ~460 nm for ZnS, 480 nm for SnS, and 240 nm for Cu. After the deposition of the precursor, the NaF layer was grown on the precursors via electron-beam evaporation at thicknesses of 0, 15, and 30 nm. The precursors were annealed with Se metals in a furnace at 590°C for 20 minutes. The final composition of the CZTSSe thin-films measured through inductively coupled plasma (ICP) was of $Cu/(Zn + Sn) \sim 0.88$ and $Zn/Sn \sim 1.05$. The metal S/Se ratio was estimated to be ~0.05, and the grain sizes of the CZTSSe thin-films were of 1–2 μm, and that of CZTSSe thin-films with the NaF layers were of 4–5 μm, as measured by field emission scanning electron microscopy (FE-SEM) (JSM-6700F).

The KPFM measurements were performed using commercial atomic force microscopy (AFM) (Nanofocus Inc. South Korea, n-Tracer). A Pt/Ir coated tip was used for to measure the topography and the surface potential. In this paper, we calibrated the tip by using a highly oriented pyrolytic graphite (HOPG) sample. The topographic and KPFM images were obtained under a non-contact mode at a resonant frequency of the probe of about 78.3 kHz. The scanning rate was set to 0.5 Hz to minimize the topological signal. The work function of the HOPG was generally known to be 4.6 eV [18]. The KPFM tool is widely used to characterize semiconductors, such as for obtaining the surface potential

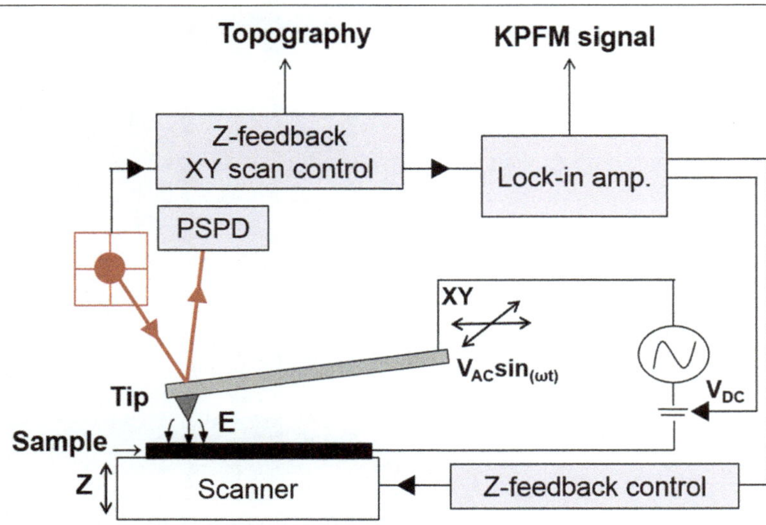

Figure 1 Schematic illustration of Kelvin probe force microscopy and surface potential measurement system.

and the work function of the sample surface at a nano-scale. From the KPFM measurements, we can obtain the contact potential difference (V_{CPD}) (i.e., the surface potential) which is different from using a metallic AFM tip and obtaining the sample work function.

The V_{CPD} can be defined as in Eq. 1. If the tip approaches the sample, then an electrostatic force forms between the metal tip and the surface of the sample. When the tip approach to the sample surface, the Fermi levels of the tip and the sample achieve a state of equilibrium. In the equilibrium state, a V_{CPD} forms as a result of the difference in the vacuum level. An external bias DC voltage (V_{DC}) nullifies the V_{CPD}, forming a feedback system, as shown in Figure 1. The V_{DC} has the same magnitude as the V_{CPD}, and the V_{CPD} has the same values as the work function difference between the tip and the sample. Furthermore, an AC voltage (V_{AC}) can also be applied to the tip at an amplitude of 1 V with a frequency of 70 kHz, together with the V_{DC}, and the V_{AC} enables the oscillation of electrical forces on the tip.

$$V_{CPD} = \frac{\phi_{tip} - \phi_{sample}}{-e} \tag{1}$$

We can consider the metal tip and the surface of the sample to form a capacitor. Thus, the electrostatic force of tip and the sample [F(z)] can be defined as in Eq. 2. The $\partial C/\partial z$ is the gradient of the capacitance, and F(z) is the equation where ΔV is the potential difference between V_{CPD} and the voltage applied to the tip. The equation for the electrostatic force applied to the tip can be divided into three terms, as shown in the first term from Eq. 3. The first term is inserted here in Eq. 2, and we can then obtain Eq. 3. Eq. 3 is divided into three parts. In the first part of Eq. 3, F_{DC} (Eq. 4) represents the static deflection of the tip that is used, and F_{ω} (Eq. 5) and $F_{2\omega}$ (Eq. 6) oscillate with frequencies proportional to ω. F_{ω} in Eq. 5 generally measures the V_{CPD}, and $F_{2\omega}$ is used for capacitance microscopy [19].

$$F(z) = -\frac{1}{2}\Delta V^2 \frac{dC(z)}{dz} \tag{2}$$

$$F = F_{DC} + F_{\omega} + F_{2\omega} \tag{3}$$

$$F_{DC} = \frac{dC}{dz}\left[\frac{1}{2}(V_{DC} \pm V_{CPD})^2\right] \tag{4}$$

$$F_{\omega} = -\frac{dC}{dz}(V_{DC} \pm V_{CPD})V_{AC}\sin(\omega t) \tag{5}$$

$$F_{2\omega} = \frac{1}{4}\frac{dC}{dz}V^2{}_{AC}[\cos(2\omega t) - 1] \tag{6}$$

3 Result and discussion

The CZTSSe thin-films deposited by sputtering method have different thicknesses in the NaF layer: 0 (a), 15 (b), and 30 nm (c). The CZTSSe thin-films contain ~4% of Na. The efficiency of the CZTSSe solar cell is almost

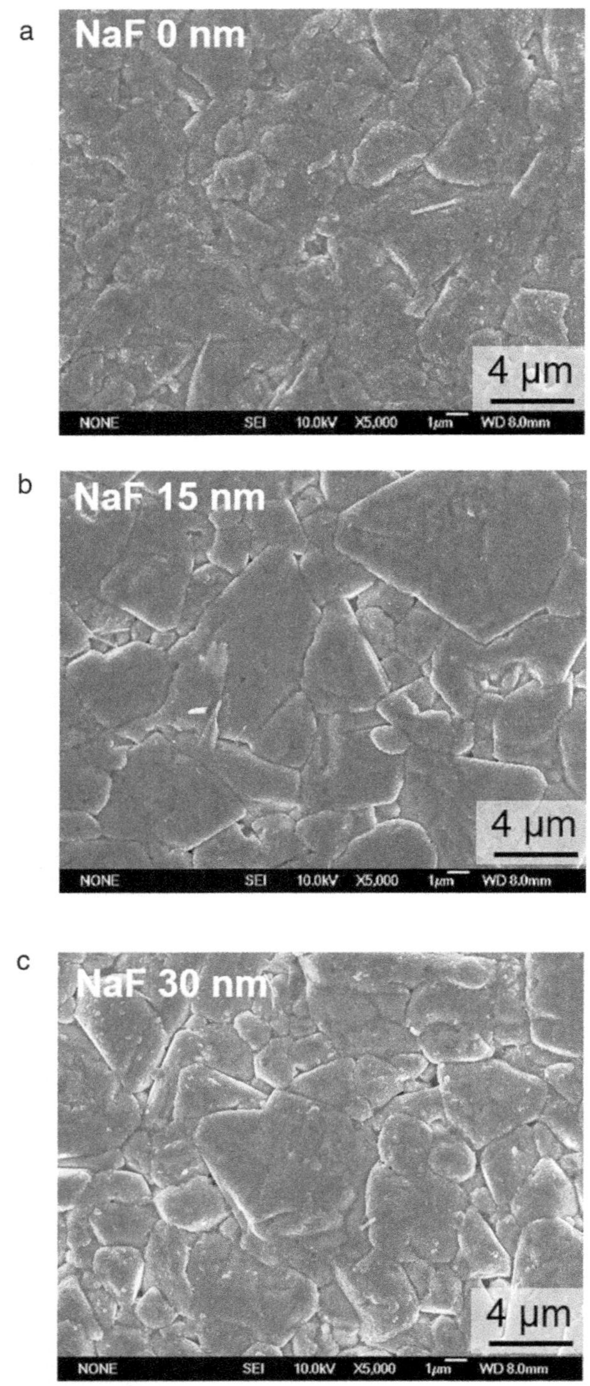

Figure 2 Scanning probe microscopy images of (a) NaF (0 nm)/CZTSSe (b) NaF (15 nm)/CZTSSe (c) NaF (30 nm)/CZTSSe thin-films deposited by sputtering method. The NaF/CZTSSe thin-film surface shows the higher grain size.

similar at ~3%. The grain size of the CZTSSe thin-films increases by adding NaF, as compared to the thin-films without an NaF layer, as shown in Figure 2. The CZTSSe thin-films without NaF presented a grain size of 1–2 μm, and that with an NaF layer was of 4–5 μm. The surface uniformity also improved by adding an NaF layer.

Figure 3 shows the topographic and the surface potential images of the CZTSSe thin-films with and without an NaF layer. The topographic images and the line profile results showed that the height of GBs in the CZTSSe thin-films with the NaF layer were reduced from 400 nm to 100 nm. We also obtained the root mean square (RMS) roughness of the CZTSSe to be 177 nm, and the films with the NaF layer had a lower roughness of 106 nm (15 nm NaF) and 112 nm (30 nm NaF). Thus, adjustments of the NaF layer thickness were observed to influence the grain growth and the formation of a uniform surface. Figures 3(d)–(f) show the surface potential images taken through KPFM measurements. The yellow region represents the positive surface potential values, and the blue region represents the negative potential values. Figures 3(e) and (f) exhibit the images of the potential mapping, and the potential near the GBs with an NaF layer increase relative to that without NaF [Figure 3 (d)]. In the CZTSSe films with NaF, the ratio of the yellow regions, especially that near the GBs, is larger than that of pure CZTSSe thin-films.

Figure 4(a)–(c) presents the line profile of the topographic, and the surface potential images are marked with a red line, as shown in Figure 3. Figures 4(d)–(f) show the total potential distribution of the CZTSSe thin-films with respect to the thickness of the NaF layers. In Figures 4 (a)–(c), the CZTSSe thin-films show that a negative potential value of about −100 mV at the half region of the GBs and a positive potential of about 110 mV at the IGs, as shown in Figure 4(a). However, the potential of the CZTSSe films with the NaF layer showed a fully positively charged potential of about 120 mV (15 nm NaF) and 380 mV (30 nm NaF) at the GB regions and a negatively charged potential of about −200 mV (15, 30 nm NaF) at the IG regions. The potential distribution was found to be related to the downward band bending because the

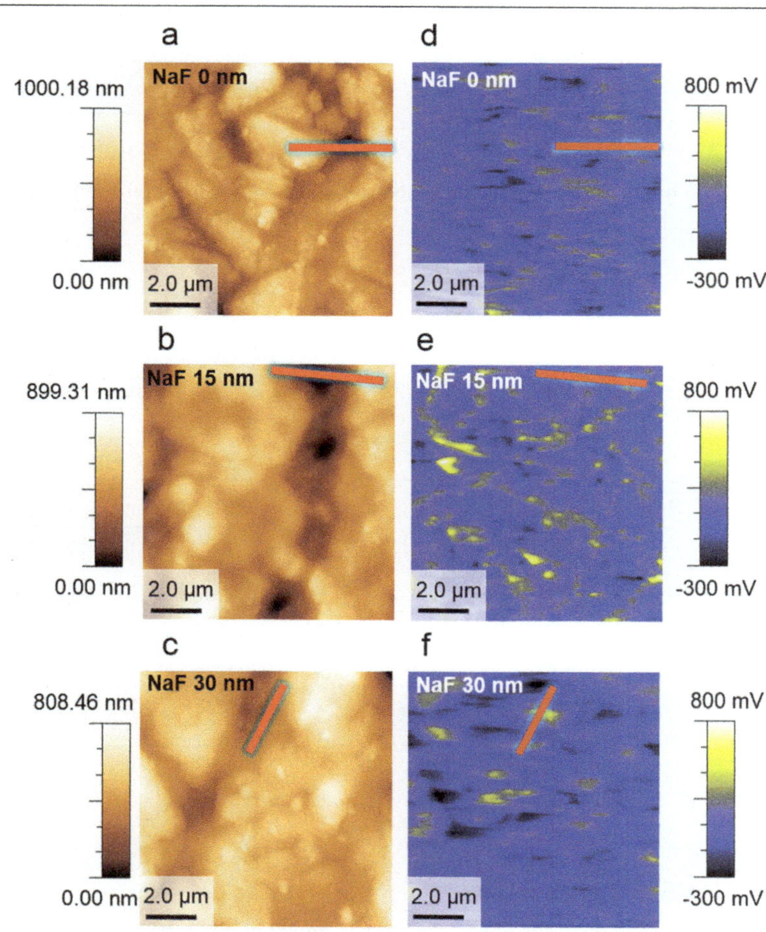

Figure 3 Topography (a) 0 nm (b) 15 nm (c) 30 nm of NaF and surface potential images (d) 0 nm (e) 15 nm (f) 30 nm of NaF of the CZTSSe thin-films. The grain size increase with increasing NaF layer and the potential value is also enhanced.

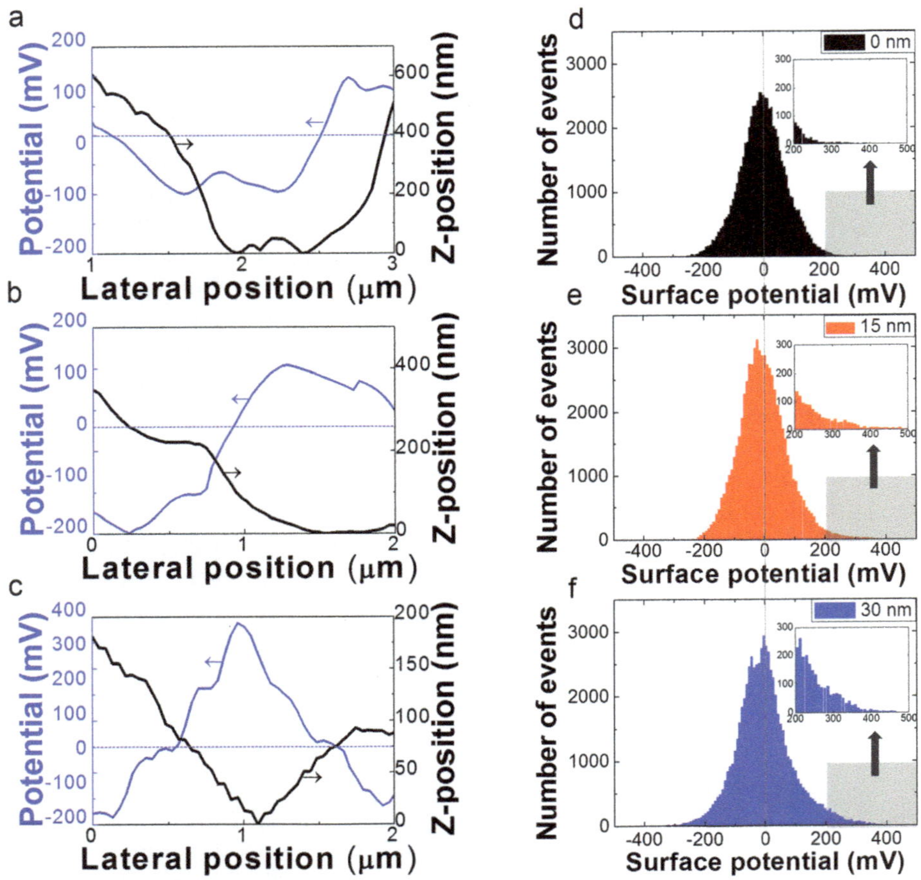

Figure 4 Line profiles of (a) 0 nm (b) 15 nm (c) 30 nm of NaF and potential distribution (d) 0 nm (e) 15 nm (f) 30 nm of NaF with CZTSSe thin-films. Line profiles along the topographic image and the surface potential as indicated red line in Figures 3. The CZTSSe thin-film with NaF shows positively charged potential of the fully GB region.

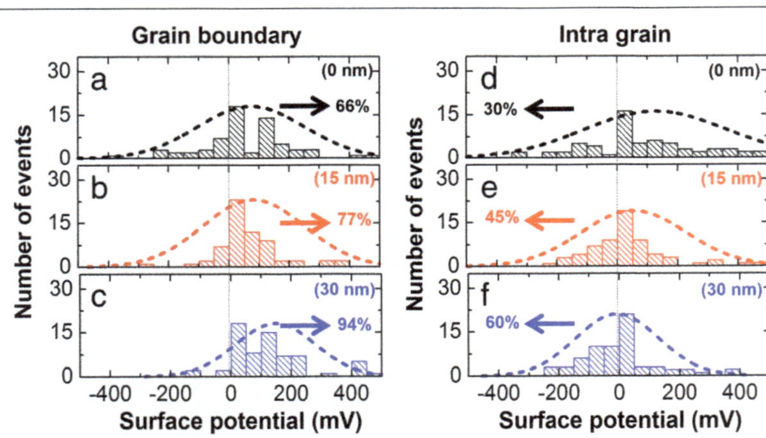

Figure 5 Statistical analysis of potential histogram at the GBs (a) 0 nm (b) 15 nm (c) 30 nm of NaF and IGs (a) 0 nm (b) 15 nm (c) 30 nm of NaF on CZTSSe thin-films. The histogram obtained based on the line profiles results of in regions of near of each sample, respectively. Higher positive potential ratio at GBs and lower negative potential ratio at IGs are exhibited in CZTSSe thin-film with NaF.

positive potential at GBs led to a reduction in the work function near the GBs in CZTSSe.

The positive potential distribution was higher as the NaF layer became thicker for the CZTSSe films, as shown in Figure 4(f). Thus, the potential value was the highest for the 30 nm NaF layer CZTSe film. In previous studies, many groups had reported benign behaviors of the GBs in polycrystalline CIGS thin-film solar cells

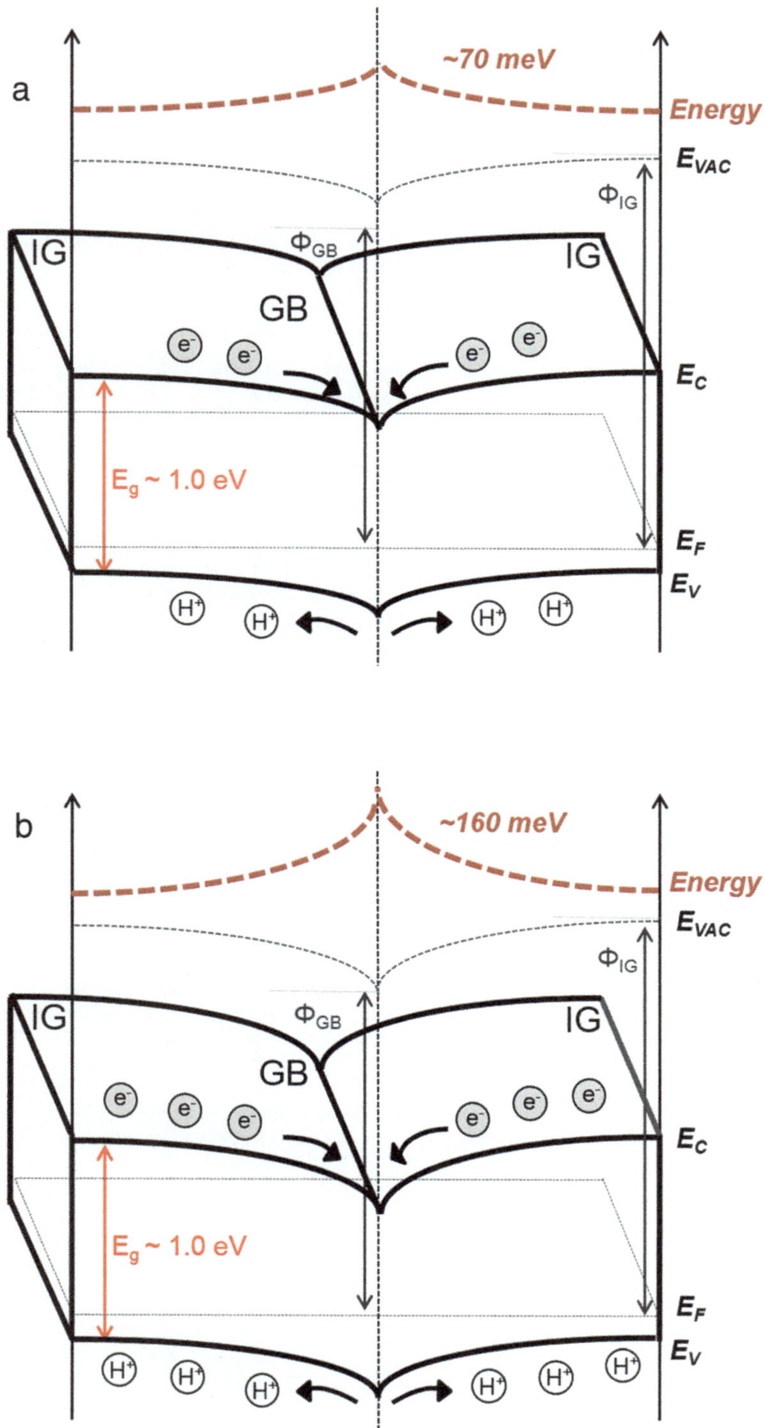

Figure 6 The schematics of band diagram around GB in (a) without NaF and (b) with NaF layer in CZTSSe thin-films. The surface potential energy increases 160 meV around grain boundary with NaF layer. In case **(b)** Electron–hole carriers more efficiently help separate at the GB.

[20]. Jiang *et al.* confirmed the presence of a positively charged potential (smaller work function) at the GBs, as compared to that at the IGs, with a formation of a local built-in potential near the GBs. This can explain how electrons are drawn to the GBs and how holes are repelled from the GB. This is beneficial for carrier collection, and it also suppresses recombination at the GBs. Therefore, the local built-in potential on GBs in the CIGS thin-films plays a critical role to achieve the desired device performance and efficiency [14,15].

Meanwhile, the GBs of the CZTSSe thin-films exhibit similar electrical properties to those of GBs in CIGS thin-films. A higher positive surface potential was measured at the GBs relative to that measured at the IGs, and a current flowed through the GBs of CZTSSe thin-films. This could improve the minority carrier collection near the GBs, and thus, the GBs are one of the essential factors that can enhance the high conversion efficiency of CZTSSe thin-film solar cells [21].

Furthermore, our group reported the presence of the positive potential ratio at the GBs and a negative potential ratio at the IGs in CZTSSe thin-film, and these are closely related to device properties, especially to the J_{SC}. The high conversion efficiency of the GBs of the CZTSSe thin-film consists of dominant, positively charged GBs and negatively charged IGs. This leads to a downward band bending at the GBs and an upward band bending at the IGs, which helps in carrier collection and suppresses the recombination at GBs that are limited to downward band-bended GBs. Thus, the J_{SC} and the shunt resistance are enhanced [16], and therefore we tried to characterize the effects of the GB on the CZTSSe film with respect to the presence of an NaF layer.

Figure 5 shows a histogram from the statistical analysis of the potential value, focusing on the GBs and the IGs in the CZTSSe with and without NaF. The results are based on the line profiles of the regions of near GBs, and the histograms exhibit a positive potential distribution at the GBs as shown in Figures 5(a)–(c) and a negative potential distribution ratio at the IGs as shown in Figures 5(d)–(f). The CZTSSe samples exhibit a lower positive potential ratio of 66% at the GBs and also a lower negative potential ratio of 30% at the IGs. However, the CZTSSe with NaF indicates a higher positive potential ratio of 77% (15 nm of NaF) and 94% (30 nm of NaF) at the GBs and a negative potential ratio of 45% (15 nm of NaF) and 60% (30 nm of NaF) at the IGs. The average potential at the GBs in the CZTSSe thin-film is of 74 mV, and that of CZTSSe with a 15 nm NaF thin-film is of 93 mV. The highest average potential of about ~160 mV is shown in the 30 nm NaF thin-films. Therefore, we conclude that the average potential at the GBs increases by adding NaF.

Figure 6(a) displays the schematics of a band diagram near the GBs in the CZTSSe thin-films, and Figure 6(b) shows the CZTSSe thin-films with a 30 nm NaF layer. The higher thickness of the NaF layer enhances the positively charged (i.e. lower work function) GBs in the CZTSSe layer. In previous studies, the Na on the surface was observed to cause surface dipoles, and thus the dipoles may increase the potential near the GBs in the CIGS surface due to the higher Na concentration around the GBs than around the IGs [13]. Thus, Na could enhance the positively charged GBs in the CIGS thin-film surface, which is a phenomenon consistent with our results. The potential at the GB increases by almost 100 meV with the NaF layer in the CZTSSe absorber, as shown in Figure 6 (b). Moreover, the ratio of the positively charged potential at the GBs and the negatively charged potential at the IG increased with the NaF layer in the CZTSSe thin-films. The electronic properties of the material should be able to efficiently collect the electrons and the holes, helping in the separation around GBs where an NaF layer was added to the CZTSSe. However, it is essential to understand the defect states and the Na passivation through the GBs in CZTSSe layer in order to identify the influence of Na in the CZTSSe solar cells.

4 Conclusion

We investigated the surface potential of CZTSSe thin-films with different thicknesses of the NaF layer by KPFM measurement. The potential value at the GBs relatively increased from 74 mV to 160 mV as the NaF thickness increased. In addition, the ratio of the positively charged GBs and that of the negatively charged IGs in CZTSSe thin-films also increased as the NaF layer increased. In these studies, the Na in CZTSSe layer was observed to have an electronically benign behaviors, efficiently collecting carriers and suppressing the recombination near the GBs.

Abbreviations

CIGS: $Cu(In,Ga)Se_2$; CZTSSe: $Cu_2(Zn,Sn)(S,Se)_4$; GBs: Grain boundaries; IGs: Intragrains; KPFM: Kelvin probe force microscopy; V_{oc}: Open-circuit voltage; J_{sc}: Short-circuit current; F.F.: Fill factor; Eff.: Photo-conversion efficiency.

Competing interests

The authors declare that they have no competing interests.

Authors' contributions

GYK, JRK, and WJ measured the electrical properties of the CZTSSe samples using scanning probe microscopy. DHS, DHK, and JKK fabricated the CZTSSe samples by sputtering with subsequent selenization. All authors read and approved the final manuscript.

Acknowledgements

This work was supported by the New & Renewable Energy of the Korea Institute of Energy Technology Evaluation and Planning (KETEP), grant funded by the Korea government, Ministry of Trade, Industry and Energy (No. 20123010010130).

Author details

[1]Department of Physics, Ewha Womans University, Seoul, Korea. [2]Daegu Gyeongbuk Institute of Science and Technology (DGIST), Daegu, Korea.

References

1. S Chen, XG Gong, A Walsh, S-H Wei, Electronic structure and stability of quaternary chalcogenide semiconductors derived from cation cross-substitution of II-VI and I-III-VI_2 compounds. Phys. Rev. B **79**, 165211 (2009)
2. TK Todorov, J Tang, S Bag, O Gunawan, T Gokmen, Y Zhu, DB Mitzi, Beyond 11% efficiency: characteristics of state-of-the-Art Cu2ZnSn(S, Se)4 solar cells. Adv. Energy Mater. **3**, 34–38 (2013)
3. JJ Scragg, J Phillip, LM Peter, Towards sustainable materials for solar energy conversion: preparation and photoelectrochemical characterization of Cu_2ZnSnS_4. Electrochem. Commun. **10**, 639–642 (2008)
4. AR Jeong, W Jo, DY Park, H Cheong, YK Seo, JH Park, JS Chung, YS Lee, Y-J Kwark, Effect of substrates on structural and optical properties of Cu-poor $CuGaSe_2$ thin films prepared by in-situ co-evaporation. Curr. Appl. Phys. **13**, 907–912 (2013)
5. D Rudmann, AF da Cunha, M Kaelin, F Kurdesau, H Zogg, AN Tiwari, G Bilger, Efficiency enhancement of Cu(In, Ga)Se2 solar cells due to post-deposition Na incorporation. Appl. Phys. Lett. **84**, 1129 (2004)
6. CM Sutter-Fella, JA Stückelberger, H Hagendorfer, FL Mattina, L Kranz, S Nishiwaki, AR Uhl, YE Romanyuk, AN Tiwari, Na assisted sintering of chalcogenides and its application to solution processed Cu2ZnSn(S, Se)4 thin film solar cells. Chem. Mater. **26**, 1420–1425 (2014)
7. T Nakada, D Iga, H Ohbo, A Kunioka, Effects of Na on Cu(in, Ga)Se2-based thin films and solar cells. Jpn. J. Appl. Phys. **36**, 732–737 (1998)
8. MA Contreras, B Egaas, D King, A Swartzlander, T Dullweber, Texture manipulation of $CuInSe_2$ thin film. Thin Solid Films **361–362**, 167–171 (2000)
9. MA Contreras, B Egaas, P Dippo, J Webb, J Granata, K Ramanathan, S Asher, A Swartzlander, R Noufi, On the role of Na and modifications to CIGS absorber materials using thin MF (M=Na, K, Cs) precursor layers. Conference Record of the Twenty Sixth IEEE Photovoltaic Specialists Conference 359–362 (1997). DOI 10.1109/PVSC.1997.654102.
10. BJ Stanbery, C-H Chang, TJ Anderson, Engineered phase inhomogeneity for CIS device optimization. Proc of the 11th Int Conf on Ternary and Multinary Compounds **152**, 915–922 (1998)
11. BJ Stanbery, S Kincal, S Kim, TJ Anderson, OD Crisalle, SP Ahrenkiel, G Lippold, Role of Na in the control of defect structures in CIS. Conference Record of the Twenty Eighth IEEE Photovoltaic Specialists Conference 440–445 (2000). DOI 10.1109/PVSC.2000.915864.
12. A Rockett, The electronic effects of point defects in $Cu(In_xGa_{1-x})Se_2$. Thin Solid Films **361–362**, 330–337 (2000)
13. Y Yan, C-S Jiang, R Noufi, S-H Wei, HR Moutinho, MM Al-Jassim, Electrically benign behavior of grain boundaries in polycrystalline $CuInSe_2$ films. Phys. Rev. Lett. **99**, 235504 (2007)
14. C-S Jiang, R Noufi, JA AbuShama, K Ramanathan, HR Moutinho, J Pankow, MM Al-Jassim, Local built-in potential on grain boundary of Cu(In, Ga)Se2 thin film. Appl. Phys. Lett. **84**, 3477–3479 (2004)
15. C-S Jiang, R Noufi, K Ramanathan, JA AbuShama, HR Moutinho, MM Al-Jassim, Does the local built-in potential on grain boundaries of Cu(In, Ga) Se2 thin films benefit photovoltaic performance of the device? Appl. Phys. Lett. **85**, 2625–2627 (2004)
16. GY Kim, AR Jeong, JR Kim, W Jo, D-H Son, D-H Kim, J-K Kang, Surface potential on grain boundaries and intragrains of highly efficient Cu2ZnSn(S, Se)4 thin-films grown by two-step sputtering process. Sol. Energy Mater. Sol. Cells. **127**, 129–135 (2014)
17. GY Kim, JR Kim, W Jo, D-H Son, D-H Kim, J-K Kang, Nanoscale observation of surface potential and carrier transport in Cu2ZnSn(S, Se)4 thin films grown by sputtering-based two-step process. Nanoscale Res. Lett. **9**, 10 (2014)
18. W Melitz, J Shen, S Lee, JS Lee, AC Kummel, R Droopad, ET Yu, Scanning tunneling spectroscopy and Kelvin probe force microscopy investigation of Fermi energy level pinning mechanism on InAs and InGaAs clean surfaces. J. Appl. Phys. **108**, 023711 (2010)
19. W Melitz, J Shen, AC Kummel, S Lee, Kelvin probe force microscopy and its application. Surface Science report **66**, 1–27 (2011)
20. RH Shin, AR Jeong, W Jo, Investigation of local electronic transport and surface potential distribution of Cu(In, Ga)Se2 thin-film. Curr. Appl. Phys. **12**, 1313–1318 (2012)
21. JB Li, V Chawla, BM Clemens, Investigating the role of grain boundaries in CZTS and CZTSSe thin film solar cells with scanning probe microscopy. Adv. Mater. **24**, 720–723 (2012)

7

Resist-free antireflective nanostructured film fabricated by thermal-NIL

Young Hun Kang[2], Jae Hyung Han[1], Song Yun Cho[2] and Choon-Gi Choi[1*]

Abstract

Resist-free antireflective (AR) nanostructured films are directly fabricated on polycarbonate (PC) film using thermal-nanoimprint lithography (T-NIL) and the moth-eye shape of AR nanostructure is elaborately optimized with different oxygen reactive ion etching conditions. Anodic aluminum oxide (AAO) templates are directly used as master molds of T-NIL for preparation of AR nanostructures on PC film without an additional T-NIL resist. AR nanostructures are well arranged with a period of about 200 nm and diameter of about 150 nm, which corresponds to those of the AAO template mold. The moth-eye AR nanostructures exhibit the average reflectance of 2% in wavelength range from 400 to 800 nm. From the results, highly enhanced AR properties with simple direct imprinting on PC film
demonstrate the potential for panel application in the field of flat display, touch screen, and solar cells.

Keywords: Antireflective films; Moth-eye; Nanostructures; Anodic aluminum oxide template; Thermal nano-imprint lithography; Oxygen reactive ion etching

1 Background

Recently, an antireflection (AR) property has been a great interest due to the demands for the enhanced performance in electronic devices such as solar cells and displays [1]. Especially, needs for AR property of plastic substrate have been increasing for the rapid growth of flexible electronic fields. Many research groups have already investigated various AR techniques such as single or multi-layer coating, surface texturization, and fabricating the periodic nanostructures [2]. AR nanostructure, so-called moth-eye structure, can efficiently reduce the light reflectance attributed to the gradual change in refractive index between the air and substrate by altering the volume fraction of structures, which in turn reduce Fresnel reflection [3]. There are many techniques available for fabricating AR nanostructure, such as electron beam lithography [4], thermal imprinting [5], hybrid nano-patterning lithography [6], dip pen nanolithography [7] and so on. However, these techniques still require high cost, complex process, and sophisticated equipments. In order to overcome these drawbacks, we employ the nanoimprint lithography (NIL) which is one of the most potential nano-scale patterning techniques with advantages of rapidness, low cost, and reproducibility. In NIL process, molds or stamps are normally produced using various materials, such as silicon, glass, and metal. However, it is difficult to prepare the periodically uniform nano-sized patterns on large area using molds from these materials. Therefore, an anodic aluminum oxide (AAO) template can be good candidate for mold material due to the advantages of the facile large scale synthesis via electrochemical method and periodical nano-sized patterns with high aspect ratio [8].

In this study, we report a facile method for the preparation of AR nanostructure on PC film using AAO template as a master mold through the T-NIL process. Furthermore, the shape of AR nanostructures is elaborately optimized with different reactive ion etching (RIE) conditions and the optical properties of the fabricated AR films are investigated.

* Correspondence: cgchoi@etri.re.kr
[1]Creative Research Center for Graphene Electronics, Electronics and Telecommunications Research Institute (ETRI), Daejeon 305-700, Korea
Full list of author information is available at the end of the article

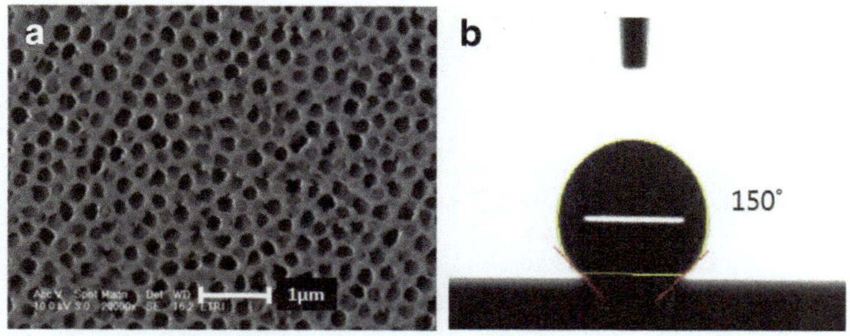

Figure 1 AAO template: (a) SEM image of the surface of AAO template and (b) contact angle of the AAO template after anti-sticking treatment.

2 Methods

2.1 Experiment

To fabricate AR nanostructures on PC film, a commercial AAO template (Whatman Inc.) fabricated by a multi-step anodising process was used as template mold. As shown in Figure 1(a), the pore diameter of AAO template was approximately 200 nm and the size of AAO template was 2×2 in^2.

Prior to T-NIL process, AAO templates were coated with the anti-adhesion layer to decrease the surface free energy of AAO for the easy separation of the AAO templates from PC film. A liquid phase deposition of *1H, 1H, 2H, 2H*-perfluorooctyl-trichlorosilane was used to form the self-assembled monolayer (SAM) based anti-adhesion layer. The formation of hydrophobic anti-adhesion layer on the surface of the AAO templates was confirmed by contact angle measurement which was determined to be 150° as shown in Figure 1(b).

T-NIL process was carried out on PC film with Nano-Imprinter (Obducat, Sweden) using AAO template mold under the various temperatures (140, 150, 160, and 170°C) and pressure (8, 10, and 12 Bar) for 20 min. A home-made PC film was used and the thickness of PC film is about 1 mm. After removing the AAO template mold, oxygen RIE process was optimized for the proper formation of AR nanostructure on PC film with the various parameters, which were oxygen flow of 30 sccm under the power of 5 W at different etching time (60, 120, and 180 s). Surface morphology of the imprinted AR nanostructures on PC film was investigated by field emission scanning electron microscope (FE-SEM) and the optical reflectance of the imprinted PC film was measured by UV-visible spectrophotometer (Hitachi U-3501). The schematic procedure for the preparation of the AR nanostructure on PC film is illustrated in Scheme 1.

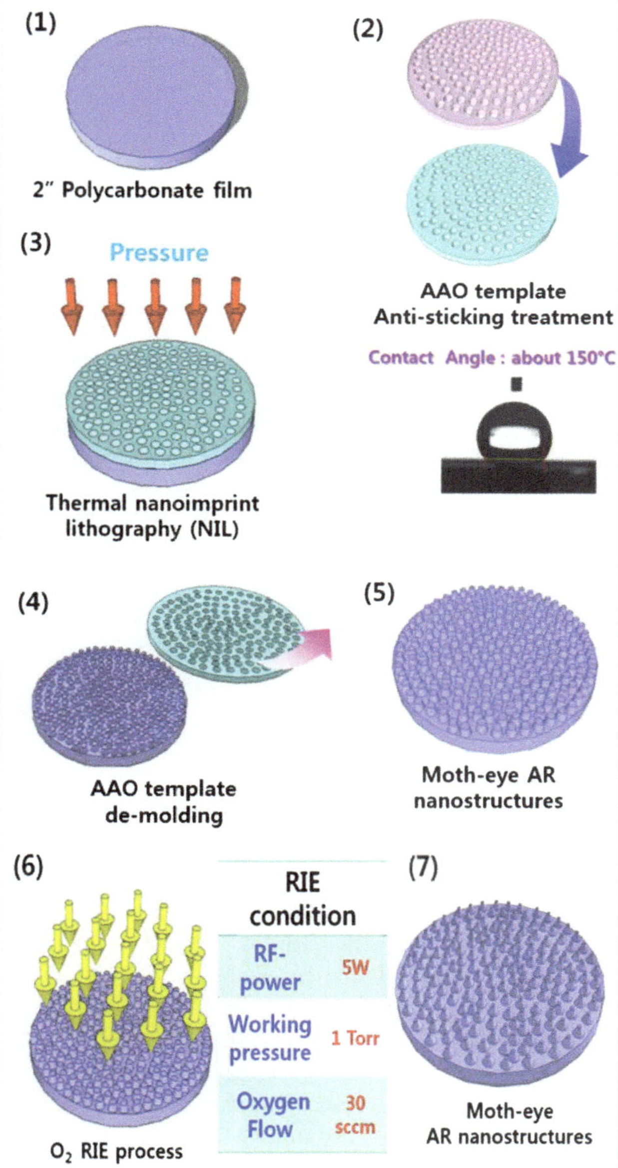

Scheme 1 Schematic diagram of T-NIL and RIE process for fabrication of AR nanostructured films.

3 Results and discussion

Figure 2 shows the FE-SEM images of the nanoimprinted AR nanostructure on PC film by T-NIL.

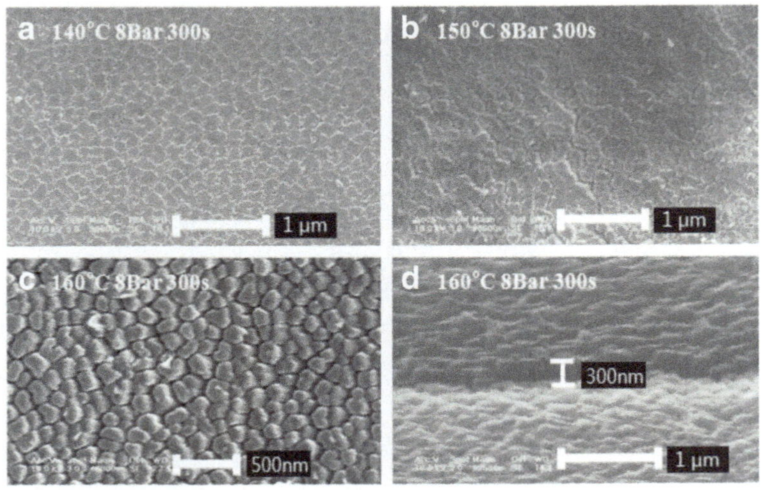

Figure 2 FE-SEM images of nanoimprinted AR nanostructures with various process temperatures: (a) 140°C, (b) 150°C, (c) 160°C, and (d) the tilted view of the imprinted AR nanostructures at temperature of 160°C.

Although the pore shape of AAO templates was clearly observed on the surface of PC films, AR nanostructure is roughly formed at a pressure of 8 bar, as shown in Figure 2(a) and (b). It is attributed that the periodic nano-sized pores of AAO templates were difficult to be filled with PC at too low process temperature. Whereas, AR nanostructure fabricated at 160°C shows a period and height of 200 and 300 nm, respectively with a diameter about 150 nm, as shown in Figure 2(c) and (d). These pattern properties on PC well correspond with those of AAO template. This result means that nanoimprinting process for AR nanostructure on PC is highly dependent on processing temperature and AR nanostructures can be completely replicated on PC film at 160°C. However, there exist the problem of separation between AAO template and PC film for high temperature process over 160°C due to the intense adhesion between inner walls of AAO nano-pores and the melted PC film, which results in the structural damage of AR nanostructures.

To evaluate the AR properties of nanoimprinted PC films, the reflectance was measured by UV-visible spectroscopy in the wavelength range from 400 nm to 1,000 nm, as shown in the Figure 3. It is clearly observed that the AR nanostructures prepared at 160°C exhibits the average reflectance of about 5%. In comparison with that of the bare PC film, the average reflectance was considerably reduced from 13% to 5% in the visible light range. At 140 and 150°C, the average reflectance of the AR nanostructures is about 7% and 9%, respectively because periodic AR nanostructures were not clearly formed and separated.

In order to optimize the shape of AR nanostructures and reduce the reflectance, AR nanostructures fabricated at 160°C by T-NIL was etched by the oxygen RIE process with the various etching time. It is observed that oxygen RIE process has great influence on the surface morphology of AR nanostructures, as shown in Figure 4. The AR nanostructures which were formed by RIE process for 60 s obviously represents the pillar shape, as shown in Figure 4(a). However, with the increase of the etching time over 60 s, AR nanostructures are collapsed due to the prolonged ion bombardment during the RIE process. At 180 s of the etching time, as shown in Figure 4(c), the period of AR nanostructures is increased and the pillar shape is changed with the decrease of height and width.

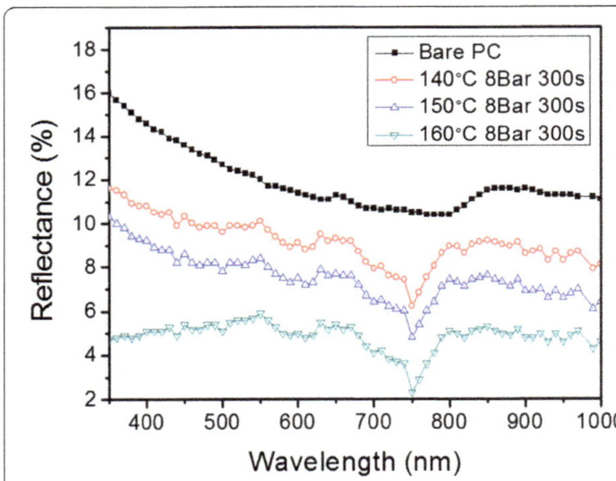

Figure 3 Reflectance of the AR nanostructures nanoimprinted on the PC film in the wavelength range of 400 nm - 1,000 nm measured by a UV-visible spectrophotometer.

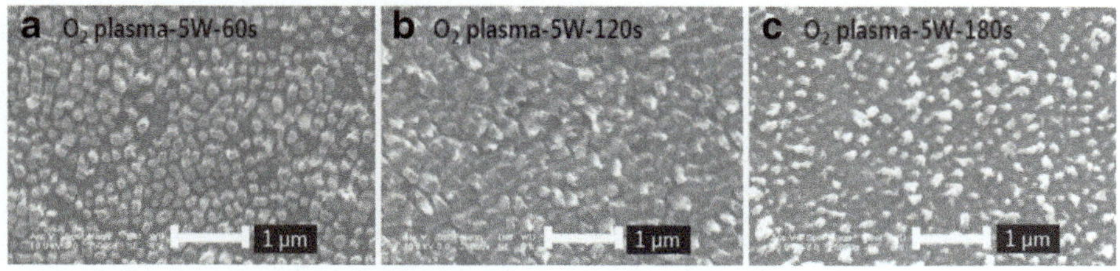

Figure 4 FE-SEM images of the AR nanostructures after O_2 RIE process with different etching time.

Figure 5 shows the reflectance of AR nanostructures after oxygen RIE process. The average reflectance of AR nanostructures after oxygen RIE process with 60 s etching time exhibits about 2% in the visible light range from 400 nm to 800 nm. In compared with that of the AR nanostructures for the etching time of 180 s, the average reflectance is also decreased from 5% to 2%. Therefore, the average reflectance of AR nanostructures with additional 60 s RIE etching time is significantly decreased from 5% to 2% in the 400–800 nm visible range in comparison with that of the nanoimprinted AR nanostructures at 160°C.

The reflectance difference between the bare PC film and PC film with oxygen RIE processed AR nanostructures is demonstrated visually in Figure 6. The film on the left-hand side is the bare PC film and the film on the right-hand side is the AR nanostructured PC film. The bare PC film has strong surface reflection because of no AR on the surface. On the contrary, no noticeable surface reflection can be observed with the AR nanostructured PC film. Therefore, it is observed that PC film with oxygen RIE processed AR nanostructures can obviously transmit the text of printed page due to the significant reduction of light reflectance. These results highlight the potential of fabricating a high performance AR film by forming moth-eye nanostructures on polymer films.

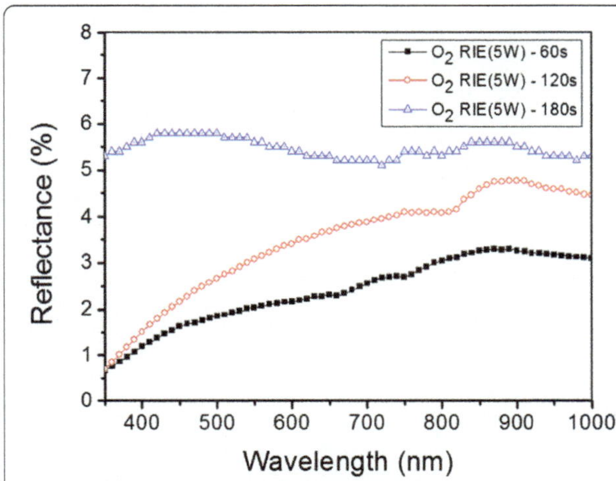

Figure 5 Reflectance of AR nanostructures on the PC film after oxygen RIE process with various etching time.

4 Conclusions

We successfully demonstrate that the resist-free AR nanostructures are directly fabricated on PC film using T-NIL and the shape of AR nanostructures is elaborately optimized with different oxygen RIE conditions. AAO templates are directly used as master molds of T-NIL for preparation of AR nanostructures on PC film without an additional T-NIL resist. AR nanostructures are well arranged with a period of about 200 nm and diameter of about 150 nm, which corresponds to those of the AAO template mold. The average reflectance of the PC film with AR nanostructures was obtained to be about 2% in the broadband wavelength range of 400–800 nm. Comparing to bare PC film and the fabricated AR film, the average reflectance of AR film is reduced to approximately 11% from 13% to 2% in the visible light region. As a result, moth-eye AR nanostructured PC film using T-NIL and oxygen RIE can efficiently reduce the light reflectance at the visible light region. We suggest that the prepared AR film is useful for panel application in the field of flat display, touch screen, and solar cells.

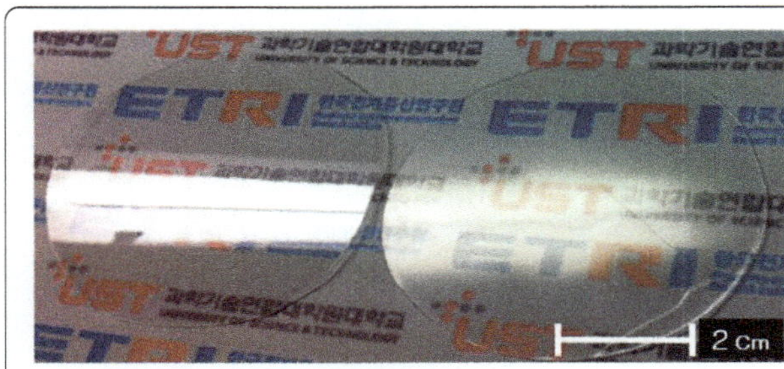

Figure 6 Digital images of the bare PC film (left) and the AR nanostructured PC film (right).

Competing interests
The authors declare that they have no competing interests.

Authors' contributions
YH Kang and JH Han completed the fabrication of the nanostructured films and the measurement of the optical properties and drafted the manuscript. SY Cho has written part of the paper. CG Choi conceived the research project and participated in the interpretation of data; he has also written part of the paper and is responsible for the revision of the paper. All authors read and approve the final manuscript.

Authors' information
Young Hun Kang is currently a senior researcher in Korea Research Institute of Chemical Technology, Daejeon, Korea. He received BS from Sungkyunkwan University in 2006 and MS from University of Science & Technology in 2009. He has recently researched development of solution processable organic & inorganic hybrid materials for application to thin film transistors and thermoelectric devices based on various printing technologies.
Jae Hyung Han is a researcher of the Creative Research Center of Graphene Electronics in ETRI, Korea. He received MS from the Department of Advanced Device Technology in University of Science and Technology (UST) in 2012.
Dr. Song Yun Cho is currently working for Korea Research Institute of Chemical Technology as a principal research scientist. He received MS in Polymer Science & Engineering from Inha University in 1998 and Ph. D. in Chemistry at the Pennsylvania State University in 2004. His recent research is specialized in the organic or oxide semiconducting materials for electronic and energy applications.
Dr. Choon-Gi Choi received Ph. D. from Université d'Orléans (France) in 1996. He is currently a director at the Creative Research Center for Graphene Electronics in Electronics and Telecommunications Research Institute (ETRI) and an adjunct professor at the Department of Advanced Device Technology in University of Science and Technology (UST), Korea. He is also an associate editor of the journal Nano Convergence with Springer publishing. His research interests are graphene electronics and optoelectronics, metamaterials, nanophotonics, nano-imprint lithography.

Acknowledgments
This research was supported by the Nano R&D Program (No. 2012-0006204) and GNT R&D Program (No. 2012-0006657) through the National Research Foundation funded by the Ministry of Science, ICT and Future Planning of Korea.

Author details
[1]Creative Research Center for Graphene Electronics, Electronics and Telecommunications Research Institute (ETRI), Daejeon 305-700, Korea. [2]Division of Advanced Materials, Korea Research Institute of Chemical Technology, Daejeon 305-600, Korea.

References
1. K Jae Kwan, P Seong-Je, K Sarah, P Hyeong-Ho, K Ki-don, C Jun-Hyuk, L Jihye, C Dae-Geun, KY Suh, J Jun-Ho, Fabrication of ZrO_2 nanopatterns for biomimetic antireflection by thermal nanoimprint lithography. Microelectron. Eng. **100**, 12–15 (2012)
2. F Joki-Korpela, J Karvinen, B Paivanranta, A Partanen, M Suvanto, M Kuittinen, TT Pakkanen, Hydrophobic and oleophobic anti-reflective polyacrylate coating. Microelectron. Eng. **114**, 38–46 (2014)
3. Y Nurhafizah Binti Abu Talip, T Jun, Fabrication of self-supporting antireflection-structured film by UV-NIL. Microelectron. Eng. **110**, 163–166 (2013)
4. Y Kanamori, E Roy, Y Chen, Antireflection sub-wavelength gratings fabricated by spin-coating replication. Microelectron. Eng. **78–79**, 287–293 (2005)
5. O Sang Soon, C Choon-Gi, K Young-Sung, Fabrication of micro-lens arrays with moth-eye antireflective nanostructures using thermal imprinting process. Microelectron. Eng. **87**, 2328–2331 (2010)
6. K Young Hun, O Sang Soon, K Young-Sung, C Choon-Gi, Fabrication of antireflection nanostructures by hybrid nano-patterning lithography. Microelectron. Eng. **87**, 125–128 (2010)
7. J Hyungil, CK Dalal, K Steven, S Raman, C Patrick Collier, Surfactant activated dip-pen nanolithography. Nano Lett. **4**(11), 2171–2177 (2004)
8. LJ Guo, Nanoimprint lithography: methods and material requirements. Adv. Mater. **19**(4), 495–513 (2007)

8

Surface effects on electronic transport of 2D chalcogenide thin films and nanostructures

Yeonwoong Jung[1], Jie Shen[1] and Judy J Cha[1,2*]

Abstract

The renewed interest in two-dimensional materials, particularly transition metal dichalcogenides, has been explosive, evident in a number of review and perspective articles on the topic. Our ability to synthesize and study these 2D materials down to a single layer and to stack them to form van der Waals heterostructures opens up a wide range of possibilities from fundamental studies of nanoscale effects to future electronic and optoelectronic applications. Bottom-up and top-down synthesis and basic electronic properties of 2D chalcogenide materials have been covered in great detail elsewhere. Here, we bring attention to more subtle effects: how the environmental, surface, and crystal defects modify the electronic band structure and transport properties of 2D chalcogenide nanomaterials. Surface effects such as surface oxidation and substrate influence may dominate the overall transport properties, particularly in single layer chalcogenide devices. Thus, understanding such effects is critical for successful applications based on these materials. In this review, we discuss two classes of chalcogenides – Bi-based and Mo-based chalcogenides. The first are topological insulators with unique surface electronic properties and the second are promising for flexible optoelectronic applications as well as hydrogen evolution catalytic reactions.

Keywords: Two-dimensional materials; Transition metal chalcogenides; Topological insulators; Bi_2Se_3; MoS_2; Surface effects

1 Introduction: renewed interest in 2D chalcogenides

Following the success of graphene [1-4], layered materials that can be peeled down to a single layer have received tremendous attention for their novel electronic, chemical, and mechanical properties. These include atom-thick, hexagonally arranged two-dimensional (2D) sheets such as hexagonal boron nitride (hBN) [5-7], silicene [8,9], and germanene [10,11], and layered oxides and chalcogenides [12] such as MoS_2 [13], WSe_2 [14], Bi_2Se_3 [15], and Bi_2Te_3 [16] to name a few. In chalcogenides, a few atomic layers are covalently bonded to make a molecular layer, and the molecular layers stack together to form crystals via the relatively weak van der Waals interaction. Van der Waals heterostructures (Figure 1) using single layer chalcogenides, graphene, and hBN as basic building blocks present exciting opportunities to tailor electronic band structures for specific applications [17].

* Correspondence: judy.cha@yale.edu
[1]Department of Mechanical Engineering and Materials Science, Yale University, New Haven, CT, USA
[2]Energy Science Institute, Yale University West Campus, West Haven, CT, USA

The electronic properties of single layer chalcogenides deviate from the bulk. For example, the topological surface state will be gapped if Bi_2Se_3 is less than 5 layers thick [18], and a single layer MoS_2 becomes a direct bandgap semiconductor instead of an indirect bandgap semiconductor [19-21]. As these chalcogenides are synthesized to single- or a few-layer thick films or nanosheets, surface effects (both desired and undesired) will be enhanced. In perfect crystals, the chalcogenide nanostructure is terminated with basal planes that are inert with saturated bonds. However, in reality, crystal defects and surface absorbates cannot be avoided, leading to surface trap sites or changes in mobility and carrier density of the sample. Thus, the unforeseen surface/parasitic effects on electron transport on 2D chalcogenide thin films and nanostructures should be studied carefully. This review will look at how surface or parasitic effects play a role in 2D chalcogenide nanomaterials. Two chalcogenide systems, Bi-based selenides (topological insulators) and Mo-based sulfides (semiconductors and hydrogen evolution catalysts), will be reviewed in detail. Detailed reviews on

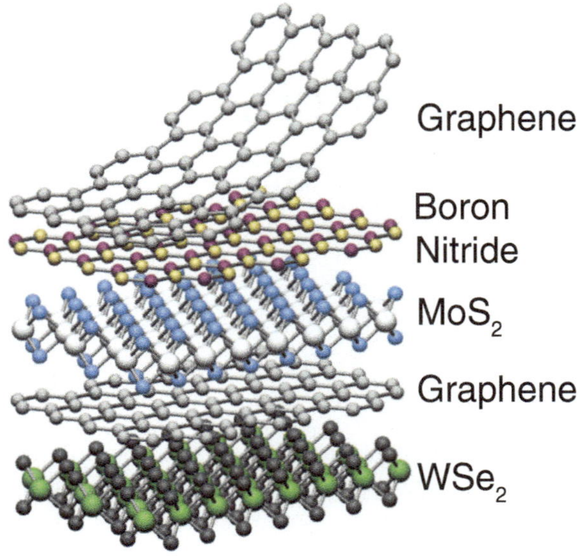

Figure 1 Van der Waals heterostructures, stacking 2D chalcogenide single layers, graphene, and hBN together to form a crystal. These heterostructures present opportunities to tailor electronic properties, not found in their bulk counterparts. Reprinted with permission from Ref 17.

synthesis and basic electronic properties of 2D chalcogenides can be found elsewhere [22-24].

2 Graphene

Before delving into the chalcogenide systems, we briefly cover the surface effects on graphene transport. We focus our attention on substrate-induced charge fluctuations and surface functionalization.

2.1 Substrate effects: SiO_x substrate, suspended graphene, and hexagonal boron nitride

Electron transport on graphene is subject to microscopic perturbations that can cause backscattering, greatly decreasing the electron mean free path as well as mobility. In the case of high quality graphene sheets without defects, impurities are induced by extrinsic sources. Scanning tunneling microscopy (STM) results show inhomogeneous electron density fluctuations caused by charge-donating impurities at the interface between graphene and SiO_2 substrate (Figure 2A) [25,26]. These substrate-induced impurities can be high enough to create standing waves in graphene due to enhanced backscattering [27]. One way to eliminate the substrate effects in graphene transport is to remove the substrate entirely. Several groups reported free-standing graphene sheets suspended over a trench [28,29]. Suspended graphene exhibits corrugation in the out-of-plane dimension, thus becoming thermodynamically stable [28]. Electron transport

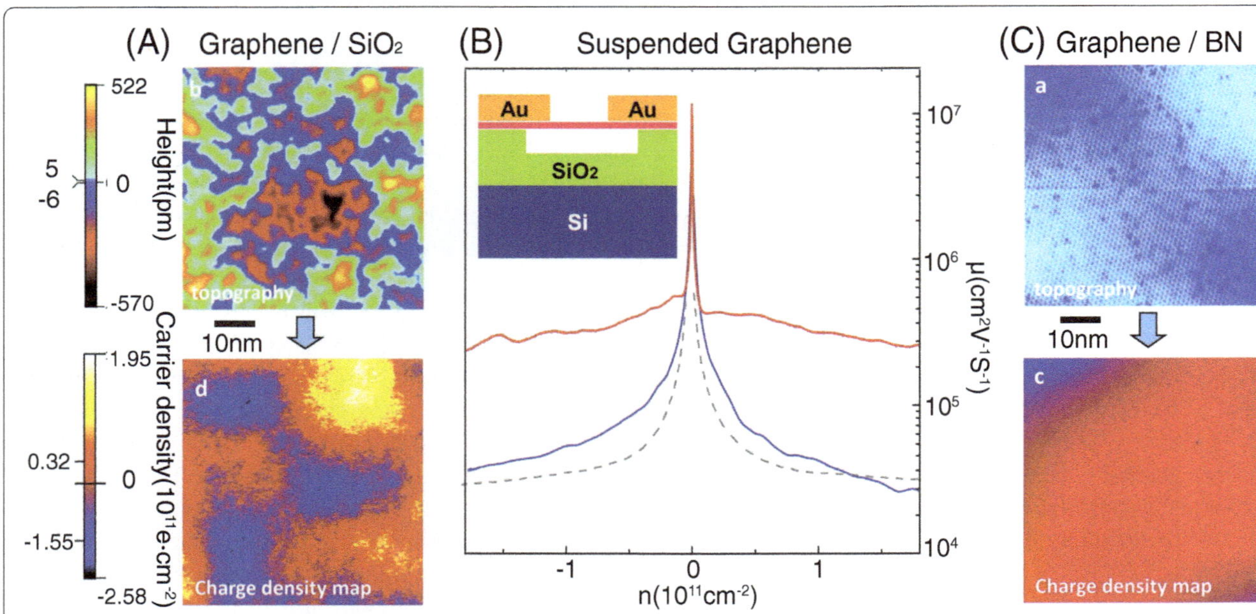

Figure 2 Substrate effects on graphene transport. (A) STM topographic (top) and charge density (bottom) maps of graphene on SiO_2. The surface is rough due to the substrate. Inhomogeneous charge density fluctuations are observed. **(B)** Mobility of graphene as a function of carrier density for graphene on SiO_2 (dotted gray), suspended graphene before (blue) and after (red) current annealing. Current annealing of suspended graphene increases mobility as surface adsorbates desorb from graphene. **(C)** STM topographic (top) and charge density (bottom) maps of graphene on hBN. Unlike the STM results of graphene on SiO_2, the topography shows a flat surface with Moiré patterns. Charge fluctuations are greatly reduced in the graphene/hBN case. **(A, C)** Reprinted with permission from Ref. 26. (B) Reprinted with permission from Ref. 29.

on the suspended graphene device (Figure 2B) shows ultrahigh electron mobility, greater than 200,000 cm^2/Vs at the electron density of ~2×10^{11} cm^{-2} [29]. This is a dramatic improvement compared to mobility values reported in graphene devices fabricated on conventional substrates [30]. Current annealing can further improve the mobility in suspended graphene devices as impurities on graphene can desorb during annealing [29]. Nearly ballistic transport can be observed in suspended devices [31]. The dramatic improvement in transport for suspended graphene devices compared to graphene/SiO_2 devices highlights how substrate or parasitic effects can completely dominate transport in these thin nanodevices.

Although suspended graphene has distinct advantages over graphene on conventional substrates in terms of transport properties, it is fragile and difficult to handle. Thus, finding an alternative substrate to SiO_2/Si while keeping the same transport properties as suspended graphene devices is essential for any successful device applications. Hexagonal boron nitride (hBN) is an appealing substrate because its surface is atomically smooth, its lattice mistmach to graphene is small (~1.7%), and it has a large bandgap (5.97 eV) [6,32]. According to STM studies [26], graphene on hBN is very flat so that microscopic Moiré patterns arising from the relative misorientation between the graphene and hBN lattices can be observed. Gate-dependent *dI/dV* spectra of graphene on hBN from STM studies exhibit a significant reduction in local microscopic charge inhomogeneity (Figure 2C). The mobility of graphene on hBN improves by ~ 3 times compared to SiO_2-supported graphene samples [32].

2.2 Surface functionalization

The surface of graphene can be functionalized with chemical species for effective Fermi level control and bandgap opening [33]. Surface functionalization includes partial plasma etching [34,35], surface absorbates [36], and covalent bonding with small molecules [37,38]. Although effective in modifying the band structure of graphene such as p- or n-type doping and bandgap opening [39-41], often these functional groups lead to increased scattering for electron transport, dramatically decreasing the mobility and the mean free path length. For room temperature applications such as graphene as transparent electrodes or conducting electrodes for Li-ion batteries [30,42-44], the degradation in mobility may not be an issue as long as the conductivity is high. However, for low-temperature transport for quantum or nanoscale effects, surface functionalized graphene may not be suitable.

3 Bi_2Se_3, Bi_2Te_3: topological insulators

Traditionally well-known thermoelectric materials, such as Bi_2Te_3 and Bi_2Se_3, have been recently classified as a new quantum matter called topological insulators [45-47]. The most unique aspect of topological insulators is the robust, spin-polarized conducting surface state that encapsulates topological insulators. The topological surface state serves as a platform to study fundamental quasiparticle phenomena such as Majorana fermions [48,49] and magnetic monopole-like behavior [50] as well as to explore future electronics such as quantum computing and spintronics [51,52]. To maximize the topological surface state for transport, often Bi_2Te_3 and Bi_2Se_3 are made into nanostructures or thin films [24]. Unfortunately, experimental studies report degradation of the topological surface state transport property due to environmental effects. Current challenges for Bi_2Se_3 and Bi_2Te_3 are minimizing these detrimental environmental effects as well as improving the intrinsic materials properties so that the topological surface states can be more easily revealed and manipulated. Here, we review three such effects.

3.1 Strain from substrates

Bi_2Se_3 and Bi_2Te_3 thin films have been successfully grown on a number of substrates such as Si, sapphire, graphene, and on mica by both molecular beam epitaxy and chemical vapor deposition [53-57]. Their topological surface states were clearly observed by angle-resolved photoemission spectroscopy [58,59]. STM studies also showed the presence of the surface states in these thin films by confirming the Dirac dispersion [60]. However, the quantum nature of the surface state was not easily revealed by transport measurements in thin films. This is mainly because the mobility of the surface state was low, as evidenced by the clear weak anti-localization effect and the lack of Shubnikov-de Hass (SdH) oscillations in transport [61,62]. Weak anti-localization occurs in diffusive regime where the electrons go through many scattering events. Thus, it indicates short mean free paths, usually around 300 – 500 nm for Bi_2Se_3 and Bi_2Te_3 [63,64]. Only a few studies show thin films exhibiting 2D SdH oscillations, which are attributed to high mobility surface electrons [65]. The reason for the low mobility of topological surface electrons in thin films is likely due to the strain from the substrate. Because Bi_2Se_3 and Bi_2Te_3 are layered materials with weak van der Waals interactions, they can be grown on substrates with a relatively large lattice mismatch. The strain induced from the lattice mismatch as well as charge fluctuations from the substrate may be the reason for the low mobility of the topological surface states.

3.2 Surface oxidation

Residual bulk carriers due to crystal defects such as Se vacancies in Bi_2Se_3 and anti-site defects in Bi_2Te_3 have hindered transport studies of topological surface states greatly. Compensation charge doping using Sb or Cu is effective in suppressing the bulk carriers [66-69]. However, it was soon realized that different size samples from the same growth batch showed different carrier densities; smaller samples showed higher carrier densities [70]. This finding hints to additional surface effects playing a role in transport. In particular, Bi_2Se_3 nanoribbon studies showed that the surface of Bi_2Se_3 was oxidized quickly by observing BiO_x and SeO_x peaks using X-ray photoemission spectroscopy [71]. The same nanoribbons showed two different carrier densities. The Hall carrier density was much higher than the density obtained from SdH oscillations. Because the Hall measurements include all carriers while the SdH measurements sample only the high mobility carriers, the discrepancy in carrier densities was attributed to additional, low-mobility carriers induced by the environmental effects including the surface oxidation. The surface oxidation induces additional scattering events for the topological surface state where the oxidized radicals can act as trap sites or induce charge fluctuations.

3.3 Fermi level pinning

Time-dependent angle-resolved photoemission spectroscopic studies show the emergence of a topologically trivial 2D electron gas forming at the surface of Bi_2Se_3 as a function of time (Figure 3) [72]. The 2D electron gas then coexists with the topological surface state, contributing to transport. This complicates the interpretation of transport studies because simply verifying the 2D nature of the carrier in transport is not sufficient to distinguish the topological insulator surface states from the trivial 2D electron gas. Indeed, a theoretical model for weak anti-localization effects in thin Bi_2Se_3 or Bi_2Te_3 films shows that contribution from the 2D electron gas can lead to uncertainty related to the value of α [73], the parameter whose value can indicate how many topologically non-trivial states exist in the thin film [74]. Earlier studies of SdH oscillations that were 2D in nature may have to be studied more carefully to distinguish the effect of the 2D electron gas [75,76]. The formation of the 2D electron gas is attributed to the Fermi level pinning. In addition, the fact that, overtime, multiple 2D electron gases emerge also hints to environmental effects playing a role.

3.4 How to reduce parasitic effects

The most logical way to reduce parasitic effects for topological insulator Bi_2Se_3 and Bi_2Te_3 has been to coat the thin films or nanostructures with a protective layer that does not interfere with the topological surface state [69]. For Bi_2Se_3, *in situ* Se coating after the growth seems particularly effective in keeping the carrier density low and mobility high. The Se layer itself is insulating, thus does not affect transport significantly. Careful Aharonov-Bohm oscillations were carried out after *in situ* Se coating on Bi_2Se_3 narrow nanowires [68]. The coating method, however, may not be feasible for proximity effect studies where topological insulators need to be interfaced directly with either superconductors or ferromagnetic insulators [77,78]. In such cases, the superconducting or ferromagnetic layers can be grown on top of topological insulators to serve as a protective layer. Care must be taken to ensure that the subsequent growth do not damage the topological insulators.

4 MoS_2 : indirect- to direct- gap semiconductor

Unlike graphene, semiconducting characteristics of 2D chalcogenides, manifested by a bandgap and high on/off

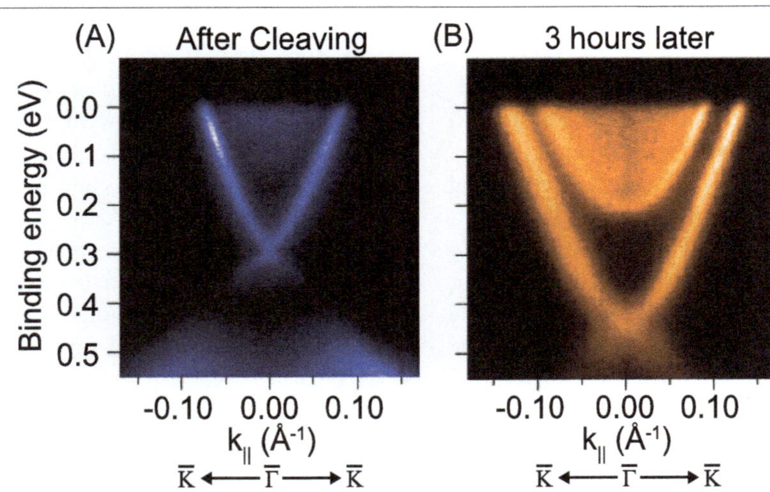

Figure 3 Band structure measurements of Bi_2Se_3 from angle-resolved photoemission spectroscopy. (A) shows the band structure of freshly cleaved Bi_2Se_3 surface. The surface state is clearly visible. **(B)** Band structure of the same surface, after left in vacuum for three hours. The Dirac point shifted down. In addition, a sharp rim is observed around the conduction band edge after three hours. This indicates emergence of a trivial 2D electron gas. Reprinted with permission from Ref. 72.

current ratio, make them attractive for electronic devices such as field-effect-transistors (FETs). In addition, the layer thickness-dependent tunable bandgap and its indirect-to-direct transition inherent to these materials suggest a novel opportunities for optical applications [19,79-81] (Figure 4A, enhanced photoluminescence in single layer MoS_2). For successful applications, charge carrier transport of the 2D chalcogenides needs to be studied carefully, which is determined by intrinsic materials properties as well as extrinsic materials preparation methods. Large uncertainty in both the intrinsic and extrinsic factors often leads to a huge variation in the performance of electronic devices based on these materials. For example, FETs based on mechanically exfoliated mono- to a few-layer thick MoS_2 generally present high carrier mobility, some approaching the theoretical limit set by phonon scattering [13,20,82-85] (Figure 4B). Meanwhile, chemically synthesized MoS_2 FETs are often limited with poor carrier mobilities, over an order of magnitude lower than the theoretical value [86-88]. Amongst a number of such variables, the surface property of the 2D chalcogenides is believed to play a critical role in their carrier transport due to the large surface areas, as in the case of graphene electronics we covered earlier. Indeed, the conductivity of few-layer thick MoS_2 is found to be highly sensitive to ambient environments [89,90], indicating the important role of its surface states for carrier transport.

In this section, we review the surface and interfacial effects on transport properties of 2D chalcogenides. We focus our discussion on MoS_2, one of the most explored 2D chalcogenides, and discuss the followings: (1) extrinsic interfacial effects on MoS_2 in contact with other medium such as metal contacts and dielectric substrates, and (2) intrinsic surface properties of MoS_2 such as surface, electronic, and crystal defects.

4.1 Extrinsic interface effects: MoS_2-metal and MoS_2-dielectrice interface

In this section, we exclusively deal with mechanically exfoliated mono-to a few-layer thick MoS_2 nanoflakes. Choice of contact metals for MoS_2 plays an important role in determining the carrier type (p- or, n-) and the performance of MoS_2 FETs. Das et al., have systematically studied various metal contacts for monolayer MoS_2 FETs, and found metals with low work functions form lower contact resistance than high work function metals [91]. As a result, very high FET mobility of ~ 700 cm^2/Vs is realized when Scandium (Sc) was used as a contact metal. The n-type characteristics were unaffected by the choice of metal contacts in this study. In contrast, Fontan et al., showed that MoS_2 FETs can display both n- or p-type behaviors depending on the choice of contact metals [92]. Gold (Au) contacts showed n-type characteristics while Palladium (Pd) contacts showed p-type characteristics. This was attributed to the local Fermi energy pinning of MoS_2 at the MoS_2/Pd interface due to the lowered Fermi energy level of Pd caused by the chemical interaction between Pd and MoS_2. A computational study found that Au forms a tunnel barrier contact while Titanium (Ti) forms a more electron injection-efficient low resistance Ohmic contact on a monolayer thick MoS_2 [93]. These studies point to the importance of characterizing the MoS_2-metal interface carefully, where transport characteristics may be dominated more by the interface effects, rather than the intrinsic MoS_2 properties.

In addition to the MoS_2-metal contact issues, dielectric substrate effects on MoS_2 are also important. A first-principle calculation has revealed that the electronic properties of a monolayer MoS_2 are greatly affected by underlying SiO_2 substrate. Presence of oxygen dangling bonds at the SiO_2/MoS_2 interface can cause

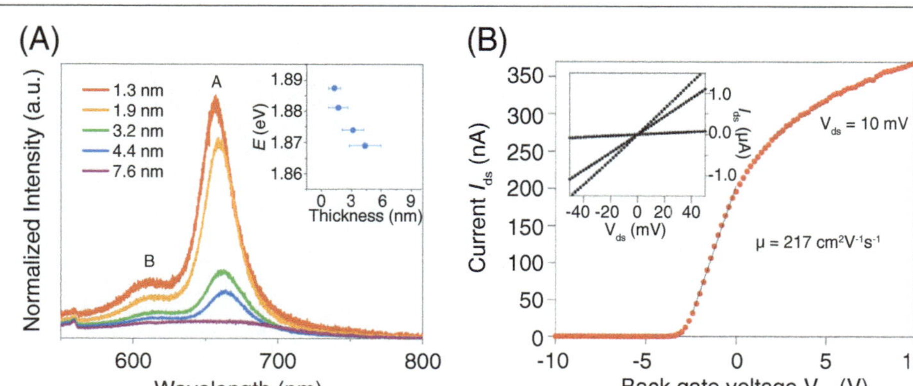

Figure 4 Optical and electrical properties of mono-to-a few layer MoS_2. (A) Photoluminescence spectra of MoS_2 thin films with varying layer thicknesses. Inset shows the energy of the exciton peak, A, as a function of layer thickness. **(B)** Room-temperature transport characteristics of a monolayer MoS_2 FET. Inset shows drain-source current vs. drain-source bias under varying back gate voltages. **(A)** Reprinted with permission from Ref 81. **(B)** Reprinted with permission from Ref 13.

the downward shift of Fermi energy below the valence band maximum, thus converting the originally n-type MoS_2 to p-type MoS_2 [94]. Using high k-materials such as aluminum oxide (Al_2O_3) or hafnium oxide (HfO_2) as gate dielectric layers instead of SiO_2 can improve the performance of MoS_2 FET without converting carrier types [91,95].

4.2 Intrinsic surface effects: structural and electronic defects in CVD- synthesized MoS_2

Zhu et al., have systematically studied the carrier transport properties of MoS_2, synthesized by chemical vapor deposition (CVD) [96]. They found that CVD-grown MoS_2 possesses a significant amount of band tail trapping states, which severely impair FET performances. Because of the trapped charges, measured mobility values largely fall short of the true band mobility predicted by theory. This finding agrees well with earlier studies of the low-temperature FET characterizations of MoS_2, which revealed highly localized electronic states and trap-assisted 2D variable hopping-range transport in MoS_2 [97,98]. The presence of intrinsic structural defects, such as point defects, dislocations, and grain boundaries, in CVD-grown monolayer MoS_2 have been confirmed by atomic resolution imaging techniques and they were correlated with localized mid-band gap states [99,100] (Figure 5A). Strong anisotropy in electron transport directions due to defects such as grain boundaries is observed [100]. In addition, sulfur vacancies in MoS_2 are found to introduce localized donor states inside the bandgap, thus are responsible for the n-type behavior of MoS_2 as well as the hopping-dominated transports [101] (Figure 5B). A theoretical study also suggests that the presence of intrinsic atomic vacancies in MoS_2 leads to localized mid-gap states that act as scattering centers, thus significantly reducing conductance [102]. The decrease of conductance depends on the type and the concentration of the defects.

Besides their effect on carrier transport efficiency, the intrinsic structural defects of MoS_2 can also affect the MoS_2-metal contact properties as well as charge carrier types. McDonnell et al., have revealed that the intrinsic defects of MoS_2 dominate the MoS_2/metal contact resistance, resulting in a low Schottky barrier independent of contact metal work functions [103]. In addition, they showed that MoS_2 can exhibit both p-type and n-type at different positions on a same sample, which is attributed to variations in the local stoichiometry of MoS_2 due to surface defects. Electrical and materials characterizations revealed that the regions of n-type possess ~1:1.8 stoichiometry in Mo:S and local S vacancies are responsible for the n-type. Meanwhile, the p-type regions possess a stoichiometry of ~1:2.3, indicating they are either S-rich (intercalates, or interstitials) or Mo-deficient. The study further emphasizes the critical role of the surface states of MoS_2 on carrier transport properties and strengthens the importance of eliminating un-wanted surface properties for improved carrier transports.

4.3 Ways to eliminate interface and surface effects

Since the surface/interface properties of MoS_2 contribute tremendously to the carrier transport in MoS_2-based electronic devices as reviewed above, the effective elimination of unwanted surface/interface effects is critical

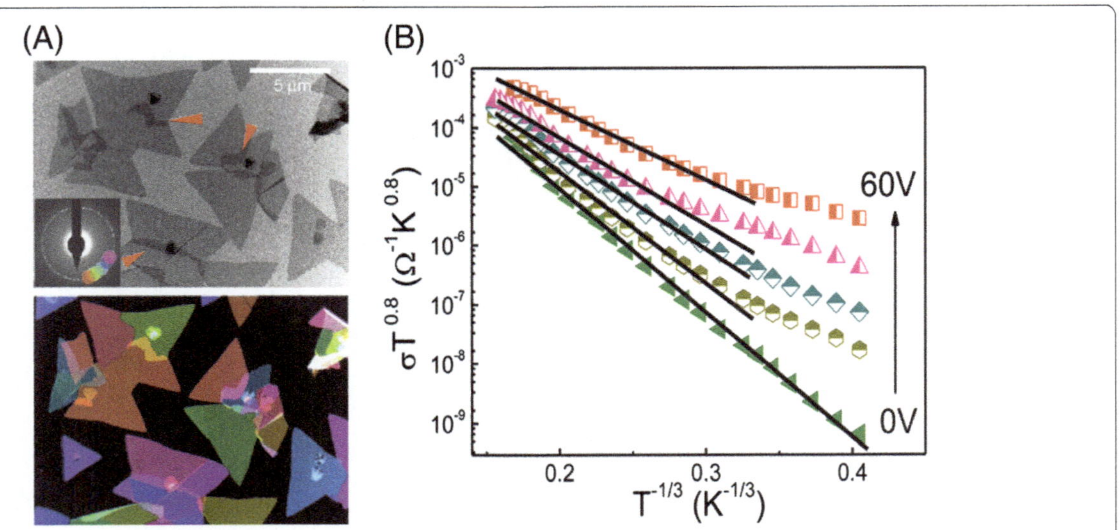

Figure 5 Intrinsic structural defects and defect dominated transport in MoS_2. (A) Bright-field transmission electron microscope (TEM) image (top) and its corresponding colored dark-field image (bottom) show irregularly shaped polycrystalline grains of MoS_2. **(B)** Temperature-dependent conductivity of a few-layer MoS_2 FET under various back gate voltages. The linear fits indicate 2-D variable-hopping transport in MoS_2. **(A)** Reprinted with permission from Ref 100. **(B)** Reprinted with permission from Ref 97.

for improving transport properties. Efforts have been focused on improving the interfacial properties - MoS_2-metal contact and MoS_2-substrate interfaces. External dopants were introduced to MoS_2 FETs to lower the Schottky barrier and contact resistance at the MoS_2-metal interface. Degenerate n-doping of MoS_2 has been realized by using Potassium, a material of high electron affinity, as a dopant [104]. Molecular doping approaches, such as using polyethyleneimine (PEI) molecules as n-dopants, have also been proven to improve FET characteristics [105,106]: ~50-70 % improvement in ON current and FET mobility. Substantial efforts have also been made in improving the MoS_2-dielectric substrate interface properties. Much of these have been focused on using better dielectric materials [83,95,107]. Bao et al., have demonstrated the highest room-temperature FET mobility of ~ 470 cm^2/Vs on polymethyl methacrylate (PMMA)-supported MoS_2 FETs, > 10 times higher than the mobility of comparable devices on SiO_2 substrates [83]. This superiority is attributed to the effective dielectric screening from the PMMA [83,108].

In comparison to the efforts to eliminate extrinsic effects on MoS_2 transport properties, not much progress has been made in improving the intrinsic properties of MoS_2 for better carrier transport. A limited number of works based on post-growth approaches have been reported. For example, surface-passivated MoS_2 FETs with passivation layers (eg. PMMA [89] or Silicon nitride (SiNx) [109]) have demonstrated better transport characteristics, attributed to the effective protection of MoS_2 against the adsorption of environmental molecules rather than the alteration of intrinsic surface/structural properties of MoS_2. More promising and fundamental approaches lie in developing effective growth schemes that could produce MoS_2 with intrinsically high surface qualities. This remains largely unexplored at present and more systematic work is needed in this direction.

5 Conclusion

The progress and renewed interest in 2D chalcogenides have been remarkable. With continued improvement in 2D chalcogenide synthesis and a myriad of possible van der Waals heterostructures with 2D chalcogenides as basic building blocks, future of 2D chalcogenides for fundamental studies as well as optoelectronic applications is promising. As we focus on mono- to a few-layer thick chalcogenide thin films and nanosheets to enhance desirable nanoscale or surface effects, it becomes increasingly important to distinguish true intrinsic materials properties from extrinsic environmental effects such as contact metal issues, surface defects such as oxidation or vacancies, and interface-induced impurities or trap sites. Systematic transport and materials characterizations are critical to reveal these parasitic effects. This review highlighted several experimental and theoretical works on extrinsic surface and interface effects on electron transport properties of Bi-based chalcogenide topological insulators and Mo-based sulfide semiconductors. Research efforts on these secondary environmental effects must be concurrent with fundamental intrinsic property studies to minimize any possible misinterpretation of data and to ensure rapid progress of the field.

Competing interests

The authors declare no competing interests.

Authors' contributions

All authors have contributed to the writing of the manuscript. All authors read and approved the final manuscript.

References

1. AK Geim, KS Novoselov, Nat. Mater. **6**, 183 (2007)
2. AK Geim, Science **324**, 1530 (2009)
3. KS Novoselov, AK Geim, SV Morozov, D Jiang, MI Katsnelson, IV Grigorieva, SV Dubonos, AA Firsov, Nature **438**, 197 (2005)
4. KS Novoselov, AK Geim, SV Morozov, D Jiang, Y Zhang, SV Dubonos, IV Grigorieva, AA Firsov, Science **306**, 666 (2004)
5. FP Bundy, RH Wentorf, J. Chem. Phys. **38**, 1144 (1963)
6. K Watanabe, T Taniguchi, H Kanda, Nat. Mater. **3**, 404 (2004)
7. L Song, Z Liu, ALM Reddy, NT Narayanan, J Taha-Tijerina, J Peng, G Gao, J Lou, R Vajtai, PM Ajayan, Advanc. Mater. **24**, 4878 (2012)
8. P Vogt, P De Padova, C Quaresima, J Avila, E Frantzeskakis, MC Asensio, A Resta, B Ealet, G Le Lay, Phys. Rev. Lett. **108**, 155501 (2012)
9. C-C Liu, W Feng, Y Yao, Phys. Rev. Lett. **107**, 076802 (2011)
10. M Houssa, E Scalise, K Sankaran, G Pourtois, VV Afanasev, A Stesmans, Appl. Phys. Lett. **98**, 223107 (2011)
11. Z Ni, Q Liu, K Tang, J Zheng, J Zhou, R Qin, Z Gao, D Yu, J Lu, Nano Lett. **12**, 113 (2011)
12. KS Novoselov, D Jiang, F Schedin, TJ Booth, VV Khotkevich, SV Morozov, AK Geim, Proc. Nat. Acad. Sci. **102**, 10451 (2005)
13. B Radisavljevic, A Radenovic, J Brivio, V Giacometti, A Kis, Nat. Nanotech. **6**, 147 (2011)
14. H Fang, S Chuang, TC Chang, K Takei, T Takahashi, A Javey, Nano Lett. **12**, 3788 (2012)
15. D Kong, JC Randel, H Peng, JJ Cha, S Meister, K Lai, YL Chen, ZX Shen, HC Manoharan, Y Cui, Nano Lett. **10**, 329 (2010)
16. D Kong, W Dang, JJ Cha, H Li, S Meister, H Peng, Z Liu, Y Cui, Nano Lett. **10**, 2245 (2010)
17. AK Geim, IV Grigorieva, Nature **499**, 419 (2013)
18. Y Zhang, K He, C-Z Chang, C-L Song, L-L Wang, X Chen, J-F Jia, Z Fang, X Dai, W-Y Shan, S-Q Shen, Q Niu, X-L Qi, S-C Zhang, X-C Ma, Q-K Xue, Nat. Phys. **6**, 584 (2010)
19. A Splendiani, L Sun, Y Zhang, T Li, J Kim, C-Y Chim, G Galli, F Wang, Nano Lett. **10**, 1271 (2010)
20. B Radisavljevic, A Kis, Nature Mater. **12**, 815 (2013)
21. W Jin, P-C Yeh, N Zaki, D Zhang, JT Sadowski, A Al-Mahboob, AM van der Zande, DA Chenet, JI Dadap, IP Herman, P Sutter, J Hone, RM Osgood Jr, Phys. Rev. Lett. **111**, 106801 (2013)
22. SZ Butler, SM Hollen, L Cao, Y Cui, JA Gupta, HR Gutiérrez, TF Heinz, SS Hong, J Huang, AF Ismach, E Johnston-Halperin, M Kuno, VV Plashnitsa, RD Robinson, RS Ruoff, S Salahuddin, J Shan, L Shi, MG Spencer, M Terrones, W Windl, JE Goldberger, ACS Nano **7**, 2898 (2013)
23. QH Wang, K Kalantar-Zadeh, A Kis, JN Coleman, MS Strano, Nat. Nanotech. **7**, 699 (2012)
24. JJ Cha, KJ Koski, Y Cui, Phys. Status Solidi (RRL) – Rapid Research Letters **7**, 15 (2013)
25. Y Zhang, VW Brar, C Girit, A Zettl, MF Crommie, Nat. Phys. **5**, 722 (2009)
26. D Rg, Y Wang, VW Brar, W Regan, H-Z Tsai, Q Wu, W Gannett, A Zettl, MF Crommie, Nano Lett **11**, 2291 (2011)

27. J Martin, N Akerman, G Ulbricht, T Lohmann, JH Smet, K von Klitzing, A Yacoby, Nat. Phys. **4**, 144 (2008)
28. JC Meyer, AK Geim, MI Katsnelson, KS Novoselov, TJ Booth, S Roth, Nature **446**, 60 (2007)
29. KI Bolotin, KJ Sikes, Z Jiang, M Klima, G Fudenberg, J Hone, P Kim, HL Stormer, Solid State Commun. **146**, 351 (2008)
30. KS Kim, Y Zhao, H Jang, SY Lee, JM Kim, KS Kim, J-H Ahn, P Kim, J-Y Choi, BH Hong, Nature **457**, 706 (2009)
31. X Du, I Skachko, A Barker, EY Andrei, Nat. Nanotech. **3**, 491 (2008)
32. CR Dean, AF Young, I Meric, C Lee, L Wang, S Sorgenfrei, K Watanabe, T Taniguchi, P Kim, KL Shepard, J Hone, Nat. Nanotech **5**, 722 (2010)
33. V Georgakilas, M Otyepka, AB Bourlinos, V Chandra, N Kim, KC Kemp, P Hobza, R Zboril, KS Kim, Chem. Rev. **112**, 6156 (2012)
34. I Childres, LA Jauregui, J Tian, YP Chen, New Journal of Physics **13**, 025008 (2011)
35. X Lu, H Huang, N Nemchuk, RS Ruoff, Appl. Phys. Lett. **75**, 193 (1999)
36. KT Chan, JB Neaton, ML Cohen, Phys. Rev. B **77**, 235430 (2008)
37. Y Xu, Z Liu, X Zhang, Y Wang, J Tian, Y Huang, Y Ma, X Zhang, Y Chen, Advanc. Mater. **21**, 1275 (2009)
38. M Fang, K Wang, H Lu, Y Yang, S Nutt, J. Mater. Chem. **19**, 7098 (2009)
39. S Niyogi, E Bekyarova, ME Itkis, H Zhang, K Shepperd, J Hicks, M Sprinkle, C Berger, CN Lau, WA deHeer, EH Conrad, RC Haddon, Nano Lett. **10**, 4061 (2010)
40. W Chen, S Chen, DC Qi, XY Gao, ATS Wee, J. Am. Chem. Soc. **129**, 10418 (2007)
41. B Guo, Q Liu, E Chen, H Zhu, L Fang, JR Gong, Nano Lett. **10**, 4975 (2010)
42. S Bae, H Kim, Y Lee, X Xu, J-S Park, Y Zheng, J Balakrishnan, T Lei, H Ri Kim, YI Song, Y-J Kim, KS Kim, B Ozyilmaz, J-H Ahn, BH Hong, S Iijima, Nat. Nanotech. **5**, 574 (2010)
43. H Wang, L-F Cui, Y Yang, H Sanchez Casalongue, JT Robinson, Y Liang, Y Cui, H Dai, J. Am. Chem. Soc. **132**, 13978 (2010)
44. Z-S Wu, W Ren, L Wen, L Gao, J Zhao, Z Chen, G Zhou, F Li, H-M Cheng, ACS Nano **4**, 3187 (2010)
45. CL Kane, EJ Mele, Phys. Rev. Lett. **95**, 146802 (2005)
46. BA Bernevig, TL Hughes, S-C Zhang, Science **314**, 1757 (2006)
47. M Konig, S Wiedmann, C Brune, A Roth, H Buhmann, LW Molenkamp, X-L Qi, S-C Zhang, Science **318**, 766 (2007)
48. F Wilczek, Nat. Phys. **5**, 614 (2009)
49. L Fu, CL Kane, Phys. Rev. Lett. **100**, 096407 (2008)
50. X-L Qi, R Li, J Zang, S-C Zhang, Science **323**, 1184 (2009)
51. D Pesin, AH MacDonald, Nat. Mater. **11**, 409 (2012)
52. I Garate, M Franz, Phys. Rev. Lett. **104**, 146802 (2010)
53. H Li, J Cao, W Zheng, Y Chen, D Wu, W Dang, K Wang, H Peng, Z Liu, J. Am. Chem. Soc. **134**, 6132 (2012)
54. W Dang, H Peng, H Li, P Wang, Z Liu, Nano Lett. **10**, 2870 (2010)
55. HW Liu, HT Yuan, N Fukui, L Zhang, JF Jia, Y Iwasa, MW Chen, T Hashizume, T Sakurai, QK Xue, Cryst. Growth Des. **10**, 4491 (2010)
56. C-L Song, Y-L Wang, Y-P Jiang, Y Zhang, C-Z Chang, L Wang, K He, X Chen, J-F Jia, Y Wang, Z Fang, X Dai, X-C Xie, X-L Qi, S-C Zhang, Q-K Xue, X Ma, Appl. Phys. Lett. **97**, 143118 (2010)
57. Q-K Xue, X Chen, Y-Y Wang, Z-C Zhang, M-H Liu, X-C Ma, L-L Wang, K He, C-Z Chang, SPIN **01**, 21 (2011)
58. D Hsieh, D Qian, L Wray, Y Xia, YS Hor, RJ Cava, MZ Hasan, Nature **452**, 970 (2008)
59. YL Chen, JG Analytis, JH Chu, ZK Liu, SK Mo, XL Qi, HJ Zhang, DH Lu, X Dai, Z Fang, SC Zhang, IR Fisher, Z Hussain, ZX Shen, Science **325**, 178 (2009)
60. P Cheng, C Song, T Zhang, Y Zhang, Y Wang, J-F Jia, J Wang, Y Wang, B-F Zhu, X Chen, X Ma, K He, L Wang, X Dai, Z Fang, X Xie, X-L Qi, C-X Liu, S-C Zhang, Q-K Xue, Phys. Rev. Lett. **105**, 076801 (2010)
61. M Liu, J Zhang, C-Z Chang, Z Zhang, X Feng, K Li, K He, W L-l, X Chen, X Dai, Z Fang, Q-K Xue, X Ma, Y Wang, Phys. Rev. Lett. **108**, 036805 (2012)
62. M Kriener, K Segawa, Z Ren, S Sasaki, S Wada, S Kuwabata, Y Ando, Phys. Rev. B **84**, 054513 (2011)
63. JJ Cha, M Claassen, D Kong, SS Hong, KJ Koski, X-L Qi, Y Cui, Nano Lett. **12**, 4355 (2012)
64. JJ Cha, D Kong, S-S Hong, JG Analytis, K Lai, Y Cui, Nano Lett. **12**, 1107 (2012)
65. AA Taskin, S Sasaki, K Segawa, Y Ando, Phys. Rev. Lett. **109**, 066803 (2012)
66. JG Analytis, RD McDonald, SC Riggs, J-H Chu, GS Boebinger, IR Fisher, Nat. Phys. **6**, 960 (2010)
67. YS Hor, A Richardella, P Roushan, Y Xia, JG Checkelsky, A Yazdani, MZ Hasan, NP Ong, RJ Cava, Phys. Rev. B **79**, 195208 (2009)
68. SS Hong, Y Zhang, JJ Cha, XL Qi, Y Cui, arXiv **1303**, 1601 (2013)
69. SS Hong, JJ Cha, D Kong, Y Cui, Nat. Commun. **3**, 757 (2012)
70. JG Analytis, J-H Chu, Y Chen, F Corredor, RD McDonald, ZX Shen, IR Fisher, Phys. Rev. B **81**, 205407 (2010)
71. D Kong, JJ Cha, K Lai, H Peng, JG Analytis, S Meister, Y Chen, H-J Zhang, IR Fisher, Z-X Shen, Y Cui, ACS Nano **5**, 4698 (2011)
72. M Bianchi, D Guan, S Bao, J Mi, BB Iversen, PDC King, P Hofmann, Nat. Commun. **1**, 128 (2010)
73. H-Z Lu, S-Q Shen, Phys. Rev. B **84**, 125138 (2011)
74. S Hikami, AI Larkin, Y Nagaoka, Progr. Theoret. Phys. **63**, 707 (1980)
75. D-X Qu, YS Hor, J Xiong, RJ Cava, NP Ong, Science **329**, 821 (2010)
76. AA Taskin, S Sasaki, K Segawa, Y Ando, Advan. Mater. **24**, 5581 (2012)
77. M-X Wang, C Liu, J-P Xu, F Yang, L Miao, M-Y Yao, CL Gao, C Shen, X Ma, X Chen, Z-A Xu, Y Liu, S-C Zhang, D Qian, J-F Jia, Q-K Xue, Science **336**, 52 (2012)
78. QI Yang, M Dolev, L Zhang, J Zhao, AD Fried, E Schemm, M Liu, A Palevski, AF Marshall, SH Risbud, A Kapitulnik, Physical Review B **88**, 081407 (2013)
79. QH Wang, K Kalantar-Zadeh, A Kis, JN Coleman, MS Strano, Nat. Nanotech **7**, 699 (2012).
80. KF Mak, C Lee, J Hone, J Shan, TF Heinz, Phys. Rev. Lett. **105**, 136805 (2010)
81. G Eda, H Yamaguchi, D Voiry, T Fujita, M Chen, M Chhowalla, Nano Lett. **11**, 5111 (2011).
82. D Jariwala, VK Sangwan, DJ Late, JE Johns, VP Dravid, TJ Marks, LJ Lauhon, MC Hersam, Appl. Phys. Lett. **102**, 173107 (2013)
83. W Bao, X Cai, D Kim, K Sridhara, MS Fuhrer, Appl. Phys. Lett **102**, 042104 (2013)
84. S Kim, A Konar, WS Hwang, JH Lee, J Lee, J Yang, C Jung, H Kim, JB Yoo, JY Choi, YW Jin, SY Lee, D Jena, W Choi, K Kim, Nat. Commun. **3**, 1011 (2012).
85. K Kaasbjerg, KS Thygesen, JW Jacobsen, Phys. Rev. B **85**, 115317 (2012)
86. KK Liu, W Zhang, YH Lee, YC Lin, MT Chang, CY Su, CS Chang, H Li, Y Shi, H Zhang, CS Lai, LJ Li, Nano Lett. 12, 1538 (2012).
87. M Amani, ML Chin, AG Birdwell, TP O'Regan, S Najmaei, Z Liu, PM Ajayan, J Lou, M Dubey, Appl. Phys. Lett. **102**, 193107 (2013).
88. W Wu, D De, S-C Chang, Y Wang, H Peng, J Bao, S-Sh Pei, Appl. Phys. Lett. **102**, 142106 (2013)
89. W Park, J Park, J Jang, H Lee, H Jeong, K Cho, S Hong, T Lee, Nanotechnology **24**, 095202 (2013)
90. H Qiu , L Pan, Z Yao, J Li, Y Shi, X Wang, Appl. Phys. Lett. **100**, 123104 (2012)
91. S Das, H-Y Chen, AV Penumatcha, J Appenzeller, Nano Lett. **13**, 100 (2013)
92. M Fontana, T Deppe, AK Boyd, M Rinzan, AY Liu, M Paranjape, P Barbara, Scientific Reports **3**, 1634 (2013)
93. I Popov, G Seifert, D Tománek, Phys. Rev. Lett. **108**, 156802 (2012)
94. IK Dolui, I Rungger, S Sanvito, Phys. Rev. B **87**, 165402 (2013)
95. H-Y Chang, S Yang, J Lee, L Tao, W-S Hwang, D Jena, N Lu, D Akinwande, ACS Nano **7**, 5446 (2013)
96. W Zhu, T Low, Y-H Lee, H Wang, DB Farmer, J Kong, F Xia, P Avouris, Nat. Commun. **5**, 3087 (2014)
97. S Ghatak, AN Pal, A Ghosh, ACS Nano **5**, 7707 (2011)
98. S Ghatak, A Ghosh, Appl. Phys. Lett. **103**, 122103 (2013)
99. W Zhou, X Zou, S Najmaei, Z Liu, Y Shi, J Kong, J Lou, PM Ajayan, BI Yakobson, J-C Idrobo, Nano Lett. **13**, 2615–2622 (2013)
100. AM van der Zande, PY Huang, DA Chenet, TC Berkelbach, Y You, GH Lee, TF Heinz, DR Reichman, DA Muller, JC Hone, Nat. Mater **12**, 554 (2013)
101. H Qiu, T Xu, Z Wang, W Ren, H Nan, Z Ni, Q Chen, S Yuan, F Miao, F Song, G Long, Y Shi, L Sun, J Wang, X Wang, Nat. Commun. **4**, 2642 (2013)
102. M Ghorbani-Asl, AN Enyashin, A Kuc, G Seifert, T Heine, Phys. Rev. B **88**, 245440 (2013)
103. S McDonnell, R Addou, C Buie, RM Wallace, CL Hinkle, ACS Nano, **8**, 2880 (2014)
104. H Fang, M Tosun, G Seol, TC Chang, K Takei, J Guo, A Javey, Nano Lett. **13**, 1991 (2013)
105. NH Pour, Y Anugrah, S Wu, X Xu, SJ Koester, Device Research Conference (DRC) 101-102, (2013)
106. Y Du, H Liu, AT Neal, M Si, PD Ye, IEEE Electr. Devices L. **34**, 1328 (2013)
107. S-W Min, HS Lee, HJ Choi, MK Park, T Nam, H Kim, S Ryu, S Im, Nanoscale **5**, 548 (2013)
108. C Jang, S Adam, J-H Chen, ED Williams, SD Sarma, MS Fuhrer, Phys. Rev. Lett. **101**, 146805 (2013)
109. DJ Late, B Liu, HS Matte, VP Dravid, CN Rao, ACS Nano **6**, 5635 (2012)

Preparation of superhydrophobic and transparent micro-nano hybrid coatings from polymethylhydroxysiloxane and silica ormosil aerogels

Saravanan Nagappan[†], Jin Joo Park[†], Sung Soo Park and Chang-Sik Ha[*]

Abstract

Superhydrophobic and transparent polymethylhydroxysiloxane (PMHOS)/silica ormosil aerogel hybrids were prepared successfully by mixing of PMHOS with various weight percentages of silica ormosil aerogels (as synthesized from methyltriethoxysilane (MTES) and methyltrimethoxysilane (MTMS) precursors) in separate seal perfume glass vials. The hybrids were spin coated on glass substrate at 1000 rpm for 60 seconds and used for further analysis. The surface morphology and chemical compositions of the hybrids were analyzed by high resolution scanning electron microscopy, high resolution transmission electron microscopy, atomic force spectroscopy, adsorption and desorption isotherm, and X-ray photoelectron spectroscopy. The transparency, thermal decomposition and static contact angle (SCA) of each sample were measured by UV-Visible spectrophotometer, TGA and drop shape analysis system, respectively. The spin coated substrates showed good superhydrophobic properties, thermal stability as well as transparency on the glass substrates.

Keywords: Superhydrophobic; Transparent; Polymethylhydroxysiloxane; Silica ormosils; Micro-nano hybrid

1 Background

Transparency of a substrate is an important factor for coating applications. Transparent substrates with several surface properties such as superhydrophilic (contact angle, CA, ≤5°), hydrophilic (CA, <90°), hydrophobic (CA ≥90°) and superhydrophobic (CA ≥150°) surfaces were prepared using various organic and inorganic hybrid materials [1-5]. These types of substrates were used widely in a range of applications such as anti-stain coating, anti-fogging, solar cell, flexible substrate fabrication, self-cleaning, and anti-icing coatings [6-11]. The coating methods such as dip coating, spin coating, spraying, and casting methods were used for the preparation of transparent substrates [12-15]. Nakajima et al. prepared transparent superhydrophobic surface by spin coating method under the sublimation of aluminum acetylacetonate (AACA) followed by fluorosilane treatment [16]. In another method, the authors also fabricated similar type of transparent superhydrophobic surface using AACA and titanium acetylacetonate (TACA) and fluorosilane [17]. The obtained superhydrophobic surface showed almost 100% transparency when the concentration of TiO_2 was lower than 20%. Meanwhile, increasing the concentration of TiO_2 would lead to reduce the transparency of the coated substrate. Xu et al. recently prepared highly transparent superhydrophobic substrates by spin coating of fluorinated silica nanoparticles on silica wafer or other substrates [18]. The spin coated substrates can produce over 95% of transparency with superhydrophobicity. Fluorine based silane precursors of low surface energy were used widely for changing the hydrophobic surface to superhydrophobic surface [19,20]. This is due to the formation of thin layers of hydrophobic low surface energy materials on the pretreated hierarchical substrates. On the other hand, this might depend on the concentration of low surface energy material used for surface modification. Increasing the concentrations of fluoro silane to other precursor

* Correspondence: csha@pnu.edu
[†]Equal contributors
Department of Polymer Science and Engineering, Pusan National University, Busan 609-735, Republic of Korea

would enhance the surface properties. In another case, long chain alkyl or aromatic compounds were also used for surface treating of hydrophobic surfaces to produce superhydrophobic surface [21,22].

Recently we developed novel bio-inspired superhydrophobic hybrid micro-nanocomposites with mesoporous, highly stable to various drying temperatures, non-stick and self-cleaning properties on any substrate [23-25]. We used lotus leaf (LL) powder, polymethylhydroxysiloxane (PMHOS), and phenyl substituted silica ormosils (PSiOr) for the preparation of superhydrophobic hybrid micro-nanocomposites in ethanol/methanol/H_2O. Meanwhile, the hybrid micro-nanocomposites suspension showed transparency lower than 70% on a glass substrate by drop or spin coating method [26]. This is due to the presence of LL powder. Superhydrophobic PMHOS powder was synthesized from polymethylhydrosiloxane (PMHS) based on the reported procedure elsewhere [23]. PMHS is a low molecular weight siloxane based material, which is used widely in many applications such as fabrication of superhydrophobic surface and materials, catalysis, transparent surface fabrications, actuators, and etc. [6,7,27-29]. This is due to ease of availability and non-toxic properties of PMHS. The introduction of functional silica ormosil aerogel to the superhydrophobic powder would enhance the surface property of the material by the formation hierarchical particles in the suspension [23].

In this work, we discussed the fabrication of transparent, and almost complete superhydrophobic surfaces by the introduction of methyltriethoxysilane (MTES) or methyltrimethoxysilane (MTMS) to the PMHOS using methanol/H_2O. The suspension that was prepared by mixing PMHOS and silica ormosil aerogel obtained from MTES or MTMS in the absence of LL powder in methanol/H_2O showed enhanced transparency on the coated substrate than the substrate prepared in the presence of LL powder as well as in methanol/ethanol/H_2O solvents [26].

2 Methods

2.1 Materials

Poly(methylhydrosiloxane) (PMHS, Mn ~1700 to 3200), methyltriethoxysilane (MTES, 99%), methyltrimethoxysilane (MTMS, 98%), oxalic acid (≥99%) and ammonium hydroxide solution (ACS reagent, 28 to 30% NH_3 basis) were acquired from Sigma-Aldrich. Sodium hydroxide (NaOH) was supplied from Junsei Chemical Co. Ltd. Anhydrous methanol (special grade) was purchased by Carlo Erba reagents Co. Ltd. Double deionized water (surface tension, 72.8 mN/m) was used in all experiments.

2.2 Synthesis of methyl substituted silica ormosil (MeSiOr) aerogel

1 mL MTES was added into various volume of methanol (4.17 mL, 7.0 mL and 9.74 mL) in different seal perfume glass vials (100 mL) followed by the addition of oxalic acid (0.5 mL of 0.001 M) solution. The solution was hydrolyzed by stirring for 30 min followed by stored at room temperature for 24 h. Ammonium hydroxide (0.61 mL of 11.2 M (28 to 30% of NH_3)) solution was added dropwise to the solution, stirred for 15 min and stored to produce gelated silica ormosil aerogel by storing at room temperature for 2 days. The gel was redispersed in 10 mL methanol and ultrasonicated for 5 min [11,23]. The obtained silica ormosil aerogel samples were called MeSiOr EX, MeSiOr EY and MeSiOr EZ, respectively, according to the volume of methanol to MTES. Similar experiment was also carried out with constant volume of MTMS, which were named MeSiOr MX, MY, and MZ, respectively. The detailed synthesis of the materials was shown in Scheme 1.

2.3 Preparation of superhydrophobic hybrid suspension

In a 30 mL glass vial 0.1 g PMHOS, 8 g methanol and 1 g deionized water were placed and stirred well for 1 h followed by the addition of various weight percentages of MeSiOr EZ suspension to the PMHOS suspension, and stirred further at room temperature for 24 h. The obtained hybrid suspensions were called PEZA, PEZB, PEZC, PEZD and PEZE. (Scheme 1 and Table 1), whereas P, EZ and (A-E) means the presence of PMHOS, MeSiOr EZ and various weight percentages of MeSiOr EZ in PMHOS. The suspensions were ultrasonicated for 2 mins and spin coated at 1000 rpm on a pre-cleaned glass substrate and cured at room temperature. Similar experiments were also carried out with various weight percentages of MeSiOr MZ, which were called PMZA, PMZB, PMZC, PMZD and PMZE, whereas P, MZ and (A-E) means the presence of PMHOS, MeSiOr MZ and various weight percentages of MeSiOr MZ in PMHOS.

2.4 Characterization

High-resolution scanning electron microscopy (HRSEM, Hitachi S-4800) and high-resolution transmission electron microscopy (HRTEM, JEM 2011 at 200 kV) were used to analyze the micro-nano surface morphology. Prior to the HRSEM measurement, the samples were coated with osmium tetraoxide (Hatfield, PA-19440), whereas, prior to the HRTEM measurement, the sample was dispersed well in ethanol and collected on a copper grid. The surface roughness of the spin coated substrates were predicted by atomic force microscopy (AFM) in tapping mode (Digital instruments nanoscope IIIa (USA) - Veeco metrology group) at a scanning rate of 1.49 Hz and on 5 μm data scales. The chemical compositions present in the samples were measured by X-ray photoelectron spectroscopy (XPS, Thermo VG Scientific (U.K), Multi Lab) using the Al Kα (hν = 1486.6 eV) and Mg Kα (hν = (hν = 1253.6 eV)

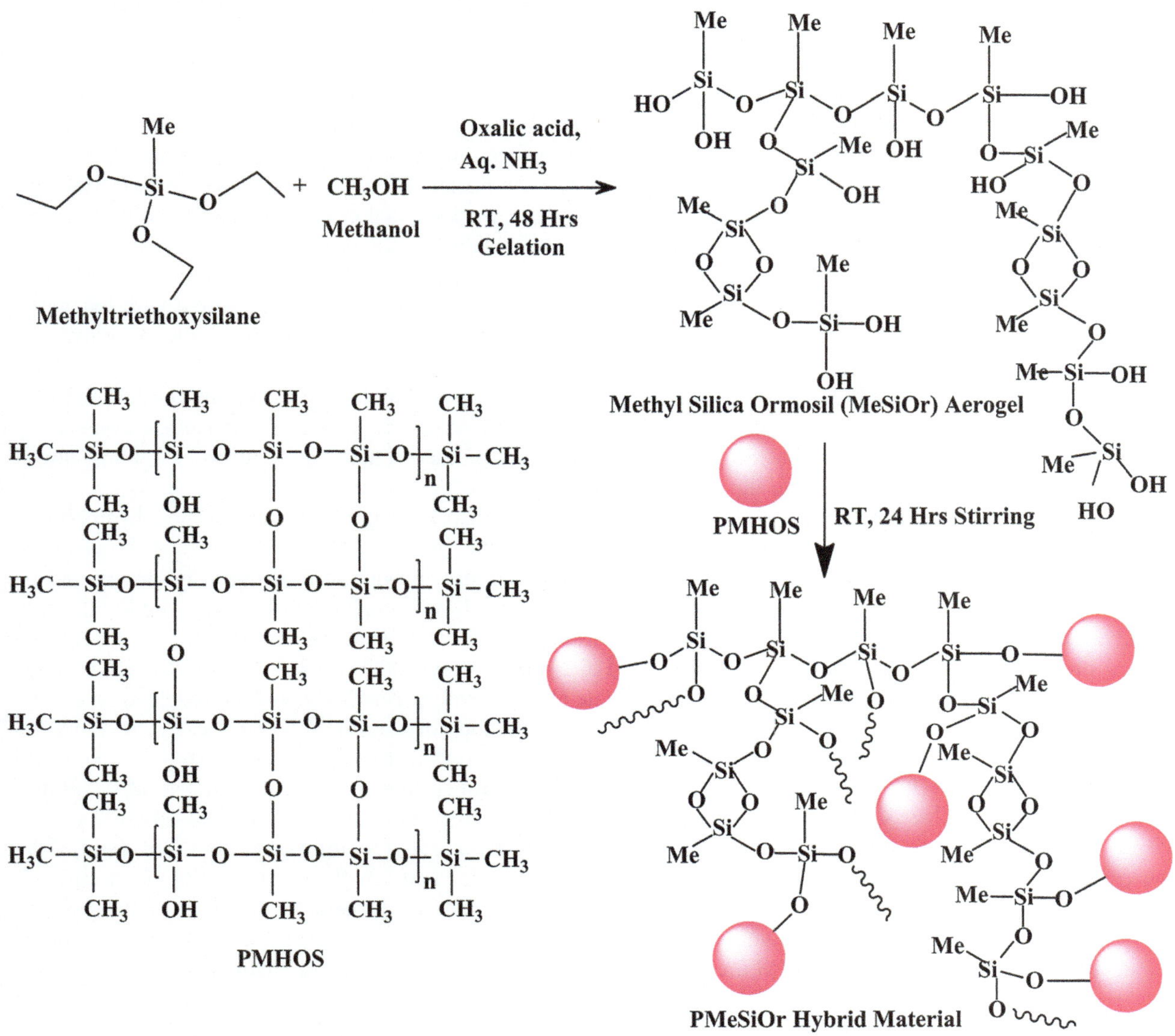

Scheme 1 Preparation of superhydrophobic hybrid material.

Table 1 Preparation of PMHOS/silica ormosil aerogel hybrids of various ratios

Sample name	PMHOS (g)	Methanol (g)	Water (g)	Silica ormosil aerogel suspension (g)
PEZA, PMZA	0.1	8	1	0
PEZB, PMZB	0.1	8	1	0.02
PEZC, PMZC	0.1	8	1	0.05
PEZD, PMZD	0.1	8	1	0.08
PEZE, PMZE	0.1	8	1	0.1

Note: P-PMHOS; EZ-MeSiOr EZ; MZ-MeSiOr MZ; (A-E)-various weight percentage of EZ or MZ based on PMHOS.

line in the binding energy analysis range of 0–600 eV in constant analyzer energy mode. The surface areas, pore volumes and pore diameters of the samples were obtained from Micromeritics ASAP 2020 V3.04 G, and calculated using the Brunauer–Emmett–Teller (BET) and Barrett–Joyner–Halenda (BJH) analysis. Thermogravimetric analysis (TGA, Q50 V6.2, Build 187, TA instruments, U.S) was performed in a nitrogen atmosphere at a heating rate at 10°C/min. The transparency of the coated substrates were measured by UV-Visible spectrophotometer (Hitachi U-2010) in the spectral wavelength range of 400–800 nm, under the conditions of 200 nm/min scanning speed, 1 nm sampling interval, and 2 nm slit width. The surface wettability was measured by drop shape analysis system (DSA 100, Krüss GmbH, Germany) at room temperature (500- μL syringe, needle diameter and length (0.5 mm and 38 mm)).

3 Results and discussion

The surface morphologies of the silica ormosil aerogel and siloxane hybrids are shown in Figure 1. The methyl functionalized silica ormosil aerogel showed densely packed silica nanoparticles on the spin coated substrate. This is due to the aggregation of silica ormosil aerogel networks at lower volume (4.2 mL) of methanol to MTES (Figure 1a). Meanwhile, porous structure was formed by increasing the volume of methanol to 7.0 mL and 9.74 mL (Figure 1b and c). Although the gels were redispersed with similar volume percentages of methanol (10 mL), the porous structure was mainly due to the gelation effect. The gels were dispersed well at higher volume percentages of methanol and reduced the surface aggregation, which creates the porous structure on the surface. The spin coated substrate also showed porous structure, which was increased by increasing the dispersibility of the silica ormosil aerogel. Similar results were also obtained for the silica ormosil aerogel obtained by MTMS (Figure 1d) [11]. Figure 1d clearly showed the formation of hierarchical surface morphology as compared with the silica ormosil aerogel surfaces obtained in the presence of MTES (Figure 1a-c). PMHOS is highly hydrolyzed and condensed solid siloxane polymer. PMHOS has hydroxyl and methyl functional groups on the surface, which makes the surface aggregation of micro-nanoparticles to form hierarchical surface. The addition of MeSiOr EZ or MeSiOr MZ suspension to the PMHOS suspension resulted in the formation of more aggregated morphology. The enhanced aggregation was mainly due to the hydroxylation and condensation of hydrophobic methyl and hydrophilic hydroxyl groups in the siloxane and silica ormosil aerogel. HRTEM images of PEZE clearly illustrate the porous and hierarchical morphology of the material (Figure 2). The spin coated substrate showed higher roughness due to the enhancement of surface aggregation. This could be confirmed further by AFM images (Figure 3). Surface roughness (Ra) and root mean square roughness (RMS) are almost similar property for the fabrication of superhydrophobic surface, though the surface property mainly depends on the Ra value than the RMS value of the coated surface. Rough surfaces sometime can show hydrophobic or hydrophilic properties if the rough surface is not properly treated with low surface energy material or not contained with other hydrophobic functional groups on the surface or at very high surface roughness (>4000 nm). Increasing the Ra or RMS value on the hydrophobic treated rough surface would also increase the surface contact angle up to an optimal level [30]. RMS also has similar property like Ra. When RMS value is less than 10 nm, the surface can show hydrophilic or hydrophobic property. MeSiOr EZ aerogel showed the surface roughness maximum (Ra) ~68.9 nm and the root mean square (RMS) roughness ~86.6 nm, whereas the Ra and RMS values were increased partially to ~95.1 nm and 115.2 nm by using MTMS under the similar conditions. The obtained results clearly showed the rough surface was created by the formation of hierarchical surface morphology. The surface roughness was enhanced further by mixing the suspensions in PMHOS. The hybrid substrate obtained by these two combinations showed the Ra and RMS values were ~166 nm and ~207.1 nm for PEZE and ~194.7 nm and 244.2 nm for PMZE, respectively. Owing to the higher Ra and RMS values of the hybrid materials, the fabricated surface showed very high surface contact angle. Three dimensional (3D) (left), two

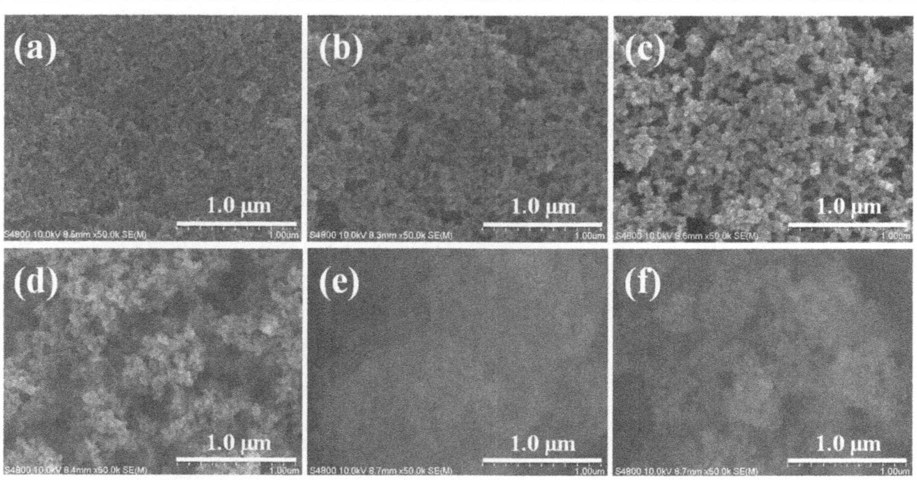

Figure 1 HR-SEM images of silica ormosil aerogels at various volume of methanol in MTES ((a) MeSiOr EX, (b) MeSiOr EY, and (c) MeSiOr EZ). (d) MeSiOr MZ from MTMS, and PMHOS/silica ormosil aerogel hybrids from MeSiOr EZ and MeSiOr MZ ((**e**) PEZE, and (**f**) PMZE). For sample notations, see Table 1.

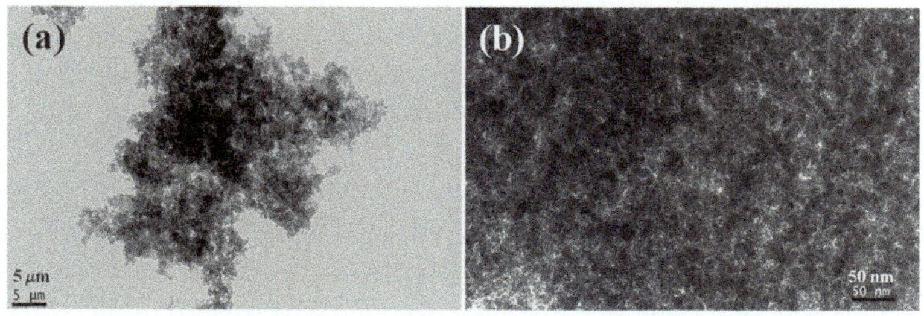

Figure 2 HR-TEM images of PMHOS/silica ormosil aerogel hybrid from MeSiOr E35 (PEZE) with (a) lower (5 µm) and (b) higher (50 nm) resolutions. For sample notations, see Table 1.

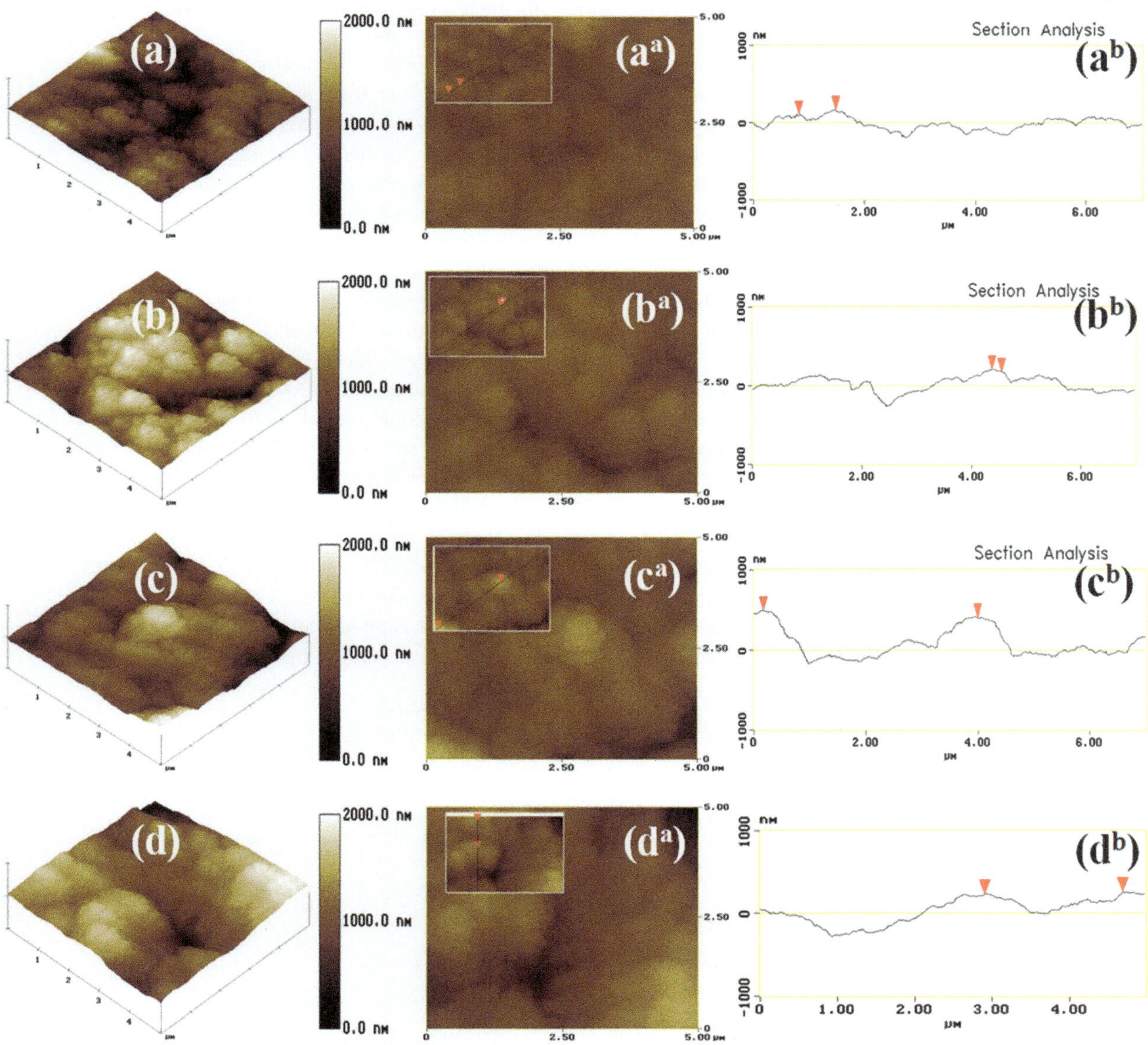

Figure 3 3D AFM images of silica ormosil aerogels (a) MeSiOr EZ, (b) MeSiOr MZ, and PMHOS/silica ormosil aerogel hybrids from MeSiOr EZ and MeSiOr MZ ((c) PEZE, and (d) PMZE). Images in the center **(a^a,b^a,c^a, and d^a)** show the 2D AFM images, while images of the right are in line profiles **(a^b,b^b,c^b, and d^b)** of the corresponding samples to the 2D AFM images of the center. For sample notations, see Table 1.

dimensional (2D) (center) views and their line profiles (right) proved the higher surface roughness created on the spin coated substrates (Figure 3).

The porous structure of the material was confirmed further by N_2 adsorption and desorption isotherm curves. The synthesized silica ormosil aerogels showed disordered mesoporous structures with type (IV) hysteresis loop (Figure 4 (a-c)). The disordered mesostructure was also obtained by hydroxylation and condensation of PMHS with surface area, pore volume and pore diameter of 410 (m^2/g), 1.0 (cm^3/g), and 25.2 nm, respectively, according to our previous study [23]. The N_2 adsorption and desorption quantity was reduced by increasing the volume of methanol to the silane precursor. Similarly, the surface areas, pore volumes, and pore diameters were reduced when the volume of methanol increased. In the mean time, the void between the adsorption and desorption curves were increased due to the creation of more porous gaps between the functional silica ormosil aerogel networks (Figure 1(a-c)) [11]. Thielemann et al. explained the effect of solvent upon the changes of surface area, pore volume and pore diameter in SBA-15 mesoporous material [31]. The surface area, pore volume and pore diameter were increased at lower volume of solvent, whereas they decreased by increasing the volume of solvent. The PEZE hybrid material also showed type (IV) hysteresis loop (Figure 4d). On the other hand, a sharp N_2 adsorption and desorption isotherm curve was observed. The detailed structural parameters of porous materials are listed in Table 2. The chemical compositions present in the functional silica ormosil aerogel were measured by XPS. The XPS patterns clearly showed the presence of carbon, oxygen and silicon in the ormosil aerogels with similar peak intensities (Figure 5a-d). The results indicate that there are partial changes occurred in the chemical compositions by increasing the volume of methanol. Meanwhile, the dispersibility of the material becomes rather uniform at higher volume of methanol. The intensity of the XPS peak was decreased by mixing the silica ormosil aerogels with PMHOS suspension, which indicates the uniform distribution and surface interactions of MeSiOr EZ or MeSiOr MZ with PMHOS (Figure 5e and f).

The thermal stability of PMHOS, according to various silica ormosil aerogels and PMHOS/silica ormosil aerogels are shown in Figure 6. The volatilization and depolymerization of methyl group in PMHOS at 500°C-600°C were already discussed in our previous study (Figure 6a) [25]. At this temperature, the superhydrophobicity of PMHOS changed to superhydrophilicity due to the partial loss of hydrophobic functional groups on the material surface. On the other hand, the thermal stability of the synthesized silica ormosil aerogel was increased gradually by increasing the volume of methanol in MTES (Figure 6b-d). The gelation can occur faster at lower volume of methanol in acid/base catalyst than at higher volumes, meaning that the aggregation of methyl group capped silicon networks reduced the free movements of siloxane chains for further condensation. This might be the reason for the reduced thermal stability of the same volume of silane precursor in the silica ormosil aerogel. On the other hand, the increase in thermal stability might be due to the formation of porous structure with complete gelation of enhanced silica ormosil aerogel networks. The thermal decompositions of the silica ormosil aerogels were occurred in two stages. The initial decomposition of the materials was occurred at ~210-220°C by the evaporation of excess water. Budunoglu et al. predicted the maximum thermal stability of the silica ormosil aerogel was ~500-600°C, when the concentration of methanol to MTMS ~35 mole fractions [11]. On the other hand, the present work illustrates the maximum degradations were occurred from ~510°C, to ~676°C, by increasing the volume of methanol to MTES. The enhanced stability was due to the presence of methyl capped functional group on the surface of silica ormosil aerogel networks, which maintains

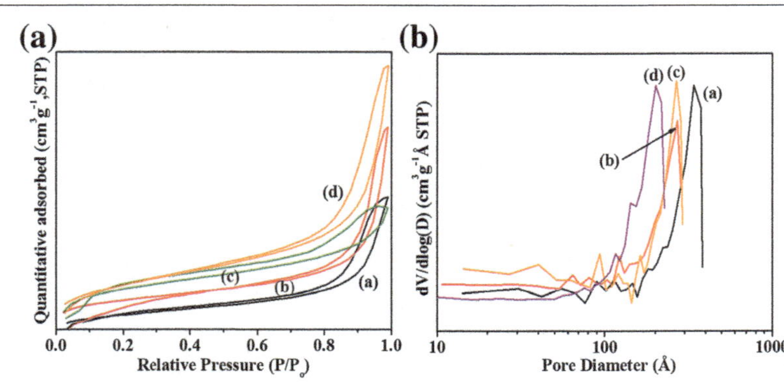

Figure 4 **(A) N_2 adsorption and desorption isotherms and (B) pore size distribution curves of silica ormosil aerogels at various volume of methanol in MTES ((a) MeSiOr EX, (b) MeSiOr EY, and (c) MeSiOr EZ), and PMHOS/silica ormosil aerogel hybrid from MeSiOr EZ (PEZE) (d).** For sample notations, see Table 1.

Table 2 BET surface areas, pore volumes, and pore diameters of silica ormosil aerogels, PMHOS and PMHOS/silica ormosil aerogel hybrid

Sample name	Surface area (m^2g^{-1})[(a)]	Pore volume (cm^3g^{-1})	Pore diameter (nm)[(b)]
MeSiOr EX	661	1.74	34.2
MeSiOr EY	559	1.55	27.3
MeSiOr EZ	529	0.5	26.8
PMHOS	413.11	1	25.2[23]
PEZE	527	1.19	20.3

Note: [(a)] – BET (Brunauer-Emmett-Teller) surface area, [(b)] calculated from the adsorption branch of the N_2 sorption curves using BJH (Barrett-Joyner-Halenda) analysis.

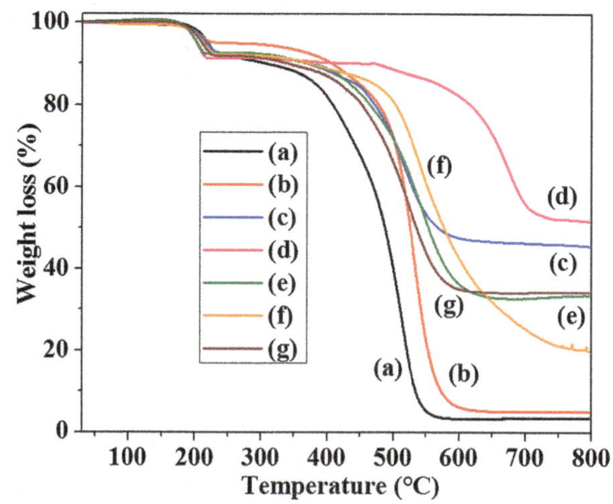

Figure 6 TGA curves of (a) PMHOS, and silica ormosil aerogels at various volume of methanol in MTES ((b) MeSiOr EX, (c) MeSiOr EY, and (d) MeSiOr EZ). (e) PMHOS/silica ormosil aerogel hybrid from MeSiOr EZ (PEZE), **(f)** MeSiOr MZ from MTMS, and **(g)** PMHOS/silica ormosil aerogel hybrid from MeSiOr MZ (PMZE). For sample notations, see Table 1.

the volatilization and decomposition of organic groups at higher temperatures. On the other hand, the materials were oxidized partially when the temperature increased to over 500°C, which may also change the surface property from superhydrophobic to superhydrophilic. The maximum degradation of the materials was due to the decomposition of capped methyl group on the surface of silica ormosil aerogel networks. Thermal stability of phenyl substituted silica ormosil was reported as high [23]. The addition of MeSiOr aerogel to the PMHOS (206°C initial; 508°C, maximum decomposition) results to the enhancement of thermal stability (218°C initial; 545°C, maximum) due to the deposition of additional hydrophobic methyl groups by silica ormosil aerogels on the siloxane networks of PMHOS (Figure 6e). The residual mass was also increased after the decomposition of PEZE, which confirmed the increased thermal stability of PMHOS by MeSiOr aerogel. Similar trends of decompositions were observed in MeSiOr MZ (216°C initial; 541°C, maximum) as well as PMZE hybrid (210°C, initial; 527°C, maximum) (Figure 6f and g).

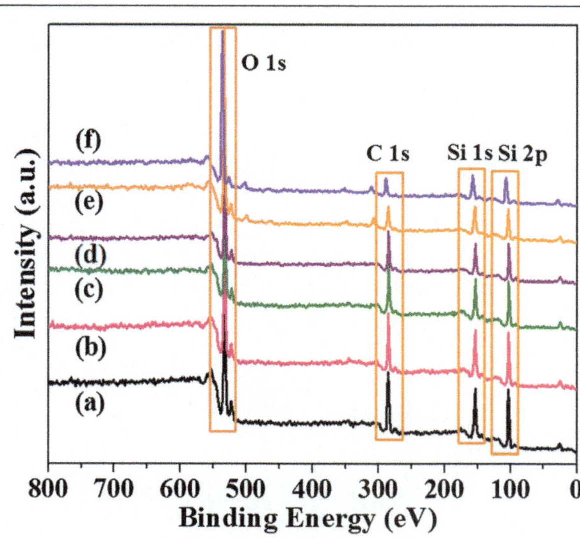

Figure 5 XP spectra of silica ormosil aerogels at various volume of methanol in MTES ((a) MeSiOr EX, (b) MeSiOr EY, and (c) MeSiOr EZ). (d) MeSiOr MZ from MTMS, and PMHOS/silica ormosil aerogel hybrids from MeSiOr E35 and MeSiOr M35 (**(e)** PEZE, and **(f)** PMZE). For sample notations, see Table 1.

The silica ormosil aerogel redispersed in methanol showed partially enhanced transparency on the spin coated substrate (Figure 7a-c). Budunoglu et al. discussed the transparency of silica ormosil aerogel obtained from MTMS under various curing temperatures [11]. At room temperature, the spin coated substrate showed ~65-75% in the visible region. On the other hand, the transparency of the coated substrate was enhanced dramatically by increasing the thermal annealing temperature from room temperature to 500°C (~90% transparency) due to reduced back reflections [11]. Similar transparent substrate was fabricated by spin coating on a glass substrate followed the evaporation of solvent at room temperature using MeSiOr suspensions (Figure 7a-c). Similarly, the spin coated superhydrophobic PMHOS glass substrate showed transparent (~85%) surface property at room temperature when dispersing in methanol (Figure 8). The transparency of the substrate was reduced partially by the addition of MeSiOr EZ suspension to the PMHOS suspension. Increasing the weight percentages of MeSiOr EZ suspension to the PMHOS suspension further reduced the transparency due to formation of hierarchical surface morphology by aggregation of PMHOS and MeSiOr EZ (Figure 8). Meanwhile, the spin coated substrate maintained reasonable transparency on the substrate. The fabrication of transparent substrates was confirmed from the optical images and UV-visible spectra (Figure 8).

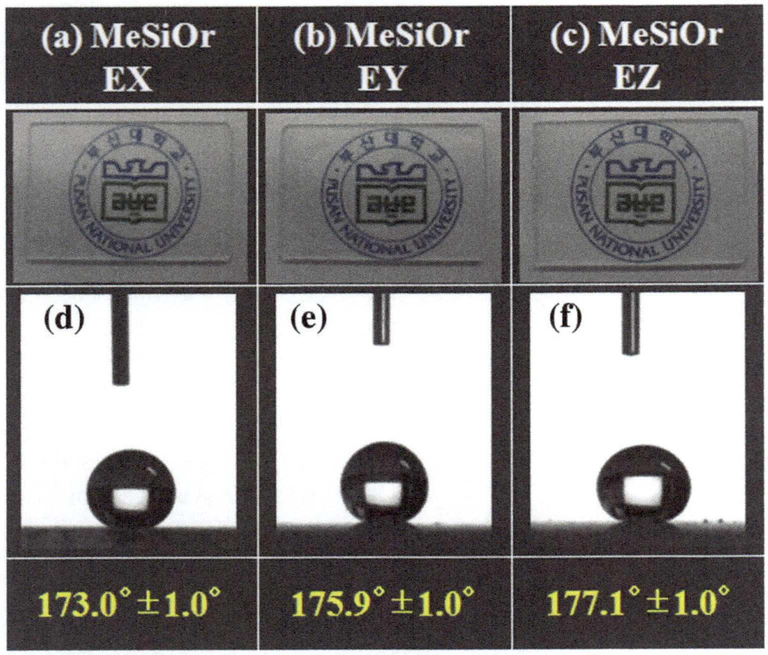

Figure 7 Optical images of silica ormosil aerogels at various volume of methanol in MTES ((a) MeSiOr EX, (b) MeSiOr EY, and (c) MeSiOr EZ) and their contact angle values (d-f). For sample notations, see Table 1.

Wettability of the substrate is an important phenomenon for using the material for self-cleaning coatings. PMHOS showed superhydrophobic property when dispersing into methanol with contact angle (CA) of 174.2° ± 1.0° (Figure 9a) [23]. Meanwhile, the silica ormosil aerogel suspension obtained from MTES or MTMS showed almost complete superhydrophobicity (~178.0° ± 1.0°) (Figure 7d-f). The addition of small amounts of MeSiOr EZ suspension or MeSiOr MZ suspension to the PMHOS suspension showed the increased contact angle on the spin coated substrate (Figure 9b and c). This is due to the aggregation of superhydrophobic silica ormosil aerogel on the PMHOS surface by hydrophobic effect, which enhanced the surface roughness of the hierarchical micro-nanoparticles. The higher surface roughness would increase the surface contact angle more preferably. HRTEM and AFM images clearly showed the hierarchical surface roughness of the hybrid material (Figures 2 and 3). Increasing the weight percentages of MeSiOr EZ suspension to the PMHOS suspension would also increase the surface

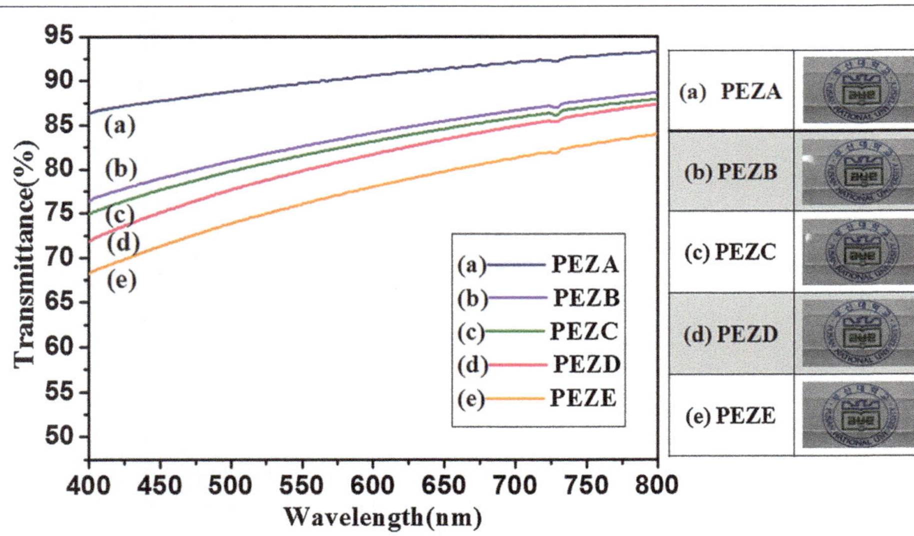

Figure 8 % Transmittance (left) with the optical images (right) of PMHOS/silica ormosil aerogel hybrids prepared with various weight percentages of MeSiOr EZ (a (0 g), b (0.02 g), c (0.04 g), d (0.08 g), and e (0.1 g)) with PMHOS.

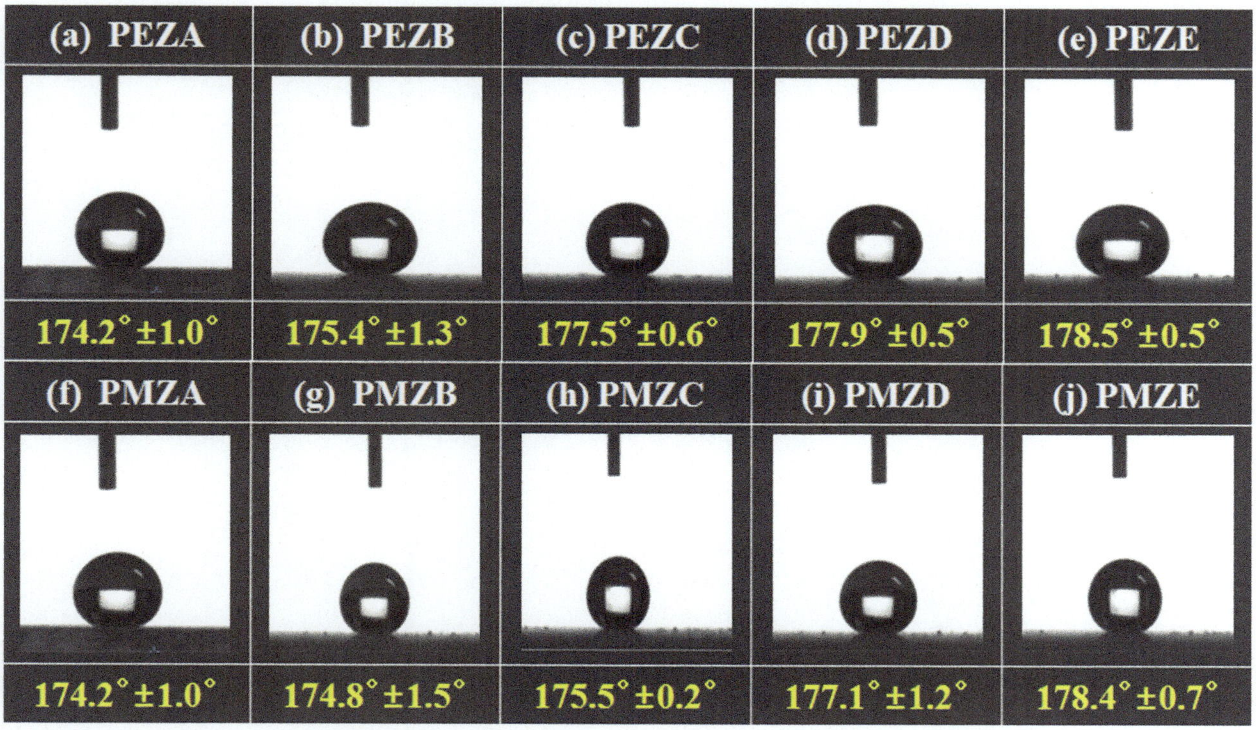

Figure 9 Static contact angles of PMHOS/silica ormosil aerogel hybrid substrates prepared at various weight percentages of MeSiOr EZ (a-e) and MeSiOr MZ (f-j) with PMHOS. For sample notations, see Table 1.

contact angle based on the mechanism of enhanced surface hydrophobicity and surface roughness (Figure 9a-e). Our previous study showed partially reduced surface contact angle (maintained the superhydrophobicity with CA over 173.5° ± 4.0°) of the similar system when using mixture of ethanol and methanol solvents for PMHOS and silica ormosil aerogel prepared from PTES [23]. In another case, we also found the decrease of surface contact angle from superhydrophobic to stable hydrophobic by the addition of hydrophilic silica nanoparticles (146.13° ± 4.0°) suspension or amino functionalized silica ormosils suspensions (136.76° ± 3.0°, and 134.0° ± 3.0°)) to the PMHOS suspension in ethanol/N,N dimethylacetamide (DMAc) solvents. The present study showed the enhanced surface contact angle with almost complete superhydrophobicity and reasonable transparency using similar type of co-solvent (methanol) for PMHOS suspension and for silica ormosil aerogel suspension. Similar surface wettability also obtained by the addition of various weight percentages of MeSiOr MZ suspension to PMHOS suspension (Figure 9f-j).

4 Conclusions

In this work, we successfully fabricated transparent substrate with superhydrophobicity on a glass substrate using various ratios and types of silica ormosil aerogels and their hybrid materials. The surface of the materials showed the formation of hierarchical surface morphology with high roughness created on the surface by the introduction of silica ormosil aerogel to the polymethylhydroxysiloxane. The synthesized silica ormosil aerogel and hybrid materials showed disordered meso structures. However, the surface areas, pore volumes and pore diameters of the silica ormosil aerogel were decreased by increasing the volume of methanol. Similarly, the transparency of the spin coated substrates also decreased partially by aggregation but with maintaining reasonable transparency. On the other hand, the thermal stability was increased by increasing the volume percentages of methanol to the silane precursors. The addition of silica ormosil aerogel to the PMHOS showed the enhanced the surface properties as well as the thermal stability of the hybrid material. The obtained results exhibit the important contribution of silica ormosil aerogels to the PMHOS for the enhancement of various properties of the hybrid material in a co-solvent by comparing with the materials obtained from ethanol/methanol/water or ethanol/DMAc/water solvents. On the other hand, the stability of the hybrid suspensions is not so good over a period of time under rest. The main reason is due to the deposition of the hybrid particles in the solvents in a few hours. Meanwhile, superhydrophobic surface can be fabricated easily at any time by simple shaking of the suspension for a few seconds followed by coating on a glass substrate.

Competing interests
The authors declare that they have no competing interests.

Authors' contribution
SN carried out the synthesis of superhydrophobic PMHOS, writing and formatting of the full manuscript according to the journal. JJP carried out the synthesis of silica ormosil aerogels and hybrid materials with the help of SN. Both authors thoroughly characterized the material using several characterization techniques. SSP and CSH were participated in this experiment and carefully revised the manuscript. All authors read and approved the final manuscript.

Acknowledgements
This study was supported by the National Research Foundation of Korea (NRF) Grant funded by The Ministry of Science, ICT & Future Planning, Korea (Pioneer Research Center Program (2010-0019308/2010-0019482), NRF-RFBR Joint Research Program (No. 2013K2A1A7076267), and Brain Korea 21 Plus Program (21A2013800002)).

References
1. S Nagappan, SS Park, CS Ha, J. Nanosci. Nanotechnol. **14**, 1441 (2014)
2. SS Latthe, H Imai, V Ganesan, C Kappenstein, A Venkateswara Rao, J. Sol-Gel Sci. Technol. **53**, 208 (2010)
3. QF Xu, JN Wang, KD Sanderson, ACS Nano **4**, 2201 (2010)
4. CH Su, J Li, HB Geng, QJ Wang, QM Chen, Appl. Surf. Sci. **253**, 2633 (2006)
5. X Deng, L Mammen, Y Zhao, P Lellig, K Müllen, C Li, HJ Butt, D Vollmer, Adv. Mater. **23**, 2962 (2011)
6. S Nagappan, MC Choi, G Sung, SS Park, MS Moorthy, SW Chu, WK Lee, CS Ha, Macromol. Res. **21**, 669 (2013)
7. S Nagappan, JJ Park, SS Park, SH Hong, YS Jeong, WK Lee, CS Ha, Compos. Interfaces **20**, 33 (2013)
8. JA Howarter, JP Youngblood, Adv. Mater. **19**, 3838 (2007)
9. NT Panagiotopoulos, EK Diamanti, LE Koutsokeras, M Baikousi, E Kordatos, TE Matikas, D Gournis, P Patsalas, ACS Nano **6**, 10475 (2012)
10. J Yang, ZZ Zhang, XH Men, XH Xu, Appl. Surf. Sci. **255**, 9244 (2009)
11. H Budunoglu, A Yildirim, MO Guler, M Bayindir, ACS Appl. Mater. Interfaces **3**, 539 (2011)
12. Y Rahmawan, L Xu, S Yang, J. Mater. Chem. A **1**, 2955 (2013)
13. L Xu, RG Karunakaran, J Guo, S Yang, ACS Appl. Mater. Interfaces **4**, 1118 (2012)
14. SA Mahadik, DB Mahadik, MS Kavale, VG Parale, PB Wagh, HC Barshilia, SC Gupta, ND Hegde, A Venkateswara Rao, J. Sol-Gel Sci. Technol. **63**, 580 (2012)
15. HS Hwang, NH Kim, SG Lee, DY Lee, K Cho, I Park, ACS Appl. Mater. Interfaces **3**, 2179 (2011)
16. A Nakajima, A Fujishima, K Hashimoto, T Watanabe, Adv. Mater. **11**, 1365 (1999)
17. A Nakajima, K Hashimoto, T Watanabe, Langmuir **16**, 7044 (2000)
18. LG Xu, JH He, Langmuir **28**, 7512 (2012)
19. JB Lin, HL Chen, T Fei, C Liu, JL Zhang, Appl. Surf. Sci. **273**, 776 (2013)
20. Y Lai, Y Tang, J Gong, D Gong, L Chi, C Lin, Z Chen, J. Mater. Chem. **22**, 7420 (2012)
21. BH Kim, DH Lee, JY Kim, DO Shin, HY Jeong, S Hong, JM Yun, CM Koo, H Lee, SO Kim, Adv. Mater. **23**, 5618 (2011)
22. D Zhong, Q Yang, L Guo, S Dou, K Liu, L Jiang, Nanoscale **5**, 5758 (2013)
23. S Nagappan, JJ Park, SS Park, WK Lee, CS Ha, J. Mater. Chem. A **1**, 6761 (2013)
24. S Nagappan, AR Sung, CS Ha, J. Biobased Mater. Bioenergy **8**, 175 (2014)
25. S Nagappan, SS Park, CS Ha, Macromol. Res. **22**, 843 (2014)
26. S Nagappan, JH Park, AR Sung, CS Ha, Compos. Interfaces **21**, 597 (2014)
27. DJ Yang, JP Li, Y Xu, D Wu, YH Sun, HY Zhu, F Deng, Microporous Mesoporous Mater. **95**, 180 (2006)
28. M Giordano, A Iadonisi, Eur. J. Org. Chem. **2013**, 125 (2013)
29. SJ Lee, M Goedert, MT Matyska, EM Ghandehari, M Vijay, JJ Pesek, J. Micromech. Microeng. **18**, 025026 (2008)
30. KL Cho, II Liaw, AHF Wu, RN Lamb, J. Phys. Chem. C **114**, 11228 (2010)
31. JP Thielemann, F Girgsdies, R Schlögl, C Hess, Beilstein J. Nanotechnol. **2**, 110 (2011)

Synthetic nanowire/nanotube-based solid substrates for controlled cell growth

Ku Youn Baik[1], Sung Young Park[2], Seon Namgung[3], Daesan Kim[4], Dong-guk Cho[5], Minju Lee[5] and Seunghun Hong[4,5*]

Abstract

The behaviour of cells can be controlled by various microenvironments such as nanostructured cell-culture substrates with controlled nanotopography and chemical properties. One of promising substrates for controlled cell growth is a solid substrate comprised of synthetic one-dimensional nanostructures such as polymer nanofibers, carbon-based nanotubes/nanofibers, and inorganic nanowires. Such nanotube/nanowire structures have a similar dimension as extracellular matrix fibers, and their nanotopography and chemical properties can be easily controlled, which expands their possible applications in controlling the growth and differentiation of cells. This paper provides a concise review on the recent applications of solid substrates based on synthetic nanowires/nanotubes for controlled cell growth and differentiation.

Keywords: Nanowire; Nanotube; Cell adhesion; Cell growth; Cell proliferation; Cell differentiation

1 Introduction

Cell behaviors, such as adhesion, proliferation and differentiation, are affected by various microenvironmental factors. Such microenvironments consist of extracellular matrices (ECM) with soluble components such as growth factors and cytokines. In order to create artificial microenvironments, biologically-derived materials have been intensively explored, since they have the advantages of being biocompatible and of having similar mechanical properties as native tissues [1-4].

Recent advances in nanotechnology present new possibilities and strategies in cell therapy and tissue engineering. Synthetic bio-inspired materials have been developed to create new controllable microenvironments rather than just to mimic *in-vivo* environments [5,6]. Among the synthetic nanomaterials, nanowires and nanotubes have a similar dimension with natural ECM components such as collagen and elastic fibers. The diameter of these nanostructures can be precisely controlled, and they can be aligned, patterned or constructed 3-dimensionally [7]. These properties facilitate mimicking microenvironments of various tissues. For example, the parallel-alignment of nanowires mimics that of ECMs in tendon and muscle, and concentric whorls in bone and mesh-like lattices of skin.

The well-investigated nanowires/nanotubes can be categorized into three gropus: 1) polymer nanofibers, 2) carbon based nanotubes/nanofibers, and 3) inorganic nanowires/nanotubes (Figure 1). These nanowire/nanotube structures have many interesting physical and chemical properties which can be easily modulated (Table 1). Polymer nanofibers are biodegradable, which makes them ideal for tissue regeneration and drug delivery [8]. Carbon-based nanostructures can be easily functionalized by chemical or physical methods to control the degree of cell adsorption [9]. Inorganic nanowires can be utilized as devices for measuring cell adhesion force [10], dynamic drug delivery system [11] and electrical cell sensors [12] due to prominent electrical and mechanical properties.

In this review, we discuss the cell engineering with solid substrates based on synthetic nanowires/nanotubes, especially polymer nanofibers, carbon based nanostructures and inorganic nanowires. The cellular adhesion, proliferation and differentiation on these substrates and various other applications will be discussed.

* Correspondence: seunghun@snu.ac.kr
[4]Department of Biophysics and Chemical Biology, Seoul National University, Seoul 151-747, Korea
[5]Department of Physics and Astronomy, and Institute of Applied Physics, Seoul National University, Seoul 151-747, Korea
Full list of author information is available at the end of the article

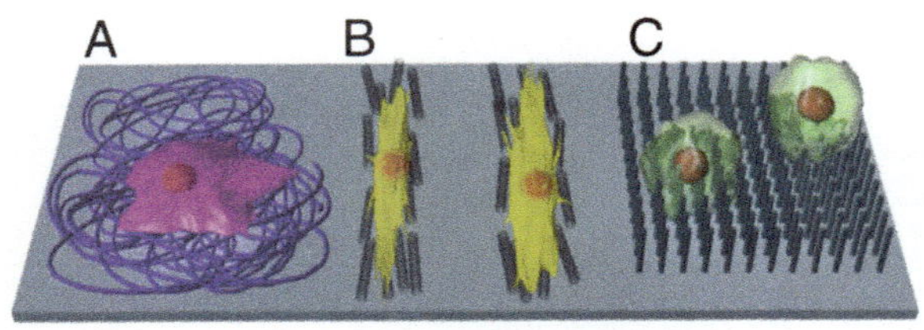

Figure 1 Schematic diagram showing the controlled cell growth on nanowire/nanotube based-substrates. Cells grown on **(A)** bundles, **(B)** laterally-aligned patterns, and **(C)** vertically-aligned structures of nanowires/nanotubes.

2 Review

2.1 Polymer nanofiber-based substrates

A polymer nanofiber is a one-dimensional nanostructure which is comprised of synthetic small molecules typically connected by covalent chemical bonds. The most important and distinguished property of polymer nanofibers compared to other synthetic nanowires is biodegradability. Taking an advantage of this property, polymer nanofibers can be utilized as an implantable substrate for various applications such as tissue regeneration, local treatment of disease, and drug delivery. Furthermore, polymer nanofibers can be functionalized with various biomaterials such as ECM molecules, ECM analogs and growth factors to enhance cellular functionality, and, thus, it has a great potential in biological and biomedical applications.

Table 1 Overview of properties of synthetic nanowires/nanotubes

	Materials	Diameter	Controllability	Biomedical applications
Polymer nanofibers	Poly (e-caprolactone) (PCL)	0.01 ~ 100 μm	Diameter	Controlled cell growth (adhesion, migration, proliferation, differentiation) [14–22]
	Polyaniline (PANi)		Structure (aligned, random)	
	Poly (L-lactic acid) (PLLA)		Surface functionalization	
	Polyethersulfone (PES)		Degradation	
	Poly (D,L-lactide-co-glycolide) (PLGA)		Elasticity	
			Composition	
Carbon-based nanostructures	Signle wall carbon nanotube (swCNT) multi wall carbon nanotube (mwCNT)	1 ~ 2/2 ~ 100 nm	Diameter	Controlled cell growth (adhesion, migration, proliferation, differentiation) [26-36]
			Structure (aligned, random, vertical)	
	Carbon nanofiber (CNF)	3.5 ~ 500 nm	Surface functionalization	Controlled cell growth (adhesion, migration, proliferation, differentiation) [39-41]
Inorganic nanowires/nanotubes	TiO_2 nanotube	10 ~ 100 nm	Diameter	Controlled cell growth (adhesion, proliferation, differentiation) [46-48]
				Orthopaedic implant [46-48]
	ZnO nanowire	10 ~ 300 nm	Diameter	Controlled cell growth (adhesion, proliferation, differentiation) [55-57]
			Structure (aligned, random, vertical)	
	GaP Nanowire	20 ~ 100 nm	Diameter	Controlled cell growth (adhesion, proliferation) [59]
				Mechanosensing [58]
	Ni nanowire	10 ~ 200 nm	Diameter	Controlled cell growth (adhesion, separation, positioning) [60,61]
			Structure (aligned, random)	
			Magnetic property	
	Au nanowire	3 ~ 100 nm	Diameter	Controlled cell growth (adhesion, differentiation) [63]
			Surface functionalization	
	Si nanowire	3 ~ 500 nm	Diameter	Mechanosensing [10]
				Electrical sensor [12,43,44]
			Structure (aligned, random, vertical)	Intracellular delivery [42]
			Surface functionalization	

2.1.1 Effect on cell adhesion

To mimic natural microenvironments for the cell control, various artificial and nanometer-scale structures with different nanotopography have been developed by many groups. For example, fibrous nanostructures can structurally and functionally mimic ECM in biological systems, and they can regulate cell responses. Several groups showed that the proliferation, migration, alignment and gene expression of cells were affected by the alignment and the dimension of nanofibers in contact with cells (Figure 2 (A-H)).

Patel *et al.* investigated the combined effects of aligned polymer nanofibers and bioactive factors on cells [13]. Polymer nanofibers were functionalized with laminin and basic fibroblast growth factor (bFGF) to imitate native ECM fibrils. In this research, aligned polymer nanofibers induced longer neurite outgrowth and faster skin cell migration compared with randomly-oriented polymer nanofibers. The cells were elongated, and they migrated along the direction of polymer nanofibers. In addition, growth factors attached to aligned nanofibers were found to be more effective than those in solution on neurite outgrowth.

Yang *et al.* also examined the effect of nanofiber alignment on the neurite outgrowth of neural stem cells (NSCs) [14]. The aligned nanofibers supported the NSC culture, and they improved the neurite outgrowth (Figure 2 (E, F)). They also found that microscale polymer fibers have no significant effect on the neurite length. Baker *et al.* showed that the mechanical properties such as stiffness and elastic modulus of polymer nanofibers based cell-laden constructs depend on the alignment of polymers but not on cell type [15]. Meniscal fibrochondrocytes (MFCs) and mesenchymal stem cells (MSCs) were tested on poly (ε-caprolactone) (PCL) polymer nanofibers. The tensile properties of aligned polymer nanofiber based cell-laden constructs were higher than those of randomly distributed one, whereas both cell-laden constructs were comparable in amount of total DNA and produced ECMs.

The diameter of polymer nanofibers is also an important factor affecting cell adhesion. Li *et al.* [16] used polyaniline (PANi) and gelatin mixed polymer with different volume ratio. They found that the diameter of PANi-gelatine polymer nanofibers decreased from 803 ± 121 nm to 61 ± 13 nm with the decrease of the volume ratio of gelatin (Figure 2 (C, D)). The H9c2 rat cardiac myoblasts cultured on randomly distributed polymer fibers showed no significant difference in proliferation, but their morphology varied with the diameter of polymer nanofibers. Cells on thicker nanofibers showed a large number of pseudo-filopodia attachment to nanofibers (Figure 2 (G)), while cells on thinner nanofibers spread randomly just as those on glass surface (Figure 2 (H)).

2.1.2 Effect on cell differentiation

Cell differentiation is the process in which a progenitor cell or a stem cell becomes a more specified cell such as a bone cell, a neural cell, a liver cell, etc. There are soluble (e.g. growth factors, cytokines) and insoluble (e.g. ECM matrix, physical environments) factors which can affect the cell differentiation. Polymer nanofibers have shown the capability of supporting various stem cell differentiations, and they can even enhance the differentiation of the stem cells compared with a conventional scaffold.

Li *et al.* investigated the *in vitro* chondrogenic differentiation of MSCs on a PCL nanofiber-based scaffold in the presence of the transforming growth factor, TGF-ß1

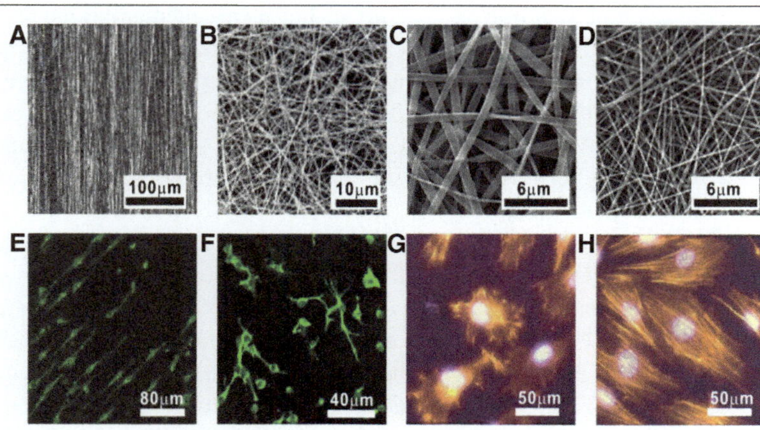

Figure 2 Polymer nanofibers for controlled cell growth. Scanning electron microscopy (SEM) images of **(A)** aligned PLLA nanofibers and **(B)** randomly-distributed PLLA nanofibers (reproduced from ref. 14 with permission, © Elsevier). SEM images of PANi-gelatin blend nanofibers with its averaged diameters of **(C)** 803 nm and **(D)** 61 nm. (reproduced from ref. 16 with permission, © Elsevier). Immunofluorescence images of neurofilaments (green) in NSCs **(E)** on aligned PLLA nanofibers and **(F)** on randomly-distributed PLLA nanofibers (reproduced from ref. 14 with permission, © Elsevier). Immunofluorescence images of actin filaments (red) and nuceli (blue) in myoblasts on the nanofibers with its averaged diameters of **(G)** 803 nm and **(H)** 61 nm (reproduced from ref. 16 with permission, © Elsevier).

[17]. Cell proliferation on polymer nanofibers was similar to that in a cell pellet culture which is a widely used cell culture system. However, the contents of sulfated glycosaminoglycan (sGAG) as a chondrogenic differentiation marker on polymer nanofibers was higher than that in a cell pellet culture. It indicates that the PCL nanofiber substrate supports a chondrogenic differentiation. MSCs were differentiated into osteoblasts on PCL nanofiber based scaffold in osteogenic differentiation medium as shown by Yoshimoto *et al.* [18] Cells migrated inside the polymer nanofiber-based scaffolds and produced an ECM of collagen after 1 week. MSCs on PCL nanofibers were also induced to differentiate along adipogenic, chondrogenic and osteogenic lineages in specific differentiation media [19]. Not only PCL but also poly (D,L-lactide-co-glycolide) (PLGA) and poly (L-lactic acid) (PLLA) support cell differentiation. Xin *et al.* reported hMSC on PLGA retained their phenotype and differentiated into chondrogenic and osteogenic cells in differentiation media [20].

Smith *et al.* examined the effect of nanofibrous scaffolds on osteogenic differentiation [21]. They prepared polymer nanofiber-based substrates in 2-D and 3-D formation. The nanofibrous structures in 2-D and 3-D enhanced the osteogenic differentiation of human embryonic stem cell (hESC)-derived osteogenic progenitor cells compared to the scaffolds without nanofibers.

The diameter of polymer nanofibers affects not only cell adhesion but also cell differentiation. Polyethersulfone (PES) nanofiber meshes with different diameter regulate proliferation and differentiation of rat neural stem cells (rNSCs) [22]. They used laminin-coated PES with diameters of 238 ± 45 nm, 749 ± 153 nm and 1452 ± 312 nm. As the diameter of polymer nanofibers decreases, cells were spread more on the nanofibers. rNSCs on 238 nm nanofiber mesh were spread on the nanofiber matrix, and they were assumed cell morphology of glial lineage. On the contrary, rNSCs on 749 nm nanofiber mesh adhered onto a single fiber, and they were elongated along a fiber axis. Those cells preferentially differentiated into neuronal lineage. However, both of cells on polymer nanofiber-based substrates showed the higher-level expression of the differentiation markers than those on a flat substrate.

Researchers also investigated the synergistic effect of polymer blends. Ku *et al.* invesgated the synergic effects of PCL and PANi blended nanofiber alignment on myoblast differentiations [23]. The differentiation of myoblasts on blended nanofibers was affacted by the PANi concentration. Kim *et al.* used commercial polymers to fabricate suitable biomaterial scaffold for the cell growth [24]. The modified PCL nanofibers incoporated within poly (ethylene oxide) (PEO) showed a good biocompatibility.

These results show that the polymer nanofiber can support multi-lineage differentiation capability of stem cells, and it is a good candidate for a scaffold in tissue regeneration or stem cell theraphy.

2.2 Carbon-based nanotubes/nanofibers

Carbon nanomaterials including carbon nanotubes (CNTs) and carbon nanofibers (CNFs) exhibit excellent mechanical and electrical properties. They have been explored in many biological applications such as bioelectronics, tissue engineering, cellular imaging and therapeutics [9,25]. The carbon nanomaterials can be advantageous for cell control because they do not emit ions and have a chemically-stable surface.

2.2.1 Carbon nanotubes

A CNT is a rolled-up graphene which is carbon atoms arranged in a planar honeycomb lattice. CNTs can be classified into single-walled CNT (swCNT) and multi-walled CNT (mwCNT) depending on the number of rolled-up graphene layers. Generally, the diameter of swCNT is about $1 \sim 2$ nm, and that of mwCNT varies from 2 nm to over 100 nm (Table 1). A CNT is mechanically strong and easy to be functionalized. The electrical and optical properties of CNTs can be used for various applicaions such as ultrasensitive biosensors. We review the recent applications of CNT-based substrates in controlling cell behaviors.

Nanoscale surface roughness of CNT-based substrates can deform cell membrane and hinder the motion of vesicles in cells. Zhang *et al.* reported that the areal coverage of vesicle motion was reduced in cells grown on a CNT layer, which was observed by total internal reflection fluorescence microscopy (Figure 3 (A)) [26]. Cellot *et al.* observed the improved postsynaptic current response of the neurons on CNTs [27]. The intimate contact between cell membranes and CNTs provided a close electrical coupling of the neurons and CNTs, which could directly stimulate neural activities (Figure 3 (B)). Furthermore, the spinal cord explants interfaced to CNT scaffolds for long-term period increased the number and length of neuronal fiber outgrowing spinal segments. This study also showed that CNT scaffolds augmented synaptic response in neurons located as far as 5 cell layers from the cell-substrate interactions.

Patterned CNTs on substrates can control the shapes or orientiations of cells. Park *et al.* succeeded in aligning hMSCs along the patterned CNTs [28]. In this work, CNTs were selectively assembled on the substrate using the pre-patterned hydrophobic SAM molecules. The selective binding of ECM proteins (e.g. fibronectin) onto CNT regions could lead the hMSCs to grow along the CNT pattern. In a similar way, Jang *et al.* showed highly directional growth of neurites along the line-shape CNT patterns after 14 days of rat hippocampal neuron culture,

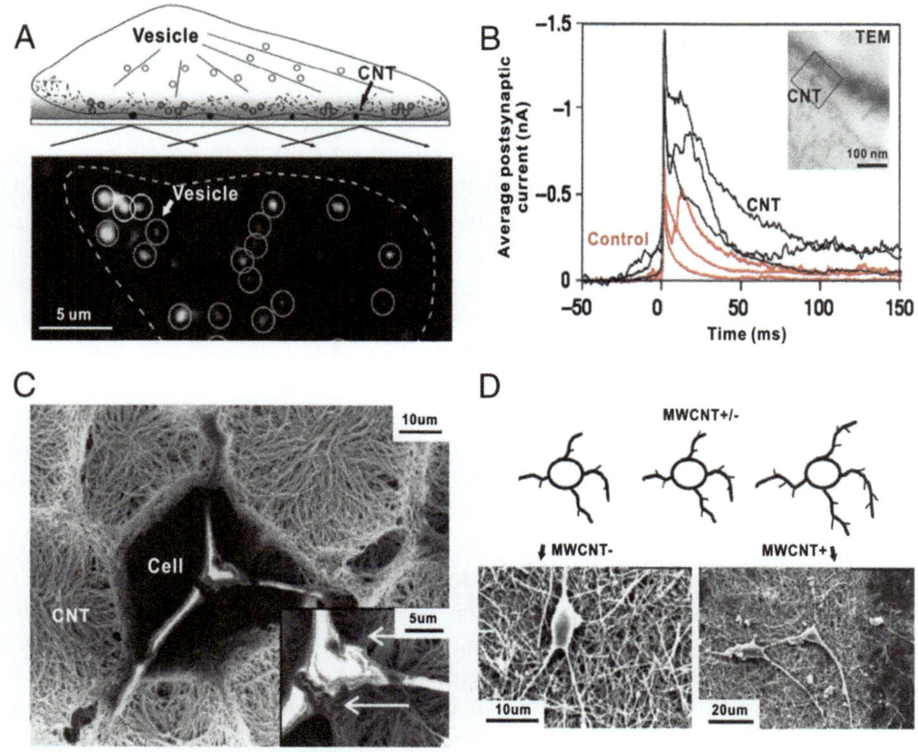

Figure 3 CNT-based substrates for controlling and monitoring cell behaviors. (A) An immunofluorescence image of secretory vesicles inside a cell on a CNT-based substrate with a schematic diagram (reproduced from ref. 26 with permission, © Wiley-VCH). **(B)** The computed waveforms of the average postsynaptic currents from the neurons on a CNT-based substrate (reproduced from ref. 27 with permission, © Nature Publishing Group). **(C)** SEM images of a fibroblast (white arrows) cultured on a 3-D CNT structure (reproduced from ref. 32 with permission, © American Chemical Society). **(D)** SEM images of neurons on positively and negatively charged mwCNTs (reproduced from ref. 33 with permission, © American Chemical Society).

which was attributed to the higher density adsorption of poly-L-lysine on the CNT patterns [29].

Previously, proteins have been known nonspecifically adsorbed well on sidewall of CNTs. Shim *et al.* demonstrated that the high affinity of proteins on CNTs could be attributed to the hydrophobic nature of CNTs or the pi-pi bonding between CNTs and proteins [30]. The adsorbed proteins mediated the interaction between cells and CNTs for a improved adhesion.

Aligned CNTs on substrates are useful in accelerating cell migration which is important in regeneration medicine. Galvan-garcia *et al.* assembled highly oriented mwCNT sheets and yarns, and they observed the directed cell growth along the CNTs [31]. In this work, the migration of fibroblasts on CNT sheets was enhanced, and the cell proliferation speed on the CNT sheets was comparable to that on glass. They also observed the directed neuronal growth along the CNT yarns, closely following the surface topography of the substrate.

A 3-D structure of CNT-film networks was explored for tissue engineering. Correa-Duarte *et al.* transformed a vertically-aligned structure of mwCNTs to a regular 3-D sieve structure by using chemically induced capillary forces (Figure 3 (C)) [32]. They successfully demonstrated that the controlled growth of a mouse fibroblast cell line (L929) in this 3-D CNT structure.

Surface functionalized CNTs provide great opportunities in controlling cell growth. Hu *et al.* reported that chemically modified CNTs affected the outgrowth and branching patterns of neurons [33]. The results showed that the positive charges induced more numerous growth cones, longer average neurite length and more neurite branching than the groups with neutral and negative charges (Figure 3 (D)). Zanello *et al.* cultured osteoblast (ROS 17/2.8) cells on the CNTs with various functional groups [34]. They found the highest cell growth on electrically-neutral swCNTs (as-produced-swCNT and swCNT-PEG). On the other hand, the lowest cell growth was observed on CNTs with charged groups (swCNT-COOH and swCNT-poly (m-aminobenzene sulfonic acid)).

CNTs can be used to transmit electrical stimulation to neural cells. Wang *et al.* developed a prototype of neural interface using vertically aligned mwCNT pillars [35]. The neurons were grown and differentiated on the hydrophilic functionalized CNT microelectrodes, and they were repeatedly excited with charge-unbalanced stimulation protocols. In this study, the CNT electrodes operated predominantly with capacitive currents without faradic reactions, which were considered ideal for neural stimulations.

CNTs were also explored to control cell differentiation. Tay *et al.* cultured hMSCs on a thin layer of carboxylic functionalized swCNTs with a vertical height of less than 100 nm [36]. In this report, they showed that the neurogenic genes were increased while osteogenic genes remained at a nominal level. This result suggests that CNT nano-roughness alone was sufficient to modulate the early stage of stem cell lineage commitment without the aid of induction media.

Some researchers performed experimental studies to reveal the mechanisms for the different differentiation behaviors of cells on CNTs. Namgung *et al.* investigated the enhanced differentiation of hMSCs on the aligned CNT networks [37]. They found that up-regulaed gene expressions of hMSCs related to the mechanotransduction pathways. These results supported that the high cytoskeletal tension of hMSCs could promote the differentiation of cells. Chen *et al.* suggested that the concentrated human bone marrow mesenchymal stem cells (hBMMSCs) following the carboxylated mwCNTs can up-regulate the secretion of growth factors [38].

2.2.2 Carbon nanofibers

Carbon nanofibers (CNFs) are graphene layers arranged as stacked cones, cups or plates, and their diameter ranges from 3.5 nm to 500 nm (Table 1) [9]. CNFs have been also explored widely in tissue engineering with the combination of polymers.

The topographical effects of CNFs were investigated in cells. McKenzie *et al.* prepared a CNF and polycarbonate urethane (PCU) composite substrate with different diameters of CNFs ranging from 60 nm to 200 nm [39]. They found that astrocytes preferentially adhered and proliferated on CNFs with the largest diameter. Price *et al.* demonstrated that this CNF-PCU substrates promoted the adhesion of osteoblasts, but the adhesions of smooth muscle cells, fibroblasts, and chondrocytes were not influenced [40].

Khang *et al.* examined osteoblast adhesion on microscale CNF patterns (30-μm-width line) on polymer substrates [41]. The patterns of CNFs on PCU substrate were prepared using an imprinting method. Osteoblasts selectively adhered and aligned on CNF patterns on PCU. The greater attractive forces between fibronectin and CNF (compared with PCU) measured by an atomic force microscope gave a plausible explanation for the selective cell adhesion.

2.3 Inorganic nanowires/nanotubes

Inorganic nanowires/nanotubes have been commonly exploited as effective tools for controlling cell behaviors by delivering biomolecules into the cells [42] and sensing intra/extracellular signals [43,44]. In addition, inorganic nanowires/nanotubes based substrates are utilized as platforms to investigate the effect of nanoscale substrate topography and size-dependent cellular responses [45–47]. Furthermore, inorganic nanowires/nanotubes have been also utilized for various other cellular applications such as controlling cell position, measuring cellular forces or sensing cellular activities. Inorganic nanowires/nanotubes could be synthesized with a variety of materials including titanium oxide (TiO_2), zinc oxide (ZnO), silicon (Si), etc. The synthesis methods include CVD [48,49], anodization [50], electrochemical deposition [51] and solvothermal growth [52]. Inorganic nanowires/nanotubes are rather easy to adjust their nanoscale surface topology and chemical properties for the control of cell behaviors.

2.3.1 Titanium oxide nanotubes

The interaction between cells and titanium oxide substrates has been extensively investigated for clinical implant applications. Recent studies showed that the nanoscale modulation of TiO_2 surface topography enhanced the efficiency in implantation. In this study, vertically-grown TiO_2 nanotubes have been used as an effective scaffold to provide various nanoscale geometries, because the diameter of TiO_2 nanotubes is easily controlled in the synthesis process.

Park *et al.* demonstrated responses of rat MSCs on vertically-oriented and highly-ordered TiO_2 nanotubes. TiO_2 nanotubes were fabricated on titanium surface by the anodization process, and the diameter of TiO_2 nanotubes were varied from 15 nm to 100 nm by controlling the applied voltage (Figure 4 (A)). They showed significantly-enhanced cellular activities such as adhesion, proliferation and differentiation on 15 nm TiO_2 nanotubes compared to 70 nm-100 nm TiO_2 nanotubes, which was attributed to the accelerated formation of integrin clustering and focal contact [45]. In other work, the similar results of the enhanced cellular activities on 15 nm TiO_2 nanotubes were shown with hematopoietic stem cells (HSCs) (Figure 4 (B) and (C)) [46].

They continued experiments with various types of cells to address the dominant factors determining cell fates on TiO_2 nanotubular enviroment [47]. Three different types of cells (endothelial cell line mlEND, hMSCs and human cord blood endothelial progenitor cells) related to blood vessel formation were studied with respect to the diameter, the crystalline structure and the chemical composition of TiO_2 nanotubes. As the diameter of TiO_2 nanotubes decreased, the cell adhesion and proliferation were enhanced regardless of the cell type. The crystalline structure and the chemical composition of TiO_2 nanotubes did not affect significantly on cell behaviors,

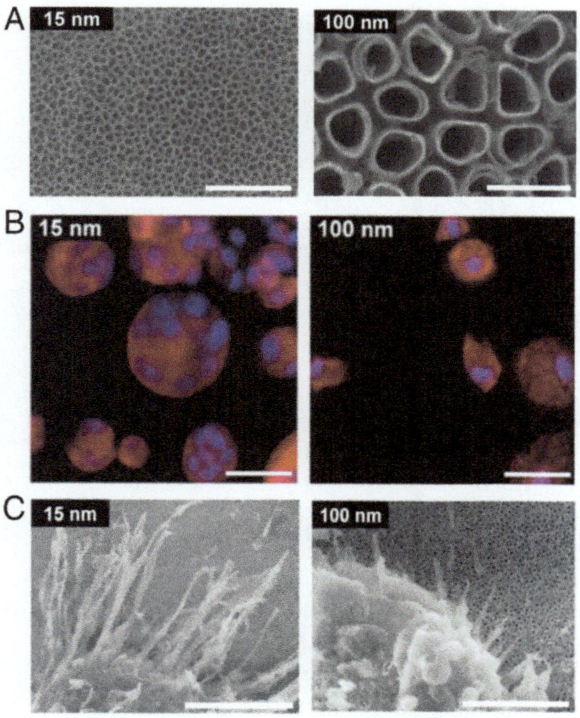

Figure 4 TiO_2 nanotubes for controlled cell growth. (A) SEM images of vertically-aligned TiO_2 nanotubes with diameters of 15 nm and 100 nm. Scale bars are 200 nm. **(B)** Immunofluorescence images of αVβ3-integrins (red) and nuclei (blue) in HSCs cultured on TiO_2 nanotubes. Scale bars are 200 μm. **(C)** SEM images of filopodia formation in HSCs on TiO_2 nanotubes. Scale bars are 3 μm (reproduced from ref. 46 with permission, © Willey-VCH).

which strongly suggests that the nanotopography should be a dominant factor for cell growth.

However, the results from Oh *et al.* were in conflict with previous experiments [50]. They observed that the differentiation of hMSCs into osteoblast-like cells was enhanced on TiO_2 nanotubes with larger diameters (70 nm to 100 nm) rather than smaller diameters (30 nm to 50 nm) in the absence of osteogenic inducing factors. These findings suggest that finely-tuned nanotopography of TiO_2 nanotubes could determine the cell fate, even though there are still controversies regarding the optimal diameter of TiO_2 nanotubes for the enhanced cell adhesion, proliferation and differentiation.

2.3.2 Zinc oxide nanowires

Zinc oxide nanowires (ZnO NWs) are also attractive nanostructures due to their semiconducting and piezoelectric properties, and they have been utilized for biosensors [53] and other biomedical applications [54]. The cell adhesion and proliferation on ZnO-based nanostructures have been also demonstrated. Lee *et al.* explored the cell adhesion and growth on vertically-grown ZnO NWs with NIH3T3 fibroblasts, umbilical vein endothelial cells and capillary endothelial cells (Figure 5 (A)) [55]. In this case, the ZnO NWs were hydrothermally grown on the uniformly distributed ZnO nano-seeds. ZnO NWs are approximately 50 nm in diameter and 500 nm in height. The cells on ZnO NWs showed weaker adhesion and lower viability than cells on a flat ZnO substrate. The authors attributed the poor adhesion and proliferation of cells on the vertically grown nanowires to the hindrance in focal adhesion formation. Simply speaking, there was not enough space for the integrin clustering on ZnO NWs, which induced poor cell adhesion and proliferation. Zaveri *et al.* also reported similar results on ZnO NWs with macrophage cells [56]. In addition, in this work, the cytotoxic effects of solubilized ZnO NWs were additionally revealed.

On the contrary, it was reported that the enhanced cell adhesion, proliferation and differentiation on ZnO nanoflowers (Figure 5 (B)) [57]. In this work, the ZnO nanoflowers were also grown by the hydrothermal methods on the micro-patterned ZnO surface. The nanoflowers were comprised of ZnO NWs with 200 ~ 300 nm in diameter and 3 ~ 4 μm in length. In this platform, osteoblasts on ZnO nanoflowers showed stronger adhesion, more DNA content and higher ALP activity than those on the flat ZnO surface. Furthermore, ZnO nanoflowers were tightly osseointegrated with the regenerated bones in the calvarial bone defects of Sprague Dawley rats. The results were attributed to the enhanced formation of lamellipodia, filopodia, and F-actin filaments on the surface of ZnO nanoflowers.

2.3.3 Other inorganic nanowires

The control of cell adhesion and viability has been also demonstrated on various other nanowires. The effects depend on the physical and chemical properties of nanowires. Of interest, the unique properties of each nanowire allow additional cellular applications. For example, dissociated sensory neurons cultured on vertically-grown epitaxial gallium phosphide nanowires (GaP NWs) showed a higher viability than planar gallium phosphide substrates (Figure 5 (C)) [58]. Interestingly, in other work, GaP NWs were utilized to measure cellular forces by analyzing the bending of GaP NWs (Figure 5 (D)) [59]. The rigidity, the high aspect ratio, and the nanoascale dimension of GaP NWs allow spatially-resolved and sensitive measurements of cellular forces.

Johansson *et al.* investigated the effect of nickel nanowires (Ni NWs) on L929 mouse fibroblasts and dissociated dorsal root ganglia (DRG) neurons [60]. They synthesized Ni NWs by the alumina template based electro-deposition method, which were 200 nm in diameter and 40 μm in length. They aligned Ni NWs using their magnetic property, and the cells were grown over the aligned Ni NWs. They showed that the cells displayed the contact guidance along the aligned Ni NWs.

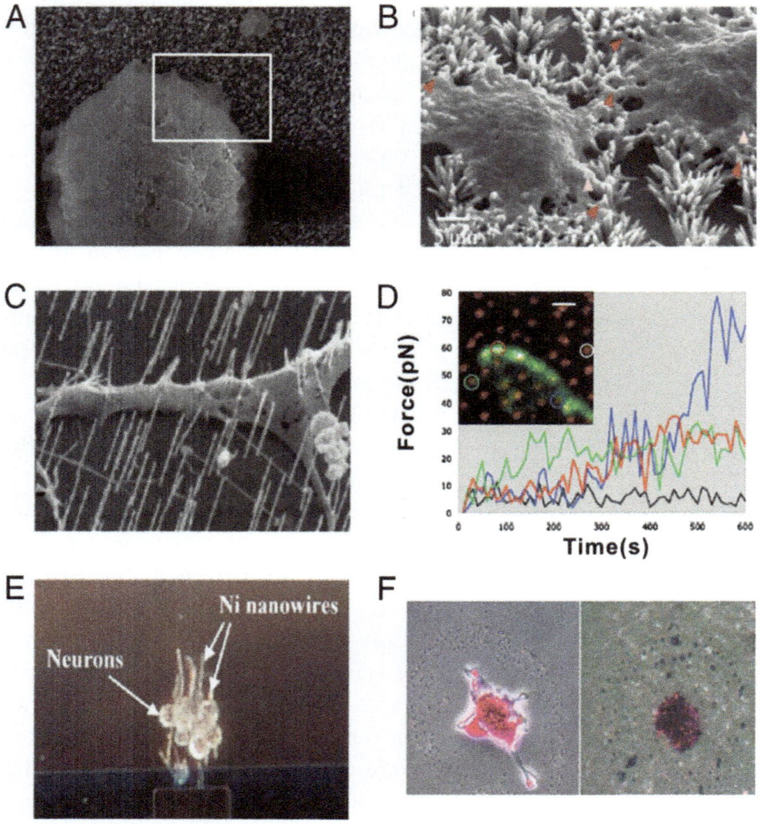

Figure 5 Various inorganic nanowires for controlling and monitoring cell behaviors. **(A)** A SEM image of NIH3T3 fibroblasts on ZnO NWs with a poor adhesion compared to the flat ZnO surface (reproduced from ref. 55 with permission, © Elsevier). **(B)** A SEM image of MC3T3-E1 osteoblasts grown on ZnO nanoflowers with an enhanced adhesion compared to the flat ZnO surface. *Lamellipodia* and *filopodia* were indicated with *pink* and *red* arrowheads, respectively (reproduced from ref. 57 with permission, © Wiley-VCH). **(C)** Neurons grown on vertically-grown GaP NWs (reproduced from ref. 59 with permission, © American Chemical Society). **(D)** The forces of a neural growth cone measured with the four GaP NWs (circles using the same color coding in the inset image). The black curve corresponds to the noise obtained by measuring the deflection of a nanowire that was not in contact with a cell (the white circle in inset image). The inset image shows neural growth cone on vertically-ordered GaP NW arrays. A scale bar is 1 μm (reproduced from ref. 58 with permission, © American Chemical Society). **(E)** An optical micrograph image of the cells internalized by Ni NWs. Ni NWs transported by an external magnetic field to the gap between the ferromagnetic electrode and the gold electrode (reproduced from ref. 62 with permission, © Springer). **(F)** Phase contrast images of differentiated stem cells on a flat RGD-treated gold surface (left) and on RGD-treated Au NWs (right). Oil red O staining was used for visualizing adipogenic differentiation (reproduced from ref. 63 with permission, © Royal Society of Chemistry).

In other works, the magnetic properties of Ni NWs were utilized for the cell separation [61] or the cell positioning (Figure 5 (E)) [62].

The adhesion and the differentiation of MSCs on gold nanowires (Au NWs) were also demonstrated by Luo *et al.* [63]. The well-known chemistry on Au surface allowed functionalization of Au NWs with self-assembled monolayers containing diverse functional groups. In this work, Au NWs were functionalized with a cell adhesive peptide (RGD- ONH2). As a result, MSCs on the functionalized Au NWs showed more fillopodia extensions compared to those on flat gold surfaces functionalized with the same RGD peptides. Moreover, in the process of adipogenic differentiation, the enhanced differentiation was observed on Au NWs (Figure 5 (F)).

Al/Al_2O_3 core/shell nanowires and patterned Si nanopillars were also studied to control the cellular responses such as cell adhesion, growth and proliferation. Lee *et al.* patterned periodic structures of Al/Al_2O_3 core/shell nanowires, and showed that the periodic patterns lead to the alignment of glial cells and dorsal root ganglia axons [64]. Similar to the Al/Al_2O_3 nanowires experiment, Bucaro *et al.* showed that Si nanopillars could control the cell polarization and alignment by controlling the radius, height and interpillar spacing of Si nanopillars [65].

3 Conclusions

In summary, we discussed various synthetic nanowires/nanotubes utilized in cell engineering. Since it is easy to control the diameter, mechanical property and surface chemistry of synthetic nanowires/nanotubes, they can be utilized to create active microenvironments to control cell behaviors. Moreover, the nanowire/nanotube-based functional devices can be used as intercellular drug delivery system and ultrasensitive biosensors for real-time detection of cell activities. Although their cytotoxicity is still under debate and requires more systematic studies, synthetic nanowires/nanotubes have great potentials not only in fundamental research but also in clinical applications such as drug screening devices, organ-on-a chips and medical tissue grafts. With the advances in nanomaterials and manipulating technique, the synthetic nanowire/nanotube based system should provide enormous opportunities in biomedical applications.

Competing interests
The authors declare that they have no competing interests.

Authors' contributions
All authors have contributed to the writing of the manuscript. All authors read and approved the final manuscript.

Acknowledgement
SH acknowledges the support from the BioNano Health Guard Research Center (H-GUARD_2013M3A6B2078961) and the NRF grant (No. 2013-007874).

Author details
[1]Department of Electrical and Biological Physics, Kwangwoon University, Seoul 139-701, Korea. [2]Center for Functional Connectomics, Korea Institute of Science and Technology, Hwarangno 14-gil 5, Seongbuk-gu, Seoul 136-791, Korea. [3]Department of Electrical and Computer Engineering, University of Minnesota, Minneapolis, MN 55455, USA. [4]Department of Biophysics and Chemical Biology, Seoul National University, Seoul 151-747, Korea. [5]Department of Physics and Astronomy, and Institute of Applied Physics, Seoul National University, Seoul 151-747, Korea.

References
1. E.S. Place, N.D. Evans, M.M. Stevens, Nat. Mater. **8**, 457–470 (2009)
2. O. Guillame-Gentil, O. Semenov, A.S. Roca, T. Groth, R. Zahn, J. Voros, M. Zenobi-Wong, Adv. Mater. **22**, 5443–5462 (2010)
3. D.O. Freytes, L.Q. Wan, G. Vunjak-Novakovic, J. Cell. Biochem. **108**, 1047–1058 (2009)
4. M.P. Lutolf, P.M. Gilbert, H.M. Blau, Nature **462**, 433–441 (2009)
5. O.Z. Fisher, A. Khademhosseini, R. Langer, N.A. Peppas, Accounts Chem. Res **43**, 419–428 (2010)
6. M.P. Lutolf, J.A. Hubbell, Nat. Biotechnol. **23**, 47–55 (2005)
7. E. Dawson, G. Mapili, K. Erickson, S. Taqvi, K. Roy, Adv. Drug Deliv. Rev. **60**, 215–228 (2008)
8. S.H. Lim, H.Q. Mao, Adv. Drug Deliv. Rev. **61**, 1084–1096 (2009)
9. P.A. Tran, L.J. Zhang, T.J. Webster, Adv. Drug Deliv. Rev. **61**, 1097–1114 (2009)
10. Z. Li, J.H. Song, G. Mantini, M.Y. Lu, H. Fang, C. Falconi, L.J. Chen, Z.L. Wang, Nano Lett. **9**, 3575–3580 (2009)
11. A. Martins, A.R.C. Duarte, S. Faria, A.P. Marques, R.L. Reis, N.M. Neves, Biomaterials **31**, 5875–5885 (2010)
12. T. Cohen-Karni, B.P. Timko, L.E. Weiss, C.M. Lieber, Proc. Natl. Acad. Sci. U. S. A. **106**, 7309–7313 (2009)
13. S. Patel, K. Kurpinski, R. Quigley, H.F. Gao, B.S. Hsiao, M.M. Poo, S. Li, Nano Lett. **7**, 2122–2128 (2007)
14. F. Yang, R. Murugan, S. Wang, S. Ramakrishna, Biomaterials **26**, 2603–2610 (2005)
15. B.M. Baker, R.L. Mauck, Biomaterials **28**, 1967–1977 (2007)
16. M.Y. Li, Y. Guo, Y. Wei, A.G. Macdiarmid, P.I. Lelkes, Biomaterials **27**, 2705–2715 (2006)
17. W.J. Li, R. Tuli, C. Okafor, A. Derfoul, K.G. Danielson, D.J. Hall, R.S. Tuan, Biomaterials **26**, 599–609 (2005)
18. H. Yoshimoto, Y.M. Shin, H. Terai, J.P. Vacanti, Biomaterials **24**, 2077–2082 (2003)
19. W.J. Li, R. Tuli, X.X. Huang, P. Laquerriere, R.S. Tuan, Biomaterials **26**, 5158–5166 (2005)
20. X.J. Xin, M. Hussain, J.J. Mao, Biomaterials **28**, 316–325 (2007)
21. L.A. Smith, X.H. Liu, J.A. Hu, P.X. Ma, Biomaterials **31**, 5526–5535 (2010)
22. G.T. Christopherson, H. Song, H.Q. Mao, Biomaterials **30**, 556–564 (2009)
23. S.H. Ku, S.H. Lee, C.B. Park, Biomaterials **33**, 6098–6104 (2012)
24. S.E. Kim, J. Wang, A.M. Jordan, L.T.J. Korley, E. Baer, J.K. Pokorski, ACS Macro Lett. **3**, 585–589 (2014)
25. K. Kostarelos, A. Bianco, M. Prato, Nat. Nanotechnol. **4**, 627–633 (2009)
26. J. Zhang, D.L. Fu, M.B. Chan-Park, L.J. Li, P. Chen, Adv. Mater. **21**, 790–793 (2009)
27. G. Cellot, E. Cilia, S. Cipollone, V. Rancic, A. Sucapane, S. Giordani, L. Gambazzi, H. Markram, M. Grandolfo, D. Scaini, F. Gelain, L. Casalis, M. Prato, M. Giugliano, L. Ballerini, Nat. Nanotechnol. **4**, 126–133 (2009)
28. S.Y. Park, S. Namgung, B. Kim, J. Im, Y. Kim, K. Sun, K.B. Lee, J.M. Nam, Y. Park, S. Hong, Adv. Mater. **19**, 2530–2534 (2007)
29. M.J. Jang, S. Namgung, S. Hong, Y. Nam, Nanotechnology **21**, 235102 (2010)
30. M. Shim, N.W.S. Kam, R.J. Chen, Y.M. Li, H.J. Dai, Nano Lett. **2**, 285–288 (2002)
31. P. Galvan-Garcia, E.W. Keefer, F. Yang, M. Zhang, S. Fang, A.A. Zakhidov, R.H. Baughman, M.I. Romero, J. Biomater. Sci.-Polym. Ed. **18**, 1245–1261 (2007)
32. M.A. Correa-Duarte, N. Wagner, J. Rojas-Chapana, C. Morsczeck, M. Thie, M. Giersig, Nano Lett. **4**, 2233–2236 (2004)
33. H. Hu, Y.C. Ni, V. Montana, R.C. Haddon, V. Parpura, Nano Lett. **4**, 507–511 (2004)
34. L.P. Zanello, B. Zhao, H. Hu, R.C. Haddon, Nano Lett. **6**, 562–567 (2006)
35. K. Wang, H.A. Fishman, H.J. Dai, J.S. Harris, Nano Lett. **6**, 2043–2048 (2006)
36. C.Y. Tay, H.G. Gu, W.S. Leong, H.Y. Yu, H.Q. Li, B.C. Heng, H. Tantang, S.C.J. Loo, L.J. Li, L.P. Tan, Carbon **48**, 1095–1104 (2010)
37. S. Namgung, K.Y. Baik, J. Park, S. Hong, ACS Nano **5**, 7383–7390 (2011)
38. Y.S. Chen, G.H. Hsiue, Biomaterials **34**, 4936–4944 (2013)
39. J.L. Mckenzie, M.C. Waid, R.Y. Shi, T.J. Webster, Biomaterials **25**, 1309–1317 (2004)
40. R.L. Price, K. Ellison, K.M. Haberstroh, T.J. Webster, J. Biomed. Mater. Res. Part A **70A**, 129–138 (2004)
41. D. Khang, M. Sato, R.L. Price, A.E. Ribbe, T.J. Webster, Int. J. Nanomed. **1**, 65–72 (2006)
42. A.K. Shalek, J.T. Robinson, E.S. Karp, J.S. Lee, D.R. Ahn, M.H. Yoon, A. Sutton, M. Jorgolli, R.S. Gertner, T.S. Gujral, G. Macbeath, E.G. Yang, H. Park, Proc. Natl. Acad. Sci. U. S. A. **107**, 1870–1875 (2010)
43. F. Patolsky, B.P. Timko, G.H. Yu, Y. Fang, A.B. Greytak, G.F. Zheng, C.M. Lieber, Science **313**, 1100–1104 (2006)
44. B.Z. Tian, T. Cohen-Karni, Q. Qing, X.J. Duan, P. Xie, C.M. Lieber, Science **329**, 830–834 (2010)
45. J. Park, S. Bauer, K. Von Der Mark, P. Schmuki, Nano Lett. **7**, 1686–1691 (2007)
46. J. Park, S. Bauer, K.A. Schlegel, F.W. Neukam, K. Von Der Mark, P. Schmuki, Small **5**, 666–671 (2009)
47. J. Park, S. Bauer, P. Schmuki, K. Von Der Mark, Nano Lett. **9**, 3157–3164 (2009)
48. K. Heo, C.J. Kim, M.H. Jo, S. Hong, J. Mater. Chem. **19**, 901–908 (2009)
49. W.I. Park, J. Yoo, J. GC Yi, Korean Phys. Soc **46**, L1067–L1070 (2005)
50. S. Oh, K.S. Brammer, Y.S.J. Li, D.Y. Teng, A.J. Engler, S. Chien, S.H. Jin, Proc. Natl. Acad. Sci. U. S. A. **106**, 2030–2135 (2009)
51. C.R. Martin, Science **266**, 1961–1966 (1994)
52. G.F. Zou, H. Li, Y.G. Zhang, K. Xiong, Y.T. Qian, Nanotechnology **17**, S313–S320 (2006)
53. P.H. Yeh, Z. Li, Z.L. Wang, Adv. Mater. **21**, 4975–4978 (2009)
54. Z. Li, R.S. Yang, M. Yu, F. Bai, C. Li, Z.L. Wang, J. Phys. Chem. C **112**, 20114–20117 (2008)
55. J.Y. Lee, B.S. Kang, B. Hicks, T.F. Chancellor, B.H. Chu, H.T. Wang, B.G. Keselowsky, F. Ren, T.P. Lele, Biomaterials **29**, 3743–3749 (2008)
56. T.D. Zaveri, N.V. Dolgova, B.H. Chu, J.Y. Lee, J.E. Wong, T.P. Lele, F. Ren, B.G. Keselowsky, Biomaterials **31**, 2999–3007 (2010)
57. J.K. Park, Y.J. Kim, J. Yeom, J.H. Jeon, G.C. Yi, J.H. Je, S.K. Hahn, Adv. Mater. **22**, 4857–4861 (2010)
58. W. Hallstrom, M. Lexholm, D.B. Suyatin, G. Hammarin, D. Hessman, L. Samuelson, L. Montelius, M. Kanje, C.N. Prinz, Nano Lett. **10**, 782–787 (2010)
59. W. Hallstrom, T. Martensson, C. Prinz, P. Gustavsson, L. Montelius, L. Samuelson, M. Kanje, Nano Lett. **7**, 2960–2965 (2007)

60. F. Johansson, M. Jonsson, K. Alm, M Kanje. Exp. Cell Res. **316**, 688–694 (2010)
61. N. Gao, H.J. Wang, E.H. Yang, Nanotechnology **21**, 105107 (2010)
62. D. Choi, A. Fung, H. Moon, D. Ho, Y. Chen, E. Kan, Y. Rheem, B. Yoo, N. Myung, Biomed. Microdevices **9**, 143–148 (2007)
63. W. Luo, M.N. Yousaf, Chem. Commun. 1237–1239 (2009)
64. J. Lee, L.K. Schwarz, C.K. Akkan, M.M. Miró, O.T. Abad, K.-H. Schäfer, M. Veith, C. Aktas, Phys. Status Solidi A-Appl. Mat. **210**, 952–956 (2013)
65. M.A. Bucaro, Y. Vasquez, B.D. Hatton, J. Aizenberg, ACS Nano **6**, 6222–6230 (2012)

11

Development of flagella bio-templated nanomaterials for electronics

Wonjin Jo[1], U Kei Cheang[2] and Min Jun Kim[2*]

Abstract

Bacterial flagella with their unique structural properties have proven to be promising bio-templates and can be exploited for the creation of nanomaterial with very high aspect ratio and surface area. Their chemically modifiable surfaces allow the flagella be modified to possess electrical/electronic properties. Their extraordinary physical properties along with the many possibilities for manipulation make them ideal systems to study for the purpose of developing nanoelectronics. First, this article reviews the characteristics of bacterial flagella and their utilization as biologically inspired templates. Next, the use of bio-templates for electronic systems such as dye-sensitized solar cell and lithium ion battery is discussed. Finally, we show the future directions for the use of flagella biotemplatednanomaterials for applications in electrical engineering fields.

Keywords: Flagella; Bio-template; Nanomaterial; Dye-sensitized solar cell; Lithium ion battery; Electronic

1 Background

The transition to renewable energy is met with the growing demand to develop new generations of electronic devices with improved efficiency and environmental performance in order to address the concerns regarding limited energy supplies and environmental issues. Moreover, the challenges in manufacturing extremely small features on mass-produced multifunctional devices incite the need to search for novel methods for the synthesis of the small scaled material integrated with high performance of functionality. As a result, the interface between nanosystems and biosystems is emerging as one of the largest and most active areas of technology to address the requirements and specifications for developing renewable energy [1,2]. The combination of bio and nano hold the promise to yield advances in the creation of new and powerful methods that enable direct, sensitive, and rapid energy generation [3,4]. In this respect, devices using on nanomaterials synthesized by biologically inspired templates can be one of the most promising platforms for direct ultrasensitive electrical detection and for building functional interfaces of core components [5]. Biological templates have self-assembled hierarchical structures and can be massively obtained in nature, allowing the massive fabrication of nanostructured materials with precise dimensions and structures. Furthermore, biologically templated nanomaterial can be easily modified to change their natural functions and properties to create new types of highly efficient electronic devices, such as dye-sensitized solar cells (DSSCs) or lithium ion batteries (LIBs), with lightweight and high performance.

Using bacterial flagella as biologically inspired templates is a relatively new approach to assemble nanoscale architectures, which can revolutionize the fabrication process and the materials used in ultra-densely integrated electronic devices. Of particular significance is the finding that it is possible to selectively metallize the flagella templates or flagella based templates to obtain electrical properties through interelement positioning at the nanometer scale [6,7]. However, even though their natural tubular nanostructure and the ease manipulation of their surface properties have already been well demonstrated, there is a lack of progress in the field of flagella-based device fabrication. To stimulate advances in using flagella as bio-templates in these forward-looking fields, this article will examine three main sections. The first section explains the characteristics of bacterial flagella to highlight their attractive properties that can be exploited for nanofabrication. Next, this article will introduce current electronic devices, including DSSCs and LIBs, built using

* Correspondence: mkim@coe.drexel.edu
[2]Department of Mechanical Engineering & Mechanics, Drexel University, 3141 Chestnut Street, Philadelphia, PA 19104, USA
Full list of author information is available at the end of the article

other bio-templates such as virus or diatom. In the last section, we will discuss recent developments in the flagella based bio-templates and present the future prospects of applying them in the electrical areas.

1.1 Characteristics of bacterial flagella

Bacterial flagella, such as those from *Escherichia coli* (*E. coli*) or *Salmonella typhimurium* (*S. typhimurium*), are organelles used for propulsion. Bacterial flagella are self-assembled helical nanostructures, approximately 20 nm in diameter and 10 μm in length. The flagella have extraordinary mechanical properties; they are extremely stiff and have an elastic modulus estimated to be around 10^{10} N/m^2 [8-10]. As a result, they have remarkable durability and stability. In addition, they have the properties to be active nanostructures. They are able to actively adapt to the environment by changing helical handedness and pitch in response to various external stimuli; this is known as polymorphic transformation. Specifically, bacterial flagella can undergo polymorphic transformations both in loaded and unloaded conditions due to chemical, electrical, thermal, mechanical, or optical stimulation. Furthermore, flagellar filaments' remarkable durability and stability allow them to withstand high temperature (up to 60°C) and extreme pH (7 ± 4) [11-13].

1.2 Structure of flagellar filaments

The flagellar filaments are polymers consisting of a single type of protein called flagellin, which have a molecular weight of about 55,000 Da, an outer diameter of 120–250 Å, and an inner diameter of 20 Å [14]. They are devoid of the amino acids cysteine or tryptophan and of any known enzymatic activity [15]. The flagellin protein is made up of four domains: D0, D1, D2, and D3 (Figure 1a). Of the four domains, D0 and D1 are essential for the assembly of the filament and for the change of conformation leading to polymorphic transformation [16]. The D3 domain is responsible for the antigenic response of bacteria [17]. As mentioned, the flagella can take on a number of polymorphic configurations. A filament with normal configuration is left-handed with a helical pitch of about 2.2 μm, a helical diameter of about 0.4 μm, and 5340 flagellin subunits per turn. The surface lattice of the filament is hexagonal, and all the subunits can be traced with 1-start, 5-start, 6-start, or 11-start helices. The 11-start helices (protofilaments) are nearly longitudinal. The filaments are helical because the subunits can bond in two different ways, with each protofilament forming a cooperative unit. Filaments with only one type of protofilament are straight. In L-type straight filaments the 11-start helices twist about 2.5° to the left, while in R-type straight filaments twist about 7° to the right. The intersubunit distance is 52.5 Å (long) for the L-type and 51.8 Å (short) for the R-type. It is thought that all the flagellins in one protofilament are of only one-bonding type, forming either a long or short protofilament. The mechanical strain in a filament is minimized if protofilaments of the same type are adjacent to one another [18,19]. This leads to the prediction of 12 polymorphic forms, 2 straight, and 10 helical, as shown in Figure 1b.

1.3 Polymorphic transformation of flagellar filaments

Given their structural makeup, filaments of different polymorphic forms can be obtained by using different flagellin types from different bacterial strains or by mixing different quantities of flagellin obtained from L-type and R-type straight filaments [23]. However, a filament's polymorphic configuration depends not only on the flagellin type, it also depends upon environmental conditions (i.e. pH, ionic strength, and temperature) as established in a series of experiments by Kamiya and Asakura [11,12,24]. Most of these experiments were performed with wild-type strains of *Salmonella*, but identical results were obtained with an *E. coli* K12 strain [25]. As an example, filaments in 0.1 M KCl buffer expose to a raise in pH from 7 to 9 changes from the normal form to the coil form, which consequentially reduces their end-to-end length by about a factor of nearly 3. The flagellar filament also have an induced dipole moment of 5×10^{-24}C·m in an electric field of $E = 10^6$ V/m [26]. For example, straight polymorphic filaments align along the field, but close-coiled forms align with the helical axis perpendicular to the field [26]. Under the appropriate condition, the filaments can even go so far as to change the handedness of the flagellar helix. Such astonishing degree of actuation opens up the possibility to utilize flagellar filaments for a broad range of applications.

1.4 Flagella depolymerization and repolymerization

The length of the flagellar filaments can be modulated using depolymerization and repolymerization processes as shown in Figure 2a; in other word, flagella can be broken down into monomers (depolymerization) and then reassembled (repolymerization) with the desired length [7,27]. This can be achieving in an in vitro stepwise procedure where the parameters of the process can be manipulated in order to control the length of the flagella. First, the flagella filaments are harvested from bacterial cells in a saturated culture by shearing the filaments off from the cells using a vortex. The filaments are then isolated form the cell bodies by differential centrifugation. A portion of the isolated flagella is sonicated to form short segments of flagella, which will be used later as seeds to initiate repolymerization, while the rest is heated to 65°C for complete depolymerization to flagellin. Next, the flagella are prepped for repolymerization by adding the flagella seeds to the flagellin solution. The flagella are

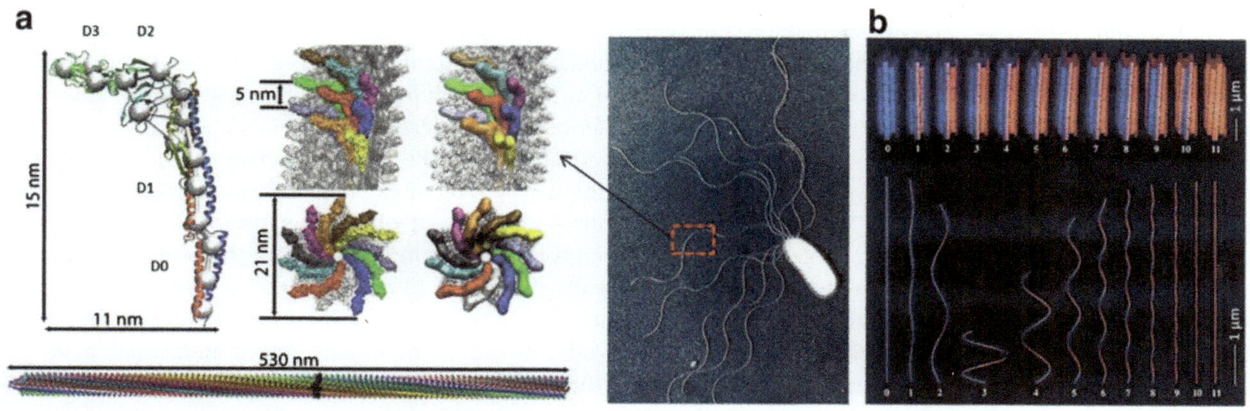

Figure 1 Structure and polymorphic transformation of flagellar filaments. (a) Single flagellin monomer and arrangement of the monomers in the filament [20]. Reprinted with permission from ref. 20, Copyright © 2006, Elsevier. **(b)** Bacterial flagellar filament polymorphic forms, shown in a curvature vs. twist plot. There are two straight forms, L (Left) and R (Right), and 10 helical forms. Of the named helical forms, normal and coiled are left-handed and the others are right-handed [21,22]. Reprinted with permission from ref. 21, Copyright © 1998, Elsevier.

left to repolymerize overnight. These filaments grow unidirectionally from one end of the seed, so it is straightforward to construct simple block copolymers by changing the types of flagellin in the supernatant fraction of the solution. When the filaments are first detached from cells by vortex, they are generally shorter than 10 µm with a broad distribution of length. However, they can be repolymerized to give a majority in the range of 10 – 25 µm, with some as long as 75 µm [22]. Once made, they can be stored for months in the polymerization buffer. Figure 2b shows optical micrograph of fluorescently-labeled flagellar filaments repolymerized from *S. typhimurium.*

1.5 Biomaterials as templates for the application of electronics

Recently, there is a fast growing interest in biological template technology, such as the use of bacteria [28,29], virus [30-32], diatoms [33,34], butterfly wings [35,36], and so on. Most biological species possess a variety of unique characteristics such as distinctive nano or microstructures. Thus, it is very desirable to utilize these naturally formed unique structures as templates to create complex inorganic structures for different applications, such as nanoelectronics. Biotemplating is an eloquent and economic way to massively synthesize hierarchically periodic

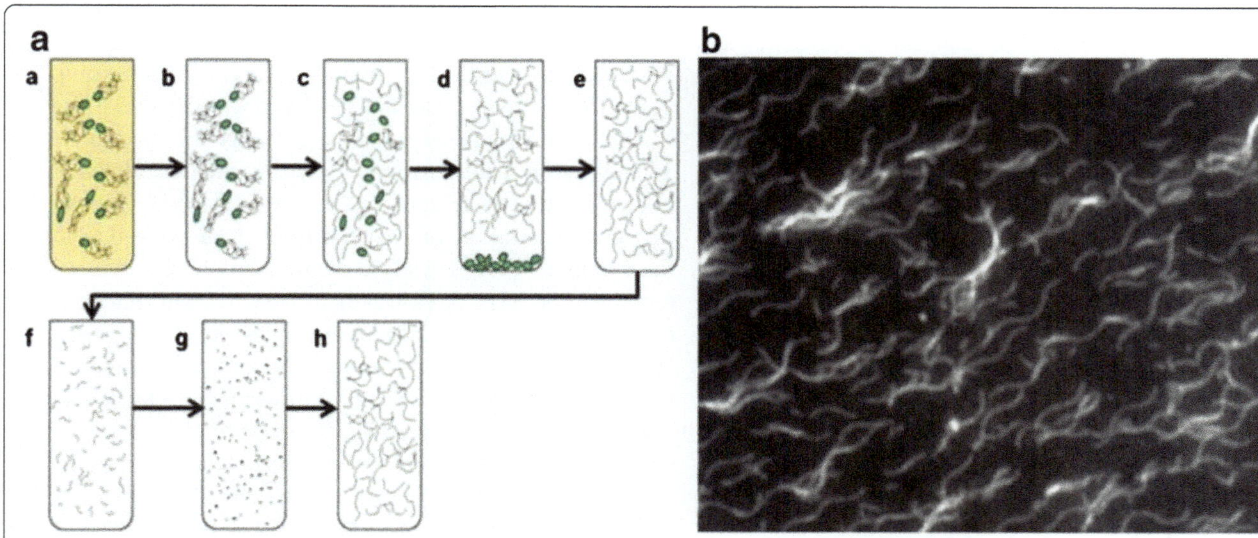

Figure 2 Flagella depolyerization and repolymerization procedures. (Left) Method for the preparation of flagellar filaments; (a) bacteria are grown for 10 hours in LB broth; (b) centrifuged and resuspended in PBS; (c) vortexed to separate flagellar filaments; (d) bacterial cell bodies are pelleted by ultracentrifugation; (e) the supernant is resuspended in PBS; (f) the flagellar filaments are broken into smaller pieces by sonication; (g) depolymerized by heating; (h) the seeds are added to the solution of flagellin and repolymerization is carried out for 24 hours. (Right) Scanning electron microscopy (SEM) images of flagellar filaments repolymerized from S. typhimurium. Reprinted with permission from ref. 7, Copyright © 2009, Elsevier.

micro and nanostructures for electrical parts without using complex and expensive processes, making the biotemplate technology ideal for the synthesis of high surface area nanomaterials. In this section, we will review some recent works on DSSC and LIB devices built using biotemplate based nanomaterials.

1.6 Dye-Sensitized Solar Cell (DSSC)

The negative impacts from existing energy source, such as environmental impacts, and the increasing scarcity of energy resources have prompted the need for cheap, renewable, and clean energy technologies. Solar energy has been a major focus of modern science and engineering, especially with the threat of global warming and fossil fuel depletion. Each year, 3×10^{24} Joules of energy radiates from the sun to the earth [37]; this manifested the idea to harness such large amount of energy through the use of solar technologies. However, existing solar technologies are expensive when compared with hydrocarbon fuels, which their limits widespread commercialization. In order to increase the appeal of solar cells, their efficient must be increased and their cost must be decreased. Over the last decade, a type of photovoltaic device called the dye-sensitized solar cell (DSSC) had emerged as a technology that can achieve reasonably high efficiency and low manufacturing cost. Unlike conventional p-n junction diode based photovoltaics, the DSSCs create photovoltaic effect by photoexcitation of dye molecules that then inject electrons into wide band gap semiconductors that are not themselves able to be photoexcited. The performance of the DSSCs can be enhanced by optimizing the structure of the nanoparticle layers (TiO_2 or ZnO) in the photoanode, the photosensitizing dye, and the electrolyte. In a typical particle based DSSC, the solar conversion efficiency is directly related to the electron percolation through the nanoparticle layers. However, the structural disorder at the contact between two crystalline NPs leads to enhanced scattering of free electrons. Therefore, it reduces electron mobility resulting in slow percolation which may lead to a decrease in efficiency [38].

Recently, much work had been performed to improve the optical path length of photoanodes by modifying surface structure or properties of photoanode [39]. Since the light harvesting efficiency is directly affected by the photoanode properties, the optimization and improvement of photoanode plays an essential role in the performance of DSSCs [40]. One approach was to replace the two dimensional flat surface with a three dimensional structure consisting of a dense array of aligned nanoscale features, such as the hierarchically self-assembled nanomaterial. The array of nanostructures created a larger surface areas compared with the flat surface; as a result, the optical path length was increased which in turn enabled rapid collection of carriers and achieved better light-accumulation effects [41]. Furthermore, the increase of the optical path length inside DSSCs increases the probability of interaction between dye molecules which is essential to yield high light absorption and to facilitate the transfer of photo-generated electrons.

To advance along the aforementioned approach, several nanotubular architectures have been investigated for the potential enhancement of electron percolation pathways, light conversion, and ion diffusion at the semiconductor-electrolyte interface [38,42,43]. Due to their lower trap density and more direct path to the current collecting electrode, arrays of nominally one-dimensional nanostructures are expected to speed up charge migration without adverse effects from recombination [44]. In an effort to further increase the power conversion efficiency, biotemplate based nanomaterials were developed in order to create paths for quick electron transfer to the conducing substrate. One such template was the M13 bacteriophage, which is a filamentous bacteriophage with a length of 860 nm and a diameter of 6.5 nm. Since the protein capsid from the M13 bacteriophage can be genetically modified with specific peptides, it allowed the formation of nanoparticles, metal oxides, or other materials [3,45]. Recently, Dang et al. integrated the single-walled carbon nanotubes (SWNTs) into photovoltaic devices by using M13 virus templates as binding agents [46]. A schematic of the fabrication process is given in Figure 3. Through the use of virus templates, SWNTs was bounded and stabilized through non-covalent binding; preserving high electron mobility by eliminating the need for surfactants and surface chemical modifications which hamper electron transfer. Additionally, the virus/SWNT complex enabled direct contact between the SWNTs and titanium dioxide (TiO_2) which in turn improved the power conversion efficiency. In addition to their applications as a binding agent, the M13 virus was also used as a sacrificial template to create an optimized interface between the electrolyte and the TiO_2 nanoparticles [47]; this created effective electron pathways (Figure 4a). In a typical DSSC system, uniform penetration of electrolytes is inadequate due to the randomly placed TiO_2 nanoparticles. As a sacrificial template, the M13 virus was first used to form interconnections with the TiO_2 nanoparticles via electrostatic binding; then the M13 virus was thermally decomposed to leave hollow channels as pathways for electrolyte penetration. Consequently, efficient electron pathways were created via the interconnected nanochannel structures and, as a result, an increased in the photo conversion efficiency was achievedover the conventional TiO_2 photoelectrodes by 35% as shown in Figure 4b.

Diatoms, unicellular photosynthetic microorganisms, are good candidates for the DSSCs application due to their unique nanostructured patterns of silica cell walls. The size of diatoms ranges from 1 to 100 μm. The diatom silica

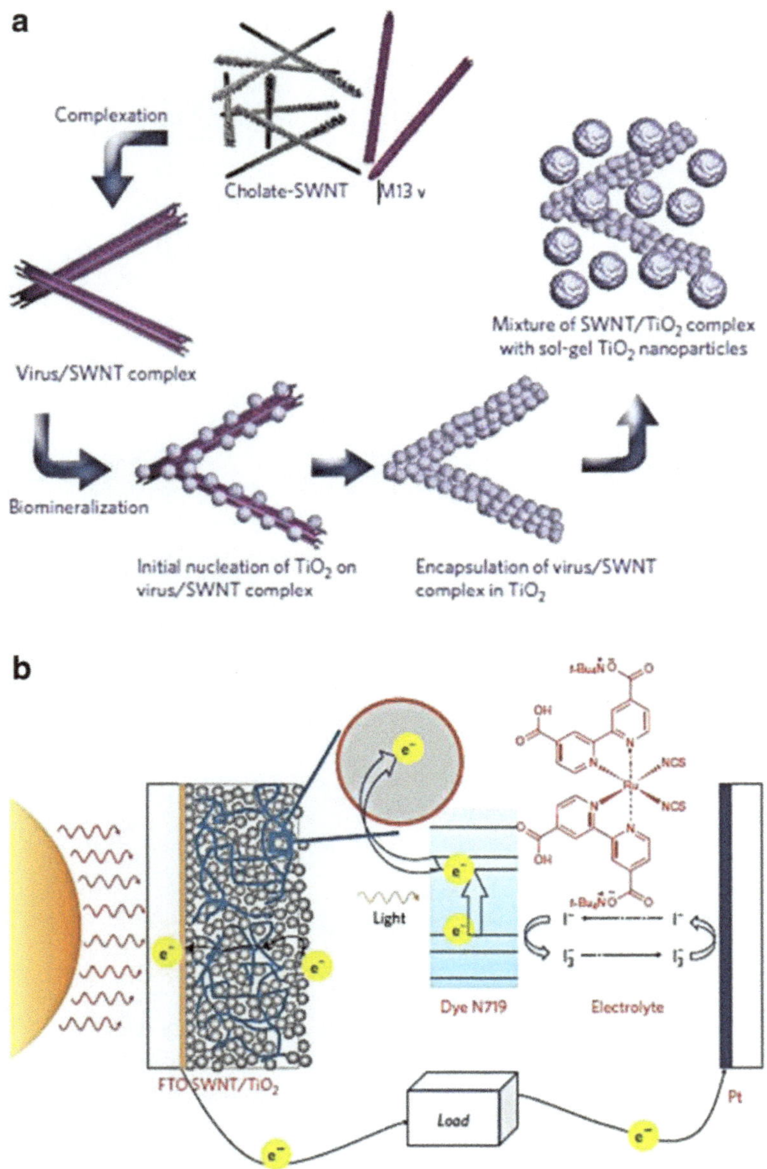

Figure 3 M13 virus-based SWNT/ TiO_2 complex was used in DSSCs. (a) Schematic diagram illustrating the fabrication process of **(a)** M13 virus-based SWNT/TiO_2 complex. **(b)** DSSCs incorporating the complex. Reprinted with permission from ref. 46, Copyright © 2011, Rights Managed by Nature Publishing Group.

walls could be chemically functionalized to interact with inorganic nanoparticles. Recently, Jeffryes et al. used diatom frustules, which are nanostructured amorphous silica skeleton, as templates for the deposition of nanostructured TiO_2 (Figure 5) [33]. The process to create nanostructured TiO_2 began with the deposition of poly-L-lysine (PLL)-TiBALDH. Next, the PLL was absorbed by the diatom silica. Finally, the PLL precipitated the soluble titanium precursor, TiBALDH, into insoluble TiO_2, resulting in the formation of amorphous titania nanoparticles. As a result, amorphous TiO_2 nanoparticles completely packed the frustule pores bringing about TiO_2loading on the frustules. The resulting diatom-TiO_2 composite material was then used in a DSSC photoanodes which effectively increased the efficiency by nearly 3 times compared with the photoanode made from only TiO_2 nanoparticles [48].

Studies of the use of butterfly wings as the biological scaffolds for the construction of electrical devices are also noteworthy for their potential uses in DSSCs. Butterfly wings exhibit periodic submicrometer structures which can be used to create effective solar collectors [49]. The

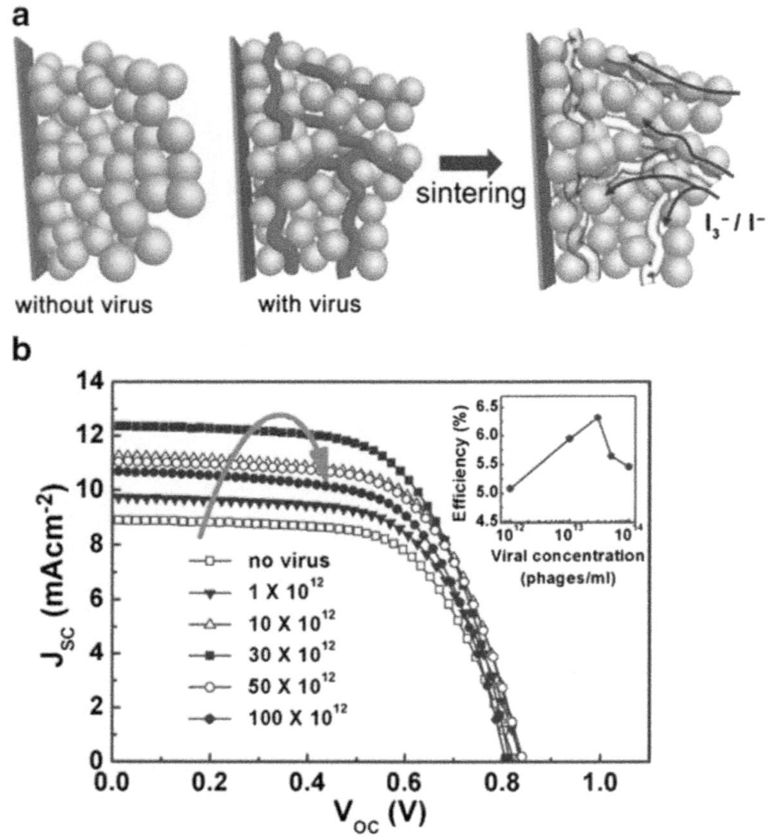

Figure 4 DSSC TiO_2 photoelectrode using M13 viruses. (a) Schematic illustration of DSSC photoelectrode fabrication process shows conventional TiO_2 nanoparticle photoelectrode and enhanced TiO_2 photoelectrode using M13 viruses as sacrificial templates. **(b)** Current density-voltage characteristics of electrodes at various virus concentrations. Reprinted with permission from ref. 47, Copyright © 2011 WILEY-VCH Verlag GmbH & Co. KGaA, Weinheim.

butterfly wing scales are approximately 150 μm long and ~50 μm wide; typically have highly ordered architectures with pores and layers. The wing scales' micro features are of great importance because they can be used to create templated microstructures that can improve the light absorptivity [35,50] by increasing the surface area. Inspired by the butterfly wing scales, Zhang et al. developed a novel titania film photoanode as described in Figure 6a [40]. Three separate photoanodes with different structures were synthesized onto a fluorine-doped tin-oxide-coated glass substrate using butterfly wings as biotemplates: Quasi-honeycomb like structure (QHS), shallow concavities structure (SCS), and cross-ribbing structure (CRS). Figure 6b presents the field emission scanning electron microscopy (FESEM) images of the synthesized QHS (upper images) and SCS (below images) structures. By investigating their wavelength-dependent absorptions and current-potential curve measurements, Zhang et al. demonstrated that the QHS titania replica photoanode has an optimal light absorptivity (Figure 6c) and higher surface area, which is extremely advantageous to light harvesting and dye absorption in the DSSCs.

1.7 Lithium Ion Battery (LIB)

To meet the demand for a clean and secure electrical energy storage unit, extensive research efforts have been made to develop rechargeable batteries with large capacity, high power, and reasonable life cycle. The most notable are the LIBs due to a number of significant advantages over competing technologies [51]. The use of lithium intercalation/alloying compounds had successfully enhanced the cycling life, safety, capacity, charge/discharge rates capability of the battery, as well as other battery characteristics such as high energy density and flexibility in shape and size. The three primary components of a LIB are the cathode, the anode, and the electrolyte. The operation of the LIB involves a electrochemical process where lithium ions migrate between the cathode and the anode while electrons flow through a close external circuit. Electrical energy is provided to the connected device via discharge where lithium ions move out of the anode (extraction) and into the cathode (intercalation). These reactions take place continuously while electrons continue to flow. Electrical energy can be restored to the battery through recharge, which is simply a reversal of the discharge process.

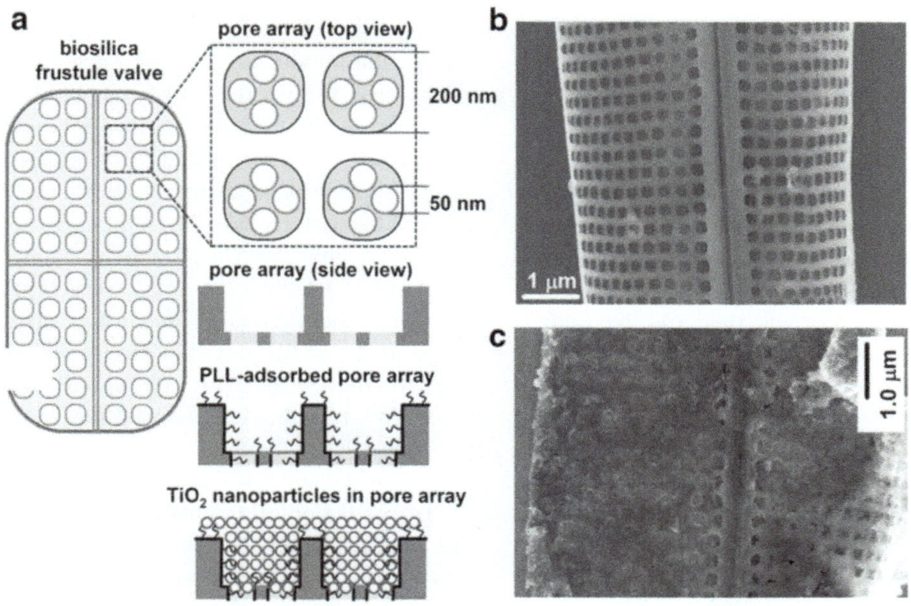

Figure 5 TiO_2 –coated diatom frusule. (a) Scheme for the PLL-mediated deposition of TiO_2 on the diatom frustule biosilica. **(b)** SEM image of bare diatom frustule. **(c)** SEM image of TiO_2 –coated diatom frusule. Reprinted with permission from ref. 48, Copyright © Cambridge University Press 2011.

Though LIBs are generally smaller and lighter than other rechargeable batteries, the growing demand for portable electronics, such as cell phones and laptops, have facilitated the need to develop even smaller batteries for the propose of scaling down portable devices. However, it is still a challenge to breach the limitations of LIBs to create smaller batteries without sacrificing performance. Fortunately, the emerging technologies to fabricate and pattern nanostructures have incited a promising solution. The high surface area offered by patterned nanostructures can significantly increase the electrode/electrolyte contact areas, improved mechanical stability, and reduced distances

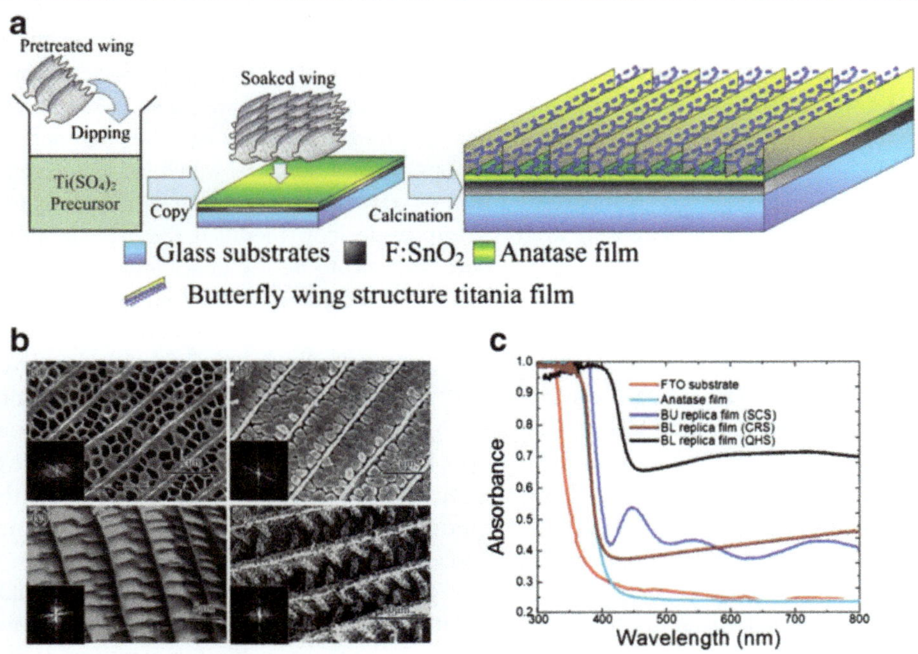

Figure 6 Photoanode template using butterfly wings. (a) Schematic illustration of the fabrication process of quasi-honeycomb structure photoanode template from butterfly wings. **(b)** FESEM images of photoanodes with different structures. **(c)** Absorption characteristics show by the wavelength-dependent absorption at the different photoanode structure. Reprinted with permission from ref. 40, Copyright © 2009, American Chemical Society.

for electron and ion transport to facilitate faster reaction kinetics. Furthermore, biogenic, bio-enabled, and bio-inspired hierarchical nanomaterials can be used to further increase reactive surface area. For instance, biological template such as viruses have proven to be extremely effect in the mass production of self-assembled hierarchically ordered structures with high surface to volume ratio. The advantages of using biotemplate technology may hold the key to a new class of electrode materials.

The tobacco mosaic viruses (TMV) are known as useful biotemplates because their size and rigid rod-like bodies which are 300 nm long and 18 nm wide and contain an internal channel of 4 nm in diameter. The biological nanostructures of viral templates with or without decoration by metal nanoparticles have been extensively studied. Royson et al. used genetically engineered TMV as robust templates for electroless deposition of nickel and cobalt [52]. The vertically patterned TMVs on a gold surface allowed for efficient and uniform metal coating, which resulted in more than 10-fold increase in available surface area. Evidently, the virus-templated electrode surfaces in a battery system showed enhanced electrode capacity, electrode life, and voltage output. Chen et al. also developed genetically engineered TMV templates to form a high capacity silicon anode with internal nickel current collector [53]. The resultant TMV/Ni/Si anodes based on high aspect ratio nanostructured surfaces produced high capacities, excellent cycle-life, and consistent rate capabilities. In these works, TMVs were genetically engineered to assign specific affinities to the targeted materials. Generally modified TMVs, which have the cysteine proteins on their surfaces, can be easily attached onto specific metal surfaces in electroless plating solutions. This is caused by the strong covalent-like interactions between the thiol groups of the cysteine protein and the metal ions. By taking advantage of the powerful binding and stabilizing properties of TMV, vanadium oxide (V_2O_5), a broadly utilized cathode material, was easily deposited on the nickel-coated TMV templates using atomic layer deposition process for the enhancement of cathode performance (Figure 7a and 7b) [54]. With the TMV-templated V_2O_5, higher capacities were achievable through the enhanced surface area resulting in the increase of active battery material loading as shown in Figure 7c and 7d. This also led to excellent electrochemical performance including high excellent rate capability and cycling stability. Like the TMV biotemplates, the M13 viruses are another potential choice for viral templates. The M13 virus templates were used to synthesize and assemble hybrid gold-cobalt oxide nanowires for advance LIB electrodes (Figure 8) [55]. Similar to the TMV templates, the M13 viruses were genetically engineered to promote strong binding affinity to cobalt ions. The virus-mediated hybrid composites generated higher initial and reversible lithium storage capacity than the pure cobalt oxide nanowires at the same current rate.

Given the importance of electrodes in battery systems, separators is also naturally a key component of LIBs since separators are directly associated with power performance and cell safety. An interesting study was performed using diatom-inspired silica coating on the surface of polyethylene (PE) separators [56]. The approach involved surface growth of a silica layer inspired by diatoms; the interfacial adhesion of the layer to the surface was inspired by mussel adhesion. The fabricated separator membranes were shown in the SEM images in Figure 9a. The silica coating increased the electrolyte wettability of the separator (Figure 9b), resulting in enhanced power, thermal shrinkage (Figure 9c), and safety.

1.8 Flagella biological templating technology

1.8.1 Flagella bio-templated nanomaterials and metallization

Compared to the viral bio-templates, bacterial flagella templates have not been demonstrated to be used to create nanomaterials for electronic devices. However, recent promising studies had shown that flagella can be a good candidate as bio-templates. In their natural state, flagella exhibit self-assembled spiral and tubular nanostructures with extremely small outer diameter (20 nm) and controllable length. Aside from having attractive structural properties, flagella are extremely stiff and highly durable; they are stable under harsh conditions such as high temperature and extreme pH. Additionally, the solvent-exposed domains (D2 and D3) of the flagellin subunits are genetically modified, enabling simple interactions with external materials via chemical or genetic processes. In the context of fabricating biologically inspired flagella templates, some studies were also focused on using genetically engineered flagella for the well-defined functional groups on their surfaces, which led to the generation of ordered arrays of nanoparticles or monodisperse nanotubes. The peptide loop modified flagella nanotubes have been demonstrated to be useful scaffolds with high affinities for metal ions including Cu(II), Au(I), Co(II), Cd(II), and Pd(II) [57] and for the immobilization of Pd and Au nanoparticles [40,58]. Genetically modified flagella have favorable in that their surfaces can be selectively functionalized to have strong affinities with different types of nanoparticles; however, the process can be complex and require expertise.

On the other hand, Kim group recently developed methods for metal oxide layer deposition on flagella surface in aqueous solution without genetic modification [7,27]. They demonstrated deposition of TiO_2 on flagella isolated from *S. typhimurium* via a biomimetic mineralization process under the temperatures of 40°C and 50°C [7]. Also, methods were develop to utilize flagella biotemplates for fabrication of silica mineralized nanotubes [27]. The process involved hydrolysis and condensation

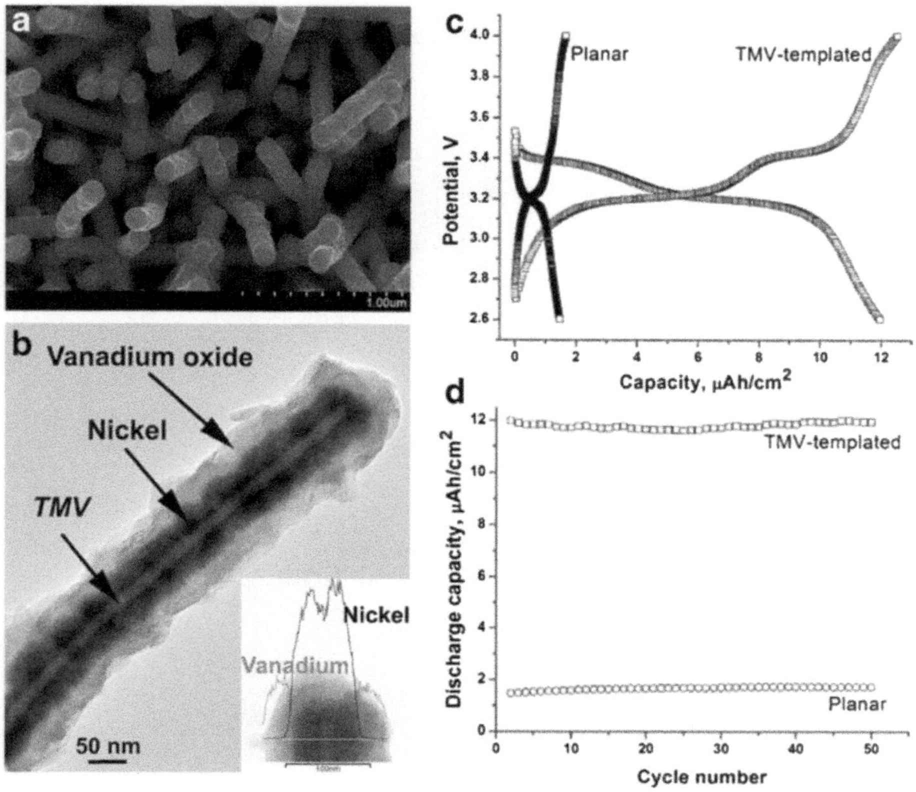

Figure 7 TMV-templated V_2O_5 electrodes. (a-b) SEM and transmission electron microscopy (TEM) image of the TMV-templated Ni/V_2O_5 core/shell cathode. **(c)** Discharge and charge curves and **(d)** the cycling performance of the cells with TMV-templated V_2O_5 electrodes. Reprinted with permission from ref. 54, Copyright © 2012, Elsevier.

reaction which was initiated by pretreatment of flagella with amino-propyltriethoxysilane, followed by the addition of tetraethoxysilane at room temperature. Moreover, the biologically derived silica nanotubes were improved and modified by decoration with gold, palladium, and iron oxide nanoparticles through wet chemical methods [6,59]. Form the TEM images shown in Figure 10, one can clearly see the silica nanotubes before (Figure 10a) and after the nanoparticle deposition (Figure 10b-10d). This method can fabricate nanotubes with unique and distinctive properties using fast and simple procedures, without the need for genetic modifications.

1.8.2 Characterization of electrical properties of flagella-templated nanotubes

The electrical characterization of the mineralized and metallized flagella based nanomaterials is a key step towards the realization of using flagella based nanomaterials as electrical components. Though very little is known about their energetics and electrical properties, the work done by Jo et al. have shed much light on the subject by characterizing the electrical properties of flagella-templated nanotubes [59]. Metallization, such as gold, palladium, or iron oxide nanoparticles, was explored and proven to effectively enhance the nanotubes' electrical conductivity (Figure 11). Given the properties of the selected metal nanoparticles, the application for the metalized nanotubes can be completely different. In particular, the electrical properties of gold incite the possibility for applications in electronics, battery electrodes, and fuel cells. Given their properties, it is possible and desirable to use nanotubes fabricated via flagella biotemplates as electrical materials.

1.8.3 Flagella templated dye-sensitized solar cell

The previous results suggest the possibility that the bacterial flagella can be selected for use in the DSSCs due to their unique properties; in particular, they have controllable length, they can change their helical form through polymorphic transformation, they can be genetically modified, and their surface can be functionalized. The functionalization of the flagella also makes the positioning of flagella nanotubes on a substrate with high precision; possibly more so than positioning nanowire arrays in current DSSCs devices. The poor control over the placement of the nanowires causes the cells to not perform as well as possible. Using bottom-up and top-down nanofabrication techniques [24], it is possible to place the flagella exactly where they are desired, and highly

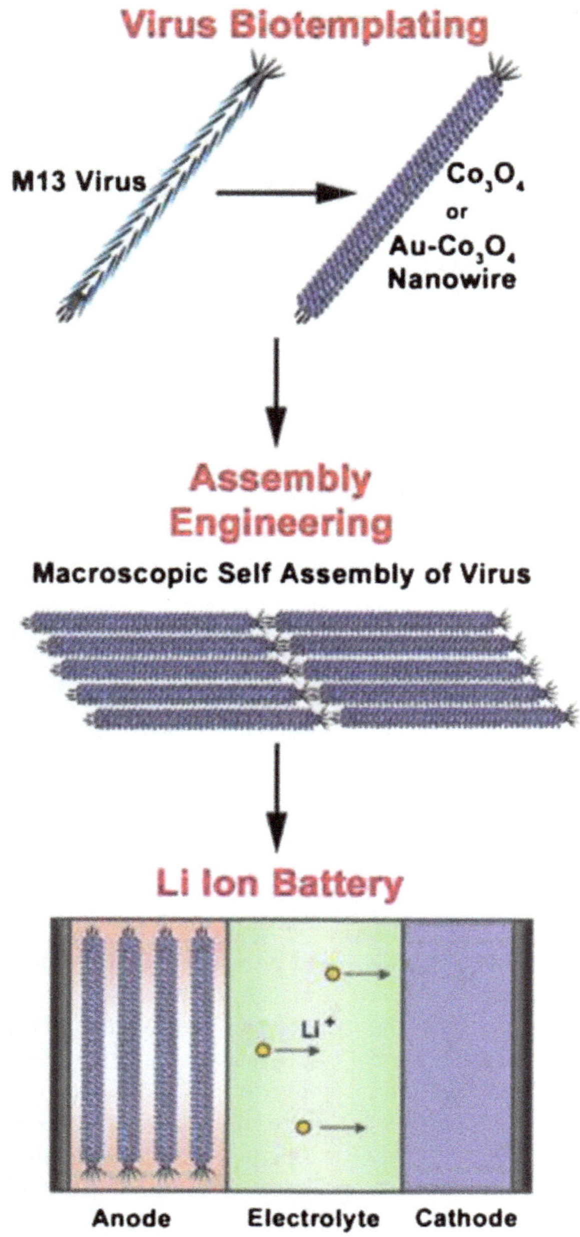

Figure 8 Schematic illustration of the synthesis and assembly of M13 virus template nanowires for negative electrode materials of LIBs. Reprinted with permission from ref. 55, Copyright © 2006, American Association for the Advancement of Science.

ordered arrays can be created, forming a *"flagellar forest"*. The flagella of *S. typhimurium* will be used in the DSSC design due to the well-defined and studied polymorphic forms.

1.8.4 Engineered flagella forest

The flagella forest is a device that harnesses arrays and patterns of flagella filaments. It consists of arrays of biotinylated flagellar filaments attached to an avidin surface treated substrate [60] in a manner where the filaments are tethered at one end. To achieve this, flagellar filaments are first functionalized at one end through biotinylation of flagella seeds followed by polymerization using non-functionalized flagellin [61]. Finally, the one-ended biotinylated flagella are introduced onto an avidinated substrate to create a flagella forest. This results in the construction of well-designed patterns of flagella arrays. The flagella forest design is shown on Figure 12a. To create a silica nanotube forest, the aforementioned flagella mineralization method can be used. Additionally, metallization of the silica nanotube forest can also be achieved.

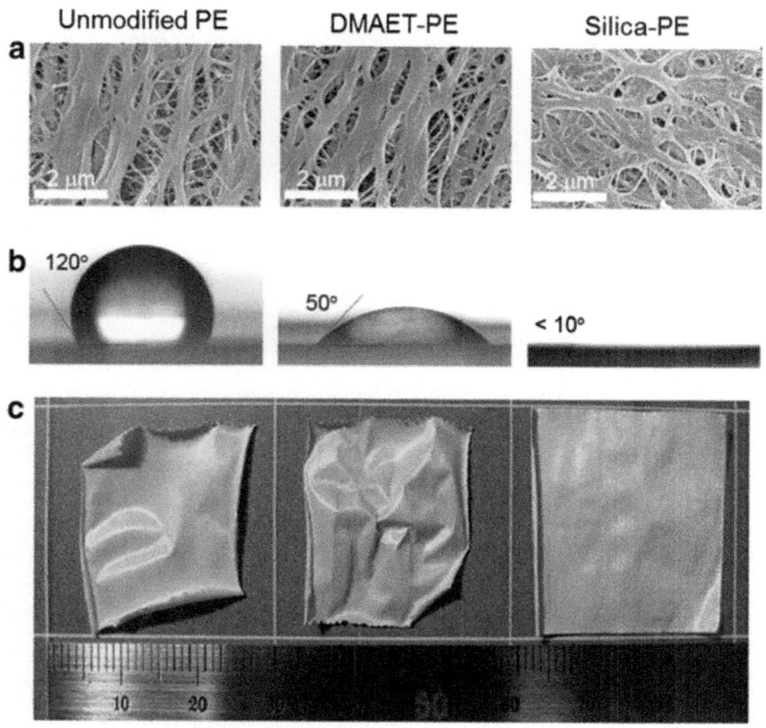

Figure 9 Diatom-inspired silica coating on the surface of polyethylene (PE) separators. (a) SEM images of separator membranes with different surface coating. **(b)** Static water contact angle measurement images shows wettability on the different separators. **(c)** Digital photographs of thermal stability of the different separators after exposure to 140°C for 1 h. Reprinted with permission from ref. 56, Copyright © 2012, American Chemical Society.

1.8.5 Flagellar nanotube forest as a dye-sensitized solar cell

Once the flagellar nanotubes are harness into a nanotube forest device, they can be readily utilized to enhance DSSCs. A schematic of the flagellar nanotube enhanced DSSC is shown on Figure 12b. This device works by photoexciting dye molecules which inject electrons into semiconducting scaffolds which then transport the electrons to the transparent conducting electrode (TCO), upon which incident light enters the device. The dye chosen for this photovoltaic device should be a ruthenium-based

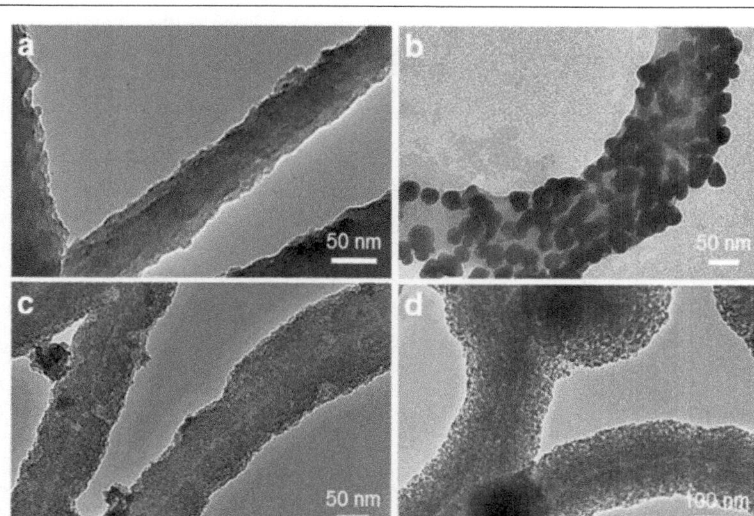

Figure 10 TEM images of silica nanotubes fabricated on flagella bio-templates. (a) Pristine silica nanotube and **(b-d)** metallized silica nanotubes by the deposition of gold, palladium, and iron oxide nanoparticles. Copyright © 2013 IOP Publishing (Ref. 59). Reproduced by permission of IOP Publishing. All rights reserved.

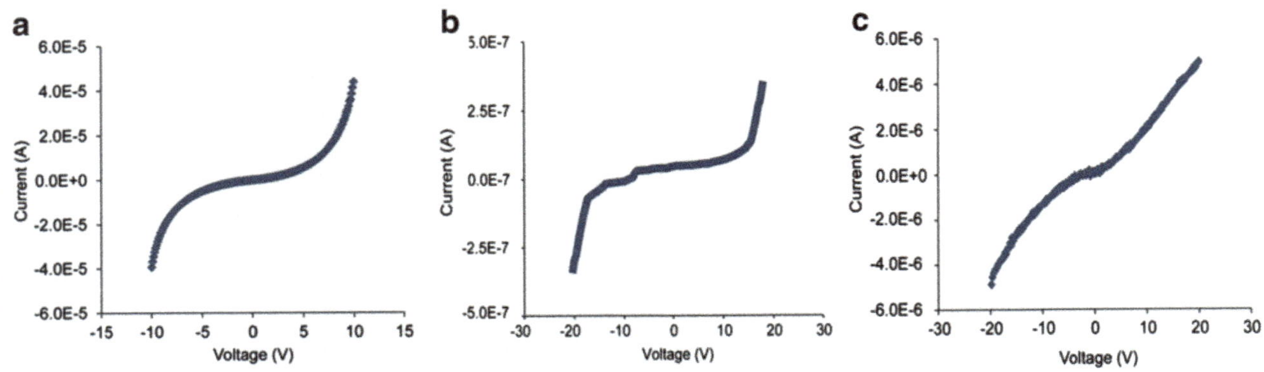

Figure 11 Current–voltage characteristics of metallized silica nanotubes fabricated from flagella bio-templates. (a) gold, **(b)** palladium, **(c)** iron oxide nanoparticles coated silica nanotubes. Copyright © 2013 IOP Publishing (Ref. 59). Reproduced by permission of IOP Publishing. All rights reserved.

dye, as demonstrated in a similar device [41]. The appropriate TCO to be used on this device is Fluorine doped Tin Oxide (FTO). The dye scaffolds used in this device will be silica nanotubes. The use of flagella as templates allows for high aspect ratio and high surface area of the dye scaffolds. This charge transfer mechanism shown in Figure 12b requires that the redox electrolyte be oxidized so that an electron can be added to the photoexcited dye in order to bring the dye back to the ground state. In the case of this solar cell, the mediator is the I^-/I_3^- redox couple, which has been shown to be the best known redox couple for this application [62]. The load to be driven by the solar cell is placed between the TCO and the counter electrode, which is on the opposite side of the solar cell from the TCO, and is generally covered by platinum. Platinum is considered due to its high electrocatalytic activity, which is necessary to reduce the redox electrolyte. This reduction of the redox electrolyte is necessary to return the oxidized redox electrolyte back to its normal state. The electrons that pass through the load and then end up on the platinum counter electrode are then transferred to the redox electrolyte which reduces the oxidized electrolyte [62].

The solar cell is assembled by dye sensitizing the TCO with the nanotube forest in a solution of the $(Bu_4N)_2Ru(dcbpyH)_2(NCS)_2$(N719) dye in dry ethanol. The platinum coated counter electrode, which is fabricated by evaporation onto an FTO electrode, and the dye-sensitized electrode are spaced apart with the use of hot melt spacers. The redox electrolyte (0.5 M LiI, 50 mM I_2, 0.5 M 4-tertbutylpyridine in 3-methoxypropionitrile) is then added to the space between the two electrodes.

In order to optimize the performance of the nanotube forest as a DSSCs, the efficiency (η) and the fill factor (FF) should be well understood. The overall efficiency is important in determining the percentage of energy from the incident light that is able to be converted into useful electrical energy. The fill factor is used in determining

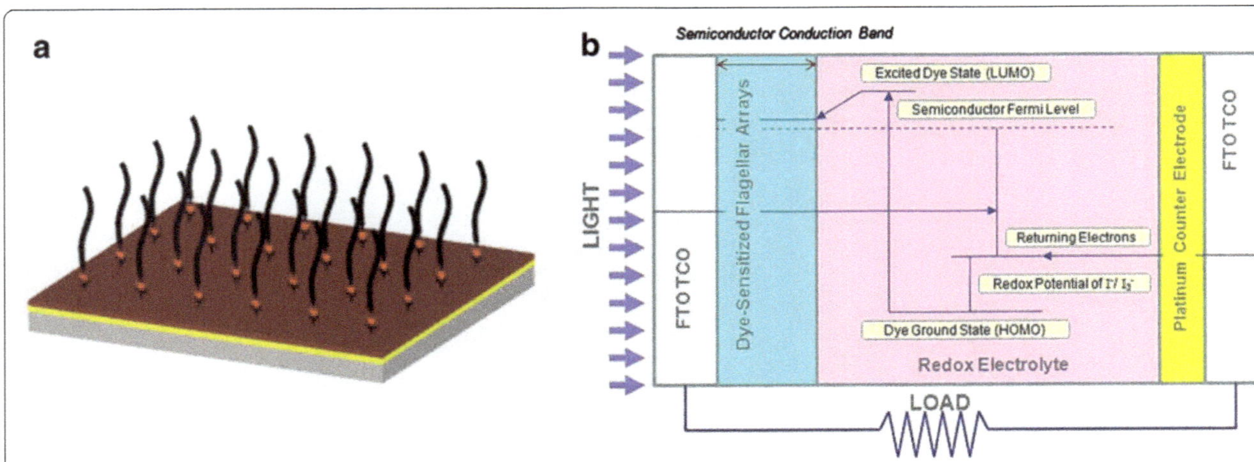

Figure 12 Flagella templated DSSC. (a) Flagellar forest assembled using avidin-biotin chemistry. **(b)** Schematic of the operation of the flagella templated DSSC.

how ideally the solar cell works as a diode, which can change depending on how well the counter electrode acts as an electrochemical catalyst, or how changing the electrolyte of the photovoltaic can affect the fill factor and be an indicator of how scalable the system is.

2 Conclusion

The current research effort and developments made towards electronic device applications have been heavily influenced by the utilization of nanomaterials synthesized from biologically inspired templates. These fascinating developments were the direct result of the advantages of using biotemplates; specifically the capability to mass produce biomaterials of uniform size, precise structure, complex morphology, and surface modification functionality. Many of these properties enhanced the fabrication method of nanomaterials and their functionalities in electronic device components. Especially, it have been said that the current technological limitations of solar cells and batteries can be amended through the use of nanostructures due to their high surface area to volume ratio. Many biotemplates had successfully been used to fabricate nanomaterials for electronic devices. The favorable results from these works have inspired the possibility to use bacterial flagella as biotemplates. The bacterial flagella can be a viable choice for biotemplates because of their unique natural properties such as tubular structures with controllable length and their polymorphic transformation. Although comprehensive studies had been done on the synthesis of high quality nanomaterials using bacterial flagella biotemplates, research on using these flagella templated nanomaterial in practical application is still in an early stage. We believe that future direction of this work should be focused on applying them to electronic devices. To move towards to realization of this work, we must optimize the manufacturing process to generate large quantity of nanomaterials with high quality in more controllable, accruable, and simple ways. Therefore, the flagella biological properties combined with synthetic chemistry have promising prospect in the fabrication of useful materials for electronic devices with high yield and functionality.

Competing interests
The authors declare that they have no competing interests.

Authors' contributions
WJ carried out acquisition of review papers, drafted the manuscript for biomaterial application sections, and reviewed critically important intelectual content. UKC involved in collecting review papers, drafted the manuscript for biomaterials sections, and revised all contents. MJK made paper conception and gave final approval of the version to be published. All authors read and approved the final manuscript.

Acknowledgement
This work was funded by Army Research Office Young Investigator Award (W911NF-10-1-0173) and National Science Foundation (CMMI-0745019).

Author details
[1]Institute for Multidisciplinary Convergence of Matter, Korea Institute of Science and Technology, 136-791 Seoul, Republic of Korea. [2]Department of Mechanical Engineering & Mechanics, Drexel University, 3141 Chestnut Street, Philadelphia, PA 19104, USA.

References
1. SW Lee, C Mao, CE Flynn, AM Belcher, Science **296**, 892–95 (2002)
2. M Knez, M Sumser, AM Bittner, C Wege, H Jeske, TP Martin, K Kern, Adv. Funct. Mater. **14**, 116–24 (2004)
3. C Mao, DJ Solis, BD Reiss, ST Kottmann, RY Sweeney, A Hayhurst, G Georgiou, B Iverson, AM Belcher, Science **303**, 213–17 (2004)
4. SE Stanca, R Eritja, D Fitzmaurice, Faraday Discuss **131**, 155–65 (2006)
5. CE Flynn, S-W Lee, BR Peelle, AM Belcher, Acta Mater. **51**, 5867–80 (2003)
6. W Jo, KJ Freedman, MJ Kim, Mater. Sci. Eng. C **32**, 2426–30 (2012)
7. WR Hesse, L Luo, G Zhang, R Mulero, J Cho, MJ Kim, Mater. Sci. Eng. C **29**, 2282–86 (2009)
8. T Atsumi, J Theor. Biol. **213**, 31–51 (2001)
9. K Gekko, Y Hasegawa, Biochemistry **25**, 6563–71 (1986)
10. F Oosawa, S Fujime, SI Ishiwata, K Mihashi, Cold Spring Harb. Sym. **37**, 85–277 (1973)
11. R Kamiya, S Asakura, J Mol, Biol. **106**, 167–86 (1976)
12. R Kamiya, S Asakura, J Mol, Biol. **108**, 513–18 (1976)
13. S Asakura, Adv. Biophys. **1**, 99–155 (1970)
14. K Yonekura, S Maki-Yonekura, K Namba, Nature **424**, 643–50 (2003)
15. K NAMBA, F VONDERVISZT, Q Rev, Biophys **30**, 1–65 (1997)
16. FA Samatey, K Imada, S Nagashima, F Vonderviszt, T Kumasaka, M Yamamoto, K Namba, Nature **410**, 331–37 (2001)
17. RRV Malapaka, LO Adebayo, BC Tripp, J Mol, Biol. **365**, 1102–16 (2007)
18. C Calladine, Nature **255**, 121–24 (1975)
19. R Kamiya, S Asakura, K Wakabayashi, K Namba, J Mol, Biol. **131**, 725–42 (1979)
20. A Arkhipov, PL Freddolino, K Imada, K Namba, K Schulten, Biophys. J. **91**, 4589–97 (2006)
21. K Hasegawa, I Yamashita, K Namba, Biophys. J. **74**, 569–75 (1998)
22. NC Darnton, HC Berg, Biophys. J. **92**, 2230–36 (2007)
23. R Kamiya, S Asakura, S Yamaguchi, Nature **286**, 628–30 (1980)
24. E Hasegawa, R Kamiya, S Asakura, J Mol, Biol. **160**, 609–21 (1982)
25. R Kamiya, H Hotani, S Asakura, Symp. Soc. Exp. Biol. **35**, 53–76 (1982)
26. M Washizu, M Shikida, S-I Aizawa, H Hotani, IEEE Ind. Applic. Soc. **28**, 1194–202 (1992)
27. W Jo, KJ Freedman, DK Yi, MJ Kim, Nanotechnology **23**, 055601 (2012)
28. Z Han, F Tongxiang, H Ting, L Xufan, D Jian, Z Di, G Qixin, O Hiroshi, Nanotechnology **20**, 085603 (2009)
29. SA Davis, SL Burkett, NH Mendelson, S Mann, Nature **385**, 420–23 (1997)
30. W Shenton, T Douglas, M Young, G Stubbs, S Mann, Adv. Mater. **11**, 253–56 (1999)
31. Y Huang, C-Y Chiang, SK Lee, Y Gao, EL Hu, JD Yoreo, AM Belcher, Nano Lett. **5**, 1429–34 (2005)
32. RJ Tseng, C Tsai, L Ma, J Ouyang, CS Ozkan, Y Yang, Nat. Nano **1**, 72–77 (2006)
33. C Jeffryes, T Gutu, J Jiao, GL Rorrer, J Mater, Res. **23**, 3255–62 (2008)
34. Y Fang, Q Wu, MB Dickerson, Y Cai, S Shian, JD Berrigan, N Poulsen, N Kröger, KH Sandhage, Chem. Mater. **21**, 10–5704 (2009)
35. WW Huang, ZL Wang, Nano Lett **6**, 31–2325 (2006)
36. Z Wang, Z Di, F Tongxiang, D Jian, G Qixin, O Hiroshi, Nanotechnology **17**, 840 (2006)
37. M Gratzel, Nature **414**, 338–44 (2001)
38. GK Mor, K Shankar, M Paulose, OK Varghese, CA Grimes, Nano Lett. **6**, 215–18 (2005)
39. Z-S Wang, H Kawauchi, T Kashima, H Arakawa, Coordin. Chem. Rev. **248**, 1381–89 (2004)
40. W Zhang, D Zhang, T Fan, J Gu, J Ding, H Wang, Q Guo, H Ogawa, Chem. Mater. **21**, 33–40 (2008)
41. M Law, LE Greene, JC Johnson, R Saykally, P Yang, Nat. Mater. **4**, 455–59 (2005)
42. JM Macák, H Tsuchiya, P Schmuki, Angew. Chem. Int. Edit. **44**, 2100–02 (2005)
43. H Jingbin, F Fengru, X Chen, L Shisheng, W Min, D Xue, W Zhong Lin, Nanotechnology **21**, 405203 (2010)

44. ABF Martinson, JW Elam, JT Hupp, MJ Pellin, Nano Lett. **7**, 2183–87 (2007)
45. CE Flynn, C Mao, A Hayhurst, JL Williams, G Georgiou, B Iverson, AM Belcher, J Mater, Chem. **13**, 2414–21 (2003)
46. X Dang, H Yi, MH Ham, J Qi, DS Yun, R Ladewski, MS Strano, PT Hammond, AM Belcher, Nat. Nano **6**, 84–377 (2011)
47. YM Lee, YH Kim, JH Lee, JH Park, N-G Park, W-S Choe, MJ Ko, PJ Yoo, Adv. Funct. Mater. **21**, 1160–67 (2011)
48. C Jeffryes, J Campbell, H Li, J Jiao, G Rorrer, Energ. Environ. Sci. **4**, 3930–41 (2011)
49. BD Heilman, IN Miaoulis, Appl. Opt **33**, 47–6642 (1994)
50. RA Potyrailo, H Ghiradella, A Vertiatchikh, K Dovidenko, JR Cournoyer, E Olson, Nat. Photon. **1**, 123–28 (2007)
51. M Armand, JM Tarascon, Nature **451**, 652–57 (2008)
52. E Royston, A Ghosh, P Kofinas, MT Harris, JN Culver, Langmuir **24**, 906–12 (2007)
53. X Chen, K Gerasopoulos, J Guo, A Brown, C Wang, R Ghodssi, JN Culver, ACS. Nano **4**, 5366–72 (2010)
54. E Pomerantseva, K Gerasopoulos, X Chen, G Rubloff, R Ghodssi, J. Power Sources **206**, 282–87 (2012)
55. KT Nam, D-W Kim, PJ Yoo, C-Y Chiang, N Meethong, PT Hammond, Y-M Chiang, AM Belcher, Science **312**, 885–88 (2006)
56. SM Kang, M-H Ryou, JW Choi, H Lee, Chem. Mater. **24**, 3481–85 (2012)
57. MT Kumara, BC Tripp, S Muralidharan, Chem. Mater. **19**, 64–2056 (2007)
58. K Deplanche, RD Woods, IP Mikheenko, RE Sockett, LE Macaskie, Biotechnol. Bioeng. **101**, 873–80 (2008)
59. W Jo, M Darmawan, J Kim, CW Ahn, D Byun, SH Baik, MJ Kim, Nanotechnology **24**, 135704 (2013)
60. E Diamandis, T Christopoulos, Clin. Chem. **37**, 625–36 (1991)
61. UK Cheang, D Roy, JH Lee, MJ Kim, Appl. Phys. Lett. **97**, 213704 (2010)
62. A Luque, S Hegedus, *Book Handbook of Photovoltaic Science and Engineering* (*Wiley*, New York, 2011)

12

Radionuclide-labeled nanostructures for *In Vivo* imaging of cancer

Won-Kyu Rhim[1†], Minho Kim[1†], Kevin L Hartman[1], Keon Wook Kang[2] and Jwa-Min Nam[1*]

Abstract

Molecular imaging plays an important role in the non-invasive diagnosis and the guiding or monitoring of disease treatment. Different imaging modalities have been developed, and each method possesses unique strengths. While a variety of molecules have been used previously in nuclear imaging, the exceptional properties of nanostructures in recent research enable the deployment of accurate and efficient diagnostic agents using radionuclide-nanostructures. This review focuses on the radionuclide labeling strategies of various nanostructures and their applications for multimodality tumor imaging.

Keywords: Nanoprobe; Molecular imaging; Radionuclide; Positron emission tomography (PET); Single photon emission computed tomography (SPECT); Multimodality

1 Introduction

Over the past decade, molecular imaging has emerged and expanded rapidly due to its great potential for understanding biological reactions, detecting diseases in early stages, providing accurate diagnosis, and visualizing pathologic processes and cellular functions at the molecular and cellular level without perturbing the living organism [1-5]. Several types of tools have been developed for molecular imaging, such as magnetic resonance imaging (MRI), computed tomography (CT), positron emission tomography (PET), single photon emission computed tomography (SPECT), contrast enhanced ultrasound (CEUS), and optical fluorescence [6]. PET, SPECT, and other radionuclide-based imaging methods have been extensively used in biomedical and clinical fields due to their excellent features of high sensitivity, easy quantitation, and strong penetration through tissue [7-10].

In the case of nuclear molecular imaging tools (PET and SPECT), radionuclide-labeled substances are used as contrast agents. Generally, radionuclides are linked via a chelating agent to targeting components, such as antibodies, antibody fragments, peptides, proteins, oligonucleotides, or receptor ligands [11-16]. Recently, nanoparticles have been noted as a remarkable radionuclide carrier due to their chemical and biological stability, high loading capacity, noteworthy physicochemical properties, and practically limitless surface-functionalization [17]. Although several functionalization methods using chelating agents have been suggested and developed [18-24], those methods have numerous limitations for nuclear molecular imaging because the radionuclides originally incorporated with chelating agents can detach from those molecules *in vivo* [25-27]. These instabilities may induce false target detection, which decreases the accuracy of diagnostic information. Therefore, functionalization methods free of chelating agents have been sought recently as a way to overcome those drawbacks [28-32].

For practical purposes, molecular imaging tools are not sufficient by themselves due to their inherent limitations. For example, while the imaging modes PET and SPECT are highly sensitive, quantitative, and show good tissue penetration depth, they are limited by their low spatial resolution [6]. Therefore, the combination of two or more molecular imaging tools, otherwise known as multimodality imaging, should be used to overcome the deficiencies of individual imaging modes to ensure the accuracy of molecular imaging applications. Moreover, the unique properties of nanostructures can be used to their advantage to lessen the amount of radionuclide used and maximize the benefits of multimodality imaging.

* Correspondence: jmnam@snu.ac.kr
†Equal contributors
[1]Department of Chemistry, Seoul National University, Seoul 151-747, Republic of Korea
Full list of author information is available at the end of the article

In this review paper, we will present radionuclide labeling strategies on various nanostructures and their applications of multimodality tumor imaging.

2 Review

Molecular imaging of living organisms involves the visualization of biological processes and cellular functions at the molecular and cellular levels [33]. This can be used for diagnosis of diseases such as cancer, neurological, and cardiovascular conditions without any invasion to the living organism [1,3]. In general, the field of molecular imaging for clinical use includes magnetic resonance imaging (MRI), computed tomography (CT), positron emission tomography (PET), single photon emission computed tomography (SPECT), contrast enhanced ultrasound (CEUS), and optical fluorescence. Taking into account sensitivity, spatial resolution, and temporal depth resolution, each molecular imaging tool has different sets of capabilities [6]. For instance, the lack of penetration depth limit of PET and SPECT, coupled with their high sensitivity (down to the picomolar range) and quantitative abilities make them powerful tools for *in vivo* imaging [34]. Molecular imaging includes the administration of imaging probes and the detection of specific signals produced from the probes. In PET, the system detects pairs of gamma rays emitted by positron-emitting radionuclides [35]. Positrons interact with nearby electrons at a distance of ~1 mm, transmutating to generate and emit 511 keV gamma photons. To enable this process, various positron-emitting isotopes are available for PET imaging. Among them, ^{18}F and ^{64}Cu are attractive radionuclides for site-specific imaging. Fluorodeoxyglucose (FDG) is the most commonly used radiotracer for tumor-targeted imaging. It is taken up by cells with a high need for glucose such as those in the heart, brain, and tumor [36]. ^{64}Cu is also a good candidate to be labeled onto biomolecules like antibodies and peptides due to its proper decay half-life, relatively low energy, and commercial availability [2]. In SPECT, computer-generated three-dimensional images of radioactive tracers are produced by detecting single gamma ray photons [37]. A single gamma photon is generated when each radionuclide decays from a higher energy state to a lower state. SPECT imaging uses gamma emitting radioisotopes such as ^{93m}Tc, ^{111}In, ^{123}I, and ^{131}I.

Nanostructures, which range in size from 1 to 100 nm, have a high surface area to volume ratio and strong, controllable binding capacity for specific molecules. This allows for varied surface modification and the multivalent binding of targets, which suggests new applications in molecular imaging and nanomedicine [38-48]. While active targeting is advantageous, the most important factor of nanostructure development for bioimaging is the enhanced permeability and retention (EPR) effect due to the leakage of neovasculature in tumor. The EPR effect is the property that only molecules and particles of specific sizes accumulate in tumor tissues. The effect is due to the abnormal angiogenesis caused by various growth factor stimulation in tumor tissues [49]. Nanostructures of 100 nm in diameter can be optimum candidates for maximizing the EPR effect *in vivo*. Moreover, modifying surface charge and flexibility allow control of nanostructure biodistribution [43,46,50]. For nanoparticles with different surface properties and multifunctional groups, proper isotope and radiolabeling methods need to be considered carefully for optimized imaging outcome. After labeling with a radionuclide, functionalized nanostructures can serve as good candidates for molecular imaging probes (PET or SPECT). The two major strategies for labeling nuclides onto nanostructures depend on whether or not chelating agents are used. The chelation-free labeling methods include self-radioactive nanostructures and the direct attachment of radionuclides onto the nanostructures, which both utilize specific interactions between the radioisotope and the nanostructure. In the earlier examples of this work, radionuclides (radioisotopes), which are used as radiotracers for nuclear molecular imaging (PET or SPECT), were combined with nanoparticles, including gold nanoparticles [18,19,51], carbon-based nanomaterials [20-22,52], and quantum dots [53,54], through a chelating agent, such as DOTA, NOTA, and others [55]. Moreover, chloramine T can act as an oxidizing agent in the synthesis of the Bolton-Hunter reagent to label radioactive iodine [56]. Although many chelating agents were tested to label radioisotopes more efficiently, DOTA is the most attractive candidate due to its FDA approval for human application.

2.1 Chelating agent-mediated nanostructures

2.1.1 Gold nanostructures

Colloidal gold nanoparticles (AuNPs) are attractive candidates as imaging probes because of their low cytotoxicity in a variety of cell and animal models [32,57-60]. Moreover, AuNPs can be easily modified with biomolecules (e.g., DNA and peptides) via thiol-Au bonds, and many other functional ligands can also be incorporated. Xie et al. reported a ^{64}Cu-chelated gold nanoshell (AuNS) structure for *in vivo* PET imaging and photothermal therapy [18]. For this, the bifunctional chelating agent p-NH2-Bn-DOTA (S-2-(4-aminobenzyl)-1,4,7,10-tetraazacyclododecane tetraacetic acid) was conjugated to bifunctional OPSS-PEG2K-NHS (opyridyldisulfide-polyethylene glycol 2000-N-hydroxysuccinimide) (Figure 1(a)). About 90% of the resulting ^{64}Cu-chelated AuNS was stable for 3 hrs in PBS solution or serum, and showed accumulation in the squamous cell carcinoma region by the EPR effect. In

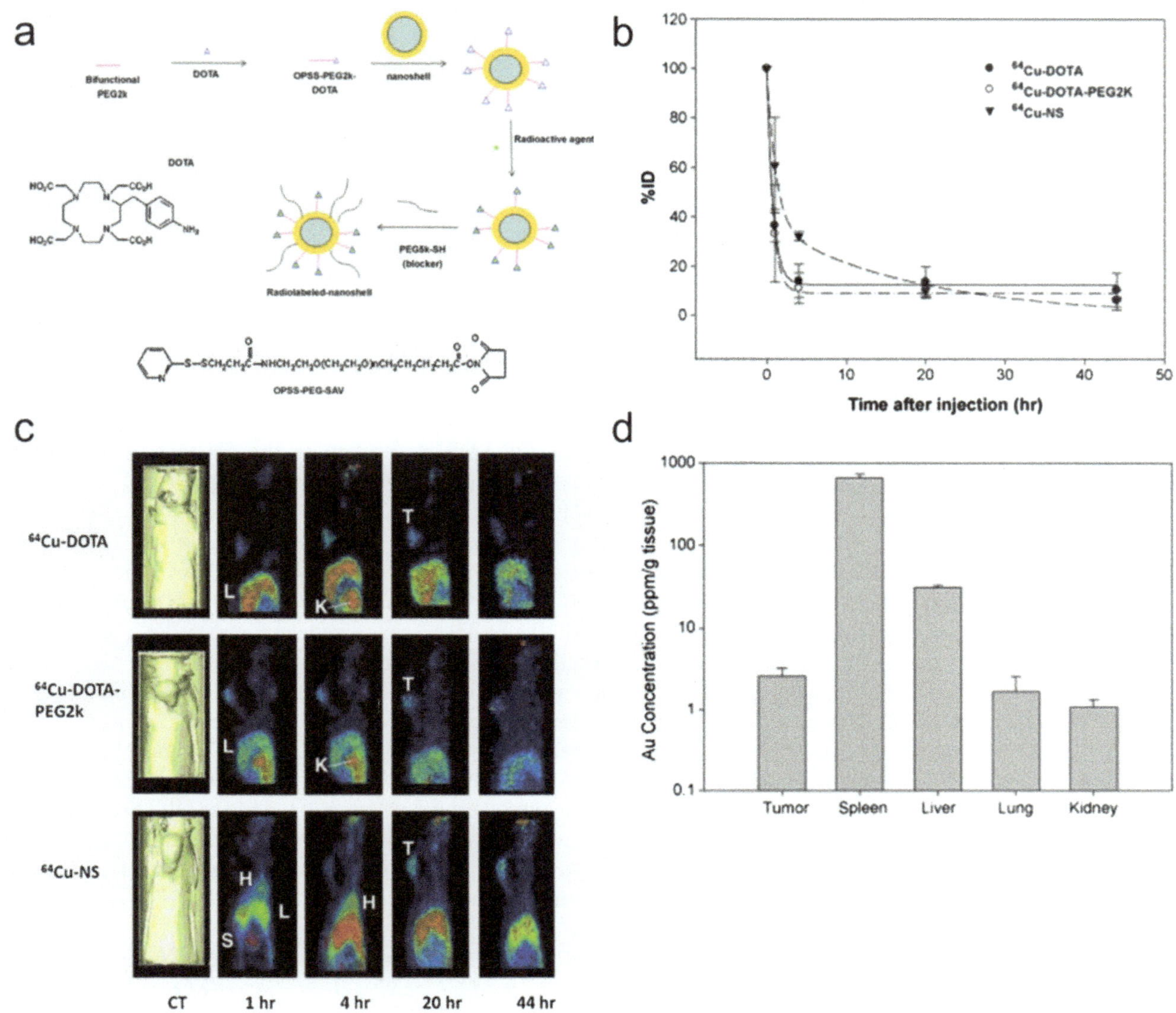

Figure 1 ***In vivo*** **PET imaging of radiolabeled gold nanoshell structure with tumor.** **(a)** Synthesis of radiolabeled AuNS conjugates. **(b)** Pharmacokinetics, expressed as the percentage of the injected dose (%ID), of ^{64}Cu-DOTA, ^{64}Cu-DOTA-PEG2K, and ^{64}Cu-NS at 0, 4, 20, and 44 h p.i. Significantly different pharmacokinetics between ^{64}Cu-NS and the two controls are evident at 4 h p.i. ($p < 0.05$). **(c)** PET images at different time points with three different probes. (^{64}Cu-DOTA, ^{64}Cu-DOTA-PEG2K, and ^{64}Cu-NS) **(d)** Biodistribution of ^{64}Cu-NS in several major organs. (Adapted with permission from Ref. 18. Copyright (2010) Elsevier).

pharmacokinetics studies, the half-life of ^{64}Cu-NS (12.76 hrs), was significantly longer than the half-lives of the ^{64}Cu-DOTA and ^{64}Cu-DOTA-PEG2K controls (0.54 and 0.52 hrs, respectively). As shown in Figure 1(b), statistical analysis proved that ^{64}Cu-NS had a much higher remaining ratio in the blood at 4 hrs post-injection (p.i.) than the two controls (32.1 ± 1.8% ID vs. 13.9 ± 6.8 and 11.0 ± 6.2% ID; $p < 0.05$) (Figure 1(c,d)). Yucai et al. furthermore reported a ^{64}Cu-labeled DOTA-PEG-Au nanocage (^{64}Cu-DOTA-PEG-AuNC) for simultaneous *in vivo* tumor targeted PET imaging and pharmacokinetics [19]. AuNCs have effective photothermal translation activity, capable of converting light into heat and increasing the local temperature. Therefore, AuNCs can act as a photothermal therapeutic agent while its properties as an imaging agent allows simultaneous monitoring of its biodistribution. DOTA was conjugated onto the AuNC surface using an NH_2-modified PEG molecule for labeling radionuclide as a PET imaging agent. The NH_2 groups were then coupled with DOTA-NHS-ester through an amide reaction via an NHS-activated ester, followed by chelation with ^{64}Cu ions. ^{64}Cu-DOTA-PEG-AuNCs showed 81.5% stability following a 24 hr incubation in serum. Due to the reduction in size and surface charge, the 30 nm diameter ^{64}Cu-DOTA-PEG-AuNCs showed improved *in vivo* pharmacokinetics with decreased RES system uptake and enhanced blood circulation compared to the 55 nm samples.

2.1.2 Carbon nanostructures

Due to their unique geometric, chemical, and physical properties, carbon nanotubes are promising candidates for nanomedicine [61-64]. SWNT can be non-covalently functionalized with phospholipid-PEG for sustained solubility and stability in human serum environments [20]. The termini of PEG on the nanotube can be furthermore conjugated to both DOTA and integrin αvβ3, with the integrin targeting the RGD (arginine-glycine-aspartic acid) peptide for ^{64}Cu labeling and tumor imaging. Liu et al. found that SWNT chelated with ^{64}Cu in this way showed high tumor accumulation over long periods (~24 hrs).

Graphene oxide (GO) is a type of planar carbon sheet that has a great potential in nanomedicine due to its interesting properties arising from the combination of an ultrahigh surface area with a modifiable planar surface chemistry [65,66]. Though toxic in some cases, the biocompatibility of graphene oxide is dependent on the functional groups displayed [67]. For example, polyethylene glycol (PEG) functionalized GO can be excreted through the urinary tract, without inducing cytotoxicity [68]. Hong et al. developed ^{66}Ga and ^{64}Cu-chelated GO for tumor vasculature imaging [21,69]. The researchers used 1,4,7-triazacyclononane-1,4,7-triacetic acid (NOTA) for chelating both ^{66}Ga and ^{64}Cu. NOTA and TRC 105 (antibody binding to CD105) were covalently conjugated to agraphene oxide sheet consisting of six-arm branched PEG molecules on the surface (Figure 2 (a)). More than 90% of labeled ^{66}Ga and ^{64}Cu remained on the GO conjugates during a 48 hrs incubation in serum conditions (Figure 2(b)). Efficient tumor uptake of radiolabeled-NOTA-GO-TRC105 occurred with a clear visualization at early time points and a stability over 48 hrs (Figure 2(c,d)). Cornelissen et al. used doxorubicin and ^{111}In-loadedgraphene oxide (GO) for tumor targeted SPECT imaging and therapy [22]. Different from Hong et al., they labeled ^{111}In using π-π stacked benzyl-DTPA onto GO. The ^{111}In achieved a labeling yield of 98% and only less than 5% of chelated ^{111}In was lost by transchelation to serum proteins during serum incubation. ^{111}In-labeled NGO-trastuzumab constructs (which consisted of an antibody against HER2) were shown to target HER2 receptors with high specificity.

2.1.3 Lipid nanostructures

Liposomes are clinically well-established drug delivery vehicles with properties including biocompatibility, high drug-to-carrier ratios, and straightforward drug encapsulation capability [70-73]. The polar and non-polar domains of the molecules comprising the liposome structure allow the stable encapsulation of both hydrophilic and hydrophobic agents. The incorporation of hydrophilic polyethylene glycol on the surface of liposomes further improves their properties of low opsonization and long circulation time [43]. Seo et al. reported a novel ^{64}Cu labeling method for liposomes [23]. The researchers conjugated a ^{64}Cu specific chelating agent, 6-[*p*-(bromoacetamido) benzyl]-1,4,8,11-tetraazacyclotetradecane-*N,N',N'',N'''*-tetraacetic acid (BAT) to an artificial lipid molecule to form a BAT-PEG-lipid, followed by successfully incorporating ^{64}Cu into liposomes at pH 5. About 88% of ^{64}Cu was encapsulated into the liposomes over the course of a 48 hr serum incubation. Different from the above method, Petersen et al. demonstrated remote loading of ^{64}Cu into a liposome using a novel inophore, 2-hydroxyquinoline [24]. By remote loading, the radionuclide becomes entrapped in the aqueous core of the liposome, rather than on the surface. For the remote loading mechanism, 2-hydroxyquinoline carries ^{64}Cu across the lipid membrane to the aqueous portion of liposome, followed by chelation into pre-encapsulating DOTA (Figure 3(a)). Using this technique, the researchers achieved very efficient loading (95.5%) and retention stability (>99%), which shows the ^{64}Cu-liposomes to be promising PET imaging agents. This method provides maximized and stable ^{64}Cu labeling efficiency. PEG conjugated liposomes can circulate for an extended time and are able to accumulate in tumor regions through the EPR effect. The ^{64}Cu-liposomes remained in the blood-pool for over 24 hrs. The ^{64}Cu-liposome activity at the tumor site increased over the 24 hr time period post injection. This result means that the number of ^{64}Cu-liposomes delivered to the tumors continues to accumulate as long as liposomes are circulating in the blood stream (Figure 3(b)).

2.2 Chelating agent-freenanostructures

2.2.1 The need for labeling strategies free of chelating agents

Although various chelating agent-mediated radioisotope labeled nanoprobes have been suggested for nuclear molecular imaging and pharmacokinetics, chelating agent-mediated functionalization methods have several limitations for nanoparticles used *in vivo* [28]. Firstly, the functionalized radionuclides using chelating agents at the surface of nanoparticles can induce the change of intrinsic surface properties of nanoparticles, such as surface charge, polarity, and degree of hydrophobicity. Basically, the physicochemical properties of nanoparticles are strongly affected by their surface conditions [74], and therefore, chelating agent-functionalized nanoparticles may show different biodistributions or pharmacokinetic properties than compared to non-functionalized nanoparticles. Secondly, radionuclides incorporated with a chelating agent may be influenced by harsh conditions *in vivo*, resulting in the detachment of radionuclides during circulation or their replacement by biological substances due to protein

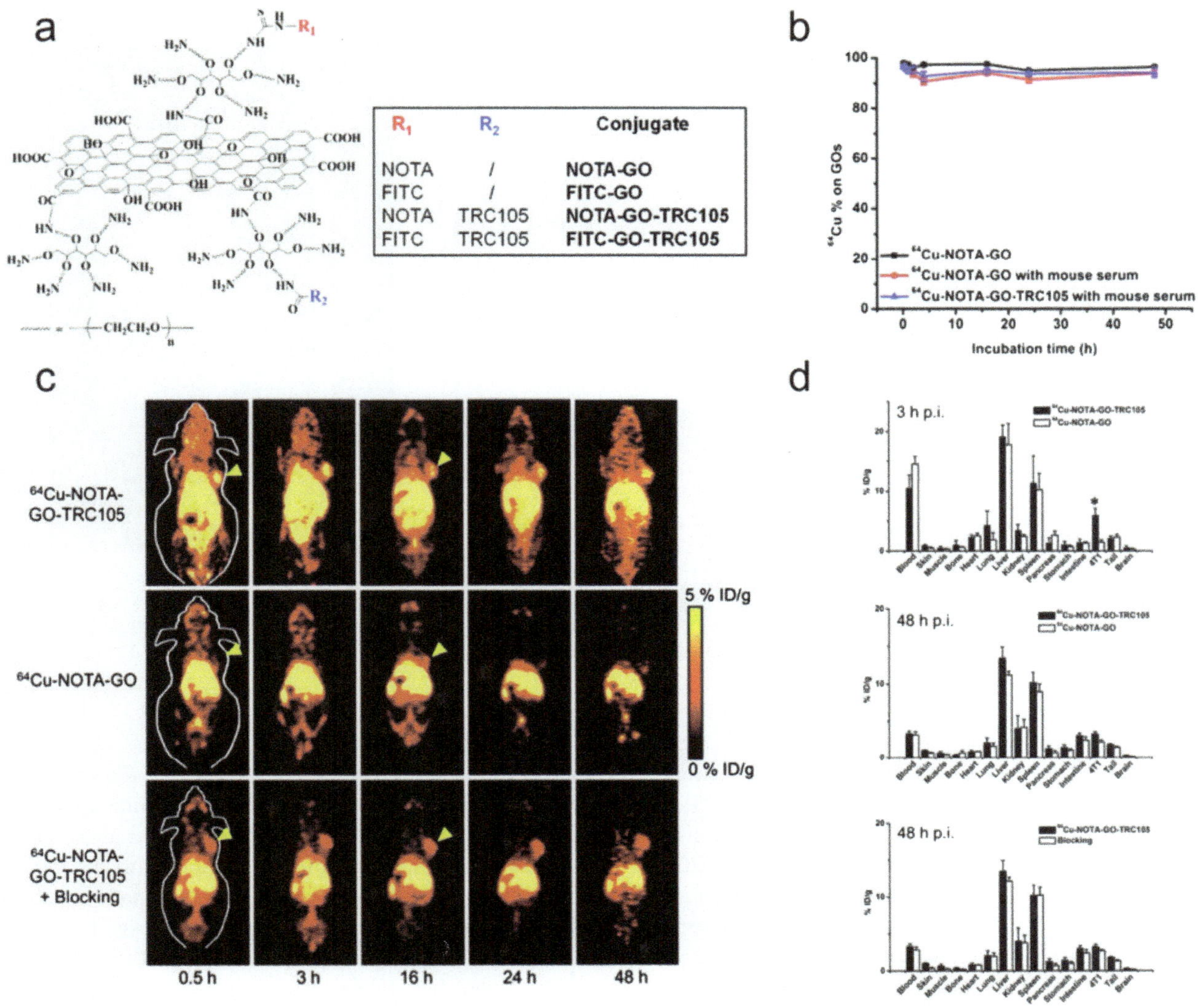

Figure 2 ***In vivo*** **targeting and PET imaging using ^{64}Cu-labeled nano-graphene.** **(a)** Schematic representation of conjugated nano-graphene. **(b)** Serum stability test of ^{64}Cu-labeled nano-graphene shows that the vast majority of ^{64}Cu remains intact on the GO after incubation in complete mouse serum at 37°C for 48 hrs. **(c)** *In vivo* PET/CT imaging of ^{64}Cu-labeled GO conjugates in 4 T1 murine breast tumor-bearing mice. Serial coronal PET images of 4 T1 tumor-bearing mice at different time points post injection (p.i.) of ^{64}Cu-NOTA-GO-TRC105, ^{64}Cu-NOTA-GO, or ^{64}Cu-NOTA-GO-TRC105 after a preinjected blocking dose of TRC105. Tumors are indicated by arrowheads. **(d)** Biodistribution studies in 4 T1 tumor-bearing mice. Biodistribution of ^{64}Cu-NOTA-GO-TRC105 and ^{64}Cu-NOTA-GO in 4 T1 tumor-bearing mice at 3 hrs and 48 hrs postinjection; $*p < 0.05$. Biodistribution of ^{64}Cu-NOTA-GO-TRC105 and ^{64}Cu-NOTA-GO-TRC105 after a blocking dose of TRC105 (denoted as blocking) in 4 T1 tumor-bearing mice at 48 h p.i. (Adapted with permission from Ref. 21. Copyright (2012) American Chemical Society).

transchelation or dehalogenation [25-27]. Accordingly, these instabilities of radionuclide-chelating agent complexes decrease the accuracy of pharmacokinetics and diagnosis data [75]. To avoid these disadvantages, researchers have been developing labeling strategies free of chelating agents. One of these studies, by Gibson et al., carried out direct radiolabelling using cyclotron accelerators to incorporate radionuclides into nanomaterials directly [76]. However, this direct incorporation method, which used neutron bombardment to generate high energy radiation, damaged the nanomaterials by causing them to aggregate.

2.2.2 Radio-labeled nanostructures free of chelating agents

Recently, nanoprobes free of chelating agents have been reported for nuclear molecular imaging [28-32]. To synthesize these, researchers have created self-radioactive nanoparticle probes using radioisotopes of metal atoms, such as ^{64}Cu or ^{198}Au, or by directly attaching radioisotopes, e.g. ^{125}I, onto the metal nanoparticle surface without a chelating agent. For example, Zhou et al. synthesized ~11 nm diameter [^{64}Cu]CuS nanoparticles for micro-PET/CT imaging and as a photothermal coupling agent for photothermal ablation, without the use of chelating agents [28] (Figure 4(a-c)). The [^{64}Cu]CuS

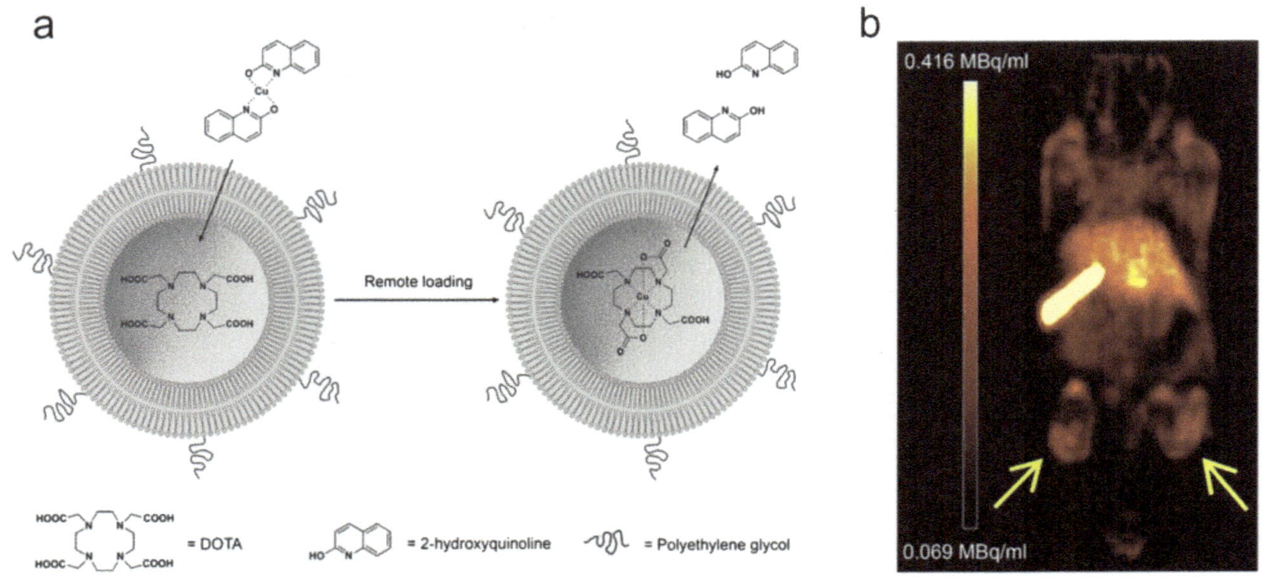

Figure 3 ^{64}Cu loaded liposome for tumor imaging using PET. (a) Remote loading of ^{64}Cu into preformed liposomes using an ionophore, 2-hydroxyquinoline. Over 99% of encapsulation efficiency was obtained with size exclusion chromatography after a remote loading method. **(b)** Coronal PET image 24 hrs post-injection of ^{64}Cu-liposomes into a mouse bearing colon adenocarcinoma (marked with arrows) on right and left flank. (Adapted with permission from Ref. 24. Copyright (2011) Elsevier).

nanoparticles, following surface PEG functionalization, showed high radiolabeling efficiency and stability in PBS or serum environments. These [^{64}Cu]CuS nanoparticles also showed high uptake in U87 human glioblastoma (U87MG) xenografts through enhanced permeability and retention effects. Furthermore, the [^{64}Cu]CuS nanoparticles strongly absorbed near-infrared (NIR) radiation (with a peak adsorption at 930 nm), and thus, they could be applied to the photothermal ablation of U87MG tumor cells upon exposure to NIR light. Therefore, these unique properties, such as small diameter, incorporation of ^{64}Cu as a radiotracer, high radiolabeling efficiency, excellent stability, and strong NIR absorption, are ideally useful for multifunctional molecular imaging and therapy. Similarly, Black et al. synthesized the radioactive ^{198}Au-doped nanostructures by directly incorporating radioactive ^{198}Au isotopes with non-radioactive ^{197}Au atoms during synthesis [29]. To use nanoparticles for*in vivo* theragnosis, the *in vivo* properties of nanoparticles including biodistribution, tumor uptake, and intratumoral distribution should be investigated and analyzed accurately. Since the *in vivo* properties of nanoparticles depend on their physicochemical characteristics including geometry, Black et al. synthesized four differently-shaped nanostructures: nanospheres, nanodisks, nanorods, and cubic nanocages, with similar sizes to confirm the shape-dependent behavior of nanoparticles using a murine EMT6 breast cancer model. They found that ^{198}Au-incorporated nanospheres showed optimal behavior with the best blood circulation, the lowest clearance by the reticuloendothelial system (RES), and the highest tumor uptake. These ^{198}Au-incorporated nanostructures have a great potential for *in vivo* molecular imaging compared to the disadvantages of radioactive nanomaterials made with chelating agents. Lately, the radioactive ^{64}Cu-alloyed [30] or ^{64}Cu-integrated [31] Au nanoparticles were synthesized for improving radiolabel stability and diagnostic accuracy, as well as PET imaging guided photothermal cancer therapy. Unlike self-radioactive nanoparticles composed of radioisotopes, Kim et al. synthesized cyclic RGD-PEGylated Au nanoparticle probes with surface conjugated radioactive ^{125}I for tumor targeting and SPECT/CT [32] (Figure 4(d-f)). The surface conjugation of ^{125}I works by the high affinity binding between the iodine ion and the Au surface [17,77-79]. This simple labeling chemistry provided high stability for ~20 hrs in serum or under various pH and high salt conditions with tumor-specific targeting for $\alpha_v\beta_3$ integrin expressing tumor cells via $\alpha_v\beta_3$-receptor-mediated endocytosis. Importantly, the ^{125}I-labeled cRGD-PEG-Au nanoprobes could target the tumor site only 10 min. after intravenous injection as observed by SPECT/CT imaging, and could be excreted from the body by renal clearance without any sign of toxicity.

2.3 Radionuclide-labeled nanostructures for multimodality tumor imaging

Generally, molecular imaging tools, such as positron emission tomography (PET), single photon emission

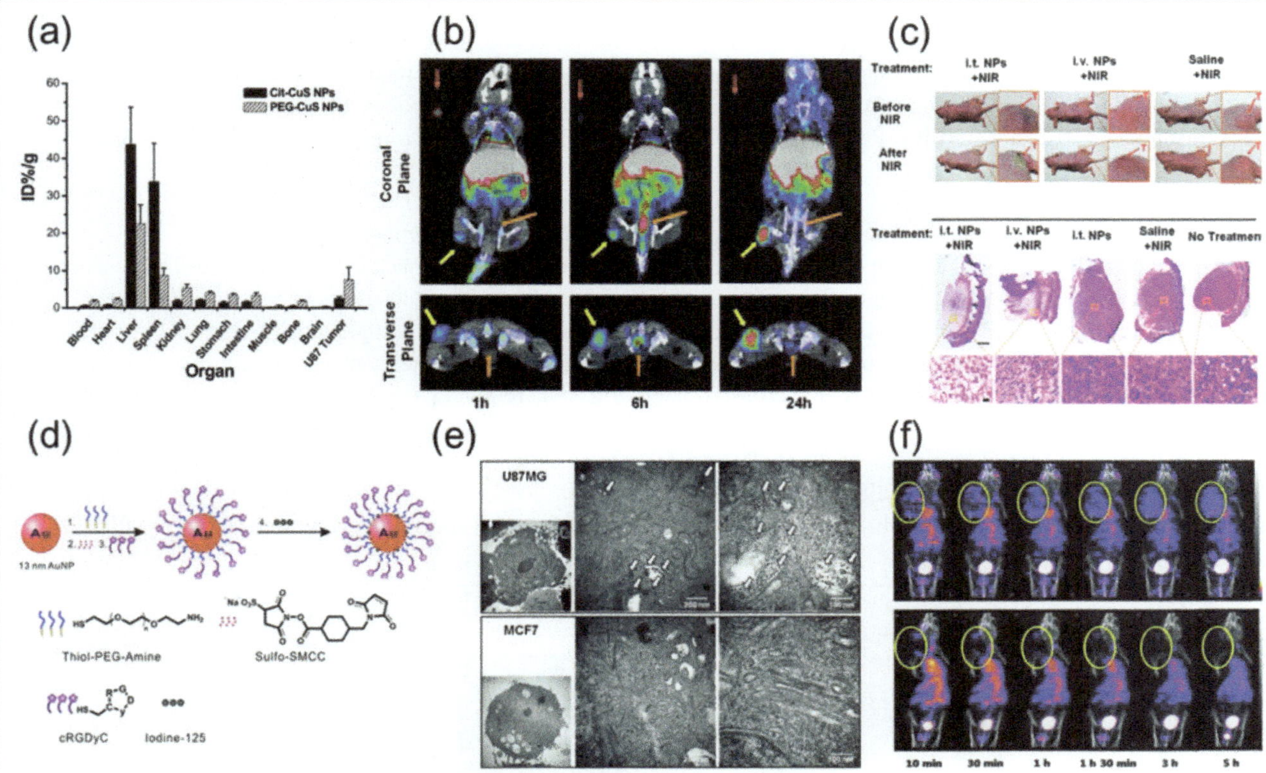

Figure 4 Chelator-free nanoprobes for nuclear molecular imaging. (a) Biodistribution of Cit-CuS NPs (n = 5) and PEG-CuS NPs (n = 5) at 24 h after i.v. injection in tumor-bearing mice. **(b)** Micro-PET/CT images of tumor-bearing nude mice acquired at 1, 6, and 24 h after i.v. injection of PEG-[^{64}Cu]CuS NPs. Yellow arrow = tumor; orange arrow = bladder, red arrow = standard. **(c)** Photographs of mice bearing subcutaneous U87 tumor before and at 24 h after NIR laser irradiation (upper panel). Representative microphotographs of tumors removed 24 h after NIR laser treatment (lower panel). Abbreviations: Cit-CuS NPs = citrate-coated CuS nanoparticles; PEG-CuS NPs = PEG-coated CuS nanoparticles; i.v. = intravenous; NIR = near-infrared. **(d)** Synthesis of cyclic RGD-PEGylated-AuNP probes labeled with radioactive iodine-125. **(e)** TEM images of cellular uptake after treated with ^{125}I-cRP-AuNPs, U87MG cells (upper panel) and MCF-7 cells (lower panel). Arrows indicate cellular uptake of ^{125}I-cRP-AuNPs in cells. **(f)** Small-animal SPECT/CT of ^{125}I-cRP-AuNPs (upper panel), and cold-form-blocked ^{125}I-cRP-AuNPs (lower panel) in nude mice bearing U87MG tumors xenografts after injection through tail vein. Yellow circles indicates tumor region (a-c adapted with permission from ref. 28. Copyright (2010) American Chemical Society & d-f adapted with permission from ref. 32. Copyright (2011) Wiley).

computed tomography (SPECT), magnetic resonance imaging (MRI), contrast-enhanced ultrasound (CEUS), and optical fluorescence, have intrinsic advantages and drawbacks simultaneously for accurate tumor imaging. For example, PET and SPECT are powerful imaging tools due to their high sensitivity, quantitative nature, and lack of a penetration limit, however, they are not perfect for single modality because of their low spatial resolution [6]. MRI enables morphological and functional imaging with no penetration limit and a high spatial resolution, however, it is limited by its low sensitivity. Optical fluorescence is widely used for molecular imaging given its low expense and high sensitivity, however, it also has several drawbacks, such as low spatial resolution, limited penetration, photobleaching, and autofluorescent emission from tissues and biomolecules. With such a wide range of capabilities, molecular imaging tools should be combined to enhance overall image quality.

2.3.1 PET(SPECT)/MR imaging

MRI is the most widely used non-invasive diagnostic technique based on the interaction of protons with surrounding molecules in the target tissue [80]. MRI provides several advantages, including high temporal and spatial resolution, soft tissue contrast, and the absence of ionizing radiation, which radionuclide-based imaging techniques, such as PET and SPECT, cannot achieve. However, MRI is limited by its low sensitivity and long acquisition time compared to PET and SPECT. Thus, as complementary techniques, PET (or SPECT) and MRI can be combined to overcome their individual limitations. The most commonly used nanoparticle-based MRI contrast agents are superparamagnetic iron oxide nanoparticles (SPIONPs), which are composed of Fe^{3+} and Fe^{2+} [81]. Usually, radionuclides are conjugated to iron oxide nanoparticles (IONPs) using chelating agents on the surface of IONPs for simultaneous dual-modal imaging techniques including PET and MRI. Lee et al. synthesized

DOTA chelating agent-mediated ^{64}Cu-labeled iron oxide nanoprobes with RGD peptide as a targeting moiety for $\alpha_v\beta_3$ integrin over-expressed U87MG tumor cells to apply PET/MR dual-modal imaging [82] (Figure 5). These probes were polyaspartic acid (PASP)-coated IO (PASP-IO) nanoparticles with a core size of 5 nm and a hydrodynamic diameter of 45 ± 10 nm, and the saturation magnetization of PASP-IO nanoparticles was about 117 emu/g of iron. With RGD conjugation, displacement competitive binding assays showed that these RGD-PASP-IO assemblies could bind specifically to integrin $\alpha_v\beta_3$ *in vitro*. Furthermore, small-animal PET, T_2-weighted MRI, and histologic analysis showed integrin-specific delivery of conjugated RGD-PASP-IO nanoparticles and prominent reticuloendothelial system (RES) uptake. Instead of ^{64}Cu, Choi et al. labeled ^{124}I on the surface of albumin-coated magnetic nanoparticles composed of $MnFe_2O_4$ via the tyrosine residue in the albumin. These probes were used for dual-modality PET/MR imaging [83]. Unlike PET/MR dual-modal imaging, Misri et al. synthesized dual-modal imaging probes for SPECT/MR by conjugating ^{111}In labeled antimesothelin antibody (mAbMB) to SPIONPs [84].

2.3.2 PET(SPECT)/optical imaging

Optical fluorescence, especially near-infrared fluorescence (NIRF) imaging technique, is most widely used for molecular imaging due to its high sensitivity, rapid scanning time with relatively inexpensive imaging instruments, and low risk to living organisms as it uses nonionizing rather than ionizing radiation. However, this imaging technique suffers from low tissue penetration, leading to poor quantification [6]. By combining fluorescence with radionuclide-based imaging techniques (PET or SPECT),

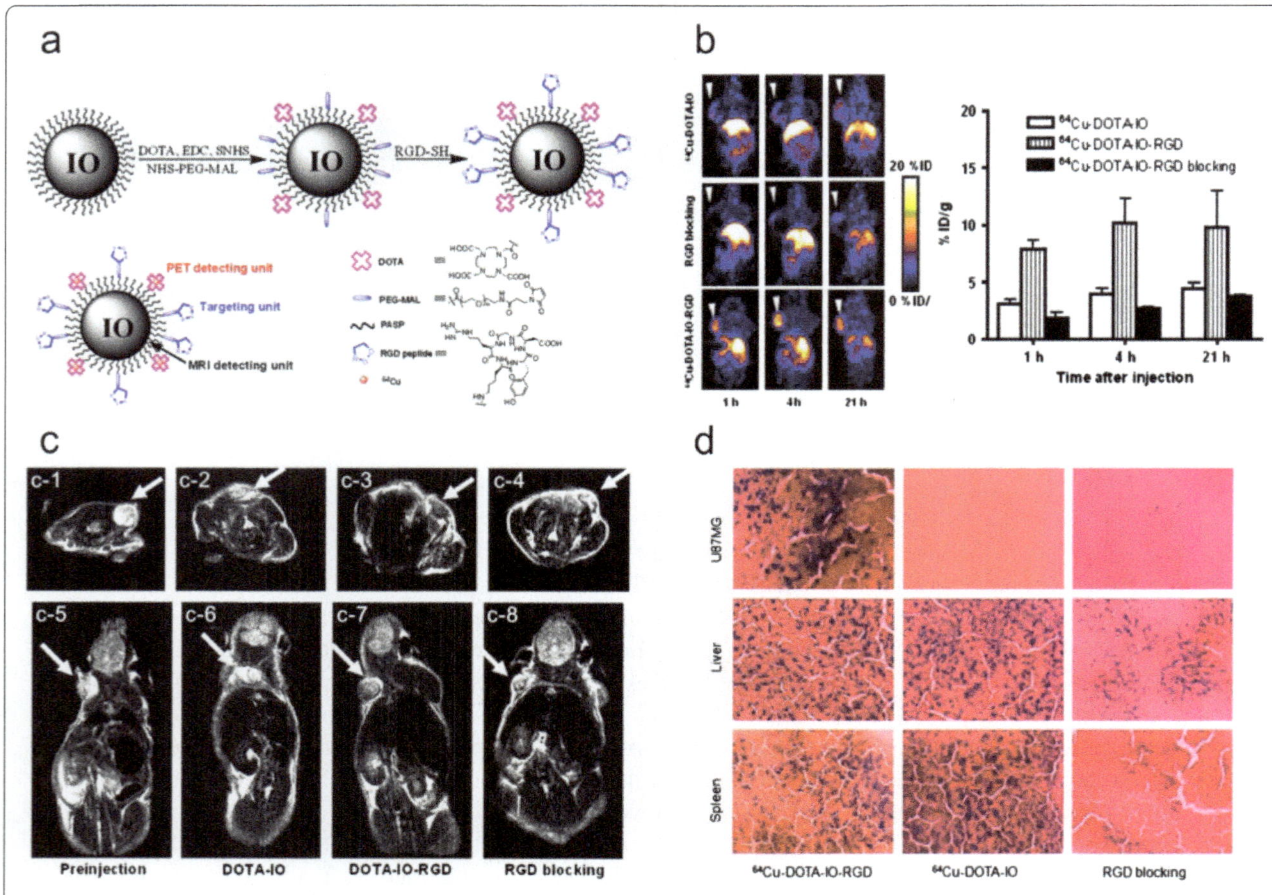

Figure 5 PET(SPECT)/MR imaging. (a) Preparation of PET/MRI dual-modality probe DOTA-IO-RGD. DOTA = 1,4,7,10-tetraazacyclododecane-*N,N′,N″,N‴*-tetraacetic acid; IO = Iron Oxide; RGD = arginine-glycine-aspartic; PEG-MAL = poly(ethylene glycol)-maleimide; PASP = polyaspartic acid. **(b)** Decay-corrected whole-body coronal PET images of human U87MG tumor-bearing nude mouse at 1, 4, and 21 h after injection of ^{64}Cu-DOTA-IO, ^{64}Cu-DOTA-IO-RGD, or ^{64}Cu-DOTA-IO-RGD (left). Time-activity curves of U87MG tumors after injection of ^{64}Cu-DOTA-IO, ^{64}Cu-DOTA-IO-RGD, or ^{64}Cu-DOTA-IO-RGD with blocking dose of c(RGDyK) (n = 3/group) (right). **(c)** T2-weighted MR images of U87MG tumor-bearing nude mice before injection of IO nanoparticles (c-1 and c-5) and at 4 h after tail-vein injection of DOTA-IO (c-2 and c-6), DOTA-IO-RGD (c-3 and c-7), and DOTA-IO-RGD with blocking dose of c(RGDyK) (c-4 and c-8). **(d)** Prussian blue-stained U87MG tumor, liver, and spleen sections after injection of ^{64}Cu-DOTA-IO-RGD, ^{64}Cu-DOTA-IO, and ^{64}Cu-DOTA-IO-RGD with blocking dose of c(RGDyK). IO was stained as blue spots in figure (magnification, 200x). (Adapted with permission from ref. 82. Copyright (2008) The Society of Nuclear Medicine and Molecular Imaging).

such dual-modal imaging can overcome the limits of imaging quality. Especially useful for imaging applications, quantum dots (QDs) are nanometer-sized inorganic fluorescent semiconductor crystals with excellent optical properties such as high quantum yield and molecular extinction coefficients, along with low photo-bleaching and composition-tunable absorption and emission spectra ranging from ultraviolet to near-infrared [45,85,86]. By introducing radionuclides using a chelating agent on the surface of QDs, PET (or SPECT) and NIRF dual-modal imaging techniques can be used. Cai et al. successfully synthesized QDs-based PET/NIRF imaging probes by conjugating QDs with cyclic RGD peptide for targeting and imaging tumors (integrin $\alpha_v\beta_3$ expressing U87MG tumor xenografts) [54] (Figure 6). In that work, DOTA was introduced as a chelating agent for incorporating radionuclides, with ^{64}Cu and CdTe QDs used for NIRF imaging. These dual-modal imaging probes showed excellent tumor uptake behavior, and a linear correlation between PET and NIRF signal intensities was confirmed. Furthermore, QD-based PET/NIRF imaging probes significantly reduced the potential toxicity and overcame the tissue penetration limit of optical imaging. In another study, Liang et al., reported SPECT/NIRF tumor imaging probes using streptavidin coated nanoparticles [87]. These streptavidin nanoparticles were conjugated with three types of molecules: biotinylated anti-Her2 Herceptin antibodiesto target a SUM190 (Her2+) tumor, biotinylated DOTA chelating agents for labeling of radionuclides (^{111}In), and biotinylated Cy5.5 fluorophores for NIRF imaging. The SPECT/NIRF imaging displayed high tumor accumulation and strong tumor-to-normal tissue contrast, and the specific tumor accumulation of radioactivity at 40 h was 21 ID%/g, which denoted a higher accumulation

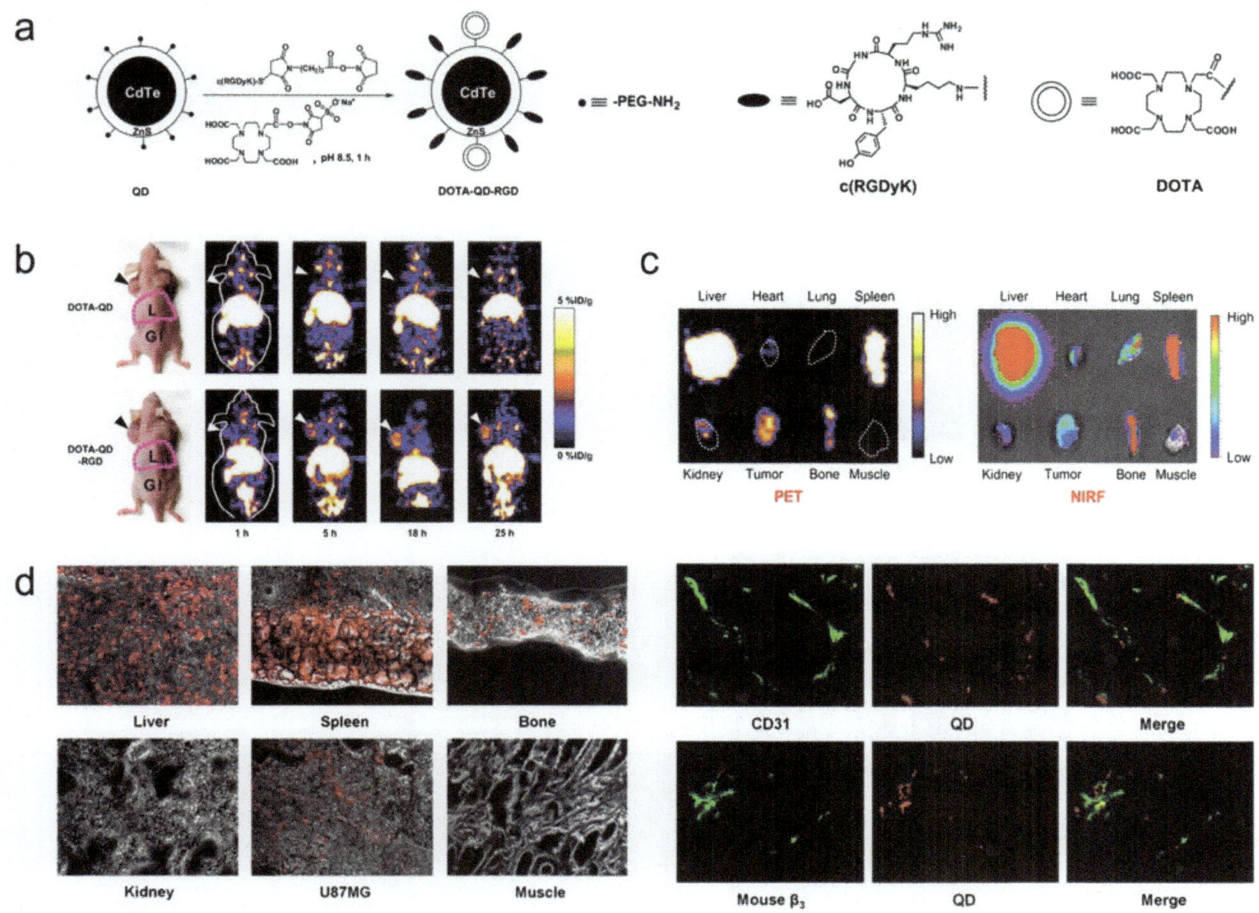

Figure 6 PET(SPECT)/optical imaging. (a) Preparation of dual functional PET/NIRF probe DOTA-QD-RGD. DOTA = 1,4,7,10-tetraazacyclododecane-*N*,*N*′,*N*″,*N*‴-tetraacetic acid; QD = Quantum Dot; RGD = arginine-glycine-aspartic acid; PEG = polyethylene glycol. **(b)** *In vivo* PET of mice bearing U87MG tumor with dual functional PET/NIRF probe. Whole-body coronal PET images of mice at 1, 5, 18, and 25 h after injection of ^{64}Cu-labeled DOTA -QD or DOTA -QD -RGD. Arrows indicate tumors. GI = gastrointestinal tract; L = liver. **(c)** *Ex vivo* PET (left) and NIRF (right) image of harvested tissues at 5 h after injection of ^{64}Cu-labeled DOTA-QD-RGD. **(d)** Histologic analysis of DOTA-QD-RGD distribution in U87MG tumors and other tissues. Overlay of bright-field and QD fluorescence images of frozen tissue slices (5 μm thick) (left). Immunofluorescence staining (CD31 and mouse β_3) of frozen U87MG tumor slices from mice injected with ^{64}Cu-labeled DOTA-QD-RGD (right). (Adapted with permission from ref. 54. Copyright (2010) The Society of Nuclear Medicine and Molecular Imaging).

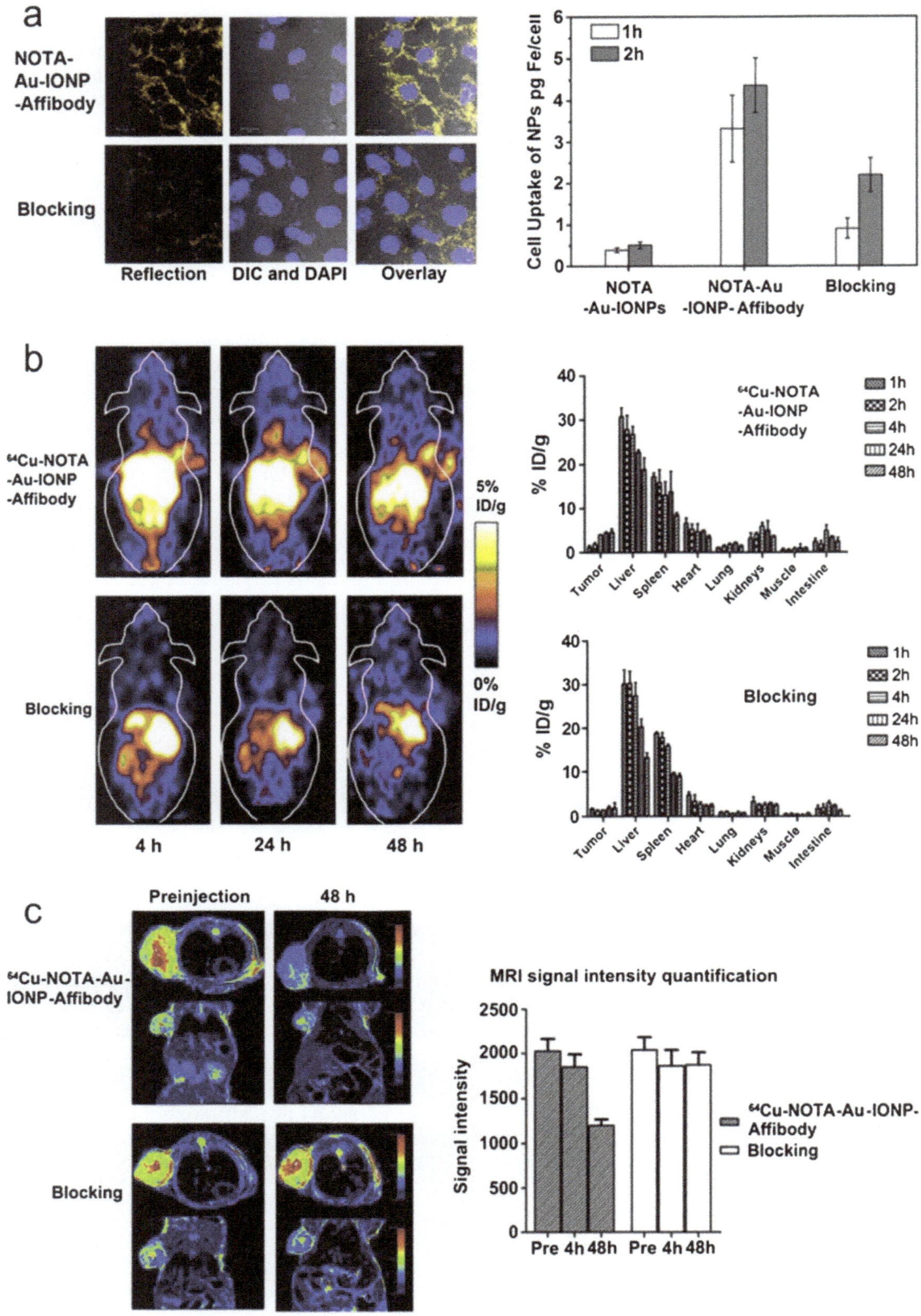

Figure 7 (See legend on next page.)

(See figure on previous page.)
Figure 7 PET(SPECT)/MR/optical imaging. (a) Reflection and confocal images of A431 cells labeled with NOTA-Au-IONP-Affibody (left). NOTA = 1,4,7-triazacyclononane-1,4,7-triacetic acid; IONP = iron oxide nanoparticle. Quantification analysis of the cellular uptake with NOTA-Au-IONP, NOTA-Au-IONP-Affibody, and Affibody blocked NOTA-Au-IONP-Affibody in A431 cells after incubation (right). **(b)** *In vivo* decay-corrected whole-body coronal PET images of A431 tumor bearing mice (n = 3/group) (left). Quantitative analysis of PET images of tissues and organs (right). **(c)** *In vivo* T2-weighted MR images of A431 tumor bearing mice (left). Quantification of MRI signal intensity before and at 4 h and 48 h after injection of ^{64}Cu-NOTA-Au-IONP-Affibody and Affibody (right). (Adapted with permission from ref. 89. Copyright (2013) Elsevier).

than all other tissues, including those from the liver, heart, kidney, spleen, and muscle.

2.3.3 PET(SPECT)/MR/optical imaging

With the easily modifiable surface chemistry of nanoparticles, two or even more imaging labels can be introduced to a single nanoparticle with multiple linking agents (ligands) or direct-incorporation. With this expansive nature, nanoprobes can be designed for imaging modalities beyond two modes. Recently, such multi-functional imaging probes containing three different imaging modalities have been developed [88-91]. Park et al. successfully synthesized optical/PET/MR hybrid nanoprobes by conjugating a PET radionuclide (^{124}I) to an MRI probe (SPIONPs) [88]. These optical/PET/MR hybrid nanoprobes could screen lymph node metastasis with high sensitivity using optical imaging and then provide accurate 3D visualization of accumulated nanoprobes with detailed anatomical information using PET and MR imaging. Yang et al. also provided Affibody modified and radiolabeled Au-IO hetero-nanostructures for tumor optical/PET/MR-based multi-modal imaging probes using NOTA as a chelating agent for a ^{64}Cu radionuclide [89] (Figure 7). Such hetero-nanoprobes can show surface-specific modification of both targeting molecules (anti-EGFR Affibody protein) and PET imaging reporters (^{64}Cu) with high efficiency and reliability. The NOTA-Au-IONP-Affibody was very stable in 10% fetal bovine serum in PBS at 37°C for 48 h with little variation of particle size. By taking advantage of the plasmonic properties of AuNPs imaging, the NOTA-Au-IONP-Affibody could be used for cell assays with a scanning confocal microscope. For both PET and MR imaging, *in vitro* and *in vivo* studies showed high specificity, sensitivity, and excellent tumor contrast in human EGFR-expressing cells and tumors. Xie et al. functionalized dopamine to the surface of IONPs for facile encapsulation into human serum albumin (HSA) matrices, and labeled both ^{64}Cu-DOTA and Cy5.5 onto the HSA coated IONPs to enable PET/MR/NIRF multi-modal imaging [90]. *In vivo* PET/NIRF/MR tri-modality imaging, and *ex vivo* analyses, indicated that the ^{64}Cu-DOTA and Cy5.5 labeled HSA-IONPs have a good retention and high extravasation rates at the tumor site. Impressively, HSA-IONPs showed a prolonged circulation half-life, high accumulation in lesions, and low uptake of the particles by macrophages due to compact HSA coating. Unlike IONP-based MR imaging, Lee et al. reported RGD peptide-conjugated multi-modal Yb^{3+}/Er^{3+} co-doped $NaGdF_4$ upconversion nanophosphors for PET/MR/optical imaging of tumor angiogenesis [91]. In their report, Gd, a component of nanophosphors, was used as a T_1-weighted MRI contrast agent rather than IONPs. The assembly was radiolabeled with ^{124}I and conjugated with a tyrosine residue of dimeric cyclic RGDyk peptide targeting integrin $\alpha v\beta_3$, and Yb^{3+}/Er^{3+} was doped to enable fluorescence imaging. These multi-modal imaging techniques are beneficial for real clinical applications, such as early diagnosis, by maximizing detection at different levels. However, the synthesis of nanostructures for multimodal imaging can present challenges of its own, as steps in synthesizing one mode may interfere with another. Thus, the synthetic procedure must be planned carefully to achieve ideal activity from each imaging mode in the finished product [92].

3 Conclusions

A variety of nanostructures have been utilized for *in vivo* diagnostic and therapeutic agents. In this review, nanostructures labeled with radionuclides were introduced for *in vivo* tumor imaging. Although many chelating agents have been used to incorporate radionuclides into nanostructures for molecular imaging, there is a need to develop radionuclide labeling strategies free of chelating agents for *in vivo* nuclear molecular imaging accuracy while maintaining biocompatibility. Moreover, multimodality molecular imaging should be used to overcome limitations in the imaging quality of a single modality. Although multimodality tumor imaging techniques have been developed remarkably over the past decade, most of those techniques have not moved beyond basic research. More effort should focus on the clinical applications of nanostructures for multimodality imaging. Given the insights that these allow into different physiological properties of tumors and disease, these nanostructures will likely to generate advancements in diagnosis and therapy across a wide range of medical fields.

Competing interests

The authors declare that they have no competing interests.

Authors' contributions

All authors have contributed to the writing of the manuscript. All authors read and approved the final manuscript.

Acknowledgement
J.-M.N. was supported by the National Research Foundation of Korea (2011-0018198) and the Pioneer Research Center Program (2012-0009586) through the NRF of Korea funded by the Ministry of Science, ICT, and Future Planning. K.W.K. was supported by National Nuclear R&D Programs through the National Research Foundation of Korea (NRF) funded by the Ministry of Education, Science and Technology (NRF-2012M2A2A7014020).

Author details
[1]Department of Chemistry, Seoul National University, Seoul 151-747, Republic of Korea. [2]Department of Nuclear Medicine, Seoul National University College of Medicine, Seoul 110-744, Republic of Korea.

References

1. TF Massoud, SS Gambhir, Genes Dev. **17**, 545 (2003)
2. M Shokeen, CJ Anderson, Acc. Chem. Res. **42**, 832 (2009)
3. R Weissleder, Science **312**, 1168 (2006)
4. HR Herschman, Science **302**, 605 (2003)
5. MA Pysz, SS Gambhir, JK Willmann, Clin. Radiol. **65**, 500 (2010)
6. M Mahmoudi, V Serpooshan, S Laurent, Nanoscale **3**, 3007 (2011)
7. SS Gambhir, Nat. Rev. Cancer **2**, 683 (2002)
8. LW Dobrucki, AJ Sinusas, Nat. Rev. Cardiol. **7**, 38 (2009)
9. AF Chatziioannou, Proc. Am. Thorac. Soc. **2**, 533 (2005)
10. ME Phelps, J. Nucl. Med. **41**, 661 (2000)
11. C Anderson, J Connett, S Schwarz, P Rocque, L Guo, G Philpott, K Zinn, C Meares, M Welch, J. Nucl. Med. Allied Sci. **33**, 1685 (1992)
12. AM Wu, PJ Yazaki, T S-w, K Nguyen, A-L Anderson, DW McCarthy, MJ Welch, JE Shively, LE Williams, AA Raubitschek, Proc. Natl. Acad. Sci. U. S. A **97**, 8495 (2000)
13. X Chen, Y Hou, M Tohme, R Park, V Khankaldyyan, I Gonzales-Gomez, JR Bading, WE Laug, PS Conti, J. Nucl. Med **45**, 1776 (2004)
14. DS Wilbur, Bioconj. Chem. **3**, 433 (1992)
15. A Roivainen, T Tolvanen, S Salomäki, G Lendvai, I Velikyan, P Numminen, M Välilä, H Sipilä, M Bergström, P Härkönen, J. Nucl. Med. **45**, 347 (2004)
16. E Mishani, G Abourbeh, Curr. Top. Med. Chem. **7**, 1755 (2007)
17. H Hong, Y Zhang, JT Sun, WB Cai, Nano. Today **4**, 399 (2009)
18. H Xie, ZJ Wang, A Bao, B Goins, WT Phillips, Int. J. Pharm. **395**, 324 (2010)
19. Y Wang, Y Liu, H Luehmann, X Xia, P Brown, C Jarreau, M Welch, Y Xia, ACS Nano. **6**, 5880 (2012)
20. Z Liu, W Cai, L He, N Nakayama, K Chen, X Sun, X Chen, H Dai, Nat. Nanotechnol. **2**, 47 (2006)
21. H Hong, K Yang, Y Zhang, JW Engle, L Feng, Y Yang, TR Nayak, S Goel, J Bean, CP Theuer, Acs Nano **6**, 2361 (2012)
22. B Cornelissen, S Able, V Kersemans, PA Waghorn, S Myhra, K Jurkshat, A Crossley, KA Vallis, Biomaterials **34**, 1146 (2013)
23. JW Seo, H Zhang, DL Kukis, CF Meares, KW Ferrara, Bioconj. Chem. **19**, 2577 (2008)
24. AL Petersen, T Binderup, P Rasmussen, JR Henriksen, DR Elema, A Kjær, TL Andresen, Biomaterials **32**, 2334 (2011)
25. CA Boswell, XK Sun, WJ Niu, GR Weisman, EH Wong, AL Rheingold, CJ Anderson, J. Med. Chem. **47**, 1465 (2004)
26. MS Cooper, MT Ma, K Sunassee, KP Shaw, JD Williams, RL Paul, PS Donnelly, PJ Blower, Bioconj. Chem. **23**, 1029 (2012)
27. KP Maresca, SM Hillier, FJ Femia, CN Zimmerman, MK Levadala, SR Banerjee, J Hicks, C Sundararajan, J Valliant, J Zubieta, WC Eckelman, JL Joyal, JW Babich, Bioconj. Chem. **20**, 1625 (2009)
28. M Zhou, R Zhang, M Huang, W Lu, S Song, MP Melancon, M Tian, D Liang, C Li, J. Am. Chem. Soc. **132**, 15351 (2010)
29. KC Black, Y Wang, HP Luehmann, X Cai, W Xing, B Pang, Y Zhao, CS Cutler, LV Wang, Y Liu, ACS Nano **8**, 4385 (2014)
30. YF Zhao, D Sultan, L Detering, SH Cho, GR Sun, R Pierce, KL Wooley, YJ Liu, Angew. Chem. Int. Ed. **53**, 156 (2014)
31. X Sun, X Huang, X Yan, Y Wang, J Guo, O Jacobson, D Liu, LP Szajek, W Zhu, G Niu, ACS Nano **8**, 8438 (2014)
32. YH Kim, J Jeon, SH Hong, WK Rhim, YS Lee, H Youn, JK Chung, MC Lee, DS Lee, KW Kang, JM Nam, Small **7**, 2052 (2011)
33. R Weissleder, U Mahmood, Radiology **219**, 316 (2001)
34. Y Xing, J Zhao, PS Conti, K Chen, Theranostics **4**, 290 (2014)
35. M Petrik, H Haas, G Dobrozemsky, C Lass-Flörl, A Helbok, M Blatzer, H Dietrich, C Decristoforo, J. Nucl. Med. **51**, 639 (2010)
36. W Chen, T Cloughesy, N Kamdar, N Satyamurthy, M Bergsneider, L Liau, P Mischel, J Czernin, ME Phelps, DH Silverman, J. Nucl. Med **46**, 945 (2005)
37. RJ Jaszczak, KL Greer, CE Floyd Jr, CC Harris, RE Coleman, J. Nucl. Med. **25**, 893 (1984)
38. J Cheon, J-H Lee, Acc. Chem. Res. **41**, 1630 (2008)
39. DA Giljohann, CA Mirkin, Nature **462**, 461 (2009)
40. H Goesmann, C Feldmann, Angew. Chem. Int. Ed **49**, 1362 (2010)
41. J Kim, JE Lee, SH Lee, JH Yu, JH Lee, TG Park, T Hyeon, Adv. Mater. **20**, 478 (2008)
42. T Lammers, S Aime, WE Hennink, G Storm, F Kiessling, Acc. Chem. Res. **44**, 1029 (2011)
43. S-D Li, L Huang, Mol. Pharm. **5**, 496 (2008)
44. Y Liu, MJ Welch, Bioconj. Chem. **23**, 671 (2012)
45. X Michalet, FF Pinaud, LA Bentolila, JM Tsay, S Doose, JJ Li, G Sundaresan, AM Wu, SS Gambhir, S Weiss, Science **307**, 538 (2005)
46. R Mout, DF Moyano, S Rana, VM Rotello, Chem. Soc. Rev **41**, 2539 (2012)
47. X Sun, Z Liu, K Welsher, JT Robinson, A Goodwin, S Zaric, H Dai, Nano Res. **1**, 203 (2008)
48. Y Xia, Nat. Mater. **7**, 758 (2008)
49. J Fang, H Nakamura, H Maeda, Adv. Drug Del. Rev. **63**, 136 (2011)
50. F Alexis, E Pridgen, LK Molnar, OC Farokhzad, Mol. Pharm. **5**, 505 (2008)
51. G Zhang, Z Yang, W Lu, R Zhang, Q Huang, M Tian, L Li, D Liang, C Li, Biomaterials **30**, 1928 (2009)
52. MR McDevitt, D Chattopadhyay, JS Jaggi, RD Finn, PB Zanzonico, C Villa, D Rey, J Mendenhall, CA Batt, JT Njardarson, PLoS One **2**, e907 (2007)
53. ML Schipper, Z Cheng, SW Lee, LA Bentolila, G Iyer, JH Rao, XY Chen, AM Wul, S Weiss, SS Gambhirl, J. Nucl. Med. **48**, 1511 (2007)
54. WB Cai, K Chen, ZB Li, SS Gambhir, XY Chen, J. Nucl. Med. **48**, 1862 (2007)
55. JD Correia, A Paulo, PD Raposinho, I Santos, Dalton Trans. **40**, 6144 (2011)
56. A Bolton, W Hunter, Biochem. J. **133**, 529 (1973)
57. AM Alkilany, CJ Murphy, J. Nanopart. Res. **12**, 2313 (2010)
58. CM Cobley, J Chen, EC Cho, LV Wang, Y Xia, Chem. Soc. Rev. **40**, 44 (2011)
59. CJ Murphy, AM Gole, JW Stone, PN Sisco, AM Alkilany, EC Goldsmith, SC Baxter, Acc. Chem. Res. **41**, 1721 (2008)
60. WK Rhim, JS Kim, JM Nam, Small **4**, 1651 (2008)
61. K Balasubramanian, M Burghard, Small **1**, 180 (2005)
62. A De La Zerda, C Zavaleta, S Keren, S Vaithilingam, S Bodapati, Z Liu, J Levi, BR Smith, T-J Ma, O Oralkan, Nat. Nanotechnol. **3**, 557 (2008)
63. H Hong, T Gao, W Cai, Nano Today **4**, 252 (2009)
64. K Kostarelos, A Bianco, M Prato, Nat. Nanotechnol **4**, 627 (2009)
65. L Feng, Z Liu, Nanomedicine **6**, 317 (2011)
66. B Tian, C Wang, S Zhang, L Feng, Z Liu, Acs Nano **5**, 7000 (2011)
67. K-H Liao, Y-S Lin, CW Macosko, CL Haynes, ACS Appl. Mater. Interfaces **3**, 2607 (2011)
68. K Yang, J Wan, S Zhang, Y Zhang, S-T Lee, Z Liu, ACS Nano **5**, 516 (2010)
69. H Hong, Y Zhang, JW Engle, TR Nayak, CP Theuer, RJ Nickles, TE Barnhart, W Cai, Biomaterials **33**, 4147 (2012)
70. WT Al-Jamal, K Kostarelos, Acc. Chem. Res. **44**, 1094 (2011)
71. PL Felgner, G Ringold, Nature **337**, 387 (1989)
72. VP Torchilin, Nat. Rev. Drug Discov. **4**, 145 (2005)
73. OC Farokhzad, R Langer, Adv. Drug Del. Rev. **58**, 1456 (2006)
74. ML Schipper, G Iyer, AL Koh, Z Cheng, Y Ebenstein, A Aharoni, S Keren, LA Bentolila, JQ Li, JH Rao, XY Chen, U Banin, AM Wu, R Sinclair, S Weiss, SS Gambhir, Small **5**, 126 (2009)
75. LA Bass, M Wang, MJ Welch, CJ Anderson, Bioconj. Chem. **11**, 527 (2000)
76. N Gibson, U Holzwarth, K Abbas, F Simonelli, J Kozempel, I Cydzik, G Cotogno, A Bulgheroni, D Gilliland, J Ponti, F Franchini, P Marmorato, H Stamm, W Kreyling, A Wenk, M Semmler-Behnke, S Buono, L Maciocco, N Burgio, Arch. Toxicol. **85**, 751 (2011)
77. WL Cheng, SJ Dong, EK Wang, Angew. Chem. Int. Ed. **42**, 449 (2003)
78. X Shao, A Agarwal, JR Rajian, NA Kotov, X Wang, Nanotechnology **22**, 135102 (2011)
79. S Singh, R Pasricha, UM Bhatta, PV Satyam, M Sastry, BLV Prasad, J. Mater. Chem. **17**, 1614 (2007)
80. AP Pathak, B Gimi, K Glunde, E Ackerstaff, D Artemov, ZM Bhujwalla, Methods Enzymol. **386**, 1 (2004)
81. Y Ling, K Wei, Y Luo, X Gao, SZ Zhong, Biomaterials **32**, 7139 (2011)
82. HY Lee, Z Li, K Chen, AR Hsu, CJ Xu, J Xie, SH Sun, XY Chen, J. Nucl. Med. **49**, 1371 (2008)

83. JS Choi, JC Park, H Nah, S Woo, J Oh, KM Kim, GJ Cheon, Y Chang, J Yoo, J Cheon, Angew. Chem. Int. Ed. **47**, 6259 (2008)
84. R Misri, D Meier, AC Yung, P Kozlowski, UO Hafeli, Nanomed.-Nanotechnol. Biol. Med. **8**, 1007 (2012)
85. WB Cai, AR Hsu, ZB Li, XY Chen, Nanoscale Res. Lett. **2**, 265 (2007)
86. IL Medintz, HT Uyeda, ER Goldman, H Mattoussi, Nat. Mater. **4**, 435 (2005)
87. M Liang, X Liu, D Cheng, G Liu, S Dou, Y Wang, M Rusckowski, DJ Hnatowich, Bioconj. Chem. **21**, 1385 (2010)
88. JC Park, MK Yu, G Il An, SI Park, J Oh, HJ Kim, JH Kim, EK Wang, IH Hong, YS Ha, TH Choi, KS Jeong, Y Chang, MJ Welch, S Jon, J Yoo, Small **6**, 2863 (2010)
89. M Yang, K Cheng, SB Qi, HG Liu, YX Jiang, H Jiang, JB Li, K Chen, HM Zhang, Z Cheng, Biomaterials **34**, 2796 (2013)
90. J Xie, K Chen, J Huang, S Lee, J Wang, J Gao, X Li, X Chen, Biomaterials **31**, 3016 (2010)
91. J Lee, TS Lee, J Ryu, S Hong, M Kang, K Im, JH Kang, SM Lim, S Park, R Song, J. Nucl. Med. **54**, 96 (2013)
92. BP Joshi, TD Wang, Cancers (Basel) **2**, 1251 (2010)

High performance electrochemical and electrothermal artificial muscles from twist-spun carbon nanotube yarn

Jae Ah Lee[1], Ray H Baughman[2] and Seon Jeong Kim[1*]

Abstract

High performance torsional and tensile artificial muscles are described, which utilize thermally- or electrochemically-induced volume changes of twist-spun, guest-filled, carbon nanotube (CNT) yarns. These yarns were prepared by incorporating twist in carbon nanotube sheets drawn from spinnable CNT forests. Inserting high twist into the CNT yarn results in yarn coiling, which can dramatically amplify tensile stroke and work capabilities compared with that for the non-coiled twisted yarn. When electrochemically driven in a liquid electrolyte, these artificial muscles can generate a torsional rotation per muscle length that is over 1000 times higher than for previously reported torsional muscles. All-solid-state torsional electrochemical yarn muscles have provided a large torsional muscle stroke (53° per mm of yarn length) and a tensile stroke of up to 1.3% when lifting loads that are ~25 times heavier than can be lifted by the same diameter human skeletal muscle. Over a million torsional and tensile actuation cycles have been demonstrated for thermally powered CNT hybrid yarns muscles filled with paraffin wax, wherein a muscle spins a rotor at an average 11,500 revolutions/minute or delivers 3% tensile contraction at 1200 cycles/minute. At lower actuation rates, these thermally powered muscles provide tensile strokes of over 10%.

Keywords: Artificial muscle; Yarn muscle; Carbon nanotube muscle; Electrochemical actuator; Thermal actuator

1 Introduction

Carbon nanotube (CNT) sheets drawn from spinnable CNT forests has been investigated for such applications as actuators [1-3], supercapacitors [4,5], solar cells [6], biofuel cells [7], and acoustic speaker [8]. There are many other potential application areas because of such properties as high electrical conductivity, high mechanical properties, optical transparency for single sheets, and high gravimetric surface area [4,9]. Zhang *et al.* developed methods to produce these sheets from spinnable CNT forest [9] and Lima *et al.* exploited these sheets for their biscrolling process, in which guest-coated CNT sheets are spun into yarn in which the guest is trapped in helical yarn corridors [10]. These hybrid yarns can contain up to 99 wt% of guest and still remain flexible and retain the original properties of the CNT host and guest materials.

CNT tensile actuators and cantilever-based bending actuators, powered by electricity, fuels, light, or heat, have been developed during the past few decades [11]. While electrostatically driven actuation of forest-drawn CNT sheets can generate giant stokes (220%) and giant elongation rates (3.7 X 10^4%/s), these ultralight aerogel-based actuators require high voltages and are not scalable in thickness to generate high forces [12]. Electrochemical actuation (by double-layer charge injection) of CNT sheets fabricated by solution processing provides very small strokes (~0.2%) [13]. Torsional and rotational motors that use single nanotubes have been demonstrated [14], but the nanotubes do not produce the actuation.

In this study, we describe methods for producing high performance CNT-based torsional and tensile artificial muscles by processes that involve inserting twist into forest-drawn CNT sheets. All of these muscles use yarn volume expansion (produced either electrochemically or by thermal expansion of a yarn guest) to reversibly generate yarn contraction and yarn untwist [1,2]. Yarn coiling, produced by inserting high twist into the CNT yarn,

* Correspondence: sjk@hanyang.ac.kr
[1]Center for Bio-Artificial Muscle and Department of Biomedical Engineering, Hanyang University, Seoul 133-791, South Korea
Full list of author information is available at the end of the article

dramatically amplifies tensile stroke and work capabilities compared with those for non-coiled yarn [2].

2 Review

2.1 Electrochemically driven liquid-state and all-solid-state CNT yarn muscles

The precursor yarns used for electrochemical muscles [1,3] were similarly made as those described for thermally powered hybrid muscles [2]. These yarns were fabricated by drawing CNT sheets from a CNT forest (Figure 1a) and inserting twist into a sheet or sheet stack to make a twisted CNT yarn. Twist insertion during forest draw is shown in Figure 1b. Figure 1c and d show, respectively, SEM images of a twisted, non-coiled single-ply yarn and a coiled single-ply yarn that has been coiled as a result of very high twist insertion. The utilized 300–400 μm high CNT forests, from which the CNT sheets were drawn, were prepared by chemical vapor deposition on an iron-catalyst-supporting silicon wafer, using acetylene gas as the carbon precursor. These forests comprised multi-walled CNTs that had an outer diameter of ~10 nm and contain approximately nine walls.

By applying a few volts potential between a twist-spun, non-coiled CNT yarn and a counter-electrode in an electrolyte bath (Figure 2a), Foroughi *et al.* [1] demonstrated that yarn artificial muscles can torsionally rotate a relatively heavy paddle to generate a rotation per length that is over 1000 times higher than for previously reported torsional muscles made from a shape-memory alloy [15], piezoelectric ceramics [16], or conducting polymer [17]. Only one-half of the pictured CNT yarn (which is tethered to prevent torsional end displacement) is submerged in the electrolyte (so that this half provides the torsional actuation, and the second half functions as a return spring to enable highly reversible torsional and tensile actuation). The paddle used to measure torsional rotation was located at near yarn center and the cantilever supporting the top end of the muscle enabled measurement of tensile actuation.

Figure 2b shows torsional rotation and tensile actuation versus time for a CNT yarn muscle (with a 40° yarn bias angle for CNT orientation with respect to the yarn axis) that uses the configuration of Figure 2a. The actuating 12 μm diameter yarn functioned as a torsional artificial muscle to provide a reversible 15,000° rotation and a maximum torsional rotation rate of 590 revolutions per minute, while rotating a paddle that was 830 times the diameter of the actuating yarn and 1800 times its mass. To achieve this result the 120 mm long muscle yarn was pulsed to +5 V (versus Ag/Ag + reference) and then to 0 V for about 5 s each (while using as liquid electrolyte 0.2 M tetrabutylammonium hexafluorophosphate ($TBA.PF_6$) in acetonitrile). The initial paddle acceleration from the Figure 2b results was $\alpha = 50$ rad/s^2 (9000°/s^2), which together with the paddles moment of inertia, provides a maximum starting torque of 1.85 N·m per kilogram of actuating yarn mass. This specific torque is similar to that achieved by large commercial electric motors (2.5 to 6 N·m/kg). The peak torsional power output (above 61 W/kg, based on actuating yarn mass) can be compared with the 300 W/kg realized by large electric motors. This torsional muscle simultaneously acts as a tensile muscle, thereby providing an additional peak power contribution of 920 W/kg by contracting against the applied 88 MPa load, which is equivalent to lifting a mass 185,000 times the mass of the actuating muscle in 1.2 s.

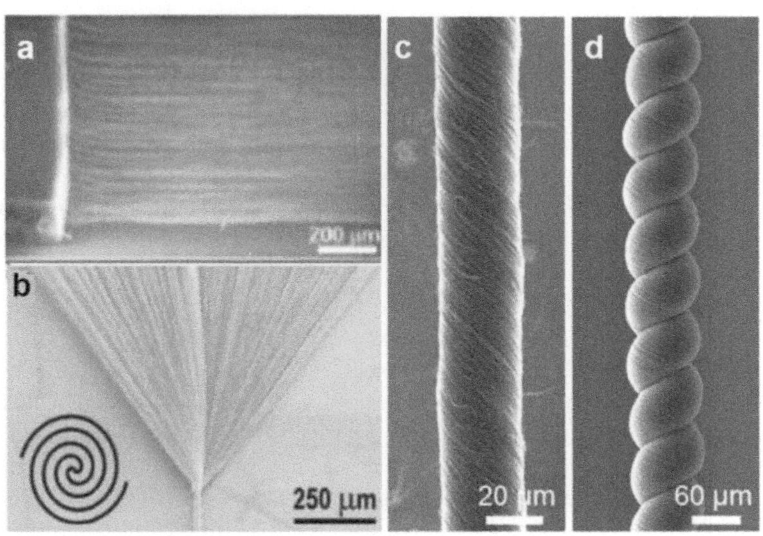

Figure 1 MWNT sheets and yarn structures for torsional and tensile actuation. SEM images of **(a)** a CNT sheet being drawn from a CNT forest (which is on the left in this image), **(b)** Fermat-type twist insertion during spinning a yarn from a CNT forest, **(c)** a single-ply, twisted, non-coiled CNT yarn, and **(d)** a single-ply, coiled yarn [2].

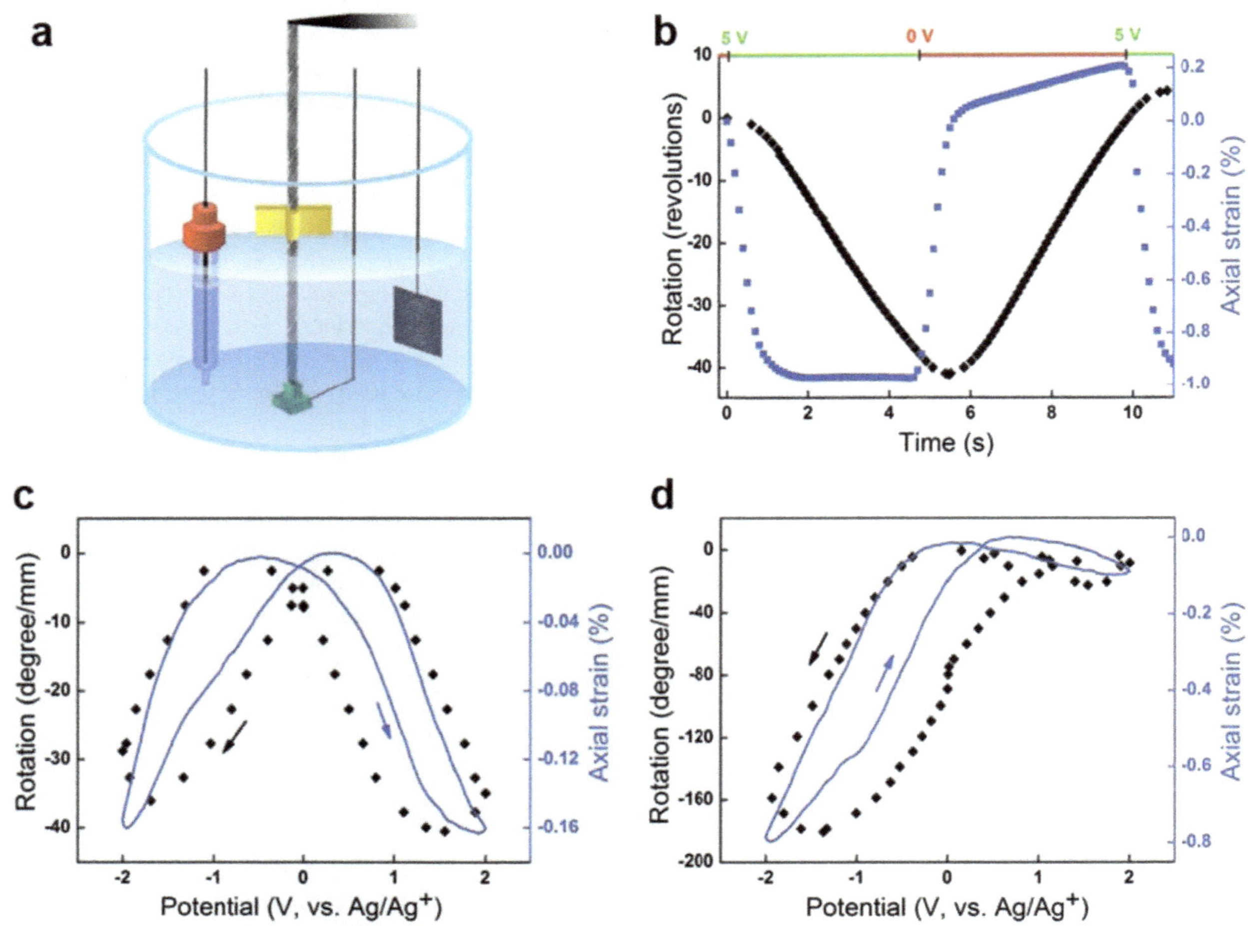

Figure 2 Configuration and results for yarn muscles electrochemically driven in a liquid electrolyte: (a) A two-end-tethered configuration for simultaneously measuring torsional and tensile actuation. **(b)** Torsional rotation (black) and axial length actuation (blue) versus time for a single-ply, twisted, non-coiled CNT yarn that is half-immersed in 0.2 M $TBA.PF_6$ in acetonitrile and pulsed to +5 V (versus Ag/Ag + reference) and then to 0 V for about 5 s each. **(c and d)** Paddle rotation and tensile actuation during a cyclic potential scan at 50 mV/s. The electrolytes were 0.1 M BMP.TFSI in propylene carbonate **(c)** and 0.2 M $TBA.PF_6$ in acetonitrile **(d)** [1].

Figure 2c and d shows that torsional actuation and yarn tensile actuation are correlated and both depend on the size of the electrolyte ion used to compensate electronic charge. When a 1-butyl-1-methylpyrrolidinium bis(trifluoromethanesulfonyl) imide (BMP.TFSI) electrolyte was used (Figure 2c), which has similar van der Waals volumes [18] of the anion (147 $Å^3$) and cation (167 $Å^3$), similar torsional and tensile actuation was shown at both positive and negative potentials. Likewise, the much large actuation during reduction than for oxidation in Figure 2d for the $TBA.PF_6$ electrolyte reflects the larger unsolvated van der Waals volume [18] for the TBA cation (293 $Å^3$) than for the PF_6 anion (69 $Å^3$).

While these advances using liquid electrolytes are fundamentally important (and use of the torsional muscles in a microfluidic circuit was demonstrated) [1], more general practical utilization is restricted by the use of a liquid electrolyte bath and the associated need for a containment system, which dramatically degrades the gravimetric and volumetric performance of the overall actuator device.

Lee *et al.* [3] eliminated these problems by showing that laterally connected anode and cathode CNT yarns could be combined in one torsional yarn muscles (Figure 3), thereby enabling all-solid-state electrochemical muscles that provide giant torsional actuation. Identical twisted, non-coiled anode and cathode yarns were made by inserting twist into forest-drawn CNT sheets, and then infiltrating these yarns with a gel electrolyte (poly(vinylidene fluoride-co-hexafluoropropylene) (PVDF-co-HFP), containing tetraethyl ammonium tetrafluoroborate ($TEA.BF_4$) with propylene carbonate (PC)). These two electrolyte-coated yarns were plied together using opposite twist directions for yarn plying than for the initially introduced yarn twist (Figure 3c, d). This use of opposite twist directions for yarn twist and yarn plying is important, since it results in a torque-balanced system when no

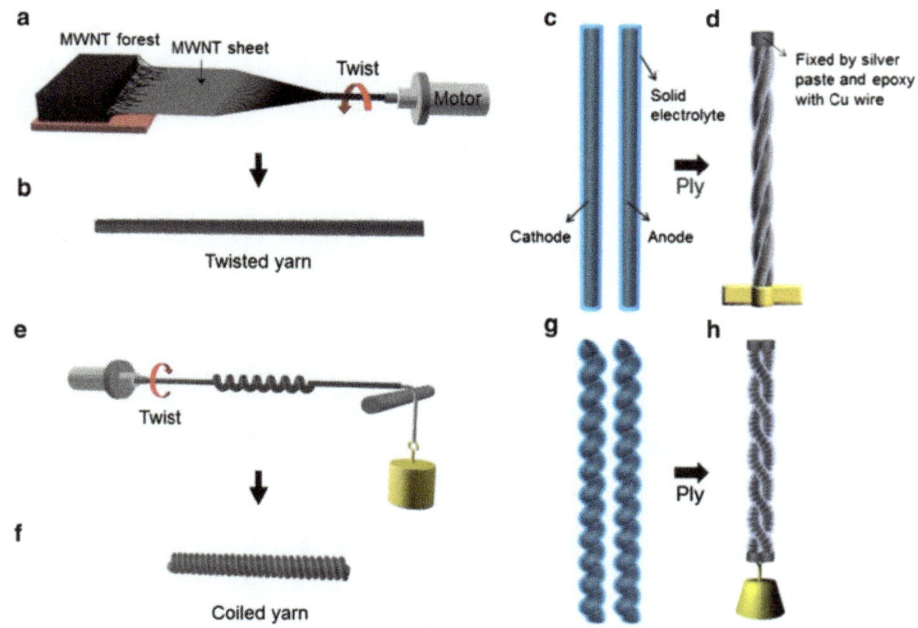

Figure 3 Illustration of yarn fabrication methods for torsional and tensile electrochemical CNT muscles. (a) A CNT sheet was pulled from a CNT forest as twist was inserted to provide yarn for the torsional muscles. Inserting twist into stacks of forest-drawn CNT sheets enabled increases in yarn diameter. **(b)** A Twisted, non-coiled CNT yarn, containing about 12,000 to 15,000 turns/m of inserted twist when a single CNT sheet was used for yarn fabrication. **(c)** Anode and cathode yarns infiltrated and coated with PVDF-co-HFP based TEA.BF_4 solid gel electrolyte. **(d)** Illustration of a plied, non-coiled yarn muscle with attached paddle for torsional actuation measurements, which had been over coated with the above electrolyte after plying two identical electrolyte-filled CNT yarns. **(e)** Illustration of the insertion of additional twist to produce yarn coiling, which was deployed to provide the fully coiled yarns used for tensile muscles. **(f)** Fully-coiled yarn used as both muscle anode and cathode for tensile muscles. **(g)** Coiled anode and cathode yarns that are infiltrated and coated with PVA-based H_2SO_4 solid gel electrolyte. **(h)** Illustration of a plied, coiled yarn muscle with attached weight for tensile actuation measurements, which had been over coated with the above electrolyte after plying [3].

voltage is applied between anode and cathode yarns, which enables reversible actuation.

Torsional actuation was obtained using the configuration of Figure 3d, where a one-end-tethered, plied, non-coiled yarn supports a paddle on the free end. Fast, reversible torsional strokes (14-53°/mm) with small hysteresis resulted when using a PVDF-co-HFP/TEA.BF_4/PC solid gel electrolyte and applied square-wave voltages between 2.5 and 5 V (Figure 4a and b). A gravimetric peak torque of 0.067 N•m/kg (3.05 nN•m) was generated for this 29 μm diameter yarn by applying a square-wave voltage of 5 V.

To obtain enhanced tensile actuation for all-solid-state electrochemical muscles, identical anode and cathode yarns in the muscles were coiled (Figure 3e and f). So that we could more conveniently investigate tensile actuation, large diameter CNT yarns were made by inserting twist into a 30-layer stack of forest-drawn sheets. After simultaneously infiltrating and coating both anode and cathode yarn with an aqueous gel electrolyte (polyvinyl alcohol (PVA) in 1 M aqueous sulfuric acid) left-handed coiled electrodes (Z twist) were plied by S twist (right-handed twist) to form a torque-balanced, two-ply, solid-state muscle (Figures 3g and h).

A reversible 0.52% tensile contraction was obtained for this electrochemical muscle (in the configuration of Figure 3h) when a square-wave voltage of 1V was applied to lift a load that provided an 11 MPa stress (Figure 4c). This load lifting capability is 27 times that of human skeletal muscle. Importantly, this electrochemically driven all-solid-state muscle provided a latched state, where charging-induced contraction was largely maintained after subsequent disconnection from the power supply (Figure 4d). In fact, 91.5% of the muscle contraction was maintained for one hour following discontinuation of the applied voltage.

2.2 Thermally and electrothermally powered CNT yarn hybrid muscles

This last category of twist-spun yarn muscles is powered by temperature changes, which can be produced by such processes as electrical heating (using the electrical conductivity of the CNTs), changes in environmental temperature, absorption of light or other electromagnetic radiation, or chemical reactions. These muscles are called hybrid CNT yarn muscles, since the thermally driven volume expansion of a guest within the twist-spun CNT yarn drives the reversible contraction

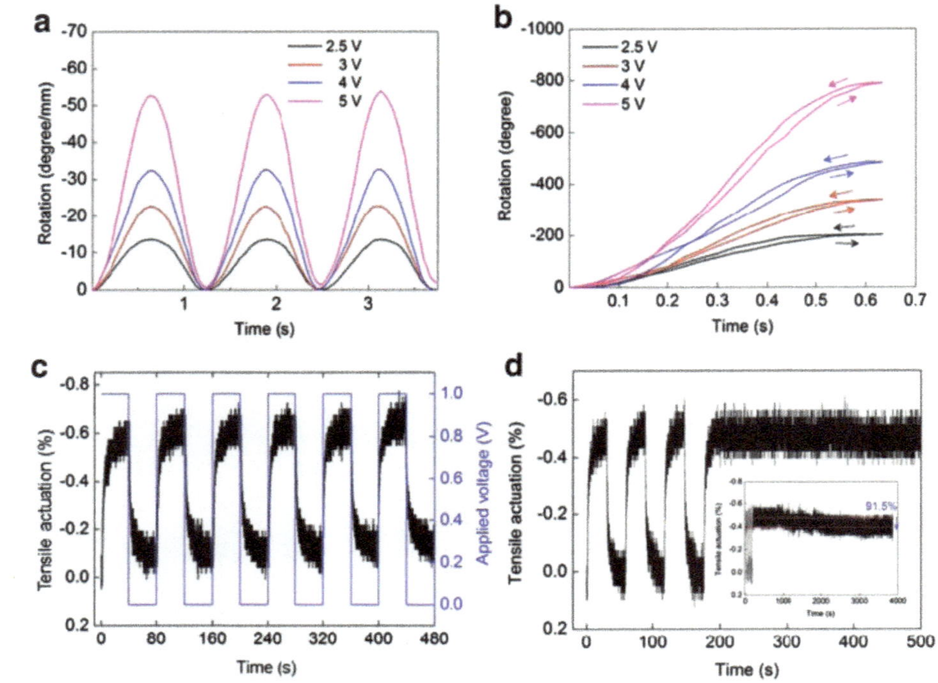

Figure 4 Torsional and tensile actuation for all-solid-state electrochemical CNT muscles. (a) Torsional rotation per muscle length, in response to different applied square-wave voltages, which were measured by paddle rotation for a two-ply, twist-spun, CNT yarn that was impregnated with PVDF-co-HFP/TEA.BF_4/PC solid gel electrolyte. The direction of yarn twist was opposite to that applied for yarn plyimg. **(b)** Data showing the largely non-hysteretic reversal of torsional actuation for the torsional muscle used in **(a)**. **(c)** Tensile actuation strain and applied voltage versus time for a two-ply yarn muscle, using polyvinyl alcohol/H_2SO_4 electrolyte, while lifting a 11 MPa load. **(d)** Tensile actuation for the tensile muscle of **(c)** during three repeated charge (1 V square wave) and discharge cycles that were followed by a hold period, where the power supply was disconnected. Inset: actuator stroke during an extended hold period where no potential was applied, showing that tensile actuation has a latched state, where actuator stroke is largely maintained for an hour without the need for input power [3].

of tensile muscles and the yarn untwist that provides torsional actuation [2]. The main function of the CNT yarn is to confine this actuating yarn guest, insure mechanical durability, enable electrical heating, and provide a helical geometry that can enable torsional actuation and amplify the effect of volume change on tensile actuator stroke.

The methods for making the twisted and coiled CNT yarns are like those above described for the electrochemical methods, and various methods can be used for incorporating high volume fractions of guest in these yarns. These methods include, for example, melt or solution infiltration (which can be followed by in situ polymerization) and biscrolling, in which the guest is deposited on a CNT sheet before twist insertion [2]. Use of paraffin wax as guest provides large yarn volume changes, and thereby large torsional and tensile strokes, so the experimental results described herein use this guest for hybrid yarn muscles.

As is true, for the twist-spun electrochemical muscles, yarn muscle structure and the type of end tethering is key for enabling large muscle strokes (tensile and torsional), as well as long muscle cycle lifetimes. One hundred tensile actuation cycles during lifting have been demonstrated for electrochemically powered a plied, coiled CNT yarn muscle (Figure 5). The tensile actuation did not significantly change during cycles. The chirality of yarn twist versus that of yarn plying is important for determining muscle performance, as is the relative chirality of possibly different segments in a yarn muscle. As a more precise refinement generalization of previous nomenclature [2], we here define a muscle yarn segment as homochiral or heterochiral according to whether or not this segment has the same or different chiralities for yarn twist and yarn plying in this segment.

Figure 6 shows the different muscle structures and configurations that have experimentally investigated for these hybrid yarns [2]. Using opposite chirality plied yarn segments (which are each homochiral), with a paddle at their interconnection (Figure 6d), maximizes initial torque on the paddle, because these segments operate additively to generate rotation. For the same two-ply homochiral yarn, but without segmentation into sections having opposite chirality, the one-end–tethered

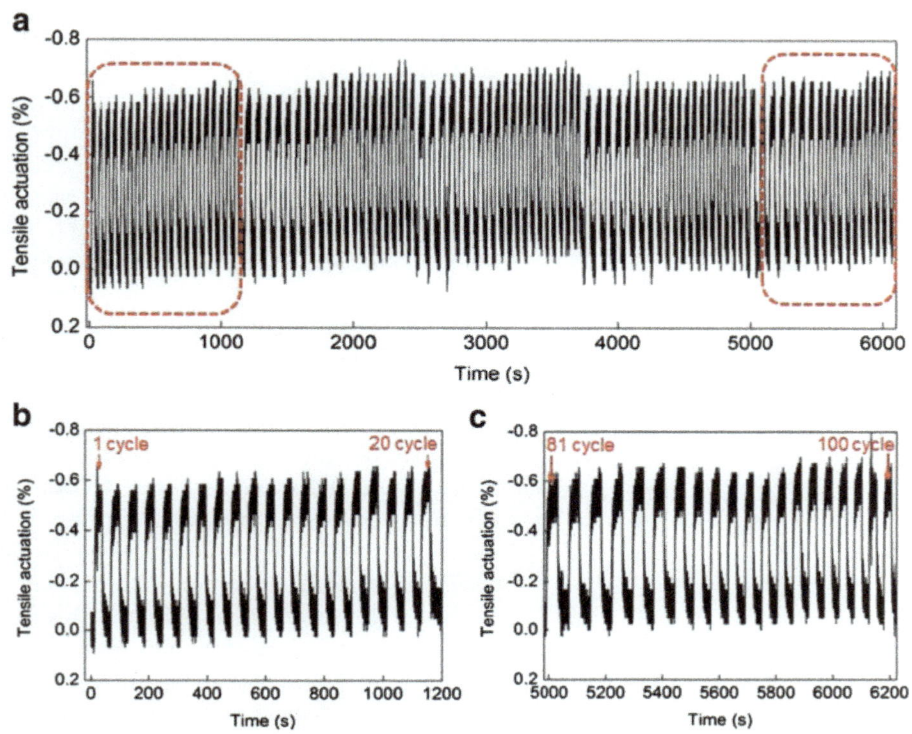

Figure 5 Cycling stability of electricochemically driven tensile actuator in solid electrolyte. (a) Tensile actuation versus time for 100 actuation cycles of the muscle of Figure 3h. **(b,c)** The first and last 20 actuation cycles of **(a)** using an expanded time scale.

configuration of Figure 6c provides twice the torsional rotation of the Figure 6d configuration, but one-half the initial torque, so both configurations provide equal torsional work capacity. Actuation of one segment in a two-end–tethered yarn (Figure 6b) generates smaller paddle rotation than for the same yarn in the configuration of Figure 6d (where two opposite chirality segments are used and both are actuated) because of the energetic cost of twisting the unactuated yarn as the actuating yarn untwists. As for the Figure 6c configuration, the Figure 6d configuration with non-plied yarn does not provide reversible actuation unless internally constrained by a solid guest, to prevent left-handed twist (Z twist) in one muscle segment from canceling right-handed twist (S twist) in the other yarn segment. Since untwist of the actuating segment is compensated by up-twist of the non-actuated segment in Figure 6a, this actuator configuration can optimize tensile contraction per actuated muscle length. Since the entire yarn untwists during actuation for the Figure 6c and d configurations (and this untwist lengthens the muscle), they do not provide maximized percent tensile contraction.

Although other guest materials can be used to obtain high performance tensile and torsional hybrid yarn muscles (some of which are more suitable for use in extremely low and high temperature ranges), the presently described results are for prototypical hybrid muscles that use paraffin wax guest. Figure 7a shows the dramatic increase in tensile actuator stroke that results from the incorporation of paraffin wax in a twist-spun, coiled CNT yarn and the figure inset shows the much smaller thermo-dimensional changes that occur for twisted, non-coiled yarns that either contain wax or are free of guest. Figure 7b shows that yarn twist insertion to produce complete coiling results in a large tensile actuator contraction when the neat yarn is heated to extreme temperatures: a 7.3% reversible tensile contraction upon heating to ~2560°C under 3.8-MPa tensile stress, corresponding to 0.16 kJ/kg of contractile work capability per yarn weight. The stroke for highly coiled, wax-filled yarn decreases at low applied stresses (Figure 7c), which is consistent with the close proximity of adjacent coils hindering contraction. This figure shows that there is an optimal amount of coiling that maximizes either stroke or work during contraction for the wax hybrid yarn. A maximum contraction of 5.6% was observed at 5.7 MPa stress for a coiled CNT yarn having intermediate twist. Adding 6.8% more twist to the coiled yarn increased the stress of maximum contraction (16.4 MPa for 5.1% strain) and the maximum measured contractile work (1.36 kJ/kg for 84 MPa load).

Figure 6 Muscle configurations for tensile and torsional actuation. Tensile load and paddle positions for **(a)** a two-end–tethered, fully-infiltrated, coiled, homochiral, single-ply yarn; **(b)** a two-end–tethered, bottom-half–infiltrated, non-coiled, twisted homochiral single-ply yarn; **(c)** a one-end–tethered, four-ply, fully-infiltrated homochiral yarn; and **(d)** a two-end–tethered, two-ply, fully-infiltrated yarn muscle comprising two homochiral segments having opposite chiralities [2]. Arrows indicate the observed direction of paddle rotation during thermal actuation. Red and green yarn-end attachments are tethers, meaning they prohibit end rotation; red attachments also prohibit translational displacement.

Fast, highly reversible torsional actuation was demonstrated for 2 million cycles for two-end–tethered, half-wax–infiltrated homochiral yarn that rotated a paddle at yarn midpoint (Figure 6b configuration). The hybrid yarn accelerated a 16.5 times-heavier paddle to a full-cycle–averaged 11,500 rotations per minute—first in one direction and then in reverse (Figure 8a). Figure 8b shows the dependence of torsional rotation on input electrical power and applied tensile load for a similar yarn that rotated a 150-times-heavier paddle for a million highly reversible cycles. The average speed was 7,600 revolutions/minute.

3 Conclusions

The reviewed work on twist-spun CNT artificial muscles demonstrates the extremely high torsional and tensile muscle performance than can be obtained by the use of twisted, non-coiled and twisted, coiled CNT yarns. The liquid-electrolyte-based twisted, non-coiled muscles (1) generate highly reversible torsional rotation (250°/mm of muscle length), which is 1000 times that for prior-art muscles, as well as a 590 revolutions/minute maximum rotation rate that was maintained for 30 full rotations and (2) a gravimetric peak torsional power output similar to that of large, high-power electric motor [1]. While the demonstrated use of these muscles for microfluidic circuits might become practical, the need for an electrolyte bath and associated confinement system is problematic (especially since it dramatically increases system weight and volume).

To eliminate this problem, our team has further demonstrated thermally powered CNT hybrid muscles and all-solid-state electrochemical CNT muscles that can be operated in air [2,3]. Paraffin-wax-filled thermally powered CNT hybrid muscles provides 2 million cycles of torsional actuation without evidence of performance decrease with cycling and an average torsional rotation speed of 11,500 rpm (20 times that of our electrochemical torsional muscles). When configured as coiled tensile muscles, these wax-hybrid muscles generated 27.8 kW/kg of power during muscle contraction, which is 80 times that for natural skeletal muscles and 30 times that we realized for our electrochemical, liquid-electrolyte CNT muscles.

We next invented all-solid-state electrochemical muscles that avoid the problems of our liquid electrolyte electrochemical muscles - since no liquid bath and associated confinement system is needed, the actuator volume and weight is only that of coupled anode and cathode muscle fibers and associated relatively minor amount of gel electrolyte [3]. The presently realized torsional muscle stroke for plied electrochemical CNT anode and cathode (53°/mm of muscle length) is giant compared with muscles other than described here, but still smaller than the 250°/mm of muscle length that we obtained for liquid-electrolyte-based muscles. Small, but useable, tensile muscle strokes were obtained (1.3% at 2.5 V and 0.52% at 1 V) when lifting loads that are ~25 times heavier than can be lifted by the same diameter human skeletal muscle. Very importantly, in contrast to the case for thermal muscles, we demonstrated the existence of a natural latched state, where stroke (contraction) can be largely maintained with little decrease for an hour when the power supply is disconnected.

There are many opportunities for further improvements in these described actuator technologies, such as in obtaining high electrical-to-mechanical energy efficiencies and large tensile strokes for the all-solid-state electrochemical muscles. More recently realized advances in both areas will be described in the future.

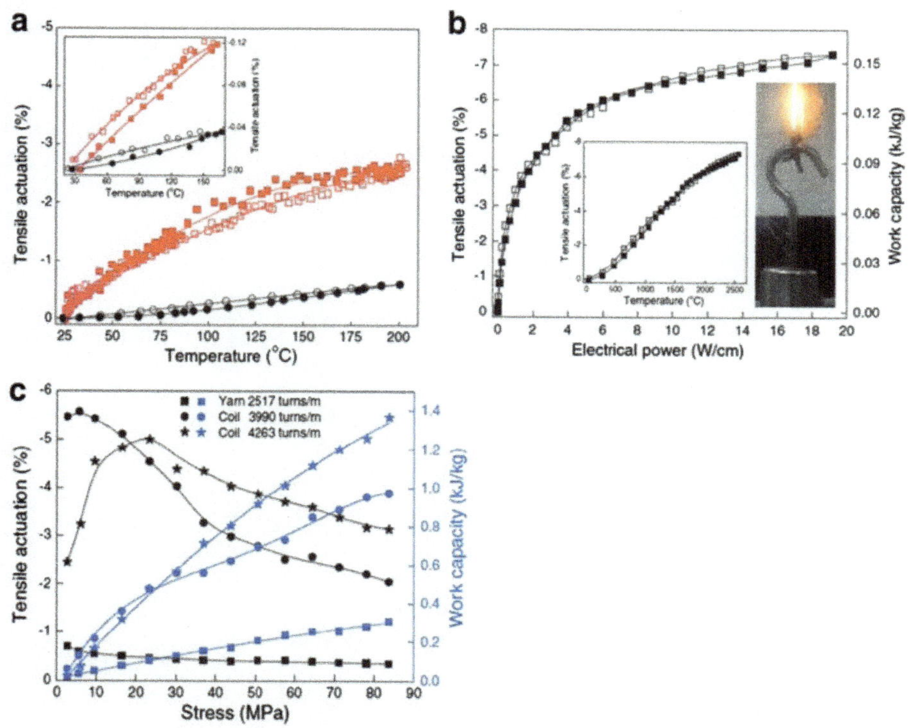

Figure 7 Thermal tensile actuation for two-end–tethered homochiral yarns. (a) Tensile actuation strain versus temperature before (black) and after (red) wax infiltration for a coiled yarn having an applied stress of 6.8 MPa. Inset: Corresponding actuation data before (black) and after (red) wax infiltration for a non-coiled yarn having an applied stress of 4.8 MPa. **(b)** Electrothermal tensile actuation strain and work capacity during contraction in vacuum as a function of applied electrical power for a neat, coiled CNT yarn. Insets: Tensile actuation versus estimated temperature for this neat yarn (left) and photograph of the incandescent yarn lifting a 10 g load. Closed symbols and open symbols in **(a)** and **(b)** are for increasing and decreasing temperature, respectively. **(c)** The stress dependence of steady-state tensile actuation and contractile work (black and blue data points, respectively) produced by Joule heating (0.19 V/cm) for a 150-mm-diameter, coiled yarn having different levels of inserted twist [2].

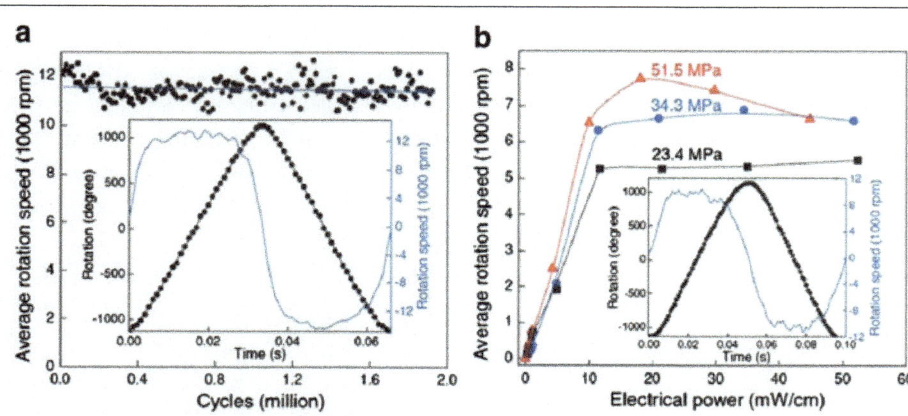

Figure 8 Torsional actuation for two-end–tethered, wax-infiltrated yarns. (a) Average rotation rate versus cycle number when excited by a 15 Hz, 40 V/cm, square-wave voltage using 50% duty cycle and 41 MPa applied load. Each point on the graph is the average speed over 120 cycles. Inset: rotation angle and rotation speed versus time for one complete cycle **(b)** Average rotation rate versus applied electrical power for different tensile loads when using the yarn of **(a)** and a heavier paddle. Inset: rotation angle and speed versus time for 51.5 MPa load [2].

Competing interests
The authors declare that they have no competing interests.

Authors' contributions
All are participated in the preparation of preparation of manuscript and discussion. All authors read and approved final manuscript.

Acknowledgments
This work was supported in Korea by the Creative Research Initiative Center for Bio-Artificial Muscle of the Ministry of Science, ICT & Future Planning (MSIP), the MSIP-US Air Force Cooperation Program (NRF-2013K1A3A1A32035592) and the Industrial Strategic Technology Program (10038599) and in the United States by Air Force Grant AOARD-13-4119, Air Force Office of Scientific Research grant FA9550-12-1-0211, and Robert A. Welch Foundation grant AT-0029.

Author details
[1]Center for Bio-Artificial Muscle and Department of Biomedical Engineering, Hanyang University, Seoul 133-791, South Korea. [2]The Alan G. MacDiarmid NanoTech Institute, University of Texas at Dallas, Richardson, TX 75083, USA.

References
1. J. Foroughi, G.M. Spinks, G.G. Wallace, J. Oh, M.E. Kozlov, S. Fang, T. Mirfakhrai, J.D.W. Madden, M.K. Shin, S.J. Kim, R.H. Baughman, Torsional carbon nanotube artificial muscles. Science **334**, 494–497 (2011). doi:10.1126/science.1211220
2. M.D. Lima, N. Li, M.J.D. Andrade, S. Fang, J. Oh, G.M. Spinks, M.E. Kozlov, C.S. Haines, D. Suh, J. Foroughi, S.J. Kim, Y. Chen, T. Ware, M.K. Shin, L.D. Machado, A.F. Fonseca, J.D.W. Madden, W.E. Voit, D.S. Galvão, R.H. Baughman, Electrically, chemically, and photonically powered torsional and tensile actuation of hybrid carbon nanotube yarn muscles. Science **338**, 928–932 (2012). 10.1126/science.1226762
3. J.A. Lee, Y.T. Kim, G.M. Spinks, D. Suh, X. Lepró, M.D. Lima, R.H. Baughman, S. J. Kim, All solid state carbon nanotube torsional and tensile artificial muscles. Nano Lett **14**, 2664–2669 (2014). 10.1021/nl500526r
4. J.A. Lee, M.K. Shin, S.H. Kim, S.J. Kim, G.M. Spinks, G.G. Wallace, R. Ovalle-Robles, M.D. Lima, M.E. Kozlov, R.H. Baughman, Hybrid nanomembranes for high power and high energy density supercapacitors and their yarn application. ACS Nano **6**, 327–334 (2012). doi:10.1021/nn203640a
5. J.A. Lee, M.K. Shin, S.H. Kim, H.U. Cho, G.M. Spinks, G.G. Wallace, M.D. Lima, X. Lepro´, M.E. Kozlov, R.H. Baughman, S.J. Kim, Ultrafast charge and discharge biscrolled yarn supercapacitors for textiles and microdevices. Nat Commun **4**, 1970 (2013). doi: 10.1038/ncomms2970
6. J. Velten, Z. Kuanyshbekova, O. Goktepe, F. Goktepe, A. Zakhidov, Weavable dye sensitized solar cells exploiting carbon nanotube yarns. Appl Phys Lett **102**, 203902–203902–3 (2013). doi:10.1063/1.4807891
7. C.H. Kwon, S.H. Lee, Y.B. Choi, J.A. Lee, S.H. Kim, H.H. Kim, G.M. Spinks, G.G. Wallace, M.D. Lima, M.E. Kozlov, R.H. Baughman, S.J. Kim, High-power biofuel cell textiles from woven biscrolled carbon nanotube yarns. Nat Commun **5**, 3928 (2014). 10.1038/ncomms4928
8. A.E. Aliev, M.D. Lima, S. Fang, R.H. Baughman, Underwater sound generation using carbon nanotube projectors. Nano Lett **10**, 2374–2380 (2010). 10.1021/nl100235n
9. M. Zhang, S. Fang, A.A. Zakhidov, S.B. Lee, A.E. Aliev, C.D. Williams, K.R. Atkinson, R.H. Baughman, Strong, transparent, multifunctional, carbon nanotube sheets. Science **309**, 1215–1259 (2005). doi:10.1126/science.1115311
10. M.D. Lima, S. Fang, X. Lepró, C. Lewis, R. Ovalle-Robles, J. Carretero-González, E. Castillo-Martínez, M.E. Kozlov, J. Oh, N. Rawat, C.S. Haines, M.H. Haque, V. Aare, S. Stoughton, A.A. Zakhidov, R.H. Baughman, Biscrolling nanotube sheets and functional guests into yarns. Science **331**, 51–55 (2011). 10.1126/science.1195912
11. D. Li, W.F. Paxton, R.H. Baughman, T.J. Huang, J.F. Stoddart, P.S. Weiss, Molecular, supramolecular, and macromolecular motors and artificial muscles. MRS Bull **34**, 671 (2009). http://dx.doi.org/10.1557/mrs2009.179
12. A.E. Aliev, J. Oh, M. Kozlov, A.A. Kuznetsov, S. Fang, A.F. Fonseca, Giant-stroke, superelastic carbon nanotube aerogel muscles. Science **323**, 1575–1578 (2009). 10.1126/science.1168312
13. R.H. Baughman, C. Cui, A.A. Zakhidov, Z. Iqbal, J.N. Barisci, G.M. Spinks, G.G. Wallace, A. Mazzoldi, D.D. Rossi, A.G. Rinzler, O. Jaschinski, S. Roth, M. Kertesz, Carbon nanotube actuators. Science **284**, 1340–1344 (1999). 10.1126/science.284.5418.1340
14. A.M. Fennimore, T.D. Yuzvinsky, W.-Q. Han, M.S. Fuhrer, J. Cumings, A. Zettl, Rotational actuators based on carbon nanotubes. Nature **424**, 408–410 (2003). 10.1038/nature01823
15. A.C. Keefe, G.P. Carman, Thermo-mechanical characterization of shape memory alloy torque tube actuators. Smart Mater Struct **9**, 665 (2000). 10.1088/0964-1726/9/5/311
16. J. Kim, B. Kang, Performance test and improvement of piezoelectric torsional actuator. Smart Mater Struct **10**, 750–757 (2001). 10.1088/0964-1726/10/4/320
17. Y. Fang, T.J. Pence, X. Tan, Fiber-directed conjugated-polymer torsional actuator: nonlinear elasticity modeling and experimental validation. IEEE/ASME Trans Mechatron **16**, 656–664 (2011). 10.1109/TMECH.2010.2049366
18. M. Ue, A. Murakami, S. Nakamura, A convenient method to estimate ion size for electrolyte materials design. J Electrochem Soc **149**, A1385 (2002). 10.1149/1.1507593

14

X-ray generation using carbon nanotubes

Richard J Parmee*†, Clare M Collins, William I Milne and Matthew T Cole*†

Abstract

Since the discovery of X-rays over a century ago the techniques applied to the engineering of X-ray sources have remained relatively unchanged. From the inception of thermionic electron sources, which, due to simplicity of fabrication, remain central to almost all X-ray applications, there have been few fundamental technological advances. However, with the emergence of ever more demanding medical and inspection techniques, including computed tomography and tomosynthesis, security inspection, high throughput manufacturing and radiotherapy, has resulted in a considerable level of interest in the development of new fabrication methods. The use of conventional thermionic sources is limited by their slow temporal response and large physical size. In response, field electron emission has emerged as a promising alternative means of deriving a highly controllable electron beam of a well-defined distribution. When coupled to the burgeoning field of nanomaterials, and in particular, carbon nanotubes, such systems present a unique technological opportunity. This review provides a summary of the current state-of-the-art in carbon nanotube-based field emission X-ray sources. We detail the various fabrication techniques and functional advantages associated with their use, including the ability to produce ever smaller electron beam assembles, shaped cathodes, enhanced temporal stability and emergent fast-switching pulsed sources. We conclude with an overview of some of the commercial progress made towards the realisation of an innovative and disruptive technology.

Keywords: X-ray; Field emission; Electron source; Carbon nanotubes; Carbon nanofibres; Nano carbons; Screen printing; Electrophoresis; Vacuum filtration; Chemical vapour deposition; Tomosynthesis; Micro focal

1 Introduction

Since the discovery of X-rays in 1895 [1], X-ray analysis and diagnostics have been some of the most widely researched areas in science and engineering. In the first quarter of the last century, almost half of the Nobel prizes awarded were connected with contributions to this field. Whilst the initial discoveries created great interest towards understanding the source of this novel light, it was only a short period before the applications of this "new light" became evident. J.J. Thomson had already established their ionising nature [2] and it was recognised that X-rays had different characteristics, initially termed qualitatively as "hard", "medium" and "soft" [3]; a classification which related to the relative absorption by soft tissue and bone. Certainly even at this early phase the implications for medical diagnostics was clear. Early examples included the identification of bullets and bone fractures in wounded soldiers. This resulted in a range of instruments, various densitometers, chiroscopes and osteoscopes [4]. Soon thereafter this lead to the analysis of the diagnostic capabilities and therapeutic benefits of this new technology [5]. The increasing importance of X-ray techniques was evident with the development of X-ray diffraction methods [6-8] which lead ultimately to evidence the structure of DNA [9]. Compton back-scattering and fluorescence provided yet another raft of valuable applications [10-13]. Nevertheless, despite the growing range of X-ray techniques, the primary area of application of X-rays has remained in medical diagnosis. Additionally, X-ray sources have proven important in a wide range of inspection technologies; for medical and analytical, to security and industrial quality control. As illustrated in Figure 1, X-ray sources can be applied to a wide range of applications, including materials analysis (X-ray fluorescence, X-ray diffraction), medical and dental diagnosis and treatment (transmission radiography, computed tomography, tomosynthesis, radiotherapy and bracotherapy), electronics inspection, food security, non-destructive testing, thickness and defect monitoring, pharmaceutical quality control, and international border security. Indeed, X-ray sources are ubiquitous and have

* Correspondence: rjp89@cam.ac.uk; mtc35@cam.ac.uk
†Equal contributors
Department of Engineering, Electrical Engineering Division, Cambridge University, Cambridge CB3 0FA, UK

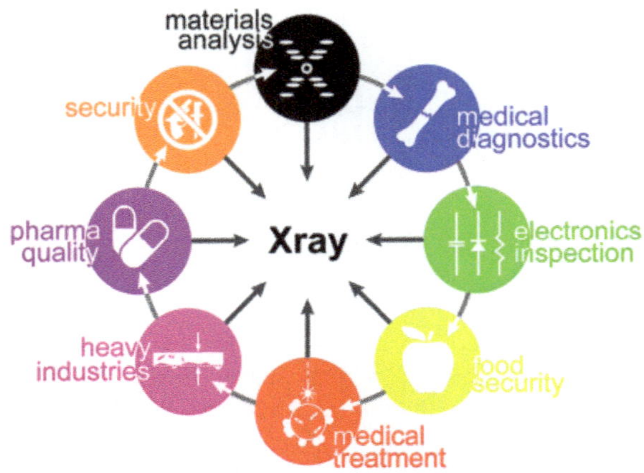

Figure 1 X-ray applications. X-ray sources find application in a wide range of applications including medical diagnostics, electronics inspection, food security and border control. Conventional X-ray sources are pervasive and the integration of carbon nanomaterials has the potential to complement the market dominance of traditional thermionic technologies.

impacted on almost every area and aspect of science and industry.

The strategies for developing new X-ray sources are based on criteria driven by the needs of current applications. Despite the apparent maturity of the technology, many critical challenges remain, including; rapid beam pulsing, dose reduction, improved image contrast, and enhancement of the spatial and temporal resolution. In this review we discuss the use of carbon nanotubes (CNTs) as a platform for emerging novel field emission X-ray sources. We detail the current state-of-the-art in CNT emitter fabrication including the electron source and the gate electrode micro-fabrication, functional enhancements including reduced turn-on electric fields and enhanced stability via the incorporation of adlayers, improved X-ray beam distribution symmetry achieved through cathode shaping, micro focal sources, pulsed emission, multi-pixel sources, and miniaturisation. We conclude by providing an overview of the commercial progress to date.

2 Review

2.1 Electron emission

Fundamental to almost all commercial X-ray sources is a source of electrons. Emitted from a cathode, these electrons are directed in a high vacuum towards an anode to which a positive voltage is applied. Emission of electrons will occur from a metal surface when subjected to a high electric field. Figures 2(a) and 2(b) show examples of a commercially available thermionic X-ray sources and their mode of operation, respectively. Figure 2(c) illustrates the approximate band diagram at the metal-vacuum interface. To stimulate appreciable electron emission the electrons must either be excited from the Fermi level over the potential barrier, or tunnel through the potential barrier. This gives rise to three principle forms of electron emission; photoemission (PE), thermionic emission (TE), and field emission (FE).

PE occurs where the metal is irradiated with an optical source; where the wavelength is selected such that it defines an energy greater than the work function (Φ) of the emitter, which typically lies in the ultra violet. As a result, electrons are excited and pass over the potential barrier. PE electron sources have a low efficiency. Much of the incident optical radiation is absorbed in the bulk material of the emitter with only a small proportion of the photon population contributing to direct emission. Although PE sources have the potential to achieve extremely fast response rates, and correspondingly high bandwidths, PE has gained very little traction in most electron emission applications as only very low emission currents are possible.

In contrast, TE can derive appreciable current densities that are capable of stimulating X-ray emission. Figure 2(a) shows examples of static and rotating anode TE X-ray sources. Significant current densities, and consequent heating therein, necessitate rotating anode and active cooling systems in TE sources. The core components of these systems are illustrated in Figure 2(b) also outlines the principle mode of operation. Here, sealed in an inexpensive and easily fabricated evacuated glass or ceramic envelope, the electrons are liberated from a metallic filament, often made of Tungsten, which is Joule heated to in excess of 1000°C [14]. Since the emission is intimately dependent on the filament temperature [15] – as increasing the emitter temperature allows for much of the electron population to pass over the surface barrier - such tubes enable analogue control over the magnitude of the emission current. In TE sources this beam current is controlled by monitoring the anode current and adjusting the inferred filament temperature using a closed-loop control system. The intrinsic finite thermal inertia of the heated coil, when coupled to the lagging response of such feedback control results in a comparatively slow time response, often several hundreds of milliseconds. In addition, care must be taken to limit the filament drive current to prevent excessive power dissipation, with subsequent damage or destruction of the filament. Issues with severe electromigration can be a significant challenge. A key design functional parameter is the focal spot size, which is related, in part, to the dimensions of the electron emission area. In the case of TE sources the physical size of the filament, typically of the order of 2 mm in outer diameter and 10 mm in length. This requires the adoption of techniques to provide first order focussing. Some TE X-ray tubes are fitted with two filaments, allowing a choice of focal spot sizes, though this is at the expense of maximum available beam current and hence photon output. In order to achieve further reduction in focal spot size, electrostatic and magnetic focussing techniques are often employed, though at the expense of source complexity and size.

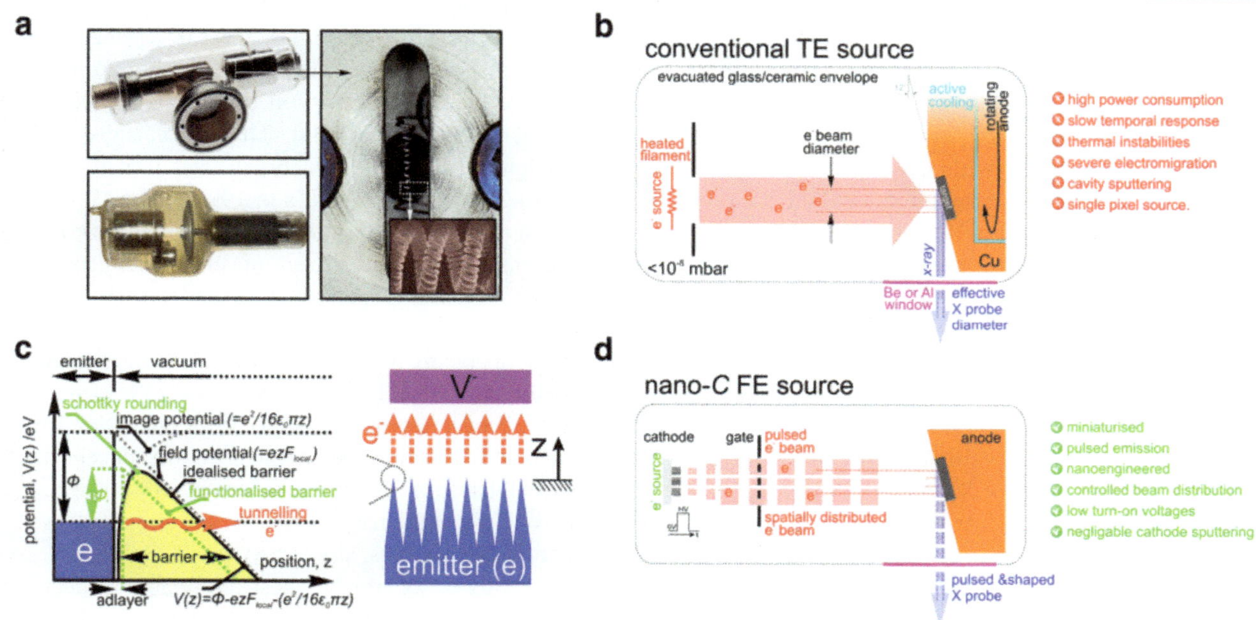

Figure 2 Thermionic and Field X-ray emission technologies. (a) Images of micro-coil and rotating anode thermionic emission (TE) electron sources. **(b)** Schematic depiction of the operation principles of a thermionic emission (TE) electron emitter based X-ray source. Not the active anode cooling. **(c)** Band diagram illustrating the routes to thermionic, photo, and field emission as means to stimulate an electron beam by overcoming the surface potential barrier. **(d)** Schematic depiction of the operation of a field emission (FE) X-ray source.

Field emission (FE) offers a number of significant benefits. Figure 2(d) illustrates a typical FE source. FE sources are often physically compact compared to their thermionic counterparts. The emission process occurs at room temperature and as such does not necessitate a direct heating element. FE is a tunnelling process and provides, as a result, near-instantaneous emission. Pulsed emission with rise times of less than 50 μs have been reported [16]. As illustrated above, electrons are emitted under the influence of a high electric field, typically of the order of a few V/μm. These electrons subsequently tunnel through the narrowed potential barrier. Traditionally the emission has been broadly described by Fowler-Nordheim theory [17], although this is now generally considered to be an incomplete representation in the case of electron emission from most nanomaterials. Nonetheless, a broadly accepted approximation of the field emission current, I, is given by;

$$I = \left(\frac{A\beta^2 V^2}{\varphi}\right) \exp\left(\frac{-B\varphi^{3/2}}{\beta V}\right) \quad (1)$$

where V is the applied voltage across the vacuum cavity, φ is the emitter work function, β is the field enhancement factor, and A and B are related to physical constants [18]. An extended analysis of the background theoretical emission mechanisms can be found in [19]. Various refinements to this model have been applied [20] to take account of the potentially elevated temperatures during emission and the extreme curvature at the nano emitters apex when whisker-like one dimensional nanomaterials are used as the field emission source. Indeed, the high electric field that is required for field emission is closely associated with the curvature and aspect ratio of the emitting material – this field enhancement factor depends upon a number of geometrical factors such as the precise form and orientation of the emitter; in general sharp whisker-like tips emit at significantly lower potentials compared to the same material that has adopted a planar morphology [21]. Such geometry-based arguments are the leading rationale for the use of high aspect ratio nanowires in field emission applications. Contrary to conventional beliefs, studies are emerging which show that the emitter morphology has a much more profound impact on reducing the turn–on and threshold electric fields compared with the effects of the emitter's work function. However, it remains challenging to attain such perturbed, high aspect ratio emitters.

Following early work by Dyke [22,23], in 1968 Spindt published details of a new method of fabricating FE arrays based on Mo conical structures. These structures were 1.5 μm high with a tip radius of 50 nm [24,25]. During the subsequent years, the use of Spindt emitters was widely adopted. They have since become common place in many electron emission systems. Spindts can be found in field emission displays [26-28] to high speed radio frequency devices [18,29,30], such as travelling wave tubes [31,32]. Extremely high current densities of up to 20 A/cm^2 have been achieved [33]. Nevertheless, the issue of developing suitably high current densities with low turn-on voltages has remained an on-going challenge, principally due to demanding requirements on the emitting material. At the time, low attainable aspect ratios that Spindt emitters offered, though better than their planar counterparts, limited emitter performance. A new material capable of forming extremely high aspect ratios was required. Self-assembly via chemical vapour deposition and the emergence of nanowires and nanotubes allowed for such high aspect ratios to be fabricated over large areas. The geometry of these new emitters allowed for a corresponding amplification in the field enhancement factor and subsequent reduction in turn-on voltage, typically by an order of magnitude.

Figure 3(a) summarises the various functional advantages of field emission (FE) sources over traditional TE and Schottky sources. The dashed orange circle depicts the characteristics of an ideal electron source. Figure 3(b) illustrates the mean turn-on and threshold electric fields, E_{on} and E_{th} respectively, and maximum current density (J_{max}) from FE sources fabricated from various materials. Figure 3(b) is structured in order of increasing dimensionality (1D, 2D to 3D/bulk) where each of these dimensional classes is further organised in terms of increasing emitter work function (Φ). The mean ($\pm 1\sigma$) work functions are shown. Each data point was acquired over multiple studies, nominally three or more, with the figure taking into account material only. The absence of Φ errors in some of the data indicates that only one such FE study has been conducted, typically due to the recent emergence of the material in question. To extract comparable data all current densities in the considered literature where normalised relative to their maximum current densities, where this normalised value, $J' = J/J_{max}$, where $0 \leq J' \leq 1$. E_{on} and E_{thr} were defined as the electric field required to emit 0.1 J' and 0.3 J', respectively. For clear assessment purposes of how the material type relates to efficient FE, various emitter morphologies were considered. Note that, on average, there is no clear increase in J_{max} or decrease in E_{on} or E_{thr} with decreasing Φ, as one would expect for a consistent view with earlier band arguments. Indeed, emitter geometry clearly smears out much, if not all of the functional benefits associated with the novel electronic properties of any given material. Nonetheless, the graphitic carbons are evidently largely unparalleled in their electron emission performance. Combined with the wide range of allotropes observed to date - including the zero-dimensional fullerenes, the one-dimensional nanotubes and nanofibres, the two-dimensional graphenes, and three-dimensional graphite - carbon truly has a morphological character perfectly matched to its impressive electronic properties.

2.2 Field emission application of CNTs

The discovery of carbon nanotubes (CNTs) is generally credited to Iijima in the early 1990s [134], although it is clear that there had been activity in this area for a considerable time prior to this [135]. This work, however; lead to a heightened interest which gave rise to some of the first studies on FE using CNTs in 1995 [136,137]. By the start of the following decade a range of applications had been identified to utilise these novel field emitters, including high resolution electron beam microscopes [138-141], flat panel displays [27,28,142-144], RF devices [18,32], electron beam lithography [145-147] and X-ray cathode emitters [148-151]. An early review [146], provides a concise overview of the state-of-the-art prior to 2004. A detailed historical perspective on the development of CNT-based FE sources is provided in [152]. Here we provide a condensed review of the progress, as it pertains to X-ray sources, since then.

CNTs have some of the highest attainable aspect ratios, high thermal conductivity, low chemical reactivity in non-oxidising atmospheres, highly parallelised *en masse* fabrication, a low sputtering cross-section, a low secondary electron coefficient, and an insensitivity to direct ion-bombdarment. CNTs are becoming increasingly inexpensive with the release of new, ever larger growth reactors. However, field emitters require ultrahigh vacuum ($<10^{-8}$ mbar) to provide stable operation. This limits their practical application as the material platform on which the emitters are fabricated largely dictates the tip robustness towards poor or compromised vacuum conditions which result in aggressive local ionization. In the case of the metallic Spindts, poor vacuum conditions caused tip degradation. As a result, much of the published work has been accomplished using demountable systems, which incorporate vacuum pumps to maintain the performance, although there are some notable exceptions [153,154]. A useful summary on this was published in 2010 [155].

It is the group of applications associated with CNT-based X-ray FE emitters [148] that is the subject of this paper, as the properties of the electron emitting CNTs offer many functional and performance advantages over conventional TE X-ray sources [156,157]. The way in which this review is structured details: shaped cathodes, micro focal sources, pulsed sources, multi-pixel sources, and miniaturised emitters. The review concludes by summarising the commercial progress to date.

Severe problems with CNT electron source fabrication, reliability, time stability, spatial uniformity, and reproducibility have beset the technology and have prevented its wide spread adoption, particularly in high beam current applications. Nevertheless, CNT-based X-ray sources were proposed following the emergence of CNTs and in 2001 Sugie *et al.* [148] were some of the first to describe the use of CNTs as a FE-based X-ray source. Here they grew

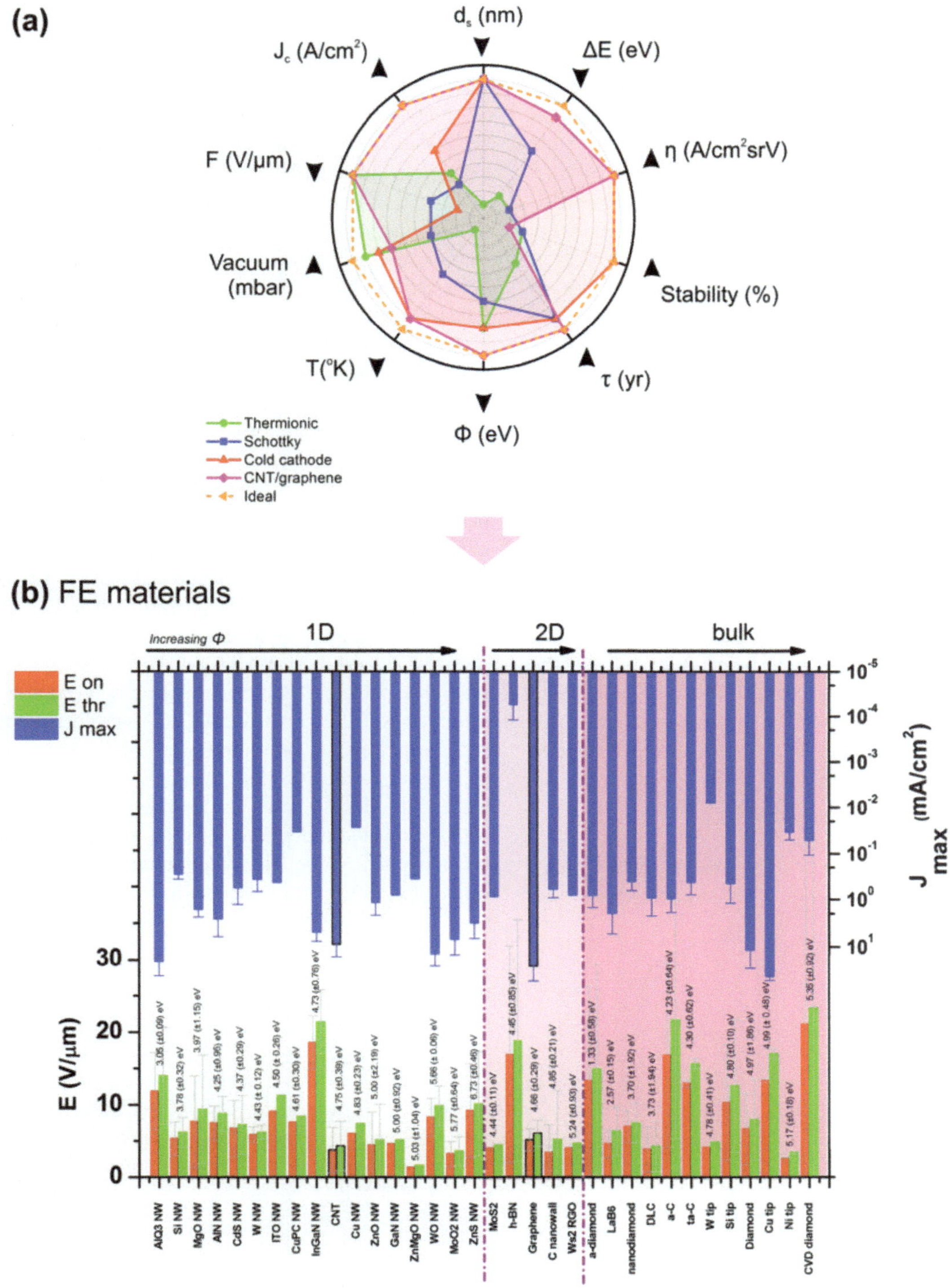

Figure 3 Comparison of common field emitting materials. (a) Polar plot for the various common-place electron emitters. CNT and graphene based field emitters out-perform such sources across most metrics, where J_{max} the current density, E_{on} is the turn-on electric field, 'Vacuum' denotes the operating vacuum, T the typical operating temperature, Φ the emitter work function, τ the lifetime, 'Stability' is the temporal stability, η the electron-optical brightness, ΔE the energy spread of the emitted electrons, d_s the virtual source size. Adapted from [34]. **(b)** Overview of the on and threshold electric fields (E_{on} and E_{thr}, respectively) and maximum current density, J_{max}, for various materials used for field emission to date, in order of dimensionality (1D, 2D and bulk) and increasing work function (Φ), including 1D nanowires - AlQ_3 [35,36], Si [37-39], MgO [40,41], AlN [42-45], CdS [46-49], W [50-52], ITO [53], CuPC [54,55], InGaN [56-58], CNTs [59-63] Cu [64-66], ZnO [67-72], GaN [73,74], ZnMgO [70,75], WO [76-79]), MoO_2 [80-82], and ZnS [37,83] -, the 2D platelets - MoS_2 [84], h-BN [85-88], graphene [89-97], C nanowall [98-100], WS_2 RGO [101] -, and the bulk materials - a-diamond [71,102], LaB6 [103-106], nanodiamond [107,108], DLC [109,110], a-C [111-113], ta-C [114-116], W [117], Si [118-120], diamond [121-125], Cu [65,66,126], Ni [127-130], and CVD diamond [112,131-133].

vertically aligned forests of CNTs on Co coated W wire, with the electron beam controlled by a counter electrode mounted 0.5 mm from the emitter. This setup produced a beam current of just 1.5 μA, but was sufficient to image onto an X-ray sensitive film a range of samples including integrated circuits. The lifetime of the cathode assembly was little more than one hour and the images exhibited severe noise artefacts. Nevertheless, this work did convincingly demonstrate the potential of CNTs in X-ray applications whilst simultaneously highlighting critical functional issues such as fluctuations in emitted beam current that needed to be addressed. Another early example of a functioning X-ray tube was produced by Haga *et al.* [150]. Catalytically synthesised CNTs and carbon nanofibres (CNFs) grown on a Pd wire were used as the FE source, although there was no counter electrode, or gate used to extract the electron beam. This device operated for several hours at 30 kV with a beam current of 50 μA. Clear images were acquired, though requiring a long and technologically unacceptable integration time of the order of minutes. The same group produced a triode-configuration (incorporating a gate electrode) which increased the beam current to about 1 mA [151]. Many more examples have now been published on the use of similar triode configurations [149,158-161]. These include designs with additional focusing electrodes for a micro focal imaging system operating at 18 kV, developed utilising 3D beam simulation tools [158].

2.3 Cathode fabrication

All X-ray tubes require the generation of a beam of electrons. This electron beam is directed towards the anode, which subsequently liberates X-rays when impacted. It is the cathode assembly that provides the source of electrons and it is the cathode design and materials which dramatically influences the resultant performance. The commercial field of vacuum electronics is now well established and was initially based on Spindt-like emitters, a comprehensive review on which is provided by Temple *et al.* [162]. The incorporation of CNTs within the cathode is the focus of this review as it is currently the subject of intense research in an attempt to enhance the electron emission, and hence, X-ray emission performance.

The first FE emitters using one-dimensional carbon allotropes were based on CNFs deposited on metallic tips [146,148,150,163]. These sought to enhance the native emission characteristics of the conventional tips by depositing CNFs, generally by CVD. Subsequent developments, however; centered on the use of planar substrates in attempts to realise more reproducible devices applicable to a wider range of electron emission applications. CNTs may be deposited on the cathode assembly using several different processes [164]. Figure 4 summarises these.

CNT thin films are readily patterned using a variety of techniques. There are numerous aspects which are yet to be well developed in the literature. Nevertheless, the attainable pattern resolution is intimately related to the orientation of the constituent CNTs within the thin film and the patterning technique employed. Aligned CNTs give higher resolution than misaligned CNTs. Wet processed thin films – such as dropped, cast, vacuum filtrated, sprayed, and screen printed – are fundamentally wet chemistry deposition techniques that can be patterned by either post deposition etching or shadow

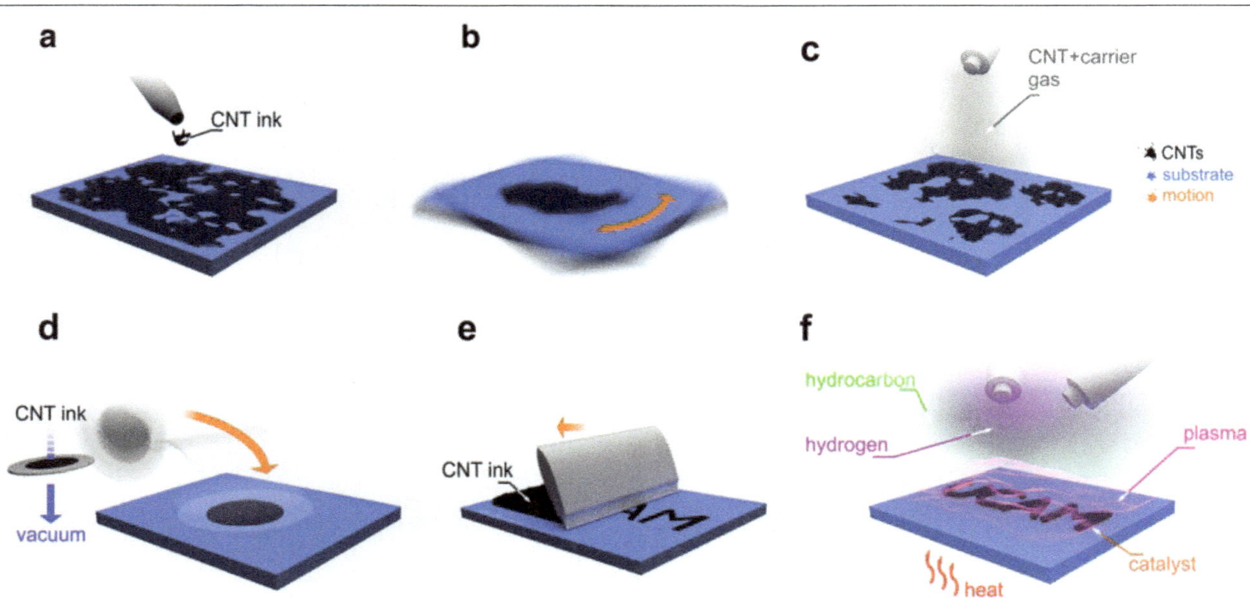

Figure 4 Common carbon nanotube thin film deposition techniques. **(a)** drop, **(b)** cast, **(c)** spray, **(d)** vacuum filtration, **(e)** screen printing, and **(f)** chemical vapour deposition.

masking to protect zones from being coated with the CNT inks during the deposition process. These shadow approaches provide a modest maximum resolution of around 100 μm as the inks often bleed beneath the masking layer, resulting in diffuse and uncontrolled edges. Without the application of external driving forces, such as magnetic or electric fields, in such patterns the CNTs are always misaligned with respect to one another. Dip pen nanolithrography and micro-ink jet printing can also be used to drop deposit CNT inks at controlled locations, with a common maximum resolution of 1 μm. Nevertheless, again, the CNTs are disordered and misaligned. In these additive approaches there remain significant challenges in preventing the nozzle from clogging during the deposition or the nanolithography tip becoming deformed. As a result inkjet printing and dip-pen nanolithography necessitates the use of very short CNTs, which dramatically compromises their usefulness in FE applications. Micro-contact printing is another additive approach to cathode patterning. Here a polymeric stamp is inked with a CNT solution and then placed in contact with a substrate which has an engineered surface hydrophobicity to ensure the CNTs adhere. Interfacial engineering marks the broadband application of this technique and often necessitates the use of self-assembled monolayers which act as adhesion promoters. Such approaches have proven useful in patterning carbon nanomaterials [165], though their spatial resolution is again limited to a few micrometers at present. In addition to the additive patterning approaches discussed, there are various subtractive means available. Most common is by way of conventional lithographically patterning the deposited CNT thin film, defining a hard mask, and oxygen plasma etching the exposed CNTs. This approach gives high resolution, typically ≤0.4 μm. Nevertheless, common to all wet chemistry approaches to CNT deposition remains the fact that the CNTs are disordered and unaligned with respect to one another.

Chemical vapour deposition (CVD) processes dramatically bolster the maximum potential resolution to around <0.1 μm by direct patterning of the physical vapour deposited catalyst whilst also allowing for the growth of individual vertically aligned CNTs. Indeed, inclusion of plasma during heating and exposure to the gaseous hydrocarbon and atomic hydrogen sources can be used to assist in the catalytic activity and alignment of the CNTs. CVD is well suited to patterning aligned CNT geometries however it is restrained, at present, to the use of high temperature (>300°C) compatible substrates, such as glass, quartz, and Si.

Another important consideration when noting the various merits of the available deposition techniques is the effective roughness and area uniformity achieved. The various deposition techniques produce films of varying thickness uniformity. CVD far exceeds the degree of uniformity of all other techniques, followed, in rank order, by casting, vacuum filtration, screen printing, and finally spray and drop casting; the latter of which typically gives coffee-stained like thin films that have a significant spatial variation in CNT density. Indeed, CVD, as demonstrated by the group of Milne, has yielded variations in surface roughness of <3% [166]. Nevertheless, for processing simplicity and related cost reductions rather than functional reasons, most CNT-based FE X-ray sources are based on wet chemistry processed thin films. What follows here summarises the fabrication details of some of the more common CNT thin film fabrication techniques, with particular emphasis on those that have been used in CNT-based FE X-ray sources.

2.3.1 Drop, cast and spray

Drop (Figure 4(a)), cast (Figure 4(b)), and spray (Figure 4(c)) approaches have proven, thus far, the fabrication method of choice by many. This is likely a direct consequence of their low cost and straightforward processing rather than any direct functional enhancements the fabrication techniques allow for. Drop, cast and spray techniques all require CNT inks. In each case, respectively, these are deposited either by direct dropping of the ink, spin coating or casting of the ink, or spray coating of the ink using a pressurised carrier gas. All these approaches initially require wet processing of the as-grown CNTs and it is this which, commonly to all, limits their usefulness and the consequent performance of the resulting X-ray source. The required high power sonication and vacuum unstable surfactants needed to form the stabilised, homogenous inks results in compromised temporal stability, deleterious hysteresis, and generally degraded emission.

2.3.2 Electrophoresis

Electrophoresis is a derivative of inking deposition. Electrophoresis involves the motion, placement, and potentially modest alignment of drop or spray deposited CNTs. Electrophoresis exploits the CNTs native anisotropic charge distribution and its interaction with an applied electric field [160,167,168]. Elsewhere it has been widely used as a means of depositing phosphor materials. It has also been applied to the deposition of bundles of SWCNTs on various substrates, such as stainless steel or doped Si [167-179]. Composite CNT films deposited by electrophoresis have been fabricated with controlled surface density with good field emission performance, current density and long-term stability under high operating voltages, which has been applied to an electron source for high-resolution X-ray imaging [160]. Wang *et al.* [169] deposited MWCNTs on Cr/Cu electrodes supported on an oxidised Si support with a glass interlayer adhesion promoter. The emitters were annealed at 480°C for 30 min following deposition to mobilise the glass interlayer and reduce the effective interfacial resistance. As with other solution-based fabrication techniques, in order to reduce the emitters turn-on bias they were taped to activate them; a threshold field of 3.5 V/μm was measured (for an emission current density of 10 $\mu A/cm^2$). Though a simple and low-cost means of defining the pixel size, electrophorectially defined emission zones only allow for relatively large cathode diameters, of the order of 2 mm, which reduces the systems functionality. Calderón-Colón *et al.* [160] employed photolithography and liquid phase electrophoretic deposition to pattern disordered and unaligned MWCNTs in a homogenous solution with $MgCl_2$ and glass frits. Here the film thickness was broadly controlled by the deposition time and the magnitude of the applied voltage, though accurate sample-to-sample reproducibility and control over the thickness, to <10 nm surface roughness, is challenging. Such techniques are comparatively low in uniformity and reproducibility, a direct consequence of the simple macro scale processing. Moreover, the common binder matrices are far from pure, making elucidation of the underlying emission mechanisms somewhat challenging, though nonetheless functional. Indeed, as with all other inking deposition techniques, the emitters do require mechanical activation for the cathodes to be of any practical use, though electrophoretic patterning is indeed rapid and simple to implement.

2.3.3 Vacuum filtration & screen printing

Many of the FE X-ray sources demonstrated have employed screen printed or vacuum filtrated CNT thin films [156,169-171] (Figure 4(d)). Here CNT solutions are deposited at the macro-scale and form highly disordered, nominally planar, spaghetti-like networks with but a few individual CNTs standing upright, at many unregistered angles relative to one another. In the case of vacuum filtration the CNT ink is poured onto a porous mixed (nitro) cellulose ester membrane, which has a partial vacuum applied on its opposite side. Such porous membranes have controlled apertures, typically around 0.2 μm in diameter. The reduced pressure stimulates the solvent within the CNT ink to pass through the membrane, whilst the membrane stops the CNTs from passing. Once the ink reservoir is depleted the thin film is rinsed with de-ionised water to remove much of the deleterious surfactants, leaving a thin CNT film on the membrane. Note, however; that much of the surfactant still remains even after extensive rinsing using deionised water. The CNT thin film can then be transferred to an arbitrary acetone-resistant substrate by applying modest pressure (40 g/cm^2) and heat (ca. 70°C) for an extended period of time (often of the order of 3 h). The membrane is then dissolved, by exposure to acetone or methanol, and the CNT thin film remains. The CNT film thickness, and hence sheet resistance and optical transparency, is controlled by adjusting the amount of CNT ink filtered through the membrane.

CNT thin film fabrication by vacuum filtration has been employed for more than a decade. It offers a rapid, low cost way to fabricate CNT thin films. Little to no infrastructure is required and the films can be processed rapidly, over large areas. Nonetheless, as in the case of drop, cast and spray, such *chemi douche* processing requires stabilised inks. CNTs experience high inter-tube van der Waals forces; they tend to agglomerate. Though this interaction has been exploited elsewhere to fabricate novel aligned nanostructured membranes [172], such agglomeration is problematic in producing homogenous CNT inks and solutions. As such, various, often sodium-based, surfactants such as sodium dodecylbenzene sulphate, sodium dodecyl sulphate, and Triton X are required to produce homogenous solutions with the CNTs well-dispersed throughout the solution. Further aggressive acid treatments and extended durations under high power ultrasonication significantly degrade the length, crystallinity and subsequent electronic character of the CNTs which necessarily limits the electron emission performance. These inks are then transferred to metallic disks using conventional screen printing methods or vacuum coated mixed cellulose ester membranes that are subsequently dissolved in acetone following transfer [173]. The morphology of these films is highly disordered. They typically lack high aspect ratio protrusions, resulting in poor field emission performance. FEs fabricated in this way have a number of intrinsic problems; chiefly that mechanical taping is required to activate the surfaces and enhance their field emission characteristics. Taping, using adhesive coated tape, increases the surface roughness of the CNT thin-films. It is a macro-scale process and affords very little reproducibility, Moreover; the necessary surfactants are usually vacuum unstable giving rise to emission profiles that deleteriously drift with time [170,174-176]. CNT inks can often have significant out-gassing [177] when even modestly heated, compromising the vacuum envelope. The resultant reduction in field emission performance and reduced current density has, as a result, prevented CNT-based pastes and ink from gaining commercial traction, though cathodes fabricated in this way appeal as the reduced reproducibility is off-set by the ease of fabrication. Such films are also somewhat dynamic and often have weak adhesion to the substrate. Their morphology shifts with time during the application of a high electric field due to the intrinsic torque induction within the CNTs due to the tip or root positioning of the growth catalyst particle.

To obviate issues of weak interfacial adhesion between the substrate and the CNTs, Kim *et al.* [154,178], using screen printing, as illustrated in Figure 4(e), employed ball-milled pastes combined with inorganic fillers including Ag and Ni alloyed nano-particles and oxides in an ethyl cellulose powder and terpineol solvent. No surfactants were used. The pastes were deposited on indium tin oxide coated glass substrates followed by 300°C post-bake to remove residual solvents. The derived slurry showed strong adhesion to the substrate, good uniformity and reproducibility with a current density of up to 350 mA/cm^2. Nevertheless, ball-milling degrades the graphitic quality of the MWCNTs, which adversely impacts the long-term time stability relative to high quality crystalline material, although these devices did show excellent high-temperature operation (up to 800°C), robustness towards harsh vacuum environments, impressive spatial uniformity, a low turn on voltage of 1 V/μm, and enhanced temporal stability relative to conventional CNT pastes, although the anode current was still degraded by around 30% under DC operation over a 100 hr test period. Another fabrication problem associated with screen printing is the limitation on the pattern resolution. The sonicated CNTs within the inks are typically of the order of 5 μm in length. These can clog the screen printing mesh resulting in a low porosity and inability to print.

Formation of stabilised, homogenous inks is central to drop, cast, spray, vacuum filtration and screen printing. The formation of these inks, as highlighted above, requires deleterious ultra-sonication and aggressive acid treatments, both of which degrade the length and electronic character of the CNTs. Though the necessary wet chemistry approaches provide a facile, rapid and inexpensive route to fabricate the emitter, the constituent CNTs are coated with deleterious vacuum unstable surfactants that can only be removed following high temperature post-deposition treatments which, if not fully removed, would otherwise cause significant outgassing during FE operation [160]. Maintaining slurry and ink consistency over time and between batches is difficult, which manifests as a reduction in device-to-device reproducibility. It is also challenging to pattern emitters fabricated in this way though screen printing (the migration of CNT inks through patterned apertures within a regular mesh), as demonstrated by Kim *et al.* Though screen printing is indeed large-area compatible, it is rather low resolution and thus limits the degree of control over the detailed design of the electron source. Oxygen plasma etching coupled to conventional lithographic techniques is another viable option though significant surface roughness of the CNT thin film can again compromise the maximum resolution. Moreover, plasma etching techniques are only applicable to non-organic substrates.

While there have been reports that electrophoresis can produce a degree of alignment [179,180] this is relative only to very randomly orientated screen printing and ink-based processes. Misalignment prevents any fabricated devices from realising the full field enhancement factor of the composite CNTs; in order to achieve this alternative fabrication methods capable of aligning, *en masse*, the CNTs must be considered, with chemical vapour deposition being the most promising method to date.

2.3.4 Chemical vapour deposition

Chemical vapour deposition (CVD) (Figure 4(f)) provides a more controllable means of growing CNTs. The CNTs self-assemble from atomic units in a highly parallelised process, which when coupled with high resolution lithographic techniques to pattern the catalyst material, allows for near nano-scale engineering of the CNTs and CNFs. CVD techniques mediate the growth of chemically untreated disordered or aligned CNT thin films depending on the substrate, catalyst and growth precursors employed. In a typical implementation, Silicon is coated with a physical vapour deposited metal catalyst which is then patterned *via* lithographic or masking techniques by either additive or subtractive process, such as magnetron sputtering, or plasma etching, respectively. The substrate is then heated to temperatures often in excess of 500°C, and the growth of the CNTs on these sites is initiated by supplying a hydrocarbon feedstock gas, such as CH_4 or C_2H_2, combined with an *a*-C etching gas species, typically H_2 or NH_3 both of which readily pyrolyse to give a constant supply of carbon and atomic hydrogen. In situ plasma can also be employed to enhance the catalysis and align the CNTs during growth. Cole *et al.* have compiled a concise overview of the CVD of nanocarbons, available elsewhere [181].

Using thermal-CVD synthesised MWCNTs directly deposited on shadow-masked Mo discs (as well as Si, fused quartz, mica, copper, and highly ordered pyrolytic graphite), with a Al/Fe (0.5/10 nm) bilayer catalyst, Sarrazin *et al.* [182] have routinely measured current densities in excess of 1 A/cm^2. Here the MWCNTs were several microns in length and randomly orientated, with most running adjacent to the substrate; CVD techniques allow for CNT alignment, though random orientation is still possible. Very little control over the type, orientation and area packing density of the CNTs was evidenced with a largely qualitative analysis presented. Nevertheless, a turn-on bias of approximately 2 V/μm was observed. Rather surprisingly, the robustness of the cathodes toward arcing was clearly shown resulting in an increase in the voltage for a given current following repeated arcing events. They suggest that either the emitting sites are not completely destroyed or that they are efficiently replaced by other nanotubes within the film. Indeed, disordered films are structurally dynamic when under the influence of a high electric field, which can augment their emission characteristics. Though film reorganisation benefits the long term stability, in that degraded CNTs are in essence replaced, the short term temporal stability is likely to be very poor. More than 100 cathodes were tested and a poor reproducibility was indeed noted, a probable consequence of the disorder and uncontrolled microscale morphology of the emitter.

As is the case for the screen printed and vacuum filtration methods, one potential problem is the degree of adhesion between the CNT and the substrate when exposed to high electric fields. Detailed control of the underlying catalysis has shown that such emitter removal concerns can be solved. Li and Cole *et al.* showed very low variation in anode current of $< \pm 0.7\%$ for constant emission over >200 hrs using an ITO/Ni (10/1 nm) bilayer catalyst which effectively enhanced the degree of adhesion [183].

Using rapid thermal CVD, Kim *et al.* [184] deposited CNTs directly on stainless steel sheets coated with TiN/Ni. The emitters showed a maximum current of 2 mA (data on the emitter area was unavailable) at an anode potential of 5 kV. The CNTs were 30 μm long, with Raman analysis suggesting a very defective material, whilst scanning electron microscopy showed vertically aligned forests with a high packing density. Though the emission showed good performance the geometry was not optimised; the dense forest results in significant shielding of the CNTs from the applied electric field – the material appears as bulk - such that the full field enhancement factor of the CNTs was not realised in this instance. Whilst individual, one dimensional nanowire and nanotube emitters have been empirically evidenced to produce the highest electric fields, the proximity of other emitters will effectively shield the field enhancement [146,185]. In a field emission device, where our interest is in the total available current density, the optimum arrangement will not be that with the highest density of emitters [186,187]. In addition, if the emission pattern of the field emitter is not uniform CVD growth can result in individual CNTs, or structures such as CNT pillar arrays and toroids [160,167,188-191]. Patterns of control electrodes may be grown, in such a way as to focus or concentrate the field from the tip.

The use of CVD to nanoengineer X-ray sources has remained in its infancy due to a number of challenges in explicating the underlying material growth. Nevertheless, following recent advances in the understanding of nanocarbon catalysis [192-194], the use of CVD-grown CNTs and CNFs in FE X-ray sources appears to be accelerating with it emerging as an emerging as an exciting candidate for viable commercialisation.

2.4 Diode & triode configurations

The first examples of X-ray emitters using FE electron sources employed diode configurations which comprised only of a cathode and an anode. However, in such devices the emission current was a function only of the anode voltage. As a result such diode configurations gave rather limited control over the magnitude of the emission current for most applications. To more accurately control the emission current, whilst also providing a degree of protection, it is now standard procedure to introduce a third electrode. This gate electrode gives rise to a triode configuration. Early in the development of CNT-based FE X-ray sources, Sugie *et al.* [148] introduced a counter-electrode wire. Since this naturally produced an anisotropic beam, the control mechanism of choice became a perforated grid, or gate electrode. The purpose of the gate is to create a local electric field that draws the electrons away from the principal emission beam. However, in practice a significant proportion of the emitted electrons will be attracted towards the gate electrode. The gate must therefore be sufficiently transparent to allow the maximum number of electrons to pass to the anode. It is normal that >50% of the cathode current is directed to the gate, thereby limiting the efficiency of such devices. However, there has been recent interesting work on improving the transparency of the gate by incorporating graphene layers. This has shown a substantial improvement in efficiency to around 60% [195] though the technique has yet to be applied to X-ray sources.

Further, the addition of a fourth electrode, mounted just above the gate, has been described [154]. This electrode had many apertures for each gate perforation, and was mounted approximately 0.2 mm above the gate electrode. The claimed transmittance of the CNT cathode electrons, to the anode was up to 80%. In addition, there are numerous examples of the addition of a further electrostatic focusing electrodes used to reduce the focal spot size. This is described in more detail below.

2.5 Demountable tubes

Many conventional X-ray tubes are constructed using sealed glass or ceramic "inserts". At this stage in the development if FE X-ray sources, many of the published results referenced in this paper relate to tubes constructed in vacuum chambers which can be directly evacuated with turbo molecular, ion, or cryo pumps. Such systems are analogous to the demountable X-ray tube [196]. There are situations where it can be of considerable benefit to have the ability to open a tube, replace damaged components, and re-evacuate it. In particular, this offers the advantage that it is possible to exchange both the anode – either for one of different target characteristics, or to replace a unit which had been degraded as a result of continuous high beam current – and cathode assembly, following failure or deterioration due to prolonged use. To date many CNT based sources have been fabricated in what are largely deemed demountable tubes. Clearly, the nature of devices developed for research and design iteration, this is the sensible choice. However some improvements – for example, miniaturisation – have seen the development of a sealed construction [150,154,197,198].

2.6 Functional enhancements

Table 1 below summarises the cumulative literature to-date on the progress towards the realisation of a functionally enhanced CNT-based FE X-ray source. Italicised values indicate the highest achieved standards at the time of publication. What follows is a detailed overview and assessment of some of the leading studies with regards to their functional enhancements, fabrication methods, and emission performance.

2.6.1 Beam profiling

Many groups have grown CNTs, by CVD or derivatives thereof, directly on conical or other shaped tips, on common Spindt-like materials [213], Si [169,205,214] or bulk metals [200]. As such, these systems do not exploit the full geometric potential of the CNTs; they simply amplify, often only marginally, the existing field enhancement factor of the tip structures and almost always result in an induced asymmetry in the electron beam profile that is challenging to accommodate with conventional electrostatic focussing. Indeed, even conventional X-ray TE cathode assemblies will, in general, produce an anisotropic focal spot, as a result of the electron beam dimensions, imperfect focusing, and the impact of the target angle. One potential route to realising symmetrical X-ray sources with shaped beam profiles is to shape the cathode such that the emergent electron beam defines, at least in part, the spatial distribution of the emergent X-ray beam. This requires patterning of individual CNTs or CNT forests into well-defined geometries. Several groups have considered the use of cathode shaping as a mean to achieve beam profiling. Most have used elliptical cathode geometries to define a circular distribution of the emergent X-ray beam [160]. Two examples are as shown in Figure 5(a,b). Shaping has also been shown to enhance the FE performance by exploiting the natively high field enhancement factor at the shaped emitter's edges. In particular, the group of Milne have shown field emission properties of individual CNFs and patterned forests [185,215,216]. No systematic studies on the effect of shaped cathode eccentricity, or other controlled geometrical parameter, on the symmetry or profiling of the emergent X-ray beam appear to have been considered to date. Indeed, the electrophoretically deposited and photolithographically patterned elliptical cathodes of Calderón-Colón *et al.* [160] showed impressive emission characteristics though no detailed assessment of the emergent beam shape was reported, making is difficult to assess the success of such beam shaping techniques (Figure 5a).

Using Si-supported CNT arrays, Ryu *et al.* demonstrated the potential of CVD in X-ray source applications [205]. A resist-assisted patterning process was used to produce a cathode assembly consisting of a gate and focusing electrode. This assembly was approximately 0.5 mm × 2.0 mm and gave an isotropic focal spot [205]. Ryu *et al.* also described a further shaping technique for post processing the as-grown vertically aligned CNTs. Here they used hydrofluoric acid to produce an array of conical pillars (Figure 5b), rather reminiscent of earlier Spindt emitters. Hydrofluoric acid, though cheap and readily available, is rather hazardous to handle and poses a significant health risk. This somewhat limits the commercial viability of the compaction technique presented. Nevertheless, they demonstrated anode currents of up to 90 mA at 7.8 V/μm, with a turn-on electric field of approximately 3 V/μm. To increase the geometric uniformity of the emitters the team used an electrical aging treatment, which degrades the taller tubes and bundles that would dominate the emission. For a high performing emitter this aging, also termed seasoning, is not desirable and CNTs of very uniform height and diameter are preferred, as evidenced by Teo *et al.* [166], to obviate the need for such post-growth treatments. Nevertheless, part of the Ryu *et al.* ageing process [205] was to remove the Ni catalyst particles at the CNTs apex, which they argued reduced the emission current. Detailed analysis and theoretical studies surmising the emission implications of the metallic nanoparticle in the emitter tip are lacking in the literature.

2.6.2 High beam current

High beam current, and therefore current density, is desirable. This gives rise to higher photon flux, and hence shorter detector integration times, in addition to improved signal-to-noise ratios. Combined with the use of gate control to pulse the X-ray source, high beam currents can be very advantageous to system designers. Recently, micro-fabricated Spindt-like emitters have been used for applications such as static tomography [217]. Here each cathode was formed from approximately 50,000 tips covering an area of about 1 mm^2. These emitters had a measured current capacity greater than that reported for CNTs, though CNTs will almost certainly exceed this once the technology matures. For current CNT-based sources the emission current density is several orders of magnitude less than that of equivalent area Spindt emitters. Nevertheless, for CNT-based emitters, current densities of up to 1 A/cm^2 have been reported [18], with studies elsewhere suggesting that individual CNTs are capable of conducting current densities

Table 1 Performance of CNT-based X-ray sources

Ref	Year	CNT deposition (substrate)	Max stable beam current (A)	|Max anode voltage| (kV)	Pulse frequency (Hz)	% drift in anode current in non-feedback mode (%I_a)	Approx. turn-on (V/μm) (@ J_a)	Focal spot size (mm)	Notes
[159]	2005	Electrophoresis	10 mA (15 V/μm)	40	DC	<5%	~2.6 (/)	0.15	Micro focus
[199]	2007	PE-CVD (W and Pd wire)	50 μA ~ (3 V/μA)	15	DC	/	/	0.05	Micro focus
[163]	2007	PE-CVD (W wire)	26 μA (5 V/μm)	40	DC	/	1.6 (10 mA/cm^2)	0.005	Micro focus
[200]	2010	PE-CVD (W and Pd wire)	0.1 mA (/)	25	DC	<15%	3.2 (/)	*0.0004*	Micro focus
[201]	2004	Solution Processed	6 mA (15 V/μm)	60	10^{7-8}	<±*1%*	~7.1 (/)	0.15 × 0.03	Micro focus high pulse rate
[149]	2002	Electrophoresis	28 mA (/)	14	10^3	±2-4%	2.0 (1 mA/cm^2)	3.2	Pulsed
[202]	2006	/	1 mA (/)	40	1	~50%	/	0.2	Pulsed multi-pixel
[203]	2009	Electrophoretic	18 mA	30	10^2	/	~5.0 (/)	0.5 × 0.3	Pulsed multi-pixel (25×1)
[171]	2011	Screen Printing	43 mA (/)	45	5	±0.2%	/	0.6	Pulsed multi-pixel (31×1)
[203]	2009	Electrophoresis	18 mA (/)	40	10^3	<7%	6.5 (/)	0.1	Multi-pixel (25×1) shaped cathode
[169]	2011	Electrophoresis	3.5 mA (7.5 V/μm)	1.4	DC	/	3.5 (10 μA/cm^2)	2.0	Multi-pixel (5×10)
[204]	2013	CVD	0.6 mA (16 V/μm)	40	DC	/	~8 (/)	/	Ballasted
[160]	2009	Electrophoresis	7.0 mA (8.5 V/μm)	50	10	±13%	5 (/)	0.1	Shaped cathode
[205]	2012	CVD	*90 mA* (7.5 V/μm)	50	DC	/	3.0 (0.1 mA)	/	Shaped cathode
[206]	2006	DC Electrophoresis	150 mA (12 V/μm)	40	10^2	/	/	0.03	Shaped cathode pulsed micro focus
[182]	2004	CVD (Mo disk)	ca. 15 mA (4.6 V/μm)	1.1	10^{-1}	/	~2.0 (/)	0.5	Miniature battery operated
[150]	2004	CVD	0.3 mA (7.0 V/μm)	30	DC	±8%	/	/	Miniature
[198]	2005	/	~40 μA (/)	/	2x10^3	~50%	/	1.0	Miniature
[207]	2012	Paste	0.6 mA (1.4 V/μm)	70	DC	±2%	/	3.7	Miniature
[208]	2013	Paste	1.2 mA (4.7 V/μm)	25	10	~ <1%	~3.5 (/)	/	Miniature
[209]	2004	CVD (Mo disk)	~2 mA (4.65 V/μm)	10	10^{-1}	/	2.2 (10 nA)	0.5	Miniature pulsed
[210]	2006	Ion bombardment (polyimide)	/	25	6x10^2	/	1.5 (1 μA/cm^2)	/	/
[148]	2001	CVD (metal wire)	1.5 μA (20 V/μm)	60	DC	±10%	/	/	/
[151]	2004	PE-CVD (W and Pd wire)	1 mA (/)	40	DC	±8%	/	2.0	/
[184]	2007	CVD	2.0 mA (0.5 V/μm)	10	DC	~50%	*~0.2 (/)*	/	/
[211]	2008	PE-CVD (Pd wire)	1.0 mA (/)	50	DC	±10%	/	3.0	/
[154,212]	2014	Screen Printing	15 mA (2.6 V/μm)	20	10	/	/	0.3	/

I_a: Anode current (A);
J_a: Anode current density (A/cm^2);
/ denotes values not given.

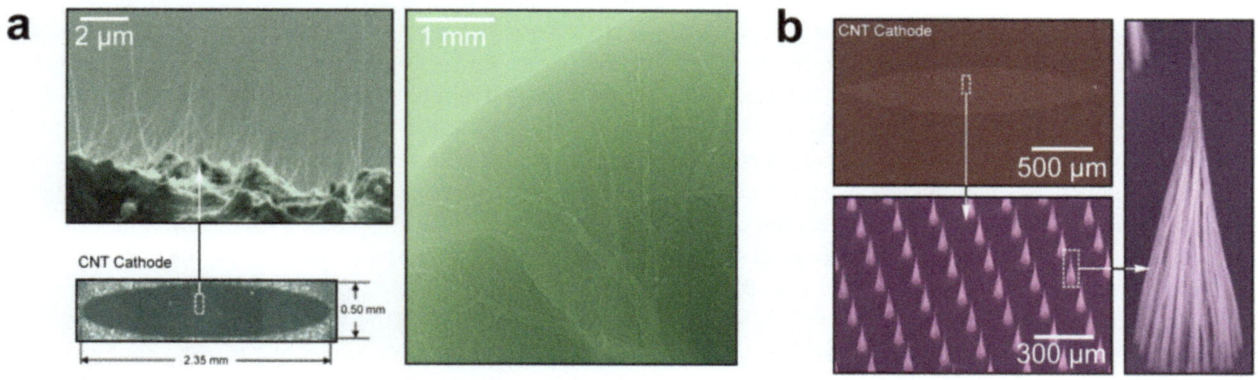

Figure 5 Shaped CNT cathodes. Control over the electron beam distribution, and subsequent symmetry of the X-ray beam, can be achieved by shaping the electron emitting areas on the cathode by conventional lithographic techniques. **(a)** A screen printed elliptical shaped electron emitter and subsequent X-ray image of a leaf [160] (Scale bars: 2 µm, 1 mm). Copyright Institute of Physics. **(b)** Scanning electron micrographs of an elliptical electron source, formed from tepee-like CNTs, demonstrated by Ryu *et al.* (Scale bar: 500 µm, 300 µm) Adapted from [205], Copyright 2012, Multidisciplinary Digital Publishing Institute.

corresponding to 5×10^6 A/cm^2 [218], which is significantly larger than any attained by Spindt-like emitters fabricated to date.

Using electrophoretically deposited SWCNTs [167], high electron beam currents have been achieved using a triode configuration by Yue *et al.* [149]. Here, the gate assembly was mounted approximately 100 µm adjacent to the CNT electron emitter resulting in a device geometry capable of sustaining a beam current of several milliamps, although this would likely prove challenging to manufacture in a practical X-ray source. In addition, they demonstrated peak pulse currents of up to 28 mA (emitted area not stipulated), without vacuum breakdown, at an anode potential of 14 kV. Using CVD grown CNT arrays, Ryu *et al.* [205] reported emission currents of up to 90 mA (emitted area not stipulated) in a diode configuration, although in a gate-controlled triode arrangement, the beam currents were substantially reduced.

Some improvement towards increasing the current density has also been achieved by controlling the emitter morphology. Toroidal CNT arrays, which have a central void [191], are one such example. By subsequent surface treatments, it is also possible to enhance the native FE characteristics of such arrays by means of the formation of nano tips, tepees and micro cones, as outlined previously [190,205]. The use of emitter forming post-treatments has also been widely investigated. One leading example is the use of conical CNTs which enhanced the beam current and stabilised the emitter geometry. Another alternative towards higher beam currents is to augment the electronic character of the CNTs, via the dry or wet deposition of various adlayers. Such adlayers adjust the interfacial characteristics at the critical emitter-vacuum interface. This often leads to a decrease in the turn-on voltage and increase the emission current density at a given anode bias. As was highlighted above, FE is highly surface sensitive; at 10^7 A/cm^2 a 1% change in the surface work function, due to intentional adlayer sputtering or indirect sputtering from reactive gas species in the vacuum cavity, can increase the current density by more than 10% [152]. Little work has been reported on the use of adlayers to enhance the emission performance of CNT-based FE X-ray sources. Nevertheless, significant advantages may be achieved by carefully designing the emitter-vacuum interface to provide high emission currents, limit adverse effects of vacuum leakage, prevent unintentional work function shifts, as well as robustness towards local ionisation and plasma etching, all of which are critical in ensuring long term stability.

2.6.3 Enhanced temporal stability

The failure mechanisms of Spindt and CNT based emitters, which are similar in many respects, have been widely studied [219], most notably by Bonard *et al.* [220]. Thermal migration, field sharpening and subsequent avalanche breakdown are perhaps the most common failure mechanism in such whisker-like emission geometries [221]. Although recent work on CNT-based emitters has shown impressive lifetimes of 7000 hours, with only a 10% emission degradation [222], it remains an on-going challenge to design and fabricate high temporally stable emitters. As noted previously, the electron emission is sensitive to the bulk emitting material and the emitters' surface chemistry. CNTs have a high sublimation temperature and high maximum current densities, making them resilient towards arcing events. Pristine CNTs are also largely inert, although when defects are added to the graphitic lattice, often through plasma processing, they are readily damaged and the formed dangling bonds bolster the CNTs emission performance, though often at the expense of the long term stability. The enhanced emission is due to the augmented surface characteristics. This effect is often noted as a shift to higher turn-on bias during hysteresis studies. Indeed, many plasma and some dopants, such as oxygen when the emitter outgases, can be particularly damaging to the CNTs. Local Ohmic (I^2R) heating can also increase the rate of emitter degradation, as can vacuum breakdown. It has been shown that the axial resistance of CNFs increases with increasing temperature. Interestingly this intrinsic property helps protect the CNFs from degradation during local heating and helps prevent thermal run-away deleteriously increasing the emission current [221]. Vacuum breakdown results in the emission of extremely high current densities which may cause a plausible local oxygen micro-plasma formation – this etches the CNTs, particularly at their apex. This shaping stimulates the formation of defects in the graphitic lattice, which preferentially emit electrons, thereby enhancing the emission, however; this results in a temporally unstable emission profile. The CNTs can be sharpened with time, which tends to increase the emission current, or can alternatively be entirely ablated, which reduces the emission current due to the reduced number of electron emitters available.

There have been various methodologies proposed which attempt to reduce temporal instabilities. Thermal annealing or electrostatic seasoning, to remove residual surfactants and non-uniform emitter profiles, are perhaps the most common and certainly the most simple and readily implemented [160]. During thermal annealing the cathode assemblies are heated to temperatures in excess of 300°C under high vacuum conditions. This out-gases the emitter, removing weakly surface bound chemisobed species. This increases the work function uniformity across the surface of the emitter. Such out-gassing techniques are also useful in emitter recovery following an arcing event. Arcing events stimulate high current flow which heats and subsequently out-gases the emitter. This out-gassing can lead to further transient arcs which, if allowed to continue, will degrade the emitter. If the emitter is initially well out-gassed any local arcs will only marginally increase the cavity pressure and the emitter will stabilise more rapidly. Annealing is also employed to enhance the pressure of the vacuum cavity, making local plasma formation increasingly unlikely. In the case of electrostatic seasoning, the extraction electric field is slowly ramped up to around 80% of the nominal emission voltage. The emitters are then left emitting for tens to hundreds of hours to increase the surface smoothness of the emitter and hence, stabilise the emission current. Such approaches are critical in achieving intrinsic emitter stability, and though feedback techniques have been employed to artificially control the stability, engineering intrinsic stability remains central to the formation of a long-term stable emitter. In feedback based systems, in the same way as conventional TE generator designs, the anode current is monitored and the extraction voltage adjusted accordingly to maintain a known, safe, emission current. Though a viable and widely adopted approach to ensuring emitter stability, the slow response times of the feedback loop cannot entirely remove transient effects, such as arcing events, and only careful design of the electron source can facilitate this.

Though CVD is certainly coming to the fore as the most reproducible fabrication technique with the finest degree of control over the emitter design, it, like other techniques, is faced with issues of tip-to-tip uniformity. Poor uniformity in effective surface roughness is known to de-stabilise the temporal stability. Indeed, wet chemistry ink approaches to emitter fabrication result in much greater surface roughness, and this exacerbates and further compromises their temporal stability. Small height variations between tips can instigate preferential emission from a small proportion of the longer CNTs, which consequently burn-out. It has been shown elsewhere that individual CNTs can in practice pass a current of several microamps [223]. Currents in excess of this threshold cause the CNT to sublime, which manifests as a temporal instability in the anode current, and subsequent X-ray emission. In the case of CVD-synthesised CNT emitters one solution – originally described for application with Spindt emitters [224] and initially proposed for general electron emission applications by [183] – consisted of integrating a ballast resistor micro-fabricated in series with the electron emitter. Here resistive deposited layers are fabricated in series with the electron emitters. The series resistance ballasts the emission, functioning as a current limiting resistance, preventing emitter sublimation, subsequently enhancing temporal stability. Yet to be applied immediately to the design of an X-ray source, the group of Milne have developed a novel field effect transistor (FET) ballasted field emission source, where each individual CNF electron source is equipped with its own dedicated FET ballast layer which is capable of limiting, and electronically controlling, the emission current to prevent emitter sublimation and significant temporal instabilities [183]. Other thin film deposition techniques have also been considered [204,225] though these too have yet to gain any significant interest.

Individual CNTs are fragile. Their propensity toward tip or root growth results in an axial asymmetry in their magnetic susceptibility which manifests as rotational torque induction when exposed to high electric fields - an effect which is typically exploited during electrophoretic alignment. However, during FE this can result in the removal and transfer of the CNTs from the cathode to the anode. This severely impacts the lifetime and stability of any field emission device into which they are incorporated. Several approaches have been taken to improve the robustness of the emitting elements by increasing the degree of adhesion between the CNTs and the supporting substrate. Bundles of vertically aligned CNTs - typically with a height of around 10 μm or less, and a diameter of 30 μm, have been shown to be one such robust assembly, that is not readily damaged or easily removed. However, FE preferentially occurs at the periphery of the patterned CNT forest or array [188,189] suggesting that if some CNTs are in fact removed the emission stability may simply be recovered by new CNTs contributing the emission current of those removed; emitter areal design can in part accommodate and engineer out temporal instabilities. Nevertheless, there is a continuing effort to engineer the CNT-support interface to enhance the degree of adhesion and hence enhance the temporal stability. In the case of ink approaches various solution additives have been exploited, such as glass frits [160], though the exact implications of such approaches with regards to the turn-on field and maximum current density remain unclear.

Elsewhere there has been significant work on the use of surface treatments to stabilise the temporal stability of the electron emission [205,226]. As discussed previously, such low work function adlayers do indeed increase the emission current density, though they also, depending on the adlayer material and means of deposition, hermetically seal the CNTs – which act as a high aspect support structure – thereby increasing the emitters stability [225].

2.6.4 Micro focal sources

FE sources have the potential for extremely high spatial modulation. By controlling, the electron emitter location at the nanoscale, coarse control over the position of the electron beam, and subsequent X-ray beam are possible. A range of focussing techniques has also been developed to reduce the electron beam focal spot size. Standard X-ray tubes have a focal spot size of typically around 1 mm diameter. For greater spatial resolution, "mini-focus" X-ray tubes have been developed, with a spot size down to 50 μm, as well as "micro focus" X-ray tubes with a spot size of ~5 μm [227]. Almost exclusively, X-ray sources with effective focal spot sizes of <10 μm, which are commercially available, employ electrostatic or magnetic focussing and require high beam current pulsing techniques, to prevent limitation in their imaging resolution by motion-induced blurring [206]. The size of the electron beam at its source and latterly as it impacts the target material, contributes, in part, to the size of the X-ray focal spot, which itself impacts on the resolution of the resultant X-ray image. A large electron beam focal spot will create a penumbra effect where the X-ray spot subtends a significant angle at the subject; this blurs the resultant image. Conventional X-ray tubes will use a lensing cup around the heated filament assembly. This, to a first order, electrostatically focuses the electrons onto the centre of the anode. Careful design of the cathode and supporting electrostatic lenses will minimise the size of the focal spot [228,229]. Additionally, further electrostatic rings or focussing coils may also be employed to redirect the flow of electrons emanating from the cathode, to further reduce the spot size, where the focal spot size is determined by measuring the Point Spread Function using standard methods [230].

In just such a way, a transmission CNT X-ray tube with a solenoid focussing unit was constructed by Heo *et al.* to produce a 5 μm focal spot [163]. Liu *et al.* developed a single electrostatic focusing lens to produce a source with a nominal focal spot size of 65 μm, which was further reduced to 35 μm by limiting the dimension of the cathode assembly [231], a concept yet to be fully exploited. A similar mini-focus tube, used for small animal CT work, was also described. This used two focussing electrodes to produce a focal spot size of about 100 μm, using a mesh gate electrode to maximise the electron emission from the cathode [232]. FE tubes with cathodes constructed from CNFs have also incorporated conventional three stage electrostatic Einzel lenses. Such systems have achieved focal spot sizes down to 40 μm [200].

2.6.5 Pulsed sources

FE sources have the potential for unprecedented temporal modulation. Compared to other one-dimensional nanomaterials, CNTs allow for near ballistic conduction making them ideally suited for such high-speed applications. Pulse X-ray sources are practically beneficial when imaging moving objects. The principle of operation is analogous to stroboscopic lighting in optical imaging. Conventional TE X-ray sources are generally not capable of rapid control; they cannot be pulsed much more rapidly than a few tens of milliseconds, without sophisticated gating mechanisms. The means of controlling the beam current in a TE source is by adjusting the filament current and hence temperature, which introduces time delays due to the finite cooling and heating time of the filament. Pulse sources are therefore generally implemented by switching the anode voltage, or by means of mechanically controlled shutters, the latter only allowing for relatively course control over the pulse shape, on period, and mark-to-space ratio, and will inevitably create a trapezoidal pulse. In the case of existing FE and TE sources, the short duration of the pulse must be compensated by increased power levels, resulting in significant engineering design constraints with regards to the resilience of the emitting material; conventional electron emitting materials degrade rapidly when used in such high power applications.

In FE X-ray sources often the means of controlling the beam current is via the gate voltage. Depending on the exact emitter geometry, this voltage can be considerably lower than the anode voltage and so at a much lower power level; it can be switched virtually instantaneously. Gate electrodes can certainly be incorporated into more traditional TE sources also, though much more control over the on/off current ratio and pulsing performance can be achieved when using FE sources as the gate and anode can be controlled simultaneously, with both responding rapidly. The result is a FE X-ray source that may readily be pulsed with a high on/off current ratio. This lowers the total thermal dissipation, reduces the total amount of emitted radiation – allowing for safer medical diagnostics, and eliminates the need for bulky mechanical components.

Most devices fabricated to date operate in either DC or low frequency (tens of Hz) mode [148,150,151,154,160, 163,169,171,182,184,159,200,202,204,205,207-209,211,212] with only a very limited set of electronically controlled devices operating at frequencies in excess of a few hundred Hz to a few kHz [149,198,203,210]. The highest electronic pulse rate achieved to date is of the order of 10^7 Hz, as reported by Cheng *et al.* [201]. Nevertheless, the pulse rate achieved by Thales [233] in their optically gated devices exceeds all other devices by a significant margin. Figure 6(a) shows the anode current and X-ray intensity temporal response by Reyes-Mena *et al.* [198]. Here the X-ray beam 0-90% turn-on time is <50 μs and the system is used to image a rotating cooling fan. Images were acquired at a constant fan angular velocity with decreasing X-ray emission pulse width. As the X-ray pulse width decreases the fan blade sharpness increases evidencing the usefulness of beam pulsing in acquiring high resolution images in moving systems. The pulsing of the gate voltage has also been shown to produce turn on times of the order of 100 μs [149,198]. At an early stage, Yue *et al.* [149], as shown in

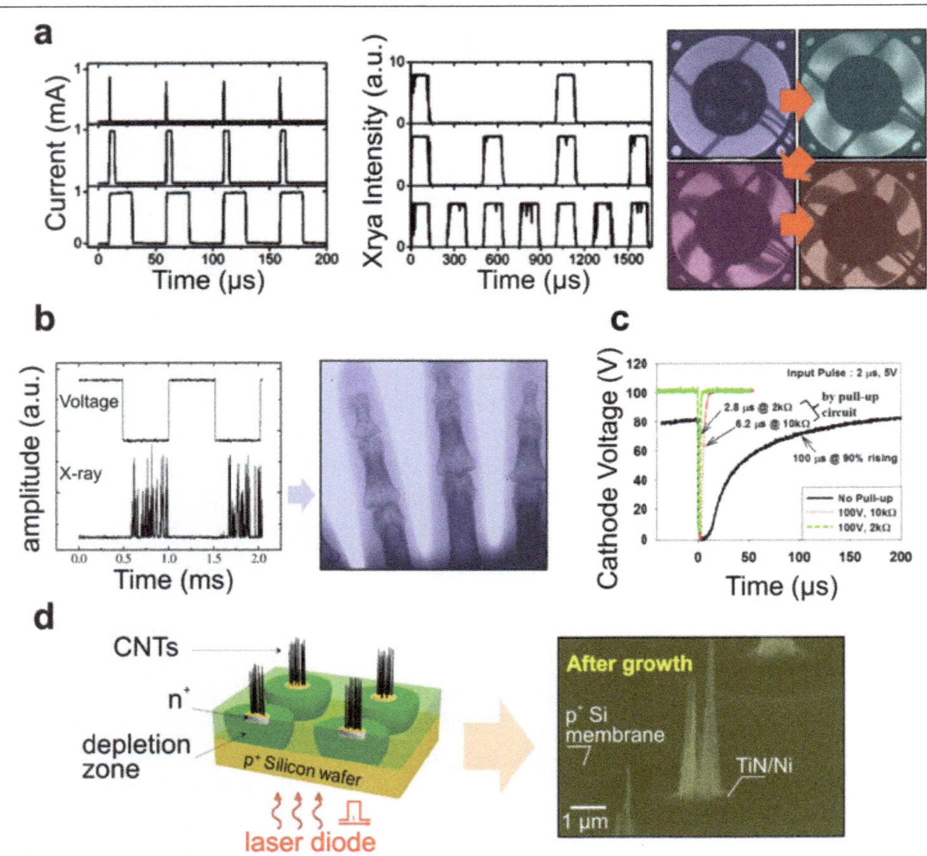

Figure 6 Pulsed Sources. (a) Anode current and X-ray intensity temporal response plots showing an X-ray beam turn-on time of <50 μs, and X-ray images of a rotating computer cooling fan acquired at a constant fan angular velocity with decreasing X-ray emission pulse width. Note the increased sharpness of the fan blades with increasingly rapid beam pulsing [198]. Copyright 2005, Joint Committee on Powder Diffraction Standards. **(b)** Bias voltage and X-ray emission temporal response plots with an acquired X-ray image of a users hand [149,156]. Copyright 2003, Elsevier and American Institute of Physics. **(c)** The time dependence of the cathode voltage with integrated pull-up resistance showing a rise time of <3 μs [154]. Copyright 2013, SPIE. **(d)** Schematic depiction of a red-source photo-gated CNF-based electron source. Each vertically aligned CNF bundle sits on top of a pin$^+$ photodiode coated with a TiN diffusion barrier coated and a Ni catalyst, as illustrated in the scanning electron micrograph taken after the CNT growth (Scale bar: 1 μm). Adapted from [233]. Copyright 2014, Thales Electron Devices.

Figure 6(b), demonstrated peak beam currents of up to 28 mA, at anode potentials of 14 kV, without degradation of the field emitters. An interesting artefact of their rapid pulsing was the chaotic and rather noisy X-ray emission, which showed an approximate periodicity of 100 μS. The noise here must certainly be reduced if a practical device is to be realised.

The limiting factor in many pulsed systems is often the capacitance of the gate assembly. Kim *et al.* [154] showed that pulsing performance could be improved and demonstrated <3 μs pulse periods, as indicated by the cathode voltage, via the use of a simple high voltage MOSFET pull-up circuit to switch the cathode, proposing that their pulsed source may find application in angiography. However it is worth noting that such a measure does not correlate with the X-ray photon flux [198]. Liu *et al.* [206], by pulsing the gate bias using a pulse generator with a constant anode bias of 40 kV, showed a stable 0.3 mA temporal response over 15 h with an approximate pulse rise time of <5 ms. Certainly electronic control over the pulsing performance has some use. Though the bandwidth is dramatically reduced, by at least an order of magnitude, the on/off ratio can be increased and the emitter fabrication simplified. Nevertheless, this is necessarily at the expense of other technological challenges, chiefly the associated RC constant of the vacuum cavity. This RC constant induces intrinsic time delays that are not experienced in the optically stimulated case. As illustrated in Figure 6(c), Kim *et al.* [154] employed 10 kΩ pull-up resistances to negate RC lag issues. The pulsing performance of arrayed sources has also been considered. Wang *et al.* [169] showed millisecond temporal response in a matrix source designed for micro radiotherapy. This had an array of 5 × 10 pixels achieving an emission current of 3 mA per pixel.

In order to obviate issues associated with the devices RC time constants, as depicted in Figure 6(d), Thales Electron Devices exploited integrated photodiodes and direct optical excitation. Vertically aligned CNFs were grown on p-i-n^+ photodiodes on a back etched Si membrane. The n+ doped areas were defined by ion implantation in a 5 μm thick intrinsic layer, whilst the p+ doped silicon wafer was subsequently thinned to obtain the 7 μm thick membrane [233]. These arrayed emitters where then irradiated, over the entire emitter area, with pulsed red laser light allowing them to attain very high bandwidth and on-off ratio of around 10, in addition to a maximum on current density of around 100 $\mu A/cm^2$.

The availability of rapidly pulsed sources opens up the potential for high-speed, real-time inspection technologies, capable of coupling high throughput, on-line manufacturing with real-time inspection. Indeed, with controlled high pulse rates the potential to perform medical diagnostics without strict patient restraints is accessed, allowing inspection of dynamic organs without image blurring. Intrinsic motion and the associated quasi-periodic respiration and heart beats, for example, allow one means of gating such high speed pulsed systems. One possible approach considered is to lock the gate electrode pulsing to the patients' cardiac or respiration signal. Such motion-induced artefacts can thus be reduced as the X-ray exposure is synchronised with a patients physiological rhythm, or indeed an objects motion [206].

2.6.6 Static tomography & tomosynthesis

X-ray based Computed Tomography (CT) imaging is fundamental to medical diagnostics. However, at present such imaging devices are affected by long scan times. Enabled by advances in computer and detection technologies, integration time issues are exacerbated when considered in relation to the concept of real-time 3D X-ray imaging. Multi-pixel sources offer one viable solution. Multi-pixel sources present a unique opportunity in next-generation inspection techniques, especially when one considers the physically small CNT FE cathodes, and the potential for nanoengineering and rapid switching. Pulsed multi-pixel sources give rise to a series of previously challenging applications typically relating to real-time medical diagnostics [234]. Tomography relies on the acquisition of large numbers of images, whilst viewing a subject at a range of angles. These systems, at present, incorporate multiple sources which are sequentially switched. This allows the system to capture an array of independent images without the necessity to physically rotate components, patients, or samples on an expensive and bulky gantry. The emerging alternative is to have multiple, distributed X-ray sources. Here, the subject remains stationary with the added benefit of more rapid data acquisition and potentially higher image quality. Indeed, such systems are largely immune to image blurring, which persists in conventional moving-source systems.

A range of different approaches to the application of tomography or tomosynthesis have been presented. The University of North Carolina, as illustrated in Figure 7(a), have demonstrated high-resolution stationary tomosynthesis using 31 individually addressed CNT X-ray sources distributed within a single large vacuum enclosure [234]. Each source had a 0.6 mm focal spot size, was operated at 28 kV at an anode current of 38 mA. Though an excellent demonstrator of the potential of CNT-based multi-pixel sources, the system is extremely large and fails to fully exploit the potential of emerging CVD techniques to pattern multiple sources on a single chip as a means to facilitate miniaturisation. Nevertheless, the system is at present one of the most advanced CNT-based FE X-ray sources available. It is based on physically separate CNT ink deposited thin films with

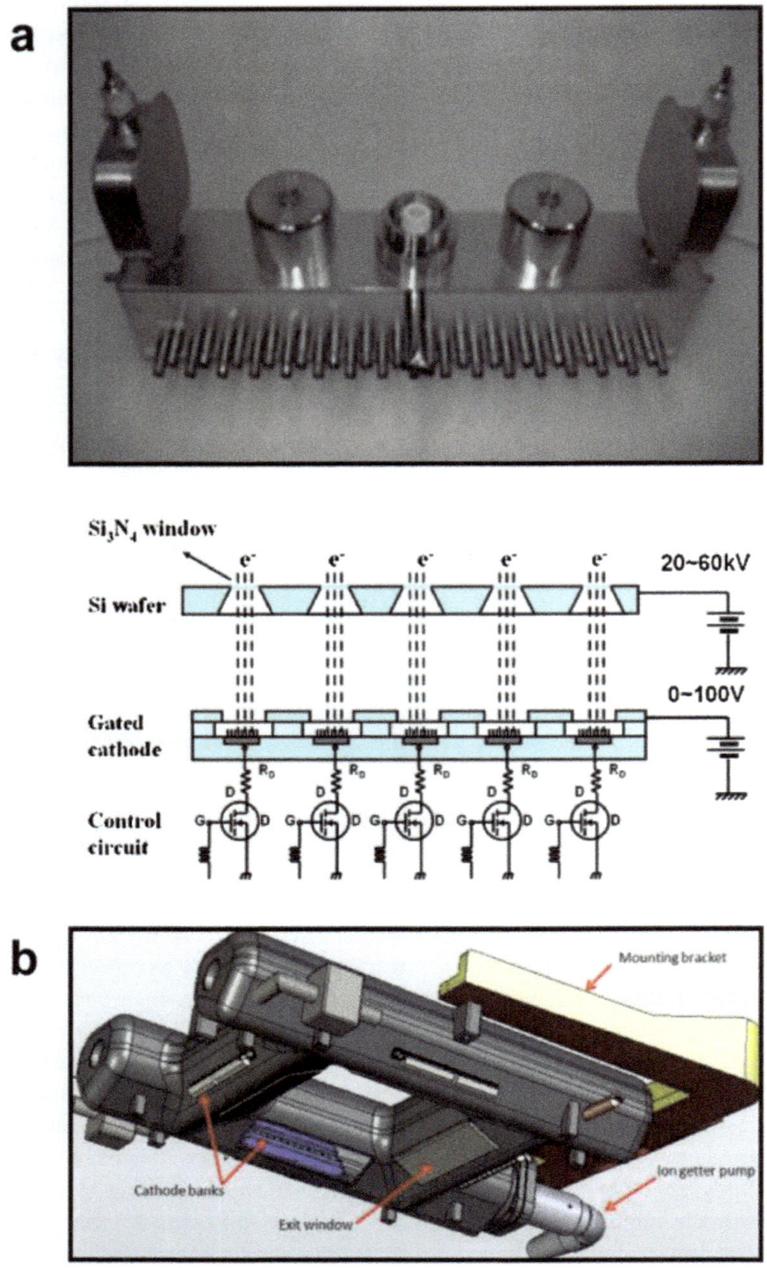

Figure 7 CNT-based tomosynthesis systems. (a) A distributed CNT tomosynthesis unit consisting of 31 individually addressable X-ray sources enclosed in a single vacuum chamber, and below a schematic the corresponding drive circuitry. Adapted from [234]. Copyright 2012, American Association of Physicists in Medicine. **(b)** An actively pumped multi-pixel X-ray source consisting of 52 individually controlled sources. Courtesy of the University of North Carolina at Chapel Hill [235].

inter-emitter pitches of the order of a few tens of centimetres. The system has also, rather impressively, demonstrated real-time 3D image reconstruction. Figure 7(b) shows another stationary computed tomography system by the same team. This system consists of 4 cathode banks, with a total of 52 separate cathode assemblies. Though rather large and limited to stationary or slow moving objects, this prototype system is the first in a new wave of CNT-based real-time 3D tomosynthesis scanners.

High spatial resolution requires a small X-ray spot size. However, practical systems also require a wide field of view in order to be able to inspect suitably large areas. Thus, to realise high resolution large area scans the X-ray source must be constructed from multiple cathodes, each with dedicated focusing electrodes. One way to maximise the field of view is to carefully engineer the vacuum chamber to maximise the total angular field of view. A 5 pixel source system has been constructed by

Zhang *et al.* [236]. This system provided sequentially acquired images over a finite angular range, each fitted with an electrostatic focusing electrode to control the size and scanning of the focal spot, which had a width of less than 300 μm. Operating at 40 kV, they showed that the electronic switching time was largely negligible with high emission reproducibility, both in FE curves and beam diameter, between the individual electron sources. Yang *et al.* demonstrated a multiple source digital breast tomosynthesis unit using an array of 25 cathodes, each individually programmed using analogue controls. Their sources employed CNT-based inks, each with a dedicated gate and an electrostatic focusing ring to control the focal spot size to approximately 200 μm [237]. In this system an impressive total angular field of view of up to 48° was realised, and the performance compared favourably with commercially available moving-source systems. A similar system, termed a multi-beam FE X-ray source, described a similar geometry [203]. A simple linear multiple source using an array of 31 CNT FE cathodes has also been constructed, for the purposes to digital breast tomosynthesis, which was shown to be capable of a total scan angle of 30° [171]. In a further extension, a digital chest tomosynthesis scanner was demonstrated [238]. This system used 75 CNT FE sources, operating at up to 80 kV with a 20° beam angle.

2.6.7 Multi-pixel sources

In the majority of the multi-pixel systems discussed the electron sources have largely been constructed using individual cathodes fabricated in linear arrays. A two dimensional geometry based on a series of linear arrays in a square or hexagonal format of physically separate detectors and sources, has been evaluated using simulation. Such a system is highly manufacturable [239], however; the inspection technique relies on close alignment of the generators and detectors which is likely challenging. Guzilov *et al.* [240] proposed circular arrays of FE emitters to emulate the classical rotary gantries used for CT, although their practical implementation has been limited to staggered linear arrays. Two dimensional arrays have a number of advantages over their linear counterparts, chief amongst which is the potential for ad hoc beam shaping of the emergent X-rays. Micro-integration of two dimensional source arrays allows for ever higher spatial resolution and subsequent beam shaping. Such micro-integrated two dimensional FE source arrays have been realised, though only few such devices have been investigated. Figure 8(a) depicts a 5 × 10 electron beam array fabricated by Wang *et al.* [169], as mentioned above. Deposited using inking techniques (Figure 8(b)), the 50 individually addressable electron beams allow for arbitrarily defined, and controllable electron, and hence X-ray beam shapes, for application in radiotherapy. An example of one such ad hoc beam shaped, electronically defined, X-ray intensity distribution is illustrated in the bottom image of Figure 8(c). A dose rate of 1.24 Gy/min at the centre of the irradiated object was obtained at an emission current of 3.0 mA [169]. High energy sources for tomosynthesis based on such multi-pixel sources, with high anode potentials of up to 160 kV, have also been demonstrated by Sprenger *et al.* [241]. This was achieved using a square array of 4 × 13 electron sources for applications including image guided radiotherapy.

2.6.8 Compact & miniature sealed tubes

X-ray generators are often bulky devices, as a result of the thermal dsspiation and subsequent cooling requirements. Certain user specifications necessitate significant miniaturisation. Miniaturised sources find use in hand-held non-destructive testing, X-ray spectroscopy, electrically controlled brachytherapy and in situ radiation therapy. Miniature TE sources have been available for many years, but have proven challenging to miniaturise [242]. Amptek has launched a series of miniature X-ray sources [243]. Sealed FE sources, on the other hand, lend themselves to these applications, and in particular are well suited for portable X-ray applications. Though difficult to manufacture, miniaturised sources are emerging on the market place. As in the case of pulsed and multi-

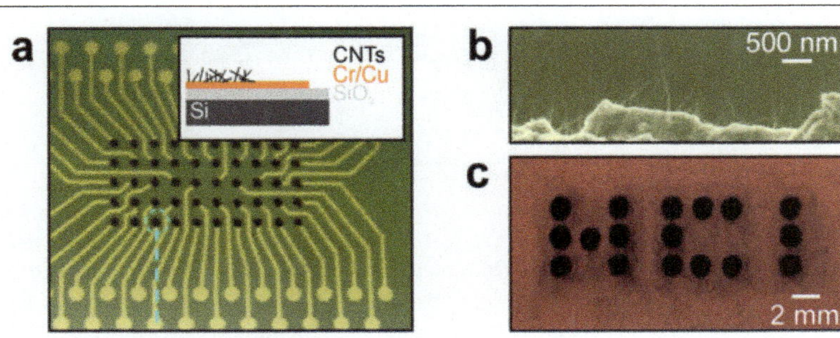

Figure 8 Multi Pixel Sources. (a) Electrophorectically deposited multiple X-ray source chip and the corresponding **(b)** scanning electron micrograph of deposited CNTs, and **(c)** he resulting electronically reconfigurable X-ray distribution. Adapted from [169]. Copyright 2011, American Institute of Physics.

pixel sources, much progress has been made towards a miniaturised CNT-based FE X-ray source though, to the best of the authors' knowledge, no such devices have, as yet, made it to market.

One of the smallest tubes fabricated, at the time of publication, as far as the authors' are aware, is that of the Electronics and Telecommunications Research Institute, Korea (Figure 9(a)). This cylindrical diode source, which is smaller than the average human finger, has an impressive outer diameter of only 6 mm and a length of 31 mm, and is based on CNTs deposited by a paste printing technique [208,244]. Developed for miniaturised X-ray diffractometry, the X-ray source of Sarrazin *et al.* [182], Figure 9(b), can support current densities of up to $1\ A/cm^2$, whilst Heo *et al.* [207], who demonstrated a 10 mm diameter and 50 mm long miniature source, as illustrated in Figure 9(c), derived around $0.12\ A/cm^2$. The sealed device can be operated up to 70 kV. The cathode was formed from sintered SWCNTs mixed with Ag nanoparticles to form a paste which was deposited on a flattened, 0.8 mm diameter, W wire. Here a conically machined Be window was magnetron sputtered with a 1.5 μm W thin film forming a transmission type X-ray target. The vacuum brazed alumina tube with Kovar electrodes and non-evaporable getters weighs just 14.5 g. The bias required to extract $10\ mA/cm^2$ was 29 kV, with a maximum current density of $0.12\ A/cm^2$, giving a nominal stability of ±2% at 50 kV, with a similar

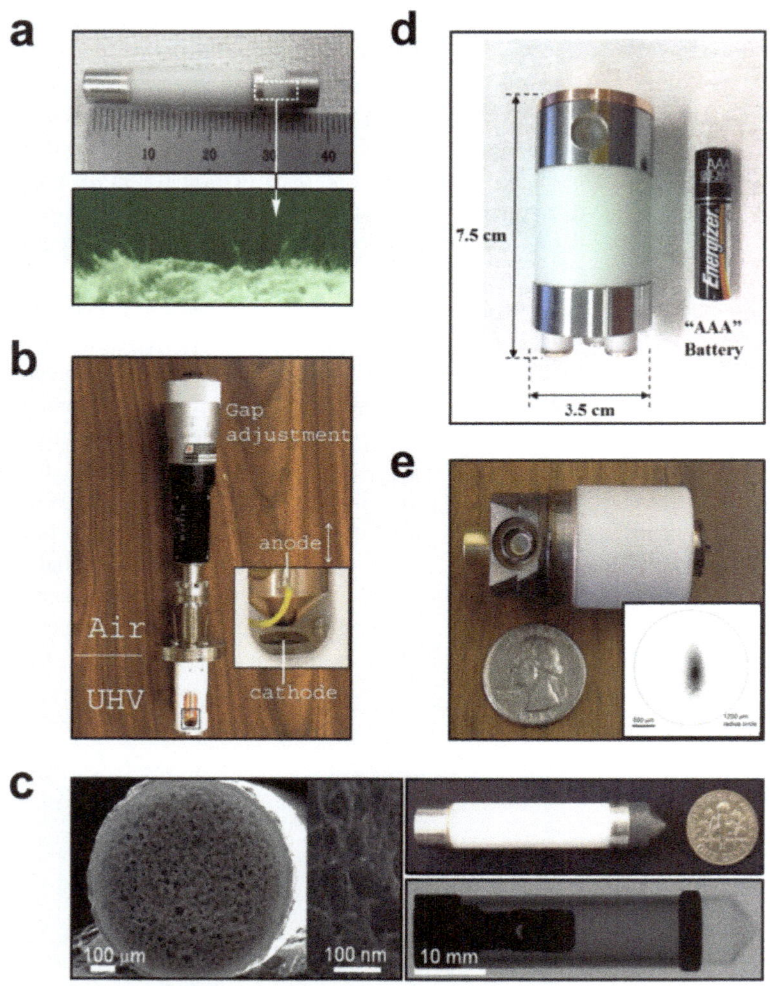

Figure 9 Miniature CNT-based FE X-ray sources. (a) A disordered miniature CNT, brazed triode X-ray source fabricated by ball milling, and firing of a CNT paste on the apex of a 0.6 mm Kovar rod, and operated at 12 kV [244]. Copyright 2013, Electronics and Telecommunications Research Institute [208]. Copyright 2013, American Institute of Physics **(b)** A miniature X-ray source fabricated by thermal CVD capable of supporting current densities of up to $1\ A/cm^2$ [182]. Copyright 2004, Joint Committee on Powder Diffraction Standards. **(c)** A sintered SWCNT paste transmission-type miniature CNT cathode X-ray source, weighting 14.5 g, and operating up to 70 kV with uniform spatial dose distribution [207]. Copyright 2012, Springer. **(d)** A 7.5 cm long selaed CNT-based X-ray source operating at 30 kV. [154]. Copyright 2013, Society of Photo-Optical Instrumentation Engineers. **(e)** An Oxford Instruments X-ray Technology Inc. miniature field emission X-ray tube with gated MWCNT 2 mm diameter cathode. Adapted from [182]. Copyright 2004, Joint Committee on Powder Diffraction Standards.

variation in the X-ray dose. Radially, the X-ray dose rate was 108 Gy/min and was broadly isotropic (20% variation) with a typical focal spot size, as defined by EU standard EN 12543–5, of the order of 3–4 mm. Serious heating prevented the source from being operated for significant lengths of time at the maximum anode bias. Kim *et al.* [154], as shown in Figure 9(d), have developed a similar brazed compact sealed source, measuring just 3.5 cm in diameter and 7.5 cm in length. This pentode device - consisting of a cathode, gate, two focusing electrodes, and an anode – had a focal spot of the order of 300 μm in diameter when operated with an anode current of 50 mA. The CNT cathode was just 0.15 cm^2 and was formed form a screen-printed and ball-milled complex CNT paste combined with inorganic fillers, oxides, metallic nano particles and a photo-initiator monomer with organic components including acrylates. The CNT films were relatively macroscopically uniform between samples, though rather perturbed and irreproducible microscopically. No active pumping system was integrated, though a non-evaporable getter was fabricated in an attempt to maintain the vacuum environment. The tube pressure was only of the order of 10^{-6} mbar upon sealing. The cavity pressure during emission was not noted. Pulse mode operation (10% duty cycle) was opted for – at 30 kV and 10 mA, to reduce thermally stimulated outgassing. Nevertheless, the tube still operated at approximately 200°C which almost certainly compromised the vacuum, even for a previously well-outgassed cathode assembly. Figure 9(e) shows the Oxford Instruments X-ray Technology Inc. miniature MWCNT FE X-ray tube with a 2 mm diameter cathode and integrated gate. The device develops a somewhat asymmetrical and diffuse focal spot with a major axis length of the order of 700 μm [182].

2.7 Commercial progress

The published information from commercial work is, at present, quite limited, and is more prevalent among smaller emerging enterprises. Of the larger international conglomerates, Thales and Thales Electron Devices have been active for just over a decade in developing CNT photocathodes for optically driven multiple X-ray sources. No compact commercial system is though, as yet, available. Due to their novel micro- and nano-fabrication techniques the Thales systems have demonstrated very fast switching times and high bandwidth, though lower on-off current ratios (~10), relative to the conventional electronic methods [245,246]. Siemens has been investigating the development of fast CNT based tomographic systems, in particular for airport security scanners [247]. This followed a period as a joint venture partner in XinRay Ltd, (North Carolina, USA) who were simultaneously developing tomographic and tomosynthesis systems for medical applications such as mammography. Xintek, and its associate company of XinRay, have also been extensively involved with the manufacture of systems for static tomosynthesis using CNT FE sources [248]. In 2004, in collaboration with NASA and Microwave Power Technology Inc., Oxford Instruments published work on their research efforts towards a MWCNT-based miniature source with a claimed 80% efficiency [182]. To demonstrate the robustness of the source they operated them at 1.5 W for >100,000 pulses, each 10s in duration (66% duty cycle). Oxford Instruments have stated an on-going commitment towards the development micro focal CNT source, which they have implemented in their Eclipse II models. These miniature tubes are based on 3 W CNT field emitters with an integrated power supply. This small package is only 160 × 38 mm, with a mass of 300 g and is battery operated [182]. Nevertheless, commercial progress remains stilted at present with the incumbent TE market dominating. FE sources with significant functional and lifetime advances are needed, with manufacturing based on readily accessible, widely applicable, and inexpensive techniques.

3 Conclusion

Here we have reviewed the current state-of-the-art in carbon nanotube-based field emission X-ray sources. Issues from fabrication to functionality have been considered. The various systems demonstrated to date have been described, highlighting some of the technological barriers facing the technology. Compared to other electron emission materials, the potential of carbon nanotubes is largely unparalleled. Their attractive morphological and electronic properties are very promising for future electron source technologies. Within the coming decade there is the distinct possibility that FE electron sources will reach a level of technological maturity where they pose a distinct and real threat to incumbent thermionic sources. The high attainable current densities, rapid pulsing and multi-pixel potential, mediated by the ability to pattern at the submicron scale, will likely result in a number of persuasive proof-of-principle prototypes that will help disrupt the incumbent, yet technologically stagnated, thermionic source market. Significant further research is needed to realise this commercial potential. Nonetheless, it remains clear that carbon nanotube-based field emission X-ray sources have great prospects. Their intrinsic manufacturability and enhanced functionality will undoubtedly be key in realising a new generation of innovative X-ray scanning techniques, applicable across many fields and industries, which have hitherto been unavailable.

Abbreviations

β: Field enhancement factor; FN: Fowler-Nordheim; I_a: Anode current (A); J: Current density (A/cm^2); FE: Field emission; PE: Photoelectric emission; TE: Thermionic emission; CNF: Carbon nanofibres; CNT: Carbon nanotubes; MWCNT: Multi-walled carbon nanotube; SWCNT: Single-walled carbon nanotube.

Competing interests

The authors declare that they have no competing interests.

Authors' contributions

RJP and MTC contributed equally. RJP and MTC wrote the manuscript. MTC designed the figures. CMC compiled the materials summary. All authors reviewed and approved the manuscript.

Authors' information

Richard Parmee: Richard Parmee is a researcher in the Electrical Engineering Division, Department of Engineering, University of Cambridge, and the William Pitt Fellow at Pembroke College. He received the MA in Electrical Sciences from the University of Cambridge is currently researching the applications of field emission using Carbon Nanotubes, for X-ray generation, at the Centre for Advanced Photonics and Electronics in Cambridge. He is an engineer in the fields of image processing, automatic X-ray inspection and electronic gauging. He has worked extensively in the areas of precision measurement, statistical quality control and end-of-line inspection, and has developed X-ray applications, equipment, image processing and detector technology for many of the largest companies in the world in the food, medical and pharmaceutical industries

Clare Melissa Collins: Clare Collins is studying for the MRes in Ultra Precision, funded by the EPSRC, at the University of Cambridge. She graduated from King's College London in 2013 with an MSci in Physics.

William Ireland Milne: William Milne was the head of the Electrical Engineering Division, Department of Engineering, University of Cambridge, from 1999–2014. He is a Professor in Electrical Engineering, and Director of the Centre for Advanced Photonics and Electronics where he heads the Electronic Devices and Materials group. He obtained his BSc from St Andrews University, Scotland, in 1970 and a PhD from Imperial College London in 1973. In his formative years he worked at the Plessey Res Co., UK, following which time, in 1976, he joined the Department of Engineering, University of Cambridge. He is a Fellow of the Royal Academy of Engineering and was awarded the 2008 Institute of Engineering and Technology's JJ Thomson medal for his work on electronic materials. He is a Distinguished Visiting Scholar at KyungHee University, Korea, a Distinguished Visiting Professor at Southeast University, China, and a Guest Professor at Shizuoka University, Japan. His research focuses on large-area silicon and carbon-based thin film materials, their growth and application of carbon nanomaterials in electronics.

Matthew Thomas Cole: Matthew Cole is the Oppenheimer Research Fellow in the Department of Engineering, University of Cambridge, and Fellow of St Edmund's College, University of Cambridge. He is based in the Centre for Advanced Photonics and Electronics and is a member of the Cambridge Graphene Centre. He gained an MEng (Hons) in Engineering Sciences from Oxford University in 2008 and a PhD in Electrical Engineering from the University of Cambridge in 2011. He is a Fellow of the Institute of Nanotechnology, a Chartered Engineer, a Chartered Physicist, and a Chartered Scientist. He has spent time as a visiting scholar at Harvard University and as a research associate at Sharp Laboratories of Europe. His interests include the heterogeneous integration of chemical vapour deposited aligned nanomaterials for electron emission applications.

Acknowledgements

The author's thank Dr C. Li, Prof. W. Lei, and Prof. B.P. Wang of Southeast University, China, for their on-going collaboration. M.T.C. thanks the Winston Churchill Trust and the Oppenheimer Research Fellowship for generous financial support. This work was supported by an EPSRC Impact Acceleration grant and a TSB Advanced Materials Feasibility award.

References

1. Roentgen, On a new kind of rays. Nature **53**, 274–276 (1896)
2. JJ Thomson, The Röntgen rays. Nature **53**, 581–583 (1896)
3. R Kienbock, Uber die Einwirkung des Rontgen-Lichtes auf die Haut. Wiener Kiin Wschr **13**, 1153 (1900)
4. L Benoist, Experimental definition of various types of X-rays by the radiochromator. C R Acad Sci Puns **135**, 225 (1902)
5. C Beck, Röntgen-ray diagnosis and therapy. Arch Roentgen Ray **9**, 110–110 (1904)
6. WL Bragg, The diffraction of short electromagnetic waves by a crystal, in *Proceedings of the Cambridge Philosophical Society*, 1913, p. 4
7. M von Laue, Concerning the detection of X-ray interferences. Nobel lecture (1915), doi: 10.1002/adom.201400238
8. HG Moseley, XCIII, The high-frequency spectra of the elements. London Edinburgh Dublin Philos Mag J Sci **26**, 1024–1034 (1913)
9. RE Franklin, RG Gosling, Molecular configuration in sodium thymonucleate. Nature **171**, 740–741 (1953)
10. AH Compton, A quantum theory of the scattering of X-rays by light elements. Physical Review **21**, 483 (1923)
11. AH Compton, On the mechanism of x-ray scattering. Proc Natl Acad Sci U S A **11**, 303 (1925)
12. R Glocker, H Schreiber, Quantitative Röntgen spectrum analysis by means of cold cathode excitation. Ann Phys **85**, 1089–1102 (1928)
13. CR Bull, R Zwiggelaar, RD Speller, Review of inspection techniques based on the elastic and inelastic scattering of X-rays and their potential in the food and agricultural industry. (J Food Eng **33**, 167–179 (1997)
14. WD Coolidge, A powerful röntgen ray tube with a pure electron discharge. Phys Rev **2**, 409–430 (1913)
15. EL Murphy, RH Good, Thermionic emission, field emission, and the transition region. Phys Rev **102**, 1464–1473 (1956)
16. W Lei, C Li, MT Cole, K Qu, S Ding, Y Zhang, JH Warner, X Zhang, B Wang, WI Milne, A graphene-based large area surface-conduction electron emission display. Carbon **56**, 8 (2013)
17. RH Fowler, L Nordheim, Electron emission in intense electric fields. Proc R Soc Lond A Math Phys Character **119**, 173–181 (1928)
18. WI Milne, KBK Teo, E Minoux, O Groening, L Gangloff, L Hudanski, J-P Schnell, D Dieumegard, F Peauger, IYY Bu, MS Bell, P Legagneux, G Hasko, GAJ Amaratunga, Aligned carbon nanotubes/fibers for applications in vacuum microwave amplifiers. J Vac Sci Technol B **24**, 345–348 (2006)
19. P Hawkes, E Kasper, *Principles of Electron Optics Vol. 2: Applied Geometrical Optics Academic* (Academic Press, San Diego, 1996), pp. 953–970
20. CJ Edgcombe, U Valdrè, Field emission and electron microscopy. Microsc Microanal **6**, 380–387 (2000)
21. J-M Bonard, T Stöckli, O Noury, A Châtelain, Field emission from cylindrical carbon nanotube cathodes: possibilities for luminescent tubes. Appl Phys Lett **78**, 2775–2777 (2001)
22. WP Dyke, J Trolan, Field emission: large current densities, space charge, and the vacuum arc. Phys Rev **89**, 799 (1953)
23. W Dyke, W Dolan, Field emission. Adv Electron Electron Phys **8**, 89–185 (1956)
24. CA Spindt, A thin-film field-emission cathode. J Appl Phys **39**, 3504–3505 (1968)
25. I Brodie, C Spindt, *Advances in Electronics and Electron Physics edited by PW Hawkes*, 83rd edn., 1992
26. CA Spindt, I Brodie, L Humphrey, ER Westerberg, Physical properties of thin-film field emission cathodes with molybdenum cones. J Appl Phys **47**, 5248–5263 (1976)
27. CA Spindt, CE Holland, I Brodie, JB Mooney, ER Westerberg, Field-emitter arrays to vacuum fluorescent display. Electron Devices IEEE Trans **36**, 225–228 (1989)
28. L Yukui, Z Changchun, L Xinghui, Field emission display with carbon nanotubes cathode: prepared by a screen-printing process. Diam Relat Mat **11**, 1845–1847 (2002)
29. DR Whaley, BM Gannon, CR Smith, CM Armstrong, CA Spindt, Application of field emitter arrays to microwave power amplifiers. Plasma Sci IEEE Trans **28**, 727–747 (2000)
30. H Makishima, S Miyano, H Imura, J Matsuoka, H Takemura, A Okamoto, Design and performance of traveling-wave tubes using field emitter array cathodes. Appl Surf Sci **146**, 230–233 (1999)
31. KL Jensen, Field emitter arrays for plasma and microwave source applications. Phys Plasmas (1994-present) **6**, 2241–2253 (1999)
32. KBK Teo, E Minoux, L Hudanski, F Peauger, JP Schnell, L Gangloff, P Legagneux, D Dieumegard, GAJ Amaratunga, WI Milne, Microwave devices - carbon nanotubes as cold cathodes. Nature **437**, 968–968 (2005)
33. EF Barasch, HP Demroff, TS Elliott, TB Kasprowicz, B Lee, T Mazumdar, PM Mcintyre, Y Pang, DD Smith, H-J Trost Gated Field-Emitter Cathodes for High-Power Microwave Applications, in *Proceedings of the 1992 Linear Accelerator Conference* (Ottawa, Ontario, Canada, 1992)
34. MT Cole, *Dry-Transfer of Chemical Vapour Deposited Nanocarbon Thins Films for Functionally Enhanced Flexible Transparent Electronics Applications* (University of Cambridge, Cambridge, 2011)

35. JJ Chiu, WS Wang, CC Kei, CP Cho, TP Perng, PK Wei, SY Chiu, Room temperature vibrational photoluminescence and field emission of nanoscaled tris-(8-hydroxyquinoline) aluminum crystalline film. Appl Phys Lett **83**, 4607–4609 (2003)
36. C-P Cho, T-P Perng, On the dendritic growth and field emission of amorphous AlQ3 nanowires. Org Electron **11**, 115–122 (2010)
37. X Fang, Y Bando, UK Gautam, C Ye, D Golberg, Inorganic semiconductor nanostructures and their field-emission applications. J Mater Chem **18**, 509–522 (2008)
38. FCK Au, KW Wong, YH Tang, YF Zhang, I Bello, ST Lee, Electron field emission from silicon nanowires. Appl Phys Lett **75**, 1700–1702 (1999)
39. Y Jr Hung, S-L Lee, LC Beng, H-C Chang, Y-J Huang, K-Y Lee, Y-S Huang, Relaxing the electrostatic screening effect by patterning vertically-aligned silicon nanowire arrays into bundles for field emission application. Thin Solid Films **556**, 146–154 (2014)
40. LA Ma, ZX Lin, JY Lin, YA Zhang, LQ Hu, TL Guo, Large-scale growth of ultra-thin MgO nanowires and evaluate their field emission properties. Physica E **41**, 1500–1503 (2009)
41. H Tan, N Xu, S Deng, Synthesis and electron emission properties of MgO nanowires. J Vac Sci Technol B **28**, C2B20–C22B23 (2010)
42. JH He, RS Yang, YL Chueh, LJ Chou, LJ Chen, ZL Wang, Aligned AlN nanorods with multi-tipped surfaces - growth, field-emission, and cathodoluminescence properties. Adv Mater **18**, 650 (2006)
43. X Ji, P Chen, J Deng, W Zhou, F Chen, Field emission properties of AlN nanostructures. J Nanosci Nanotechno **12**, 6531–6533 (2012)
44. J Pelletier, D Gervais, C Pomot, Application of wide-gap semiconductors to surface ionization: work functions of AlN and SiC single crystals. J Appl Phys **55**, 994–1002 (1984)
45. W Sun, Y Li, Y Yang, Y Li, C Gu, J Li, Morphology inducing selective plasma etching for AlN nanocone arrays: tip-size dependent photoluminescence and enhanced field emission properties. J Mater Chem C **2**, 2417–2422 (2014)
46. PG Chavan, MA More, DS Joag, SS Badadhe, IS Mulla, Photo-enhanced field emission studies of tapered CdS nanbelts, in *Proceedings 27th International Vacuum Nanoelectronics Conference* (IEEE, Engelberg, Switzerland, 2014), pp. 83–84
47. P Chavan, R Kashid, S Badhade, I Mulla, M More, D Joag, CdS nanowires: ultra-long growth and enhanced field emission properties. Vacuum **101**, 38–45 (2014)
48. T Ge, L Kuai, B Geng, Solution-phase chemical route to branched single-crystalline CdS nanoarchitectures and their field emission property. J Alloys Compd **509**, L353–L358 (2011)
49. Y Lin, Y Hsu, S Lu, S Kung, Non-catalytic and template-free growth of aligned CdS nanowires exhibiting high field emission current densities. Chem Commun, 2391–2393 (2006); doi: 10.1039/b604309g
50. YH Lee, CH Choi, YT Jang, EK Kim, BK Ju, NK Min, JH Ahn, Tungsten nanowires and their field electron emission properties. Appl Phys Lett **81**, 745–747 (2002)
51. S Wang, Y He, X Fang, J Zou, Y Wang, H Huang, PMFJ Costa, M Song, B Huang, CT Liu, PK Liaw, Y Bando, D Golberg, Structure and field-emission properties of sub-micrometer-sized tungsten-whisker arrays fabricated by vapor deposition. Adv Mater **21**, 2387 (2009)
52. KS Yeong, JTL Thong, Field-emission properties of ultrathin 5 nm tungsten nanowire. J Appl Phys **100**, 114325 (2006)
53. N Wan, J Xu, G Chen, X Gan, S Guo, L Xu, K Chen, Broadband anti-reflection and enhanced field emission from catalyst-free grown small-sized ITO nanowires at a low temperature. Acta Mater **58**, 3068–3072 (2010)
54. AS Komolov, EF Lazneva, SA Pshenichnyuk, AA Gavrikov, NS Chepilko, AA Tomilov, NB Gerasimova, AA Lezov, PS Repin, Electronic properties of the interface between hexadecafluoro copper phthalocyanine and unsubstituted copper phthalocyanine films. Semiconductors **47**, 956–961 (2013)
55. WY Tong, ZX Li, AB Djurisic, WK Chan, SF Yu, Field emission from copper phthalocyanine and copper hexadecafluorophthalocyanine nanowires. Mater Lett **61**, 3842–3846 (2007)
56. S-H Jang, J-S Jang, Schottky barrier characteristics and carrier transport mechanism for ohmic contacts to strained p-Type InGaN/GaN superlattice. Electrochem Solid State Lett **13**, H403–H405 (2010)
57. TH Seo, KJ Lee, TS Oh, YS Lee, H Jeong, AH Park, H Kim, YR Choi, E-K Suh, TV Cuong, VH Pham, JS Chung, EJ Kim, Graphene network on indium tin oxide nanodot nodes for transparent and current spreading electrode in InGaN/GaN light emitting diode. Appl Phys Lett **98**, 251114 (2011)
58. F Ye, XM Cai, XM Wang, EQ Xie, The growth and field electron emission of InGaN nanowires. J Crsy Growth **304**, 333–337 (2007)
59. J Lee, Y Jung, J Song, JS Kim, G-W Lee, HJ Jeong, Y Jeong, High-performance field emission from a carbon nanotube carpet. Carbon **50**, 3889–3896 (2012)
60. JM Bonard, JP Salvetat, T Stockli, L Forro, A Chatelain, Field emission from carbon nanotubes: perspectives for applications and clues to the emission mechanism. Appl Phys A Mater **69**, 245–254 (1999)
61. DD Nguyen, Y-T Lai, N-H Tai, Enhanced field emission properties of a reduced graphene oxide/carbon nanotube hybrid film. Diam Relat Mat **47**, 1–6 (2014)
62. M Shiraishi, M Ata, Work function of carbon nanotubes. Carbon **39**, 1913–1917 (2001)
63. H Ago, T Kugler, F Cacialli, WR Salaneck, MSP Shaffer, AH Windle, RH Friend, Work functions and surface functional groups of multiwall carbon nanotubes. J Phys Chem B **103**, 8116–8121 (1999)
64. K Zheng, X Li, X Mo, G Chen, Z Wang, G Chen, Enhanced field emission and patterned emitter device fabrication of metal-tetracyanoquinodimethane nanowires array. Appl Surf Sci **256**, 2764–2768 (2010)
65. O Auciello, JC Tucek, AR Krauss, DM Gruen, N Moldovan, DC Mancini, Review of synthesis of low-work function Cu-Li alloy coatings and characterization of the field emission properties for application to field emission devices. J Vac Sci Technol B **19**, 877–883 (2001)
66. T Glatzel, H Steigert, S Sadewasser, R Klenk, MC Lux-Steiner, Potential distribution of Cu(In, Ga)(S, Se)(2)-solar cell cross-sections measured by Kelvin probe force microscopy. Thin Solid Films **480**, 177–182 (2005)
67. S Ding, H Cui, W Lei, X Zhang, B Wang, Stable field emission from ZnO nanowires grown on 3D graphene foam, in *Proceedings 27th International Vacuum Nanoelectronics Conference* (IEEE, Engelberg, Switzerland, 2014), pp. 178–179
68. F Jamali Sheini, DS Joag, MA More, Field emission studies on electrochemically synthesized ZnO nanowires. Ultramicroscopy **109**, 418–422 (2009)
69. X Sun, Designing efficient field emission into ZnO using morphological and electronic design techniques, low-threshold, high-emission current densities have been obtained with ZnO. SPIE. Nanotechnology (2006)
70. H Tampo, H Shibata, K Maejima, TW Chiu, H Itoh, A Yamada, K Matsubara, P Fons, Y Chiba, T Wakamatsu, Y Takeshita, H Kanie, S Niki, Band profiles of ZnMgO/ZnO heterostructures confirmed by Kelvin probe force microscopy. Appl Phys Lett **94**, 242107 (2009)
71. YH Yang, CX Wang, B Wang, NS Xu, GW Yang, ZnO nanowire and amorphous diamond nanocomposites and field emission enhancement. Chem Phys Lett **403**, 248–251 (2005)
72. ZP Zhang, WQ Chen, YF Li, J Chen, Defect-Assisted Field Emission fron ZnO nanotrees, in *Proceedings 27th International Vacuum Nanoelectronics Conference* (Elsevier, Engelberg, Switzerland, 2005), pp. 78–79
73. DKT Ng, MH Hong, LS Tan, YW Zhu, CH Sow, Field emission enhancement from patterned gallium nitride nanowires. Nanotech **18**, (2007); doi: 10.1088/0957-4484/18/37/375707
74. YQ Wang, RZ Wang, MK Zhu, BB Wang, B Wang, H Yan, Structure and surface effect of field emission from gallium nitride nanowires. Appl Surf Sci **285**(Part B), 115–120 (2013)
75. R Yousefi, FJ Sheini, MR Muhamad, MA More, Characterization and field emission properties of ZnMgO nanowires fabricated by thermal evaporation process. Solid State Sci **12**, 1088–1093 (2010)
76. A Agiral, JGE Gardeniers, *On-Chip Tungsten Oxide Nanowires Based Electrodes for Charge Injection*, 2010
77. K Huang, Q Pan, F Yang, S Ni, D He, The catalyst-free synthesis of large-area tungsten oxide nanowire arrays on ITO substrate and field emission properties. Mater Res Bull **43**, 919–925 (2008)
78. Y Kojima, K Kasuya, T Ooi, K Nagato, K Takayama, M Nakao, Effects of oxidation during synthesis on structure and field-emission property of tungsten oxide nanowires. Jpn J Appl Phys **46**, 6250–6253 (2007)
79. S Yue, H Pan, Z Ning, J Yin, Z Wang, G Zhang, Amazing ageing property and in situ comparative study of field emission from tungsten oxide nanowires. Nanotech **22**, (2011); doi: 10.1088/0957-4484/22/11/115703
80. A Khademi, R Azimirad, AA Zavarian, AZ Moshfegh, Growth and field emission study of molybdenum oxide nanostars. J Phys Chem C **113**, 19298–19304 (2009)
81. J Zhou, Large-area nanowire arrays of molybdenum and molybdenum oxides: synthesis and field emission properties. Advanced Mater **15**, 1835–1840 (2003)
82. J Liu, Z Zhang, C Pan, Y Zhao, X Su, Y Zhou, D Yu, Enhanced field emission properties of MoO2 nanorods with controllable shape and orientation. Mater Lett **58**, 3812–3815 (2004)
83. X Fang, Y Bando, G Shen, C Ye, UK Gautam, PMFJ Costa, C Zhi, C Tang, D Golberg, Ultrarine ZnS nanobelts as field emitters. Adv Mater **19**, 2593 (2007)

84. RV Kashid, DJ Late, SS Chou, Y-K Huang, D Mrinmoy, DS Joag, MA More, VP Dravid, Enhanced field-emission behavior of layered MoS2 sheets. Small **9**, 2730–2734 (2013)
85. T Yamada, T Masuzawa, T Ebisudani, K Okano, T Taniguchi, Field emission characteristics from graphene on hexagonal boron nitride. Appl Phys Lett **104**, 5 (2014)
86. T Yamada, T Ebisudani, K Okana, T Masuzawa, Y Neo, H Mimura, T Taniguchi, Field emission characteristics of graphene/h-BN structure, in *Proceedings 27th International Vacuum Nanoelectronics Conference* (IEEE, Engelberg, Switzerland, 2014), pp. 182–183
87. AB Preobrajenski, AS Vinogradov, N Martensson, Monolayer of h-BN chemisorbed on Cu(111) and Ni(111): the role of the transition metal 3d states. Surf Sci **582**, 21–30 (2005)
88. S Ohtani, T Yano, S Kondo, Y Kohno, Y Tomita, Y Maeda, K Kobayashi, Electron emission from h-BN films codoped with Mg and O atoms. Thin Solid Films **546**, 53–57 (2013)
89. W Chen, Y Su, H Chen, S Deng, N Xu, J Chen, Temperature Dependence of the Field Emission from Monolayer Graphene, in *Proceedings 27th International Vacuum Nanoelectronics Conferernce* (IEEE, Engelberg, Switzerland, 2014), pp. 36–37
90. T Hallam, MT Cole, WI Milne, GS Duesberg, Field emission characteristics of contact printed graphene fins. Small **10**, 95–99 (2014)
91. Z-S Wu, S Pei, W Ren, D Tang, L Gao, B Liu, F Li, C Liu, H-M Cheng, Field emission of single-layer graphene films prepared by electrophoretic deposition. Adv Mater **21**, 1756–1760 (2009)
92. C Wu, F Li, Y Zhang, T Guo, Field emission from vertical graphene sheets formed by screen-printing technique. Vacuum **94**, 48–52 (2013)
93. A Malesevic, R Kemps, A Vanhulsel, MP Chowdhury, A Volodin, C Van Haesendonck, Field emission from vertically aligned few-layer graphene. J Appl Phys **104**, 084301 (2008)
94. UA Palnitkar, RV Kashid, MA More, DS Joag, LS Panchakarla, CNR Rao, Remarkably low turn-on field emission in undoped, nitrogen-doped, and boron-doped graphene. Appl Phys Lett **97**, 063102 (2010)
95. L Li, W Sun, S Tian, X Xia, J Li, C Gu, Floral-clustered few-layer graphene nanosheet array as high performance field emitter. Nanoscale **4**, 6383–6388 (2012)
96. S Pandey, P Rai, S Patole, F Gunes, G-D Kwon, J-B Yoo, P Nikolaev, S Arepalli, Improved electron field emission from morphologically disordered monolayer graphene. Appl Phys Lett **100**, 043104 (2012)
97. L Jiang, T Yang, F Liu, J Dong, Z Yao, C Shen, S Deng, N Xu, Y Liu, H-J Gao, Controlled synthesis of large-scale, uniform, vertically standing graphene for high-performance field emitters. Adv Mater **25**, 250–255 (2013)
98. ATH Chuang, J Robertson, BO Boskovic, KKK Koziol, Three-dimensional carbon nanowall structures. Appl Phys Lett **90**, 123107 (2007)
99. D Banerjee, S Mukherjee, KK Chattopadhyay, Synthesis of amorphous carbon nanowalls by DC-PECVD on different substrates and study of its field emission properties. Appl Surf Sci **257**, 3717–3722 (2011)
100. YH Wu, BJ Yang, BY Zong, H Sun, ZX Shen, YP Feng, Carbon nanowalls and related materials. J Mater Chem **14**, 469–477 (2004)
101. CS Rout, PD Joshi, RV Kashid, DS Joag, MA More, AJ Simbeck, M Washington, SK Nayak, DJ Late, Superior field emission properties of layered WS2-RGO nanocomposites. Sci Rep **3**, (2013); doi: 10.1038/srep03282
102. M-C Kan, J-L Huang, JC Sung, K-H Chen, B-S Yau, Thermionic emission of amorphous diamond and field emission of carbon nanotubes. Carbon **41**, 2839–2845 (2003)
103. A Yutani, A Kobayashi, A Kinbara, Work functions of thin LaB6 films. Appl Surf Sci **70–71**(Part 2), 737–741 (1993)
104. RP Menaka, S Ghosh, AK Ganguli, Manipulating the anisotropy and field emission of lanthanum hexaboride nanorods. RSC Adv **2**, 7875–7885 (2012)
105. M Jha, R Patra, S Ghosh, AK Ganguli, Vertically aligned nanorods of lanthanum hexaboride with efficient field emission properties. Sol State Comms **153**, 35–39 (2013)
106. J Xu, G Hou, H Li, T Zhai, B Dong, H Yan, Y Wang, B Yu, Y Bando, D Golberg, Fabrication of vertically aligned single-crystalline lanthanum hexaboride nanowire arrays and investigation of their field emission. NPG Asia Mater **5**, e53 (2013)
107. S Raina, WP Kang, JL Davidson, Field emission from nanodiamond grown with 'ridge' type geometrically enhanced features. Diam Relat Mat **17**, 790–793 (2008)
108. B-R Huang, S Jou, T-C Lin, Y-K Yang, C-H Chou, Y-M Wu, Field emission property of arrayed nanocrystalline diamond. Diam Relat Mat **20**, 314–317 (2011)
109. HS Jung, HH Park, SS Pang, SY Lee, The structural and electron field emission characteristics of pulsed laser deposited diamond-like carbon films with thermal treatment. Thin Solid Films **355**, 151–156 (1999)
110. H Zanin, PW May, MHMO Hamanaka, EJ Corat, Field emission from hybrid diamond-like carbon and carbon nanotube composite structures. ACS Appl Mater Interfaces **5**, 12238–12243 (2013)
111. D Banerjee, A Jha, KK Chattopadhyay, Low-temperature synthesis of amorphous carbon nanoneedle and study on its field emission property. Physica E **41**, 1174–1178 (2009)
112. RK Tripathi, OS Panwar, AK Srivastava, I Rawal, S Chockalingam, Structural, nanomechanical, field emission and ammonia gas sensing properties of nitrogenated amorphous carbon films deposited by filtered anodic jet carbon arc technique. Talanta **125**, 276–283 (2014)
113. J Robertson, Mechanisms of electron field emission from diamond, diamond-like carbon, and nanostructured carbon. J Vac Sci Tech B **17**, 659–665 (1999)
114. A Hart, BS Satyanarayana, WI Milne, J Robertson, Effect of surface treatment and back contact material on field emission from tetrahedral amorphous carbon. Diam Relat Mat **8**, 809–813 (1999)
115. A Ilie, A Hart, AJ Flewitt, J Robertson, WI Milne, Effect of work function and surface microstructure on field emission of tetrahedral amorphous carbon. J Appl Phys **88**, 6002–6010 (2000)
116. WI Milne, Field emission from tetrahedrally bonded amorphous carbon. Appl Surf Sci **146**, 262–268 (1999)
117. M Setvín, J Javorský, D Turčinková, I Matolínová, P Sobotík, P Kocán, I O ťádal, Ultrasharp tungsten tips—characterization and nondestructive cleaning. Ultramicroscopy **113**, 152–157 (2012)
118. KL Ng, J Yuan, JT Cheung, KW Cheah, Electron field emission characteristics of electrochemical etched Si tip array. Sol State Comms **123**, 205–207 (2002)
119. ZX Pan, JC She, SZ Deng, NS Xu, Si Tip Arrays with Ultra-Narrow Nanoscale Charge Transfer Channel, in *Proceedings 27th International Vacuum Nanoelectronics Conference* (IEEE, Engelberg, Switzerland, 2014), pp. 74–75
120. SX Chen, JJ Li, CZ Gu, Ieee, Field Emission From the Composite Structure of Silicon Tips and Vertical Carbon Nanotubes, in *2007 7th Ieee Conference on Nanotechnology, Vol 1–3*, 2007, pp. 367–370
121. JW Glesener, HB Lin, AA Morrish, Field emission from cleaved diamond fibers. Thin Solid Films **308**, 204–208 (1997)
122. H-F Cheng, K-Y Teng, H-C Chen, G-C Tzeng, C-Y Tang, IN Lin, Bias enhanced nucleation and growth processes for improving the electron field emission properties of diamond films. Surf Coat Technol **228**, S175–S178 (2013)
123. F Zhao, Z D-d, W S-l, C G-a, Z R-t, Diamond-like carbon decoration enhances the field electron emission of silicon nanowires. Surf Coat Technol **228**, S349–S353 (2013)
124. K Panda, KJ Sankaran, BK Panigrahi, N-H Tai, IN Lin, Direct observation and mechanism for enhanced electron emission in hydrogen plasma-treated diamond nanowire films. ACS Appl Mater Interfaces **6**, 8531–8541 (2014)
125. P Abbott, ED Sosa, DE Golden, Effect of average grain size on the work function of diamond films. Appl Phys Lett **79**, 2835–2837 (2001)
126. X Song, CX Zhao, JQ Wu, J Chen, Field emission from copper micro-cones formed by ion bombardment of copper substrate using an oxide masking, in *25th International Vacuum Nanoelectronics Conference (IVNC)* (IEEE, Jeju Island, South Korea, 2012), pp. 376–377
127. J Wang, L Wei, L Zhang, J Zhang, H Wei, C Jiang, Y Zhang, Controlled growth of nickel nanocrystal arrays and their field electron emission performance enhancement via removing adsorbed gas molecules. Crystengcomm **15**, 1296–1306 (2013)
128. E Le Shim, SB Lee, E You, CJ Kang, KW Lee, YJ Choi, Rapid Fabrication of Wafer Scale Patterned Nickel Nanocone Arrays for Field Emission Applications, in *25th International Vacuum Nanoelectronics Conference (IVNC)* (IEEE, Jeju Island, South Korea, 2012), pp. 198–199
129. SW Joo, AN Banerjee, Field emission characterization of vertically oriented uniformly grown nickel nanorod arrays on metal-coated silicon substrate. J Appl Phys **107**, 114317 (2010)
130. T Nakane, K Sano, A Sakai, A Magosakon, K Yanagimoto, T Sakata, Field-emission barrier height of an alloy tip: possible effect of surface segregation. J Vac Sci Tech A **15**, 1563–1567 (1997)
131. J Wang, T Ito, Improved field emission characteristics of nano-structured carbon films deposited on polycrystalline CVD diamond. Diam Relat Mat **16**, 364–368 (2007)
132. JB Cui, J Ristein, M Stammler, K Janischowsky, G Kleber, L Ley, Hydrogen termination and electron emission from CVD diamond surfaces: a combined secondary electron emission, photoelectron emission microscopy, photoelectron yield, and field emission study. Diam Relat Mat **9**, 1143–1147 (2000)
133. O Gröning, Properties and characterization of chemical vapor deposition diamond field emitters. Solid-State Electron **45**, 929–944 (2001)

134. S Iijima, Helical microtubules of graphitic carbon. Nature **354**, 56–58 (1991)
135. A Oberlin, M Endo, T Koyama, Filamentous growth of carbon through benzene decomposition. J Crystal Growth **32**, 335–349 (1976)
136. LA Chernozatonskii, YV Gulyaev, ZJ Kosakovskaja, NI Sinitsyn, GV Torgashov, YF Zakharchenko, EA Fedorov, VP Val'chuk, Electron field emission from nanofilament carbon films. Chem Phys Lett **233**, 63–68 (1995)
137. A Rinzler, J Hafner, P Nikolaev, L Lou, S Kim, D Tomanek, P Nordlander, D Colbert, R Smalley, Unraveling nanotubes: field emission from an atomic wire. Science **269**, 1550–1553 (1995)
138. N de Jonge, Y Lamy, K Schoots, TH Oosterkamp, High brightness electron beam from a multi-walled carbon nanotube. Nature **420**, 393–395 (2002)
139. S Akita, H Nishijima, Y Nakayama, F Tokumasu, K Takeyasu, Carbon nanotube tips for a scanning probe microscope: their fabrication and properties. J Phys D **32**, 1044 (1999)
140. M Mann, K Teo, W Milne, T Tessner, Direct growth of multi-walled carbon nanotubes on sharp tips for electron microscopy. NANO **1**, 35–39 (2006)
141. H Nishijima, S Kamo, S Akita, Y Nakayama, KI Hohmura, SH Yoshimura, K Takeyasu, Carbon-nanotube tips for scanning probe microscopy: preparation by a controlled process and observation of deoxyribonucleic acid. Appl Phys Lett **74**, 4061–4063 (1999)
142. W Milne, K Teo, Growth and characterization of carbon nanotubes for field-emission-display applications. J Soc Inf Display **12**, 289–292 (2004)
143. W Choi, D Chung, J Kang, H Kim, Y Jin, I Han, Y Lee, J Jung, N Lee, G Park, Fully sealed, high-brightness carbon-nanotube field-emission display. Appl Phys Lett **75**, 3129–3131 (1999)
144. N Lee, D Chung, I Han, J Kang, Y Choi, H Kim, S Park, Y Jin, W Yi, M Yun, Application of carbon nanotubes to field emission displays. Diamond Relat Mater **10**, 265–270 (2001)
145. KBK Teo, M Chhowalla, GAJ Amaratunga, WI Milne, P Legagneux, G Pirio, L Gangloff, D Pribat, V Semet, VT Binh, WH Bruenger, J Eichholz, H Hanssen, D Friedrich, SB Lee, DG Hasko, H Ahmed, Fabrication and electrical characteristics of carbon nanotube-based microcathodes for use in a parallel electron-beam lithography system. J Vac Sci Tech B **21**, 693–697 (2003)
146. J Nd, J-M Bonard, Carbon nanotube electron sources and applications. Philos Transact Math Phys Eng Sci **362**, 2239–2266 (2004)
147. SM Vieira, KB Teo, WI Milne, O Gröning, L Gangloff, E Minoux, P Legagneux, Investigation of field emission properties of carbon nanotube arrays defined using nanoimprint lithography. Appl Phys Lett **89**, 022111 (2006)
148. H Sugie, M Tanemura, V Filip, K Iwata, K Takahashi, F Okuyama, Carbon nanotubes as electron source in an x-ray tube. Appl Phys Lett **78**, 2578–2580 (2001)
149. GZ Yue, Q Qiu, B Gao, Y Cheng, J Zhang, H Shimoda, S Chang, JP Lu, O Zhou, Generation of continuous and pulsed diagnostic imaging x-ray radiation using a carbon-nanotube-based field-emission cathode. Appl Phys Lett **81**, 355–357 (2002)
150. A Haga, S Senda, Y Sakai, Y Mizuta, S Kita, F Okuyama, A miniature x-ray tube. Appl Phys Lett **84**, 2208–2210 (2004)
151. S Senda, M Tanemura, Y Sakai, Y Ichikawa, S Kita, T Otsuka, A Haga, F Okuyama, New field-emission x-ray radiography system. Rev Sci Instrum **75**, 1366–1368 (2004)
152. F Charbonnier, Developing and using the field emitter as a high intensity electron source. Appl Surf Sci **94**, 26–43 (1996)
153. Y Iwai, T Koike, Y Hayama, A Jyouzuka, T Nakamura, Y Onizuka, M Miyoshi, H Mimura, Field emission characteristics of inflamed graphite at high temperature and its applications, in *Technical Digest of IVNC*, 2012, pp. 274–275
154. J-W Kim, J-W Jeong, J-T Kang, S Choi, J Choi, S Ahn, S Y-h, A digital compact x-ray tube with carbon nanotube field emitters for advanced imaging systems, in *SPIE Medical Imaging* (International Society for Optics and Photonics, Nishikawa, 2013), pp. 861–866, From Conference Volume 8668, Medical Imaging 2013: Physics of Medical Imaging, Robert M. Nishikawa; Bruce R. Whiting, Lake Buena Vista (Orlando Area), Florida, USA | February 09, 2013
155. Y Saito, *Carbon Nanotube and Related Field Emitters: Fundamentals and Applications* (Wiley-VCH Verlag, Weinheim, 2010)
156. Y Cheng, O Zhou, Electron field emission from carbon nanotubes. Comptes Rendus Physique **4**, 1021–1033 (2003)
157. I Oxford, *Cold Cathode vs Thermionic X-ray Tubes [Online]*, 2000. http://wigner.elte.hu/koltai/science/pub/Which_Source_Cold_cathode_or_Thermionic.pdf
158. HY Choi, WS Chang, HS Kim, YH Park, JU Kim, Acquisition of X-ray images by using a CNT cold emitter. Phys Lett A **357**, 36–41 (2006)
159. J Zhang, Y Cheng, YZ Lee, B Gao, Q Qiu, WL Lin, D Lalush, JP Lu, O Zhou, A nanotube-based field emission x-ray source for microcomputed tomography. Rev Sci Instrum **76**, 094301–094304 (2005)
160. X Calderón-Colón, H Geng, B Gao, L An, G Cao, O Zhou, A carbon nanotube field emission cathode with high current density and long-term stability. Nanotech **20**, 325707 (2009)
161. G Pirio, P Legagneux, D Pribat, K Teo, M Chhowalla, G Amaratunga, W Milne, Fabrication and electrical characteristics of carbon nanotube field emission microcathodes with an integrated gate electrode. Nanotechnology **13**, 1 (2002)
162. D Temple, Recent progress in field emitter array development for high performance applications. Mat Sci Eng R Rep **24**, 185–239 (1999)
163. SH Heo, A Ihsan, SO Cho, Transmission-type microfocus x-ray tube using carbon nanotube field emitters. Appl Phys Lett **90**, 183109 (2007)
164. WI Milne, Q Zhang, *Advances in Nanodevices and Nanofabrication* (Pan Stanford Publishing, Singapore, 2012)
165. MT Cole, T Hallam, WI Milne, GS Duesberg, *Field Emission Characteristics of Contact Printed Graphene Fins. Small*, 2013
166. KBK Teo, SB Lee, M Chhowalla, V Semet, VT Binh, O Groening, M Castignolles, A Loiseau, G Pirio, P Legagneux, D Pribat, DG Hasko, H Ahmed, GAJ Amaratunga, WI Milne, Plasma enhanced chemical vapour deposition carbon nanotubes/nanofibres - how uniform do they grow? Nanotech **14**, 204–211 (2003)
167. B Gao, GZ Yue, Q Qiu, Y Cheng, H Shimoda, L Fleming, O Zhou, Fabrication and electron field emission properties of carbon nanotube films by electrophoretic deposition. Adv Mater **13**, 1770–1773 (2001)
168. AR Boccaccini, J Cho, JA Roether, BJ Thomas, E Jane Minay, MS Shaffer, Electrophoretic deposition of carbon nanotubes. Carbon **44**, 3149–3160 (2006)
169. S Wang, X Calderon, R Peng, EC Schreiber, O Zhou, S Chang, A carbon nanotube field emission multipixel X-ray array source for microradiotherapy application. Appl Phys Lett **98**, 213701–213703 (2011)
170. J Li, W Lei, X Zhang, X Zhou, Q Wang, Y Zhang, B Wang, Field emission characteristic of screen-printed carbon nanotube cathode. Appl Surf Sci **220**, 96–104 (2003)
171. F Sprenger, X Calderon, E Gidcumb, J Lu, X Qian, D Spronk, A Tucker, G Yang, O Zhou, *Stationary digital breast tomosynthesis with distributed field emission x-ray tube*, 2011. pp 79615I-79615I-79616
172. MT Cole, M Doherty, R Parmee, P Dawson, WI Milne, Ultra-broadband polarisers based on metastable free-standing aligned carbon nanotube membranes. Adv Opt Mater (2014), doi: 10.1002/adom.201400238
173. M Qian, T Feng, K Wang, H Ding, Y Chen, Q Li, Z Sun, Field emission of carbon nanotube films fabricated by vacuum filtration. Physica E **43**, 462–465 (2010)
174. Y Cho, H Song, G Choi, D Kim, A simple method to fabricate high-performance carbon nanotube field emitters. J Electroceram **17**, 945–949 (2006)
175. YS Shi, C-C Zhu, W Qikun, L Xin, Large area screen-printing cathode of CNT for FED. Diam Relat Mat **12**, 1449–1452 (2003)
176. J-Y Pan, C-y Chen, Y-L Gao, C-C Zhu, Improved field emission characteristics of screen-printed CNT-FED cathode by interfusing Fe/Ni nano-grains. Displays **30**, 114–118 (2009)
177. H-Y Shin, W-S Chung, KH Kim, Y-R Cho, B-C Shin, Effects of bonding materials in screen-printing paste on the field-emission properties of carbon nanotube cathodes. J Vac Sci Technol B **23**, 2369–2372 (2005)
178. J-W Kim, J-W Jeong, J-T Kang, S Choi, S Ahn, Y-H Song, Highly reliable field electron emitters produced from reproducible damage-free carbon nanotube composite pastes with optimal inorganic fillers. Nanotechnology **25**, 065201 (2014)
179. K Yamamoto, S Akita, Y Nakayama, Orientation and purification of carbon nanotubes using ac electrophoresis. J Phys D **31**, 34–36 (1998)
180. K Yamamoto, S Akita, Y Nakayama, Orientation of carbon nanotubes using electrophoresis. Japanese J Appl Phys **35**, L917 (1996)
181. MT Cole, C Collins, R Parmee, C Li, WI Milne, NanoCarbon Electron Emitters: Advances & Applications, in *Chemical Functionalisation of Carbon Nanomaterials: Chemistry & Applications - Structure & Synthesis* (Taylor & Francis, Philadelphia, 2015)
182. P Sarrazin, D Blake, L Delzeit, M Meyyappan, B Boyer, S Snyder, B Espinosa, Carbon-nanotube field emission x-ray tube for space exploration XRD/XRF instrument. Adv X Ray Anal **47**, 8 (2004)
183. MT Cole, C Li, Y Zhang, SG Shivareddy, JS Barnard, W Lei, B Wang, D Pribat, GAJ Amaratunga, WI Milne, Hot electron field emission *via* individually transistor-ballasted carbon nanotube arrays. ACS Nano **6**, 3236–3242 (2012)
184. HS Kim, D Quang, JH Kim, HJ Lee, DM Yoon, SS Shin, JW Ha, KJ Lee, YG Hwang, CH Lee, Field-emission electron source using carbon nanotubes for X-ray tubes. J Korean Phys Soc **52**, 3 (2008)
185. MT Cole, KBK Teo, O Groening, L Gangloff, P Legagneux, WI Milne, Deterministic cold cathode electron emission from carbon nanofibre arrays. Sci Rep **4**, (2014); doi: 10.1038/srep04840

186. O Gröning, O Küttel, C Emmenegger, P Gröning, L Schlapbach, Field emission properties of carbon nanotubes. J Vac Sci Technol B **18**, 665–678 (2000)
187. L Nilsson, O Groening, C Emmenegger, O Kuettel, E Schaller, L Schlapbach, H Kind, J Bonard, K Kern, Scanning field emission from patterned carbon nanotube films. Appl Phys Lett **76**, 2071–2073 (2000)
188. S Fujii, H S-i, H Machida, H Kawai, K Ishida, M Katayama, H Furuta, T Hirao, K Oura, Efficient field emission from an individual aligned carbon nanotube bundle enhanced by edge effect. Appl Phys Lett **90**, 153108 (2007)
189. JL Killian, NB Zuckerman, DL Niemann, BP Ribaya, M Rahman, R Espinosa, M Meyyappan, CV Nguyen, Field emission properties of carbon nanotube pillar arrays. J Appl Phys **103**, 064312 (2008)
190. P Rai, DR Mohapatra, KS Hazra, DS Misra, SP Tiwari, Nanotip formation on a carbon nanotube pillar array for field emission application. Appl Phys Lett **93**, (2008). 131921-131921-131923
191. JL Silan, DL Niemann, BP Ribaya, M Rahman, M Meyyappan, CV Nguyen, Carbon nanotube pillar arrays for achieving high emission current densities. Appl Phys Lett **95**, 133111 (2009)
192. K Celebi, MT Cole, KBK Teo, HG Park, Observations of early stage graphene growth on copper. Electrochem Solid State Lett **15**, K1–K4 (2012)
193. K Celebi, MT Cole, JW Choi, F Wyczisk, P Legagneux, N Rupesinghe, J Robertson, KBK Teo, HG Park, Evolutionary kinetics of graphene formation on copper. Nano Lett **13**, 967–974 (2013)
194. H Kim, C Mattevi, MR Calvo, JC Oberg, L Artiglia, S Agnoli, CF Hirjibehedin, M Chhowalla, E Saiz, Activation energy paths for graphene nucleation and growth on Cu. ACS Nano **6**, 3614–3623 (2012)
195. C Li, MT Cole, W Lei, K Qu, K Ying, Y Zhang, AR Robertson, JH Warner, S Ding, X Zhang, B Wang, WI Milne, Highly electron transparent graphene for field emission triode gates. Adv Funct Mater **24**, 1218–1227 (2014)
196. TB Rymer, PG Hambling, Demountable X-ray tube for crystallography. J Sci Instrum **29**, 192 (1952)
197. S Heo, H Kim, J Ha, S Cho, A vacuum-sealed miniature X-ray tube based on carbon nanotube field emitters. Nano Res Lett **7**, 1–5 (2012)
198. A Reyes-Mena, C Jensen, E Bard, DC Turner, K Erdmann, Q Qiu, B Gao, J Lu, O Zhou, Miniature X-ray tubes utilizing carbon-nanotube-based cold cathodes. Adv X Ray Anal **48**, 204–209 (2005)
199. Y Sakai, A Haga, S Sugita, S Kita, S-I Tanaka, F Okuyama, N Kobayashi, Electron gun using carbon-nanofiber field emitter. Rev Sci Instrum **78**, 013305 (2007)
200. W Sugimoto, S Sugita, Y Sakai, H Goto, Y Watanabe, Y Ohga, S Kita, T Ohara, A fine-focusing X-ray source using carbon-nanofiber field emitter. J Appl Phys **108**, 044507 (2010)
201. Y Cheng, J Zhang, YZ Lee, B Gao, S Dike, W Lin, JP Lu, O Zhou, Dynamic radiography using a carbon-nanotube-based field-emission X-ray source. Rev Sci Instrum **75**, 3264 (2004)
202. J Zhang, G Yang, YZ Lee, S Chang, JP Lu, O Zhou, Multiplexing radiography using a carbon nanotube based x-ray source. Appl Phys Lett **89**, 064106 (2006)
203. X Qian, R Rajaram, X Calderon-Colon, G Yang, T Phan, DS Lalush, J Lu, O Zhou, Design and characterization of a spatially distributed multibeam field emission x-ray source for stationary digital breast tomosynthesis. Med Phys **36**, 4389–4399 (2009)
204. Y Sun, DA Jaffray, TW Yeow, The design and fabrication of carbon-nanotube-based field emission X-Ray cathode with ballast resistor. IEEE T Electron Dev **60**, 7 (2013)
205. J Ryu, J Kang, K Park, Carbon nanotube electron emitter for X-ray imaging. Mater **5**, 2353–2359 (2012)
206. Z Liu, G Yang, YZ Lee, D Bordelon, J Lu, O Zhou, Carbon nanotube based microfocus field emission x-ray source for microcomputed tomography. Appl Phys Lett **89**, 1–3 (2006)
207. SH Heo, HJ Kim, JM Ha, SO Cho, A vacuum-sealed miniature X-ray tube based on carbon nanotube field emitters. Nano Res Lett **7**, 5 (2012)
208. J-W Jeong, J-T Kang, S Choi, J-W Kim, S Ahn, Y-H Song, A digital miniature x-ray tube with a high-density triode carbon nanotube field emitter. Appl Phys Lett **102**, 023504 (2013)
209. P Sarrazin, D Blake, L Delzeit, M Meyyappan, B Boyer, S Snyder, B Espinosa, Carbon-nanotube field emission X-ray tube for space exploration XRD/XRF instrument. Adv X Ray Anal **47**, 232–239 (2004)
210. TT Tan, HS Sim, SP Lau, HY Yang, M Tanemura, J Tanaka, X-ray generation using carbon-nanofiber-based flexible field emitters. Appl Phys Lett **88**, 103105 (2006)
211. S Kita, Y Watanabe, A Ogawa, K Ogura, Y Sakai, Y Matsumoto, Y Isokane, F Okuyama, T Nakazato, T Otsuka, Field-emission-type X-ray source using carbon-nanofibers. J Appl Phys **103**, 064505 (2008)
212. J-W Kim, J-W Jeong, J-T Kang, S Choi, S Ahn, Y-H Song, Highly reliable field electron emitters produced from reproducible damage-free carbon nanotube composite pastes with optimal inorganic fillers. Nanotech **25**, (2014); doi: 10.1088/0957-4484/25/6/065201
213. C-K Park, J-P Kim, S-J Yun, S-H Lee, J-S Park, Field emission properties of carbon nanotubes grown on a conical tungsten tip for the application of a microfocus x-ray tube. Thin Solid Films **516**, 304–309 (2007)
214. K Teo, M Chhowalla, G Amaratunga, W Milne, G Pirio, P Legagneux, F Wyczisk, D Pribat, D Hasko, Field emission from dense, sparse, and patterned arrays of carbon nanofibers. Appl Phys Lett **80**, 2011–2013 (2002)
215. V Merkulov, D Lowndes, Y Wei, G Eres, E Voelkl, Patterned growth of individual and multiple vertically aligned carbon nanofibers. Appl Phys Lett **76**, 3555–3557 (2000)
216. MK Debe, *Field emission device having nanostructured emitters*. Google Patents, 1998
217. PR Schwoebel, JM Boone, *Stationary x-ray source for tomographic medical imaging*, 2006. 61423N-61423N-61423
218. E Minoux, O Greening, KBK Teo, SH Dalai, L Gangloff, JP Schnell, L Hudanski, IYY Bu, P Vincent, P Legagneux, GAJ Amaratunga, WI Milne, Achieving high-current carbon nanotube emitters. Nano Lett **5**, 2135–2138 (2005)
219. R Forman, Evaluation of the emission capabilities of Spindt-type field emitting cathodes. Appl Surface Sci **16**, 277–291 (1983)
220. J-M Bonard, C Klinke, KA Dean, BF Coll, Degradation and failure of carbon nanotube field emitters. Phys Rev B **67**, 115406 (2003)
221. WI Milne, KBK Teo, GAJ Amaratunga, P Legagneux, L Gangloff, JP Schnell, V Semet, VT Binh, O Groening, Carbon nanotubes as field emission sources. J Mater Chem **14**, 933–943 (2004)
222. C Spindt, I Brodie, L Humphrey, E Westerberg, Physical properties of thin-film field emission cathodes with molybdenum cones. J Appl Phys **47**, 5248–5263 (2008)
223. M Doytcheva, M Kaiser, J Nd, In situ transmission electron microscopy investigation of the structural changes in carbon nanotubes during electron emission at high currents. Nanotech **17**, 3226 (2006)
224. S Itoh, Current status of the Spindt-type field emitter. Proc ISDEIV **2**, 875–876 (2006)
225. C Li, Y Zhang, M Mann, P Hiralal, HE Unalan, W Lei, BP Wang, DP Chu, D Pribat, GAJ Amaratunga, WI Milne, Stable, self-ballasting field emission from zinc oxide nanowires grown on an array of vertically aligned carbon nanofibers. Appl Phys Lett **96**, 143114 (2010)
226. Y Liu, L Gao, A study of the electrical properties of carbon nanotube-NiFe < sub > 2</sub > O < sub > 4</sub > composites: effect of the surface treatment of the carbon nanotubes. Carbon **43**, 47–52 (2005)
227. J Madsen, Focal spot size measurements for microfocus X-ray sets. NDT Int **22**, 292–296 (1989)
228. W Ehrenberg, WE Spear, An electrostatic focusing system and its application to a fine focus X-Ray tube. Proc Phys Soc B **64**, 67 (1951)
229. BM Moores, P Brubacher, Focal spot studies and electron focusing in a demountable X-ray tube. Phys Med Biol **19**, 605 (1974)
230. ASfTaM Philadelphia, Standard test method for measurement of focal spots of industrial X-Ray tubes by pinhole imaging. Annual Book ASTM Stand (2010). ASTM-E-1165-12, Volume 03.03; doi: 10.1002/adom.201400238
231. Z Liu, J Zhang, G Yang, Y Cheng, O Zhou, J Lu, Development of a carbon nanotube based microfocus x-ray tube with single focusing electrode. Rev Sci Instrum **77**, 054302–054306 (2006)
232. S Sultana, X Calderón-Colón, G Cao, O Zhou, J Lu, *Design and characterization of a carbon-nanotube-based micro-focus x-ray tube for small animal imaging*, 2010. pp 76225G-76225G-76229
233. P Legagneux, P Ponard, L Gangloff, S Xavier, C Bourat, J Martinez, CS Cojocaru, A Gohier, JP Mazellier, JP Schnell, D Pribat, KBK Teo, Carbon nanotube photocathodes for optically driven multiple X-ray sources. http://www.phantomsnet.net/imaginenano/Abstracts/Nanospain2011_Legagneux_Pierre_pierre.legagneux@thalesgroup.com_Imaginenano2011_Abstract_Legagneux.pdf
234. X Qian, A Tucker, E Gidcumb, J Shan, G Yang, X Calderon-Colon, S Sultana, J Lu, O Zhou, D Spronk, F Sprenger, Y Zhang, D Kennedy, T Farbizio, Z Jing, High resolution stationary digital breast tomosynthesis using distributed carbon nanotube x-ray source array. Med Phys **39**, 2090–2099 (2012)
235. SX Chang, O Zhou, *Carbon Nanotubes Field Emission X-ray for Research and Clinical Applications in Radiation Oncology* (University of North Carolina, Chapel Hill). http://amos3.aapm.org/abstracts/pdf/77-22629-310436-91429.pdf [Online], (accessed 19/08/2014, 2014)

236. J Zhang, G Yang, YZ Lee, Y Cheng, B Gao, Q Qiu, JP Lu, O Zhou, *A multi-beam x-ray imaging system based on carbon nanotube field emitters*, 2006. pp 614204-614204-614208
237. G Yang, R Rajaram, G Cao, S Sultana, Z Liu, D Lalush, J Lu, O Zhou, *Stationary digital breast tomosynthesis system with a multi-beam field emission x-ray source array*, 2008, pp. 69131A–69131A
238. J Shan, P Chtcheprov, AW Tucker, YZ Lee, X Wang, D Foos, MD Heath, J Lu, O Zhou, *Stationary chest tomosynthesis using a CNT x-ray source array*, 2013. pp 86680E-86680E-86612
239. E Quan, DS Lalush, *Three-dimensional imaging properties of rotation-free square and hexagonal micro-CT systems*, 2009. pp 72580I-72580I-72510
240. IA Guzilov, KV Kuzmich, OY Maslennikov, EV Smirnova, PV Minakov, AY Poroykov, AT Rakhimov, B Seleznev, W Sen, Multi beam X-ray tube with the field emitter on the base of nanocrystalline graphite for computer tomography, in *Vacuum Electronics Conference, 2009. IVEC '09. IEEE International*, 2009, pp. 289–291
241. F Sprenger, X Calderon-Colon, Y Cheng, K Englestad, J Lu, J Maltz, A Paidi, X Qian, D Spronk, S Sultana, G Yang, O Zhou, *Distributed source x-ray tube technology for tomosynthesis imaging*, 2010, pp. 76225M–76225M
242. Source 1 X-ray. (2000). http://www.source1xray.com/index-2.html
243. Amptek miniature X- Ray source. (2013), http://www.amptek.com/products/mini-x-ray-tube/
244. J-T Kang, J-W Kim, JW Jeong, S Choi, J Choi, S Ahn, Y-H Song, Analysis of failure in miniature X-ray tubes with gated carbon nanotube field emitters. ETRI J **35**, 1164–1167 (2013)
245. J Mazellier, *Carbon nanotubes based photocathodes for multisource X-Ray generators*, 220–223. http://www.nsti.org/procs/Nanotech2012v1/3/X2.123
246. P Legagneux, Carbon nanotube cathodes for microwave and X-ray electron tubes, in *Vacuum Electron Sources Conference and Nanocarbon (IVESC), 2010 8th International* (IEEE, New York, 2010), pp. 86–86
247. Siemens, *Siemens CNT X-ray [Online]*, 2010. http://ragingbull.com/forum/topic/857725
248. Xintek, *Nanotechnology Innovations [Online]*, 2014. http://www.xintek.com/cnt-x-ray-sources/

15

Charge transfer between photosynthetic proteins and hematite in bio-hybrid photoelectrodes for solar water splitting cells

Greta Faccio[1], Krisztina Gajda-Schrantz[2,3,4], Julian Ihssen[1], Florent Boudoire[2,5], Yelin Hu[2,6], Bongjin Simon Mun[7], Debajeet K Bora[2], Linda Thöny-Meyer[1] and Artur Braun[2*]

Abstract

Functionalization of the hematite photoanode with the photosynthetic light antenna protein C-phycocyanin (PC) can yield substantial enhancement of the photocurrent density. Photoelectrochemical cells with bio-hybrid electrodes from photosynthetic proteins and inorganic semiconductors have thus potential for the use in artificial photosynthesis. We investigate here processing routes for the functionalization of hematite photoanodes with PC, including *in situ* co-polymerization of PC with enzymatically-produced melanin, and using a recombinant PC genetically engineered to carry a hexa-histidine tag (αHisPC). First, the effect of the immobilisation of PC on the electrode morphology and photocurrent production is evaluated. Then, the electronic charge transfer in dark and light conditions is assessed with electrochemical impedance spectroscopy and valence band (VB) X-ray photoemission spectroscopy. The relative shift of the VB spectrum towards the Fermi energy E_F upon illumination is smaller for the more complex processed coating and virtually disappears for αHisPC immobilised with a melanin film. Optimal conditions for protein immobilisation are determined and the dark currents benefit most from the most advanced protein coating processes.

Keywords: Artificial photosynthesis; Antenna protein C-phycocyanin; Hematite photoanode; Impedance spectroscopy; Valence band (VB) X-ray photoemission spectroscopy

1 Background

Hematite, a ferric oxide with 2.1 eV optical band gap energy, is considered a low cost and environmentally benign photoanode material for solar hydrogen production by water splitting in photoelectrochemical cells (PEC). The research and development of photoelectrodes for PEC is focused on the tailoring of the electronic structure and the microstructure, and recently also on the design of hetero structures. The functionalization of a photoelectrode with light harvesting proteins was recently demonstrated for hematite using C-phycocyanin (PC) [1], which constitutes basically a dye sensitization.

Dye sensitization [2] is an established technology for the extension of the light absorption range of inorganic photoelectrodes for solar energy conversion. To be fully functional, the electric charge generated in the dye upon light absorption needs to be transferred to the photoelectrode, i.e. the interface between the dye and the photoelectrode must warrant proper charge transfer for electrons or holes.

We recently developed a process which enhances the photocurrent of hematite thin-film electrodes by immobilising light-harvesting proteins such as PC on the surface [1]. So far, the functionalization has been performed with the commercially available PC from *Spirulina* sp. which contains the α- and β-subunits of the protein. In the present study, we evaluate the simpler α-subunit of PC from *Synechocystis* sp. PCC6803 which is recombinantly produced in *E. coli* and engineered to carry an N-terminal hexa-histidine tag (αHisPC) [3]. The electronic charge transfer is investigated by electrochemical impedance spectroscopy (EIS) and with valence band X-ray photoemission spectroscopy (VB PES).

* Correspondence: Artur.Braun@alumni.ethz.ch
[2]Laboratory for High Performance Ceramics, Empa, Swiss Federal Laboratories for Materials Science and Technology, Überlandstrasse 129, 8600 Dübendorf, Switzerland
Full list of author information is available at the end of the article

2 Methods

2.1 Materials

1,1′-Carbonyldiimidazol (CDI, >97%), 1,4-dioxane (99.8%), 2,2′-azino-bis(3-ethylbenzothiazoline-6-sulfonic acid) diammonium salt (ABTS, >98%), monopotassium phosphate (1 mol l^{-1}), hydrogen dipotassium phosphate (1 mol l^{-1}), potassium hydroxide (KOH, 95%), iron nitrate nonahydrate (>98%), tetrahydrofuran (THF, >99.9%), sodium chloride (NaCl, 5 mol l^{-1}), L-tyrosine (>99%), Tris–HCl (>99%) and C-phycocyanin isolated from *Spirulina* sp. (99.9%) were purchased from Sigma-Aldrich. Oleic acid (>60%) and L (+)-tartaric acid (>99.5%) were purchased from Fluka. Tyrosinase from *Verrucomicrobium spinosum* was produced in *E. coli* and proteolitically activated as described in [4].

2.2 Production and purification of C-phycocyanin from *Synechocystis* sp. PCC6803

The α-subunit of PC from *Synechocystis sp.* PCC6803 was produced in an N-terminally His-tagged form according to [3] and purified with a modified protocol from [5]. Briefly, strain *E. coli* DH5α was co-transformed with plasmids pAT101 and pBD414V and transformants were selected for resistance to both kanamycin (50 μg ml^{-1}) and spectinomycin (100 μg ml^{-1}). Cultures were routinely set up in TB medium (500 ml volume), incubated at 37°C under constant shaking (180 rpm), and expression was induced by adding IPTG (final conc. 1 mM). Cells were harvested by centrifugation 17 h post-induction and stored at −20°C. Upon thawing and re-suspension in 100 mM potassium phosphate buffer with pH 7.5 (5 ml g^{-1} wet cell weight), lysozyme (final conc. 1 mg ml^{-1}) and a protease inhibitor mix (Complete Protease Inhibitor Cocktail Tablets, Roche) were added. The cell suspension was then subjected to a freeze-thaw cycle at −20°C. Cells were further disrupted by sonication, and the cell-free extract, isolated by ultracentrifugation, was loaded onto a HisTrap column (1 ml, GE Healthcare Life Sciences) for purification by immobilized metal ion affinity chromatography (IMAC). αHisPC was eluted with imidazole (conc. 200 mM) in 100 mM potassium phosphate (pH 7.5). αHisPC-containing fractions were concentrated and desalted with a Vivaspin ultrafiltration device (10000 cut-off, Sartorius). The purified αHisPC was stored at −20°C for further use. Protein concentrations of αHisPC solutions were determined spectrophotometrically with a Synergy Mx spectrophotometer (Biotek, Switzerland) and using an extinction coefficient of 140000 M^{-1} cm^{-1} at 620 nm. Protein purity was assessed spectrophotometrically as the ratio of the absorbance at 280 nm and at 620 nm.

2.3 Hematite films and single crystals

Hematite films were prepared as described previously [6]. Fluorine doped tin oxide (FTO) coated glass slides (12 mm × 30 mm × 2 mm, TEC-8 from Hartford Glass Inc.) were automatically dip coated (DipMasterTM-50, Chemat Technology Inc., USA) with the precursor solution and dried at 75°C for 5 min on a hot plate. The samples were then further annealed for 30 min at 500°C in an air vented muffle furnace. Dip coating and annealing were repeated three times so as to obtain a film of four layers of hematite, which has typically a thickness of 500 to 600 nm. Prevalence of the hematite phase was confirmed with X-ray diffraction (Siemens D5000).

The thickness of the hematite films was determined with a mechanical profilometer (XP-1 Stylus Profiler by Ambios Technology). The morphology of the pristine and protein coated hematite films was investigated with a Philips XL 30 scanning electron microscope (SEM, Oxford instruments).

Hematite single crystals with (0001) orientation were purchased from SurfaceNet GmbH, Rheine, Germany. They had a size of 5 mm × 5 mm × 1 mm size and were Epi surface polished.

2.4 Coating of hematite with proteins and melanin

Similarly to [1], the hematite thin films were first dip-coated with a 0.001% agarose solution, let dry, and second with ~100 μl cm^{-2} of a 30 mg ml^{-1} CDI solution prepared in dioxan. Protein immobilisation was carried out by coating the hematite surface with a melanin-forming mixture containing PC or αHisPC (concentration 0.2 mg ml^{-1}), L-tyrosine (concentration 0.1 mg ml^{-1}), and tyrosinase (50 μg ml^{-1} concentration) in 0.05 M phosphate buffer saline (PBS; 160 g l^{-1} NaCl, 4 g l^{-1} KCl, 28.88 gl^{-1} Na_2HPO_4, 4.4 g l^{-1} KH_2PO_4, pH 7.2-7.4) for 3 h. The surface was dried between all the steps, eventually rinsed with PBS and distilled water, and then stored at 4°C in the dark until the actual measurements were performed. Because the tyrosinase preparation contains traces of trypsin, the protease inhibitor phenylmethanesulfonyl fluoride (PMSF) was added to the coating mixture at a 1 mM final concentration. The thickness of the films treated with melanin alone, or with melanin plus PC or with melanin plus αHisPC was below 1000 nm. All samples were prepared in either duplicate or triplicate and average values are reported throughout this paper.

In addition to the aforementioned thin hematite films on FTO-coated glass slides, three (0001) hematite single crystals were coated with:

1) Melanin plus αHisPC: 50 μl solution containing αHisPC, L-tyrosine and tyrosinase in PBS (concentrations as above) after pre-treatment of the hematite single crystal with agarose and CDI solution (as described above).

2) PC only: 50 μl solution containing PC, after pre-treatment of the hematite single crystal with agarose and CDI solution (as described above).

3) Melanin plus PC: 50 μl solution containing PC and the melanin-forming mixture (as described above), after

pre-treatment of the hematite single crystal with agarose and CDI solution (as described above).

2.5 Photoelectrochemical assessment of the films

For the assessment of the photoelectrochemical properties of the photoanode, we largely followed the recommendations provided in [7], when not otherwise stated in this manuscript. Linear voltammetry in the dark and under illumination was conducted using a specifically designed spectro-photoelectrochemical cell (known as EPFL cappuccino cell) made of polyether ether ketone (PEEK) and a potentiostat (Voltalab80 PGZ 402). The hematite photoanode was connected as the working electrode in a three electrode configuration. A platinum sheet of 5 mm × 5 mm was used as counter electrode and an Ag/AgCl (with saturated KCl) electrode was used as reference electrode. The electrolyte was 0.05 M PBS. Sunlight was simulated by a 1 Sun Oriel Lamp (L.O.T.–Oriel AG) with a quartz filter, corresponding to AM 1.5 global standard solar spectrum. The bias potential was scanned from 0 mV to 1000 mV. All shown results are the arithmetic average of data of at least three consecutive measurements. Electrochemical impedance spectra were recorded with the Voltalab80 PGZ 402 using the same cell in the frequency range from 100 kHz to 100 mHz with 20 data points per decade and a perturbation amplitude of 10 mV. Spectra of hematite films coated on FTO glass and of variously prepared PC films on hematite were recorded. Dark and light conditions as described above were applied during the impedance measurements.

2.6 Photoemission spectroscopy

X-ray photoemission (PES) spectra were recorded at beamline 8–1 at Stanford Synchrotron Radiation Lightsource, Menlo Park, CA. Specifically, the spectra were recorded under Fe 2p resonant condition with 53.2, 54.4 and 55.8 eV photon energy, so that the valence band spectra were particularly representing the Fe relevant features. The PES spectra were measured from the surface finished (0001) hematite single crystals that were coated with PC in the aforementioned different processing routes. The measurements of PES were carried out under ultrahigh vacuum condition, i.e. 10^{-10} mbar base pressure. At the time of experiment, the films could be considered dry by ambient conditions. A 1.5 AM solar simulator (HAL-302 Solar Simulator, 350–750 nm, Asahi Spectra, Japan) was used to illuminate the samples during the PES experiments.

3 Results and Discussion

We have recently observed that the photocurrent density of a hematite electrode increases when the light-harvesting protein phycocyanin (PC) is copolymerized with melanin on its surface. In the present work we investigate the influence of PC concentration, the time of incubation with the melanin-forming mixture on the photocurrent. In addition to the commercially available PC preparation, we test the effect of a recombinantly produced form of PC on the performance of the photoelectrochemical cell (PEC). Each sample is measured prior to and after the immobilization of proteins under pH-neutral conditions in PBS, if not mentioned otherwise.

3.1 Production of recombinant αHisPC

The α-subunit of PC from *Synechocystis sp.* PCC6803 is produced in recombinant form in *E. coli* with a two-plasmid system [3]. Carrying an N-terminal Hexahistidine tag, αHisPC is purified by immobilized metal ion affinity chromatography (IMAC). The solution containing the purified αHisPC has a vivid blue colour. Differently from PC that has a maximum absorbance peak at 620 nm, the purified αHisPC shows maximum optical absorbance at 626 nm (Figure 1).

Glazer et al. reported that the peak of maximum absorbance of C-phycocyanin when composed of both α- and β-subunits results from the sum of the absorbance peaks of both subunits. The α-subunit absorbs at a longer wavelength than the β-subunit, e.g. the α-subunit has an absorbance maximum at 620 nm and the β-subunit at 608 nm in C-phycocyanin from *Synechococcus* sp. [9].

Similarly to what has been reported for PC [8], incubation under alkaline conditions, e.g. in the presence of 1 M KOH, absorbance in the 600–620 nm region is rapidly annulled and a novel absorbance peak at 750 nm is detected (Figure 1).

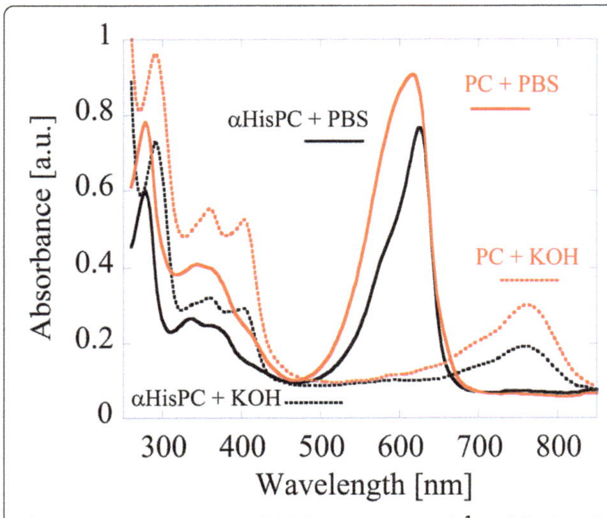

Figure 1 UV-vis spectra of PC (conc. 0.2 mg ml^{-1}, red line) and αHisPC (conc.0.2 mg ml^{-1}, black line) in the native form in PBS pH 7.4 (continuous lines) and in denatured form in 1 M KOH (dashed lines). Data for PC are reproduced from [8].

3.2 Effect of protein-melanin coating on the photocurrent

In order to establish optimal parameters for the coating process, three protein concentrations spanning two orders of magnitude (0.02, 0.2 and 2 mg ml^{-1}) and three reaction times (1.5, 3 and 4.5 h) are tested. The highest enhancement in photocurrent density was achieved at concentrations between 0.2 mg ml^{-1} to 2 mg ml^{-1} of PC and a reaction time of 3 h to 4.5 h. For hematite thin films coated with PC, the highest relative enhancement in photocurrent density is 23% and is obtained at a concentration of 2 mg ml^{-1} and a reaction time of 3 h, when measured in KOH. Under identical conditions, a higher value is obtained with samples coated with αHisPC at a concentration of 0.2 mg ml^{-1} that give a 55% increase in photocurrent density.

Figure 2 shows a typical I-V (photocurrent density) curve of the pristine and the protein coated electrodes in KOH (Figure 2a) and PBS (Figure 2b) electrolytes. The protein layer causes the higher photocurrent density. The higher photocurrent densities can be attributed to the panchromatic light harvesting action of protein and the electronic transfer to the hematite surface. The photocurrent onset potential for both samples is virtually the same vs. Ag/AgCl., i.e. ~0 V in KOH and 600 mV in PBS, when we account for the effect of the different pH on the Nernst potential.

We expect that the extreme pH conditions with the use of 1 M KOH electrolyte (~ pH 13–14) leads to melanin dissolution and structural changes in PC causing denaturation. Yet we measure in KOH a significant increase in photocurrent density (left panel, Figure 2a). A possible explanation might be found in the newly occurring absorbance feature at 750 nm for PC and αHisPC, when measured in the strongly alkaline environment (Figure 1). While the proteins may become denatured, the chromophores may still be intact and absorb light [10] and produce electron–hole pairs which add to the photocurrent. Moreover, the recombinant monomeric αHisPC appears to be a valid alternative to the commercial PC preparation. The 55% increase measured for αHisPC is significantly higher than the 8% increase measured for analogous samples prepared with PC (all measured in 1 M KOH). We thus speculate that PC and αHisPC couple differently with the surface, thus conferring different charge transfer. As a possible explanation, the presence of the Histag on αHisPC might provide a better conducting bonding to the hematite surface as histidine residues are known to coordinate iron ions and the Histag has been shown to bind iron in some proteins [11]. For the evaluation of the effect of melanin alone, samples lacking any photosynthetic protein were prepared and analysed as a control experiment. Compared to the hematite films with coatings containing PC or αHisPC and melanin, no significant increase in photocurrent was observed. Therefore the key factor for the increase in photocurrent is the biliprotein, and not the melanin. Based on these results, a 0.2 mg ml^{-1} protein concentration and 3 h enzymatic reaction time was used for further experiments.

3.3 Surface morphology of phycocyanin-melanin coated samples

Electron microscopy was used for the closer visual inspection of the hematite coated with melanin and different amounts of PC. Electron micrographs of hematite coated with melanin and PC show the formation of strands of protein-melanin complexes (Figure 3). Since the same concentration of melanin was present in all samples, strand formation depends on the protein concentration. At the lower 0.2 mg ml^{-1} protein concentration, SEM micrographs of samples coated with PC (Figure 3) shows formation of stretched dense formations in tight contact with the hematite surface, and this could play a role in the enhancement of the photocurrent. A significant strand formation is visible in samples with a PC concentration of 2 mg ml^{-1} rather than at

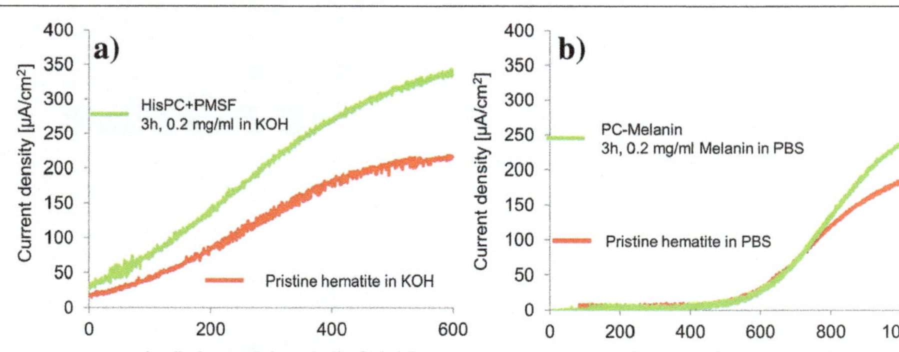

Figure 2 Photocurrent density before and after immobilization of αHisPC (a) and PC (b). Pristine hematite before (red bottom line) and after immobilization with melanin (green top line) of 0.2 mg ml^{-1} PC or αHisPC in the presence of 1 mM PMSF, for 3 h and measured in 1 M KOH (left panel) and in 0.05 M PBS (right panel).

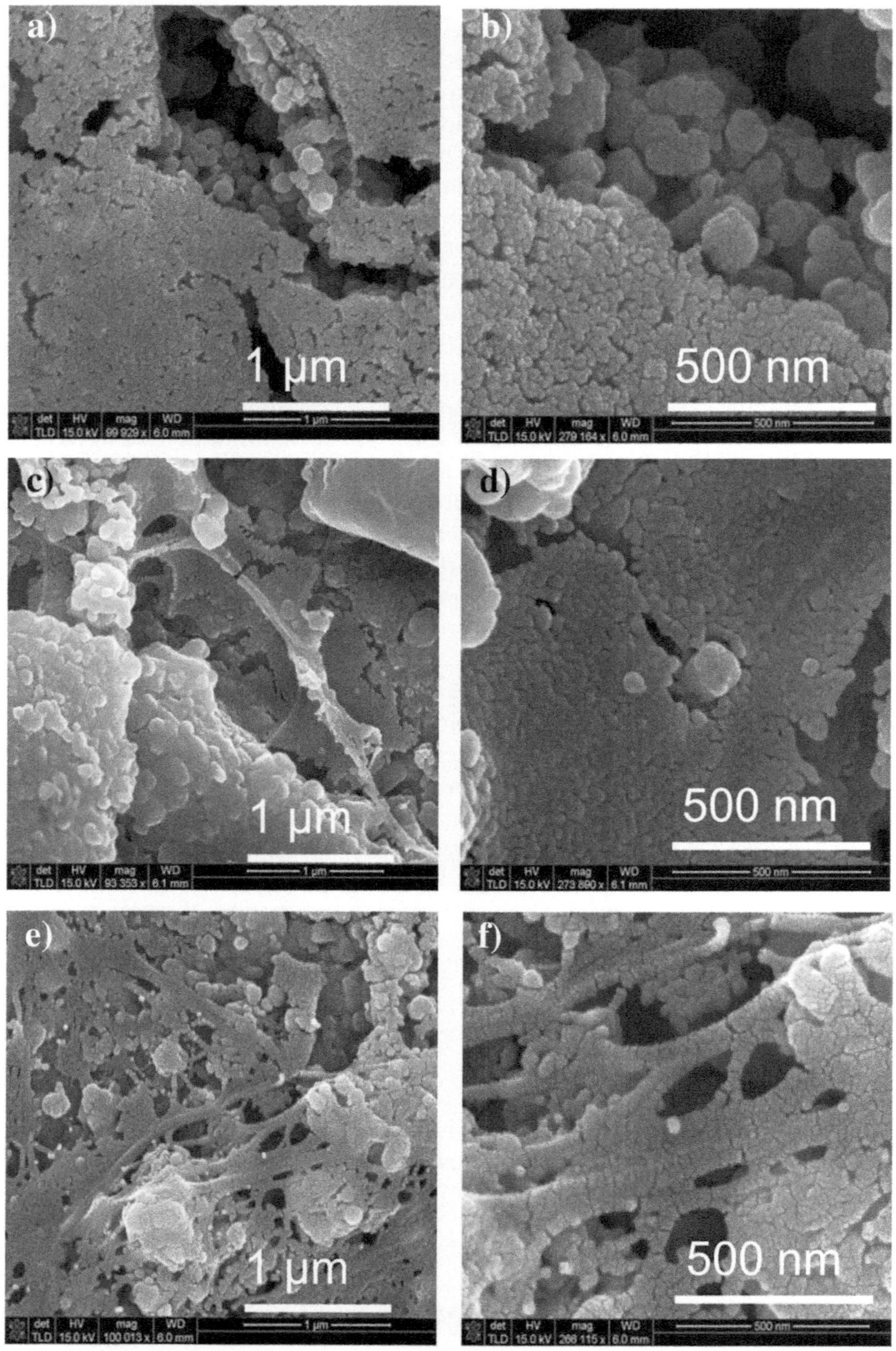

Figure 3 Scanning electron micrographs of hematite films coated with different amounts of PC, e.g. incubation for 3 h with a melanin-forming mixture containing 0.02 (a, b), 0.2 (c, d) and 2 (e, f) mg ml^{-1} incubated from top to bottom row. The scale bar is 1 μm (left) or 0.5 μm (right).

0.2 mg ml^{-1}. In addition to the protein strains, spherical granules with a diameter of 200 nm are observed (Figure 3). These are likely due to melanin which tends to organise in such granules when isolated from natural sources [12]. The pattern observed differs from that reported by Costa et al. [13] who studied natural and synthetic melanin and their interaction with Fe (III) by SEM. However, such heterogeneous kind of geometry was also described in [14] using laser light scattering which also showed how enzymatically-produced melanin can be largely hydrated and can have non-uniform density. However, the highest enhancement in current density was measured for samples with the 0.2 mg/ml PC concentration.

3.4 Charge transfer in PC-melanin coated hematite films

Important for the functionality of bio-inorganic interfaces in sensors or energy converters is not only their physical immobilization but also sufficient charge transfer between the organic component and the inorganic component, i.e. the electrode of the current collector. Charge transfer between electrodes and proteins or even whole cells has been studied for various conducting substrates, for example for GaAs and photosystem I [15], gold and blue copper protein azurin [16], carbon nanotubes [17] and for stainless steel [18]. It is clear from numerous examples, that electrons can be directly transferred from and to proteins which are in contact with inorganic surfaces, including hematite. Impedance spectroscopy can be a valuable analytical tool for the assessment of bio-organic interfaces [19]. Dheilly et al. [20] used electrochemical impedance spectroscopy (EIS) to probe the dynamics of the charge transfer of bacterial biofilms attached to electrodes in an aqueous environment. We apply EIS for studying the interaction between the photosensitizing protein (PC) and hematite using different coating procedures under dark and illuminated conditions using PBS as electrolyte. For the reader we illustrate in Figure 4 the sequence of the various synthesis and coating steps for the assembly of the electrode.

The schematic bio-hybrid electrode architecture in Figure 5a) shows one possible impedance model circuit for our protein-semiconductor assembly (following a suggestion of Ben-Yoav et al. [22]), along with an energy diagram for the solid state processes and the processes in the electrolyte. The water splitting reaction is based on electron holes h^+ which come from the hematite photoanode: $2\ H_2O + 4\ h^+ \rightarrow O_2 \uparrow + 4H^+$. The excited PC molecule generates also an electron–hole pair. The electron moves from the PC to the hematite. Since the highest occupied molecule orbital (HOMO) energy level of PC is above the VB of hematite, the electron hole cannot transfer from PC to hematite. Rather, the hole h^+ from the PC can form with the aqueous electrolyte hydroxyl radicals ($^{\bullet}HO$) upon excitation with visible light >470 nm wavelength [23]. This would balance the charge transfer via the Type I mechanism [24]. This constitutes an electron acceptor. Backflow of electrons to the electrolyte via the surface would be prevented by the $^{\bullet}HO$ radicals. Since the electron does not get freedom to recombine with holes at the hematite surface and back injected to electrolyte, the functionalization with PC increases the photocurrent density.

Figure 6 shows the impedance spectra of the pristine hematite film coated on FTO glass, measured in PBS in the dark (Figure 6a) and under 1.5 AM solar simulated light (Figure 6b). The spectra were recorded with bias ranging from 0 mV to 1000 mV in steps of 200 mV at

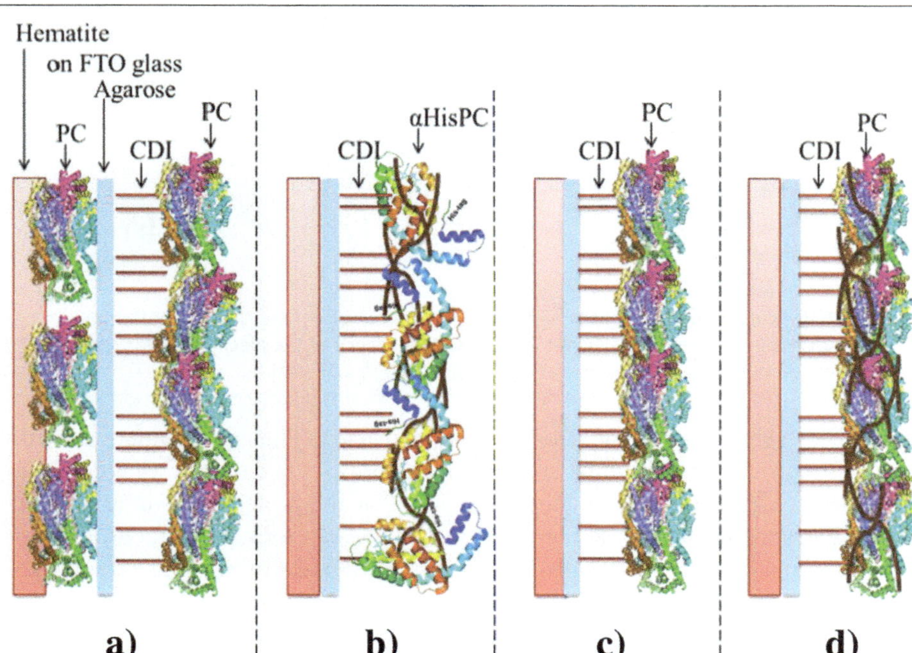

Figure 4 Schematic of the applied coating procedures for PC and αHisPC covalently attached to hematite via a chemical crosslinker, e.g. CDI. Differently from Bora [21], as shown here for reference in a), the hematite surface is directly coated with agarose and melanin using the chemical crosslinker CDI. Distances are not in proportion. Additionally, a layer containing melanin and αHisPC (**b**, upper panel, PDB ID: 4F0T), PC (**c** upper panel), or PC plus melanin (**d** upper panel, PDB ID: 1HA7) was applied.

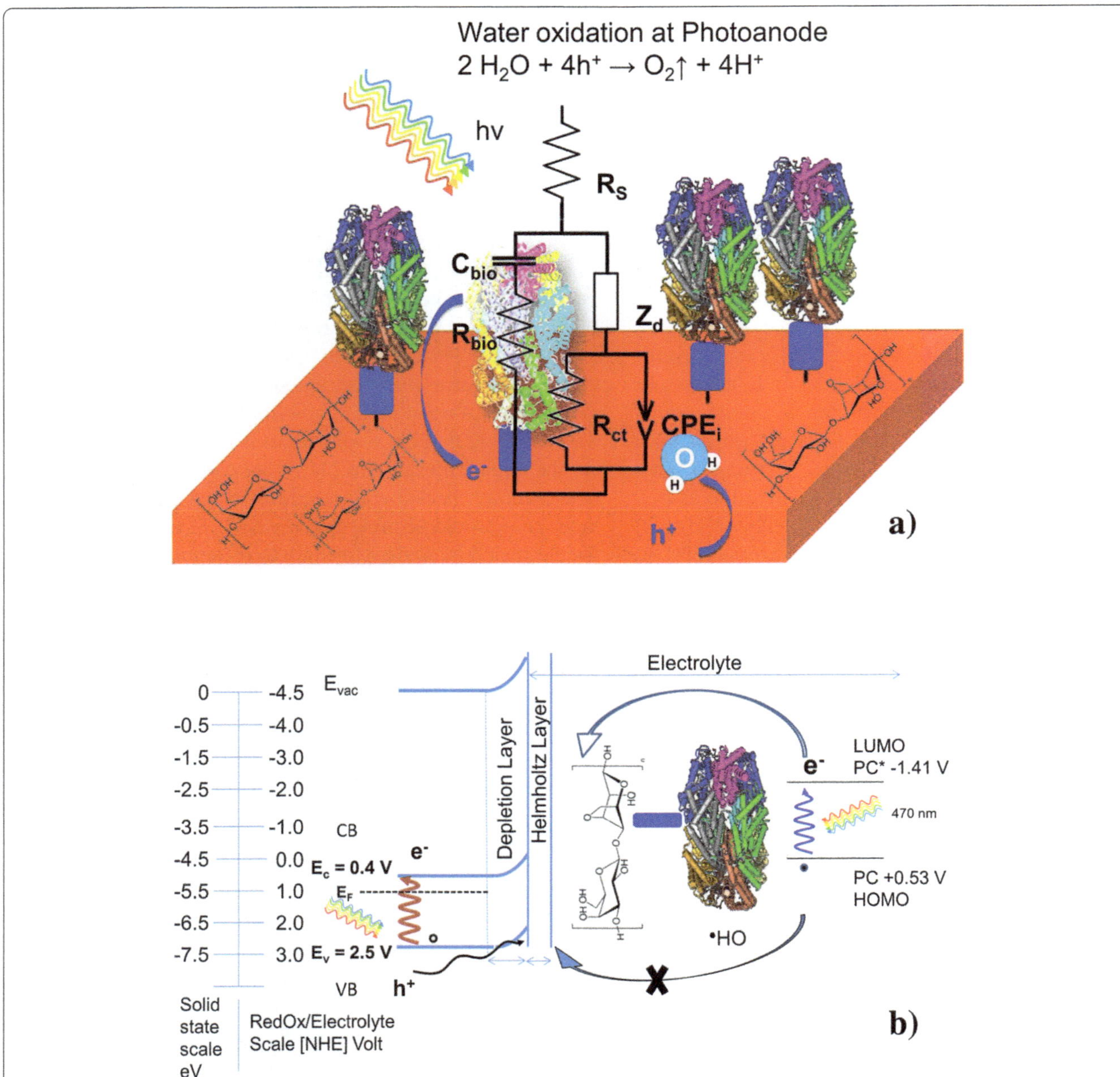

Figure 5 Top a) - The semiconducting hematite interfacing the electrolyte is represented by the charge transfer resistance R_{ct} and a constant phase element CPE_i and a Warburg impedance element Z_d. R_s is the resistance of the electrolyte. Inspired by reference [22]. Bottom **b)** – Energy diagram for the PC in aqueous electrolyte, linked with CDI and agarose to hematite.

frequencies ranging from 100 kHz (low impedance point near origin) to 0.1 Hz (high impedance point at end of spectrum). The impedance spectra of pristine hematite in dark fit in the impedance window posed by axes with 10 kOhm. Under illumination, the spectra obtained under 600 mV to 1000 mV bias show a convolution of two semicircles (Figure 6b) for the spectrum obtained at 800 mV. The first semicircle has a significantly smaller radius than the second one. The radius of the larger of both semicircles scales systematically with the applied DC bias and thus represents the charge transfer resistance.

For the hematite film coated with protein, the impedance measured in dark condition is below 5 kOhm (Figure 7a), revealing that the protein layer (αHisPC coated with the enzymatic melanin formation procedure) improves charge transfer between electrolyte and hematite. Note that we have plotted the data in Figures 6 and 7 in the same scaled window for easier comparison.

Moreover, upon illumination of the protein film, the impedance decreases dramatically (see the very small box in the lower left corner in the impedance spectrum in 7b) to far below 1 kOhm. Here it is sufficient to plot the impedance in an isometric window of 300 Ohm to

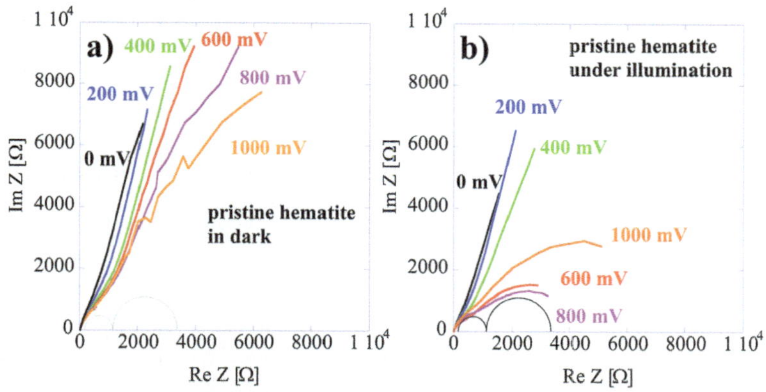

Figure 6 EIS spectra (Nyquist plots) of pristine hematite in dark a) and under illumination b) for DC bias from 0 mV to 1000 mV. The spectra are plotted on the same isometric scale.

present virtually the entire impedance spectra when measured under light condition. The lowest impedance is found for 800 mV bias. The curvature of the spectra in the inset of Figure 7b) shows the clear tendency towards semicircles as the direct current (DC) bias approaches the water splitting potential. We interpret this as increased charge transfer of the system during illumination and increased DC bias. The protein coating improves the conductivity and charge transfer between the hematite and the electrolyte also under illumination significantly. The impedance for pristine and protein coated hematite increases again in the transition from 800 mV to 1000 mV. This is, where typically considerable oxygen evolution by water oxidation takes place.

Charge transfer across interfaces depends on the electronic structure of these interfaces. A method for the assessment of the electronic structure of surfaces and interfaces is X-ray and photoelectron spectroscopy (XPS) and photoemission spectroscopy (PES). In Brizzolara et al. [25] the conclusion about covalent attachment of the proton pump protein bacteriorhodopsin via genetic substitution of cysteine for serine (S35C) was based partially on the XPS core level spectra of sulfur, i.e. via detection of the chemical shift of the sulphur core level spectrum. An early valence band (VB) XPS study on a protein (D-luciferin) is presented in Wada et al. [26], where XPS helped sketch a model for the luminescence. In this study, we find that the position of the VB spectrum changes depending on the type of phycocyanin coating process applied to hematite.

Figure 8 shows a set of VB spectra which were recorded with photoemission spectroscopy (PES) with photon energies close to the Fe 3p resonance edge, ~54 eV. The PES was employed in resonance so as to identify the contribution of the Fe ions in the VB density of states (DOS). The spectra were recorded from three different samples, Figure 8a) PC coated on hematite single crystal, Figure 8b) PC and melanin (PC + Mel) co-polymerized on hematite single crystal, and Figure 8c) His-tagged PC and melanin (αHisPC + Mel) on hematite single crystal.

We observe that, under dark condition (8a), the intensity maximum of the spectrum from the PC film shifts

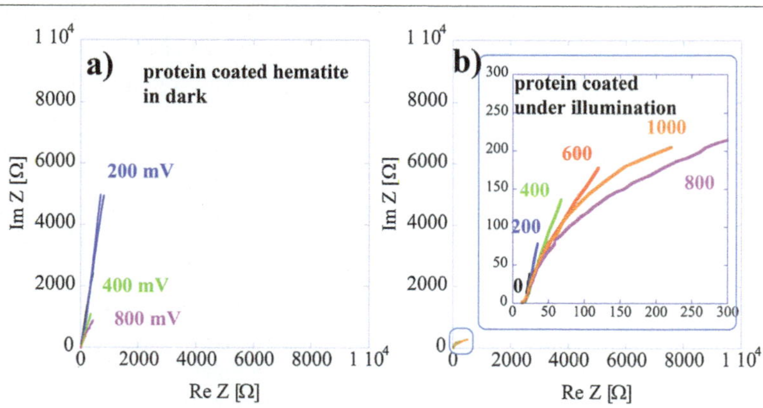

Figure 7 EIS spectra (Nyquist plots) of protein coated hematite recorded in dark a) and under illumination b) for DC bias from 0 mV to 1000 mV. The inset on the right shows the largest magnification. Note the very small window in the lower left corner, showing the very small overall impedance of the illuminated protein film. Coating procedure: enzymatic, αHisPC plus melanin (condition B in Figure 4, upper panel) [8].

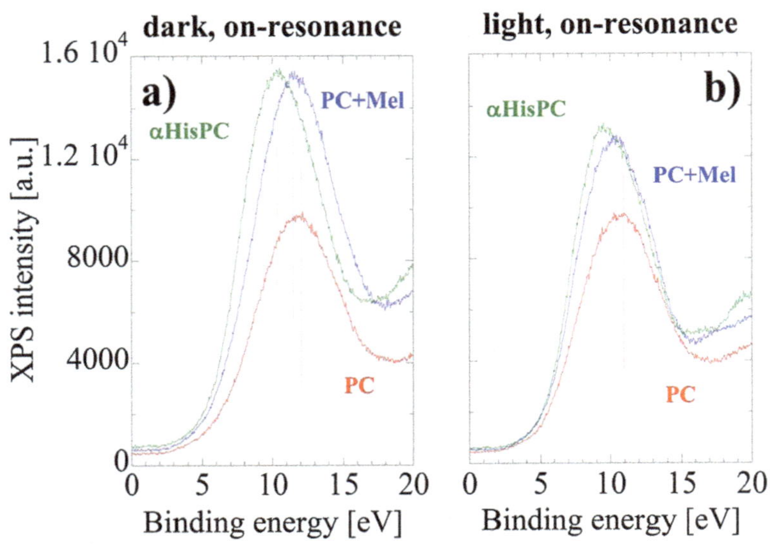

Figure 8 Comparison of three different coatings (condition B-D in Figure 4, upper panel) on hematite in Fe3p on-resonance condition at 54.4 eV photon energy under dark (a) and illuminated (b) conditions. With more advanced film processing technology, the VB PES spectra shifts closer towards the Fermi level (0 eV binding energy).

from 12 eV to just below 11.5 eV towards the Fermi level, when PC was co-polymerized with melanin (Figure 8a). With αHisPC + Mel on hematite, the intensity maximum shifts to 10.3 eV. A similar trend is shown in the case of illumination with the solar simulator (Figure 8b). We notice that with increasing improvement of film processing technology, the maxima of the VB spectra move closer to the Fermi level E_F, reflecting the changes in the electronic structure with increased DOS near E_F. The charge transfer between hematite and phycocyanin might be improved in αHisPC as histidines are known to coordinate to iron ions. At this time it is not clear whether the spectral shift towards the E_F is a result of hole doping originating from the PC + Mel in the films. In addition, the His-tag presumably improves entrapment of the His-tagged protein into the melanin network, as the formation of histidine-tyrosine bonds have been observed upon tyrosinase-catalysed crosslinking [27].

It is clear, however, that there is a correlation between improved electric transport, increased photocurrent and increased DOS in the VB near E_F. Although literature on the electronic structure of the interfaces of biological macromolecules and inorganic materials is scarce, some pioneering studies have been published. A PES and NEXAFS study on the electronic structure of an ultra-thin film of the surface layer of *Bacillus sphaericus* deposited on Si wafers showed how the highest occupied and lowest unoccupied molecular orbitals (HOMO and LUMO) constitute the band gap of the assembly [28]. In an extension of that study, these authors showed with resonant PES how charge transport evolved from torsion effects in the protein [29].

We want to turn to another observation in our PES spectra. When subjecting the protein-hematite assembly to 1.5 AM simulated solar light, the spectral weight of the intensity range near the O2p bonding peak decreases and shifts by up to 0.7 eV toward the Fermi level (0 eV binding energy).

We interpret this shift as the hole-doped DOS resultant from the exposure of visible light. With the solar lighting on, the electron–hole pairs can be created in the conduction band (CB) and the holes stays near the surface generating a p-type DOS. A similar trend of binding energy shift is observed when silicon is doped [30]. Our PES studies on proteins are in principal not different from studies on synthetic organic molecules and polymers which are frequently used in solar cells and light emitting diodes. A comprehensive review on such VB spectroscopy studies is presented by Koch et al. [31].

Among our protein functionalization processes, the coating with PC only (Figure 9a) is the least complex one. For the comparison of the PES spectra in Figure 9, we determine the position of the VB by extrapolation of the spectral flank near E_F towards zero intensity – as a simple and practical metric. The intercept is then by our definition the VB position. The VB position of the PES spectra shifts from 6 eV to 5 eV when the sample is illuminated with the solar simulator (Figure 9a). This 1 eV shift thus originates from the illumination effect, i.e. a charge carrier (electron hole pair) generation.

When we turn to the PC + Mel coated hematite (Figure 9b), the VB position shifts from 5.6 eV to 5 eV, i.e. the shift is only 0.6 eV upon illumination. In the case of the HisPC coating (Figure 9c), there is no shift of the VB spectrum

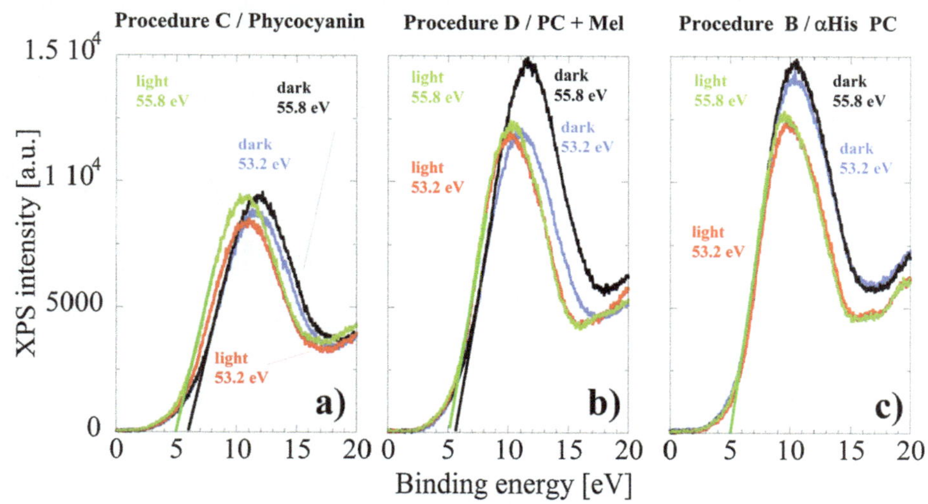

Figure 9 Comparison of VB PES spectra for the three different protein coatings (procedures C, D, and B from left to right from Figure 4). The steep straight lines intersecting near 5 eV show the VB position. The sample coated with PC and melanin (panel **b**), procedure D in Figure 4) shows 0.6 eV shift on VB position while PC only (panel **a**), procedure C in Figure 4) shows 1.0 eV shift under illumination. In the case of αHisPC (panel **c**)), no shift in the VB position is observed.

when the light is switched on. Interestingly, for all films, the position of the VB in the illumination condition is 5 eV. Hence, it appears that light has no influence on the charge transfer as far as the interface of the protein film with the hematite is concerned.

In addition to the electric charge transfer, which is the scope of this paper, fluorescence resonance energy transfer (FRET) could hypothetically occur between phycocyanin and hematite and thus increase the photocurrent density. The efficiency of the FRET mechanism is inversely proportional to the sixth power of the distance between light donor and light acceptor, making it extremely sensitive to distances and virtually negligible for small distances. Due to the architectures of our bio-hybrid electrodes, where agarose, CDI and His tag virtually constitute "spacers" between PC and hematite, it is difficult to perceive that FRET could occur across these spacers.

4 Conclusions

The immobilization of photosynthetic proteins on hematite, in particular C-phycocyanin, yields a substantial increase of the photocurrent density which can be assigned to electron hole pair formation in the protein by light absorption in a wavelength range that exceeds the absorption range of the hematite. Refined immobilization strategies like co-polymerization of PC with melanin or genetic engineering to attach a His-Tag to PC successively increase the photocurrent. Analysis of the charge carrier dynamics of the bio-hybrid electrode assembly under photoelectrochemical conditions shows that upon illumination the enhanced charge transfer and DC bias are substantially amplified (factor 10) when the hematite film is coated with the light harvesting protein. This, however, seems to have not the same strong effect on the photocurrent density (only factor 2). A clear and systematic dependency of spectral shift of the valence band is observed in the dark and in the illuminated condition. While the His-tag strategy clearly improves the charge transfer between PC and hematite in the dark, no further enhancement is observed under illumination condition. This might be due to the fact that the illumination generates electron hole pairs, but this should have no effect on the charge transport characteristic of the covalent linkages between protein and hematite.

Competing interests

The authors declare that they have no competing interests.

Authors' contributions

GF synthesized and processed samples and wrote major part of the manuscript. KGS synthesized and processed samples and measured current densities. JI organized the work, inspired the use of melanin and provided intellectual input on protein functionalization. FB measured the protein samples with photoemission spectroscopy at SSRL and ALS. YH measured the protein samples with photoemission spectroscopy at SSRL and ALS and analyzed the impedance spectra and arranged for SSRL beamtime. BSM measured the protein samples with photoemission spectroscopy at SSRL and ALS and discussed the spectra. DKB provided intellectual input in. LTM organized the work and provided intellectual input. AB organized the work, inspired the protein functionalization, provided intellectual input, did data analysis on photoemission spectra, arranged for ALS beamtime and wrote major part of the manuscript. All authors read and approved the final manuscript.

Acknowledgements

This study was funded at large by the VELUX Foundation under project no° 790 (Biomimetic photoelectrochemical cells for solar hydrogen generation: Bio-PEC) for K. G.-S. and G.F. G.F. was also financially supported by the TYROMAT project (COFUND action between Empa Postdocs programme and FP7: People) and COST TD1102 Action Project by SBFI C13.003. Financial support by the Strategic Korean-Swiss Cooperative Program in Science and

Technology within the project "Spectroscopy on photoelectrochemical electrode materials: SOPEM" for A.B. and NRF-2013K1A3A1A14055158 for B.S.M. is gratefully acknowledged. Financial support by the Swiss National Science Foundation for F.B. (project # 200021–137868 Reaction–diffusion processes for the growth of patterned structures and architectures: bottom up approach for photoelectrochemical electrodes) and for Y.H. (project # 200021–132126 Defects in the bulk and on surfaces and interfaces of metal oxides with photoelectrochemical properties: In-situ photo-electro-chemical and resonant x-ray and electron spectroscopy studies) and R'Equip # 121306 is gratefully acknowledged.
Portions of this research were carried out at the Stanford Synchrotron Radiation Lightsource, a national user facility operated by Stanford University on behalf of the U.S. Department of Energy, Office of Basic Energy Sciences. The solar simulator was provided by Jinghua Guo, Advanced Light Source in Berkeley. Florian Häussler (Empa Dübendorf) assisted with the sample preparation at Empa. We are grateful to Prof. Wendy Schluchter from the Department of Biological Sciences of the University of New Orleans for providing the plasmids for the production of αHisPC in this study.

Author details
[1]Laboratory for Bioactive Materials, Empa, Swiss Federal Laboratories for Materials Science and Technology, Lerchenfeldstrasse 5, 9014 St. Gallen, Switzerland. [2]Laboratory for High Performance Ceramics, Empa, Swiss Federal Laboratories for Materials Science and Technology, Überlandstrasse 129, 8600 Dübendorf, Switzerland. [3]Department of Inorganic and Analytical Chemistry, Szeged University, Dóm tér 7, H-6720 Szeged, Hungary. [4]Institute of Chemistry, Research Group of Environmental Chemistry, University of Szeged, Rerrich Béla tér 1, H-6720 Szeged, Hungary. [5]Department of Chemistry, University of Basel, Spitalstrasse 51, 4056 Basel, Switzerland. [6]Laboratory for Photonics and Interfaces, Ecole Polytechnique Federale de Lausanne, Rue Cantonale, 1021 Lausanne, Switzerland. [7]Department of Physics and Photon Science, GIST School of Physics and Chemistry, GIST College, Ertl Center for Electrochemistry and Catalysis, 123 Cheomdan-gwagiro (Oryong-dong), Buk-gu, Gwangju 500-712, Korea.

References

1. Bora DK, Braun A, Constable EC: Energy Environ. Sci. 2013, **6**(2):407–425.
2. Anderson S, Constable EC, Dare Edwards MP, Goodenough JB, Hamnett A, Seddon KR, Wright RD: *Nature* 1979, **280**(5723):571–573.
3. Tooley AJ, Cai YP, Glazer AN: Proc. Natl. Acad. Sci. 2001, **98**(19):10560–10565.
4. Fairhead M, Thoeny-Meyer L: *New Biotechnol* 2012, **29**(2):183–191.
5. Taurino I, Reiss R, Richter M, Fairhead M, Thoeny-Meyer L, De Micheli G, Carrara S: Electrochim. Acta. 2013, **93**:72–79.
6. Hu Y, Bora DK, Boudoire F, Haeussler F, Graetzel M, Constable EC, Braun A: J. Renew. Sust. Ener. 2013, **5**(4):043109.
7. Chen Z, Jaramillo TF, Deutsch TG, Kleiman-Shwarsctein A, Forman AJ, Gaillard N, Garland R, Takanabe K, Heske C, Sunkara M, McFarland EW, Domen K, Miller EL, Turner JA, Dinh HN: J. Mater. Res. 2010, **25**(1):3–16.
8. Ihssen J, Braun A, Faccio G, Gajda-Schrantz K, Thöny-Meyer L: **Light harvesting proteins for solar fuel generation in bioengineered photoelectrochemical cells.** Curr. Protein. Pept. Sci. 2014, **15: in press.**
9. Glazer AN, Fang S, Brown DM: J. Biol. Chem. 1973, **248**:5679–5685.
10. Berns DS, MacColl R: Chem. Rev. 1989, **89**:807–825.
11. The Recombinant Protein Handbook: **Protein amplification and simple purification.** *Amersham Biosciences* 2002, 41–58. 18-1142-75.
12. Watt AA, Bothma JP, Meredith P: *Soft Matter* 2009, **5**(19):3754–3760.
13. Costa TG, Younger R, Poe C, Farmer PJ, Szpoganicz B: Bioinorg. Chem. Appl. 2012, **2012**:712840.
14. Bridelli MG: Biophys. Chem. 1998, **73**(3):227–239.
15. Frolov L, Rosenwaks Y, Richter S, Carmeli C, Carmeli I, Phys J: Chem. C. 2008, **112**:13426–13430.
16. Chi Q, Farver O, Ulstrup J: *PNAS* 2005, **102**(45):16203–16208.
17. Deng S, Jian G, Lei J, Hu Z, Ju H: *Biosens Bioelectron* 2009, **25**:373–377.
18. Omanovic S, Roscoe SG: Langmuir 1999, **15**(23):8315–8321.
19. Owino IOK, Sadik OA: *Electroanalysis* 2005, **17**(23):2101–2113.
20. Dheilly A, Linossier I, Darchen A, Hadjiev D, Corbel C, Alonso V: Appl. Microbiol. Biotechnol. 2008, **79**(1):157–164.
21. Bora DK, Rozhkova AA, Schrantz K, Wyss PP, Braun A, Graule T, Constable EC: Adv. Funct. Mater. 2012, **22**(3):490–502.
22. Ben-Yoav H, Freeman A, Sternheim M, Shacham-Diamand Y: Electrochim. Acta. 2011, **56**(23):7780–7786.
23. Zhang S, Xie J, Zhang J, Zhao J, Jiang L: Biochim. Biophys. Acta. 1999, **1426**(1):205–211.
24. Foote CS: Photochem. Photobiol. 1991, **54**(5):659.
25. Brizzolara RA, Boyd JL, Tate AE, Vac J: Sci. Technol. A. 1997, **15**(3):773–778.
26. Wada N, Shitaba R, Shibazaki M, Suzuki N: Photochem. Photobiol. 1989, **49**(4):513–518.
27. Hellman M, Mattinen M-L, Fu B, Buchert J, Permi P: J. Biotechnol. 2011, **151**(1):143–150.
28. Vyalikh DV, Danzenbacher S, Mertig M, Kirchner A, Pompe W, Dedkov YS: Phys. Rev. Lett. 2004, **93**(23):238103.
29. Vyalikh DV, Maslyuk VV, Blueher A, Kade A, Kummer K, Dedkov YS: Phys. Rev. Lett. 2009, **102**(9):098101.
30. Sezen H, Suzer S: J. Chem. Phys. 2011, **135**:141102.
31. Koch N: phys. stat. sol. RRL. 2012, **6**(7):277–293.

16

Graphene nanoribbon devices at high bias

Melinda Y Han[1] and Philip Kim[2*]

Abstract

We present the electron transport in graphene nanoribbons (GNRs) at high electric bias conduction. When graphene is patterned into a few tens of nanometer width of a ribbon shape, the carriers are confined to a quasi-one-dimensional (1D) system. Combining with the disorders in the system, this quantum confinement can lead into a transport gap in the energy spectrum of the GNRs. Similar to CNTs, this gap depends on the width of the GNR. In this review, we examine the electronic properties of lithographically fabricated GNRs, focusing on the high bias transport characteristics of GNRs as a function of density tuned by a gate voltage. We investigate the transport behavior of devices biased up to a few volts, a regime more relevant for electronics applications. We find that the high bias transport behavior in this limit can be described by hot electron scattered by the surface phonon emission, leading to a carrier velocity saturation. We also showed an enhanced current saturation effect in the GNRs with an efficient gate coupling. This effect results from the introduction of the charge neutrality point into the channel, and is similar to pinch-off in MOSFET devices. We also observe that heating effects in graphene at high bias are significant.

PAC codes: 73.23.-b; 85.35.-p

1 Background

The discovery of graphene [1] has enabled intense fundamental and applied research activities in this novel two-dimensional (2D) carbon based electronic system. Electron transport in graphene is substantially different from that of conventional 2D electronic systems owing to the linear energy dispersion relation near the charge neutrality point (Dirac point) in the electronic band structure [2,3]. This unique band structure is fundamentally responsible for the distinct electronic properties of carbon nanotubes (CNTs) [4]. When graphene is patterned into a narrow ribbon, and the carriers are confined to a quasi-one-dimensional (1D) system, we expect the opening of an energy gap. Earlier theoretical work showed that this energy gap depends on the width and crystallographic orientation of the graphene nanoribbon (GNR) [5-7], similar to CNTs. The first experimental work on GNRs [8-10] has shown that a transport gap can indeed be open up by patterning graphene into nanometer size ribbons or constrictions. The resulting transport gap formation can be most simply attributed to quasi-onedimensional (1D) confinement of the carriers, which induces an energy gap in the single particle spectrum [5-7,11-13]. Detailed experimental studies of disordered graphene nanoribbons (GNRs) [14-20], however, suggest that this observed transport gap may not be a simple band gap. In an effort to explain these experimental results, various theoretical explanations for the transport gap formation in disordered graphene nanostructures have been proposed, including models based on Coulomb blockade in a series of quantum dots [21], Anderson localization due to edge disorder [22-24], and a percolation driven metal-insulator transition [25]. In order to distinguish between these different scenarios, systematic experiment including treatment of both disorder induced localization and electron-electron interaction is required.

In our recent experiment [20], we carried out systematic studies of the scaling of the transport gap in GNRs of various dimensions. From this scaling of several characteristic energies with GNR width (W) and length (L), we find evidence of a transport mechanism in disordered GNRs based on hopping through localized states whose size is close to the GNR width. We found that At the charge neutrality point, a length-independent transport gap forms whose size is inversely proportional to the GNR width. In particular, we found that in this gap, electrons are localized, and charge transport exhibits a transition between thermally activated behavior at higher temperatures and variable range hopping at lower temperatures. By varying

*Correspondence: pk2015@columbia.edu
[2]Department of Physics, Columbia University, New York, NY 10027, USA
Full list of author information is available at the end of the article

the geometric capacitance, we find that charging effects constitute a significant portion of the activation energy.

Extending this earlier work, in this review, we examine the electronic properties of lithographically fabricated GNRs with widths in the tens of nanometers. Here we investigate the transport behavior of devices biased up to a few volts, a regime more relevant for electronics applications. We will first address characteristics of graphene at high bias which are not specific to graphene nanoribbons, then we address GNRs at high bias specifically. Graphene devices operated at high source-drain bias show a saturating $I - V$ characteristic. This decrease in conductivity at high applied electric field is described by carrier velocity saturation due to optical phonon emission. This result is analogous to the high bias results obtained CNTs. In a well known experiment, Yao *et. al.* [26] found that current in metallic single wall carbon nanotubes saturates at high electric field. Their result is explained in terms of zone-boundary optical phonon emission from high energy electrons. At high electric fields, a steady-state population is developed between right and left moving charge carriers with a maximum energy difference corresponding to the phonon energy $\hbar\Omega = 160$ meV, leading to a saturated current of $(4e/h)/(\hbar\Omega) \approx 25\ \mu$A. A slightly different behavior was reported in semiconducting single wall carbon nanotubes by Chen and Fuhrer [27]. In these devices, current does not saturate completely, and the transport is described by an electric field dependent carrier velocity. The authors fit their data with a model based on a carrier velocity that saturates to a constant value at high electric field and a carrier density dependent on the local potential along the device. They find a saturation velocity of 2×10^7 cm/s in their device. These results demonstrate the feasibility of 1D GNR devices for electronic applications with a proper bandgap engineering.

2 Graphene nanoribbon fabrication

GNRs used in this study were fabricated by lithgrapically patterned structure from mechanically exfoliated graphene. The process flow is outlined in Figure 1. Briefly, we begin with exfoliated graphene, fabricate metal electrodes using standard electron beam (e-beam) lithography procedures, pattern an etch mask using an negative e-beam resist, and etch away unprotected graphene using an oxygen plasma etch. An atomic force microscope (AFM) image of a finished device is shown in Figure 2.

Once a suitable piece of graphene has been deposited and identified using the procedure described above, the next step is to electrically contact the graphene with metal electrodes using e-beam lithography. We begin by spinning on a layer of poly(methyl methacrylate) (PMMA) e-beam resist and baking on a hotplate at 180°C for 2 minutes. Then we use e-beam lithography to write a 2 mm by 2 mm grid of alignment marks at roughly the location of the graphene, and develop in a solution of methyl isobutyl ketone:isopropal alcohol (MIBK:IPA) 1:3 for 5–10 seconds. This quick development leaves alignment mark "holes" in the PMMA, which we use for alignment in the following e-beam lithography step, eliminating the need for metal alignment mark deposition or another PMMA spin step. Electrodes are patterned in this PMMA layer with e-beam lithography, using an optical image of the sample with the alignment mark holes for design and alignment. Thermal evaporation is then used to deposit 1–2 nm of chrome and 25–50 nm of gold, and the chip is placed in acetone overnight at room temperature for lift-off (Figure 1(b)).

Once the graphene has been successfully contacted with Cr/Au electrodes, we create an etch mask to define the nanoribbons. A negative tone e-beam resist, hydrogen silsesquioxane (HSQ) (2% in MIBK) is spun on to the chip (at 4000 rpm, for a typical film thickness of 14 nm). We use HSQ as the resist for this step because a negative resist is ideal for creating a small etch mask, and because with HSQ we can obtain small feature sizes. The etch mask is written at a relatively high e-beam dose (1300 μC/cm^2 for the ribbons in our 30 keV system, with lower doses for larger features) and developed in a solution of 0.26N tetramethylammonium hydroxide (TMAH) in water for 1 minute (Figure 1(c)).

After defining the etch mask, the graphene is ready to be etched. The device is exposed to oxygen plasma in a Technics reactive ion etcher (RIE) with 200 mTorr O_2 at 50 W for 5–10 seconds. These conditions etch graphene at a rate of about one layer per second, so that unprotected single layer and few-layer graphene are etched away cleanly (Figure 1(d)). The finished device (Figure 2) is then ready to be wirebonded and measured.

The devices measured in this experiment are back-gated and dual-gated etched graphene devices. Graphene devices often fail or change drastically and irreversibly when the current density per unit width exceeds a threshold of $\sim$ 2 mA/μm. We operate the device at currents below this threshold. Current-voltage characteristics at varying gate voltages were measured for 17 ribbon devices with a range of widths and lengths, and three "wide" devices with $W = 200$ nm, in order to compare to the behavior of non-ribbon devices.

3 Saturating behavior fits a velocity saturation model

Figure 3 shows a plot of current vs source-drain bias for varying gate voltages in a back-gated device. We focus here on the curves taken at densities far from the charge neutrality point, such as the curve singled out in Figure 4. Here we see that at low bias the slope of the curve is constant, and at high bias the curve turns down, approaching a linear behavior with a reduced slope.

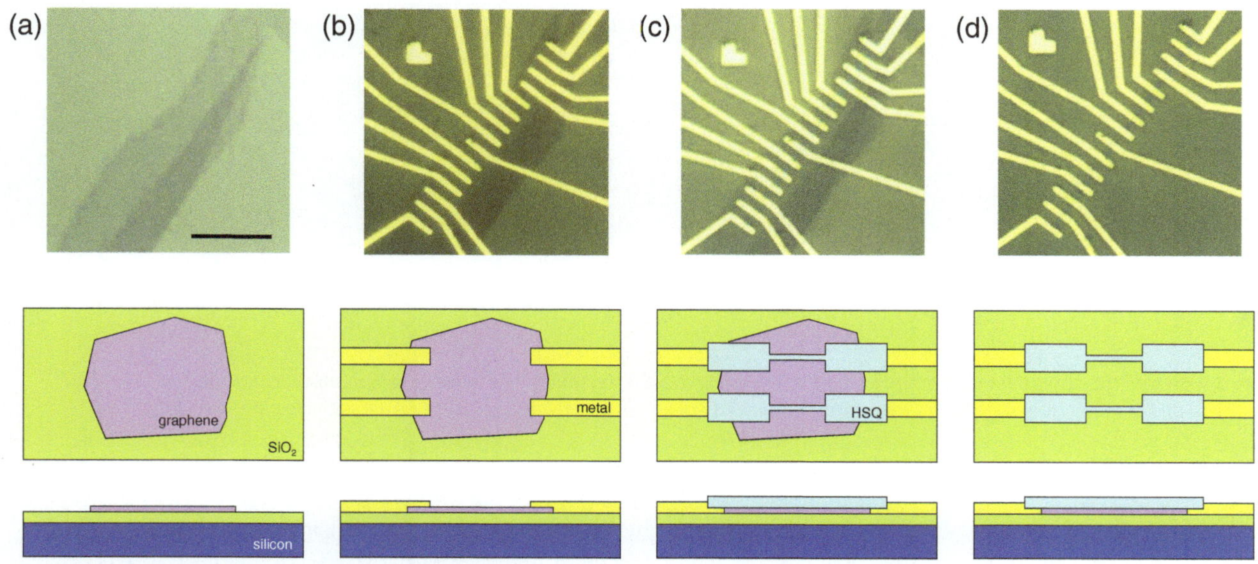

Figure 1 Process flow for graphene nanoribbon fabrication. Optical microscope image (top row), cartoon top view (middle row), and cartoon side view (bottom row), for each of four major processing steps. **(a)** Graphene deposition on Si/SiO_2 substrate. **(b)** E-beam lithography fabrication of metal electrodes. **(c)** Patterning of negative e-beam resist etch mask. **(d)** Removal of unprotected graphene by oxygen plasma etching. Scale bar in optical image is 20 μm, all four optical images have the same scale.

To describe this saturating decrease in conductivity, we propose a model based on an electric field-dependent carrier velocity $v_d(E)$ of the form:

$$v_d(E) = \left(\frac{1}{\mu_0 E} + \frac{1}{v_{sat}}\right)^{-1} \tag{1}$$

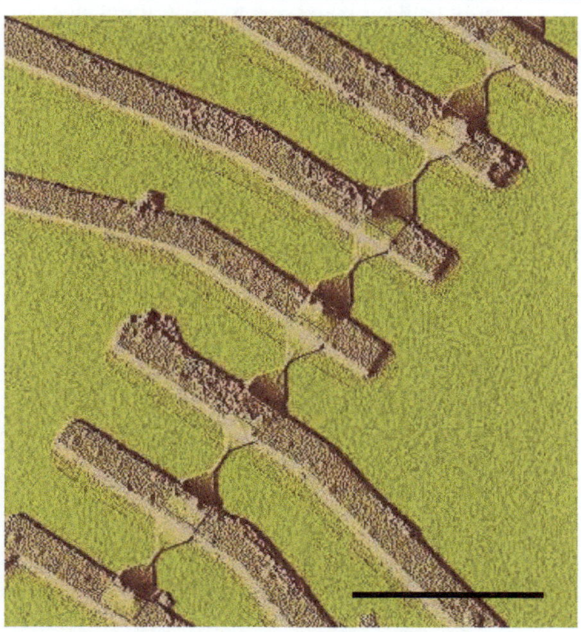

Figure 2 Atomic force microscope (AFM) image of the device in Figure 1. The ribbons in this image correspond to the lower six ribbons in the optical image in Figure 1(d). Scale bar is 4 μm.

where μ_0 is the low field mobility and v_{sat} is a phenomenologically introduced saturation velocity. The total current through the device is given by

$$I = jW = -nev_d W \tag{2}$$

We assume that the capacitance to the back gate dominates in determining the charge density in the channel, so that

$$ne = C_g\left(V(x) - (V_g - V_{CNP})\right) = C_g(V - V_0) \tag{3}$$

where $V = V(x)$ is the potential at position x along the channel, and we have defined $V_0 \equiv V_g - V_{CNP}$. Using the relation $E = dV/dx$, we have

$$I^{-1} = -\frac{1}{WC_g(V - V_0)}\left(\frac{1}{\mu_0 dV/dx} + \frac{1}{v_{sat}}\right) \tag{4}$$

Rearranging terms and integrating gives the current

$$I = WC_g(V_0 - V_b/2)\frac{\mu_0 V_b/L}{1 + \mu_0 V_b/v_{sat}L} \tag{5}$$

In its limiting forms, Equation 5 for the current qualitatively gives the behavior seen in Figure 4. At low V_b, current is linear in V_b with a conductivity $WC_g V_0 \mu_0/L$, determined by the low field mobility, as expected. At high V_b, current is again linear in V_b, but now with a conductivity of $WC_g v_{sat}/2$ and an offset determined by the

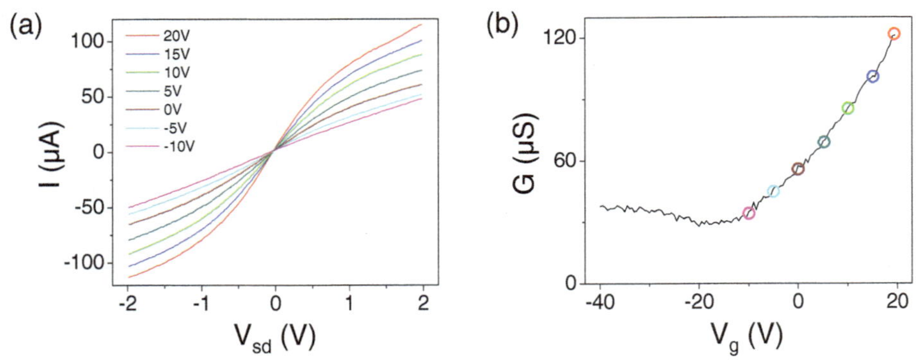

Figure 3 High bias transport data from a ribbon with W=70 nm and L=500 nm measured at room temperature in vacuum. (a) Current vs. source drain bias at varying gate voltages V_g, as shown in the legend. **(b)** Conductance vs. gate voltage for the same device at a source drain bias of V_{sd} =200 meV.

gate voltage. At low fields, the variation in carrier density is small and the linear I–V results from the linear form of $v_d(E) \approx \mu_0 E$ in this regime. At high fields, v_d approaches a constant value v_{sat}, and the linear dependence of the carrier concentration on V_b is responsible for an I–V characteristic approaching linear behavior. Note this is in contrast to the case of carbon nanotubes, where there are a set number of conducting channels, so that the current saturates with the drift velocity.

The expression in Equation 5 for $I = I(V_b)$ was fit to the I–V characteristics in Figure 3; the result is shown in Figure 5. For ribbon devices, the geometry is not well approximated by a parallel plate capacitor, so the gate capacitance was calculated numerically.For the device in Figure 5, the capacitance was calculated to be 47.5 nF/cm^2 using a numerical calculation based on the finite element method. The model fits well for curves taken at densities far at high carrier densities, and begins to break down for curves measured near the charge neutrality point, as seen in Figure 5 for $V_g = -10$ V. This fit has two free parameters, v_{sat} and μ_0. For this dataset, this model gives μ_0 values between 400 and 600 cm^2/Vs, compared to the value of 700 cm^2/Vs from low bias sweeps of G–V_g.

The values of v_{sat} obtained from this fit are plotted against V_g in Figure 6(a). In Figure 6(b), we plot v_{sat} against the inverse of the Fermi energy

$$E_F = \hbar v_F \sqrt{\pi C_g (V_g - V_{CNP})} \tag{6}$$

Converting V_g to E_F involves the value of V_{CNP}, which commonly drifts throughout measurement due to changes in adsorbed molecules and positions of trapped

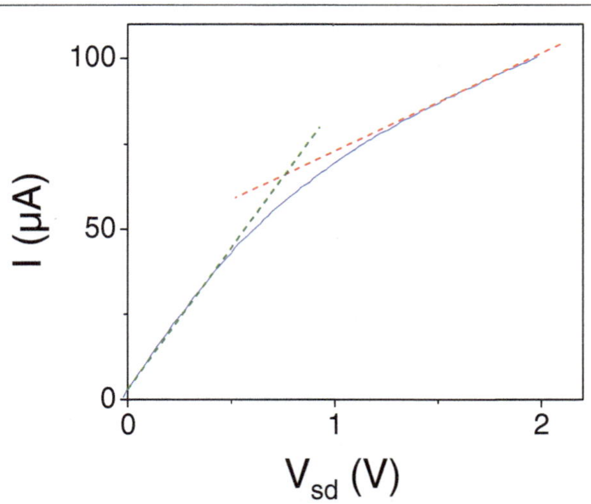

Figure 4 Current-voltage characteristic for the curve with $V_g - V_{CNP}$ = -30V shown in Figure 3(a). The distinct slopes at low and high V_{sd} resulting from the saturation behavior are highlighted with green and red dashed lines, respectively.

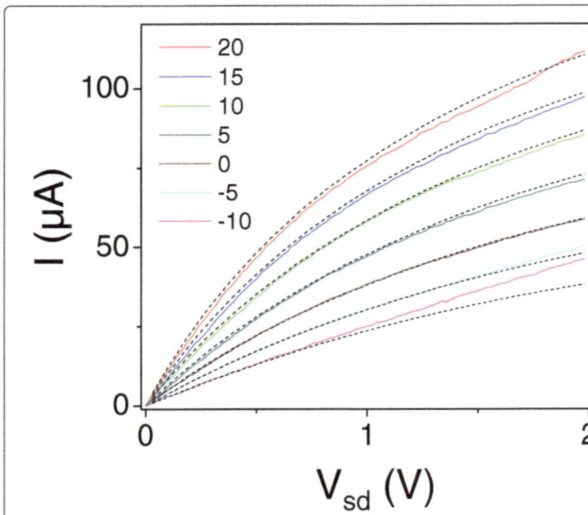

Figure 5 Fit of the current model in Equation 5 to the data in Figure 3. The legend gives the gate voltage for each sweep, dashed lines are fits to each sweep.

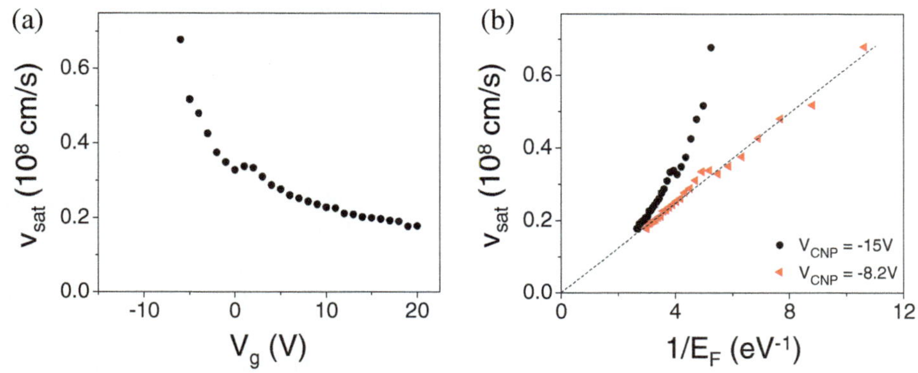

Figure 6 Saturation velocity values extracted from the fits in Figure 5. (a) Saturation velocity v_{sat} as a function of the gate voltage V_g. **(b)** v_{sat} vs the inverse of the Fermi energy E_F. Black circles and red triangles correspond to a conversion of V_g to E_F using $V_{CNP} = -15$ V and $V_{CNP} = -8$ V, respectively. Dashed line is a linear fit to this data.

charges. Here the black circles correspond to conversion of V_g to E_F using $V_{CNP} = -15$ V, the same value used in Equation 5 for the original fit. Red triangles represent a conversion to E_F using $V_{CNP} = -8$ V so that the a linear fit of v_{sat} vs. E_F^{-1} intersects the origin.

In order to understand the inverse relationship between v_{sat} and E_F, we seek a physical understanding of the electric field dependent carrier velocity, or drift velocity, in Equation 1. This expression corresponds to scattering by optical phonons, which would produce an electric field dependent mean free path. By Matthiessen's rule, mean free paths add as

$$\frac{1}{l} = \frac{1}{l_{sc}} + \frac{1}{l_{op}} \tag{7}$$

where l is the total mean free path and l_{sc} is the mean free path for elastic impurity scattering and quasi-elastic acoustic scattering, and l_{op} is the mean free path for optical phonon emission. If electrons are immediately scattered upon reaching the optical phonon energy, so that

$$l_{op} = \frac{\hbar\Omega}{eE} \tag{8}$$

where E is the electric field and Ω is the relevant optical phonon frequency, then the mobility μ is given by

$$\frac{1}{\mu} = \frac{1}{\mu_0} + \frac{E}{v_{sat}} \tag{9}$$

This form of the mobility results in the expression for the drift velocity $v_d = \mu E$ given in Equation 1. For electrons and holes in graphene, which have a constant carrier velocity of v_F, drift velocity can be understood as the time averaged velocity of carriers when scattering is taken into account.

From the above calculation we see that our phenomenological velocity saturation model can be understood in terms of a picture where electrons scatter by optical phonon emission upon reaching the phonon energy $\hbar\Omega$ under the influence of the applied electric field. With this in mind, we derive an expression for current density using a different approach, in order to gain insight into our measured values for the saturation velocity. Current density is given by

$$\vec{j} = -e \int d\vec{k} D_k \vec{v}(\vec{k}) g(\vec{k}) \tag{10}$$

where $D_k = 2/(2\pi)^2$ is the density of electronic states in k-space, $\vec{v}(\vec{k}) = v_F$ is the electron velocity, and $g(\vec{k})$ is the distribution function. In the relaxation time approximation, we have

$$g(\vec{k}) = g^0(\vec{k}) - e\vec{E} \cdot \vec{v}(\vec{k}) \tau(\epsilon(\vec{k})) \left(-\frac{\partial f}{\partial \epsilon}\right) \tag{11}$$

where $g^0(\vec{k})$ is the equilibrium distribution function, τ is the relaxation time, and f is the Fermi-Dirac distribution function. For a device with its length in the x direction, we seek $\vec{j} = j\hat{x}$, so we consider only $\vec{E} = E\hat{x}$, and

$$\vec{E} \cdot \vec{v}(\vec{k}) = Ev_F \cos\theta \tag{12}$$

where θ is the angle between $d\vec{k}$ and $\vec{E}$. We assume that electrons are immediately scattered upon reaching the energy threshold for phonon emission, giving

$$\tau = \frac{\hbar\Omega}{eEv_F} \tag{13}$$

So that for Equation 10 we have

$$j = e \int \frac{dk}{\pi^2} v_F \cos^2\theta \hbar\Omega \left(-\frac{\partial f}{\partial \epsilon}\right) |_{\epsilon=\hbar v_F k} \tag{14}$$

In polar coordinates

$$\begin{aligned} j &= e\int_0^{\infty}\frac{dk}{\pi^2}\int_0^{2\pi} d\theta v_F \cos^2\theta \hbar\Omega\delta(\hbar v_F k - E_F) \\ &= \frac{e}{\pi}\Omega\frac{E_F}{\hbar v_F} \end{aligned} \quad (15)$$

At high fields, we assume $j = nev_{sat}$ and use $E_F = \hbar v_F\sqrt{\pi n}$ to obtain

$$\frac{v_{sat}}{v_F} = \frac{\hbar\Omega}{E_F} \quad (16)$$

Using this expression with $v_F = 10^8$ cm/s [2,3], we obtain a value of $\hbar\Omega = 62.0$ meV from the linear fit (dashed line) in Figure 6(b). This is well below the value of the longitudinal zone-boundary phonon for graphene, which has $\hbar\Omega = 200$ meV [28]. We suggest that our measured phonon energy corresponds to the SiO_2 surface phonon energy $\hbar\Omega = 55$ meV [29-31], although we note that values measured in other ribbon devices of different geometries vary widely (from $\approx$ 22 meV to $\approx$ 120 meV), possibly due to discrepancies in determining the relevant device geometry, the corresponding capacitance, and the position of the charge neutrality point.

4 Top-gated graphene devices show an enhanced current saturation effect

In dual-gated devices, we observe a velocity saturation behavior similar to that the back-gated device behavior described above. However, we also see an enhanced current saturation at certain gate voltage combinations, as first reported in Reference [32]. Figure 7 shows current-voltage characteristics and corresponding conductance-gate voltage sweeps for a dual-gated device with $W = 35$ nm and $L = 2$ μm. At combinations of V_{bg} and V_{tg} near the charge neutrality point, we see a "kink" in the I–V curve, where the current first begins to flatten out, then turns upwards again. Figure 8 highlights this behavior in one I–V curve from the same device. This effect is specific to top-gated devices, where the strong capacitive coupling allows the bias voltage to dominate the carrier density in the channel.

The "kink" effect in graphene is similar to pinch-off in traditional MOSFETs, where a strong bias voltage pulls the quasi-Fermi level at one end of the channel into the charge-depleted bandgap. In graphene, where there is no bandgap, this results in a transition within the channel from one carrier type (electrons or holes) to the other. In a device that is n-type, as in Figure 8(c)(I), a positive source-drain voltage (applied to the source) depletes the electron density in the channel near the source (II). At sufficiently strong positive bias voltage, the bias voltage begins to pull holes into the channel, so a region of the channel is at charge neutrality and contributes a large resistance (III). As bias is further increased, hole density at the source also increases, so conductivity increases again (IV). In Reference [32] we showed that wide plateaus in current could be achieved when this "kink" effect is made to coincide with velocity saturation.

5 Heating effects can overcome transport gap at high bias

The results discussed above come from graphene nanoribbons measured at high bias, but the key features of the data, saturation velocity at strong electric fields and the "kink" effect in the current for top-gated devices, are also seen in wide graphene devices [32]. This leads to the question, how are nanoribbons different from wide, non-ribbon devices when operated at high bias? Here we present the preliminary results of a comparison between dual-gated ribbons and wide devices and so far find no

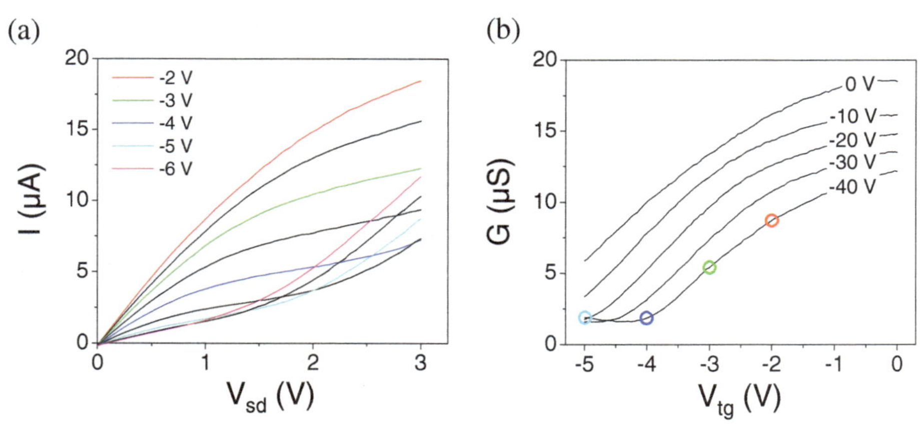

Figure 7 Current-voltage characteristics for a dual-gated device with W=35 nm and L=2 μm measured at room temperature in vacuum. **(a)** I–V_{sd} at constant V_{bg} = -40 V and varying V_{tg}. Select curves are shown in color and have V_{tg} values as noted in the legend, the black curves fall between these curves at 0.5 V increments. **(b)** G-V_{tg} at varying V_{bg} values, as noted on each curve, measured with V_{sd} = 1 meV. Colored dots correspond to the gate voltage positions where the colored curves in Figure 7 (a) were measured.

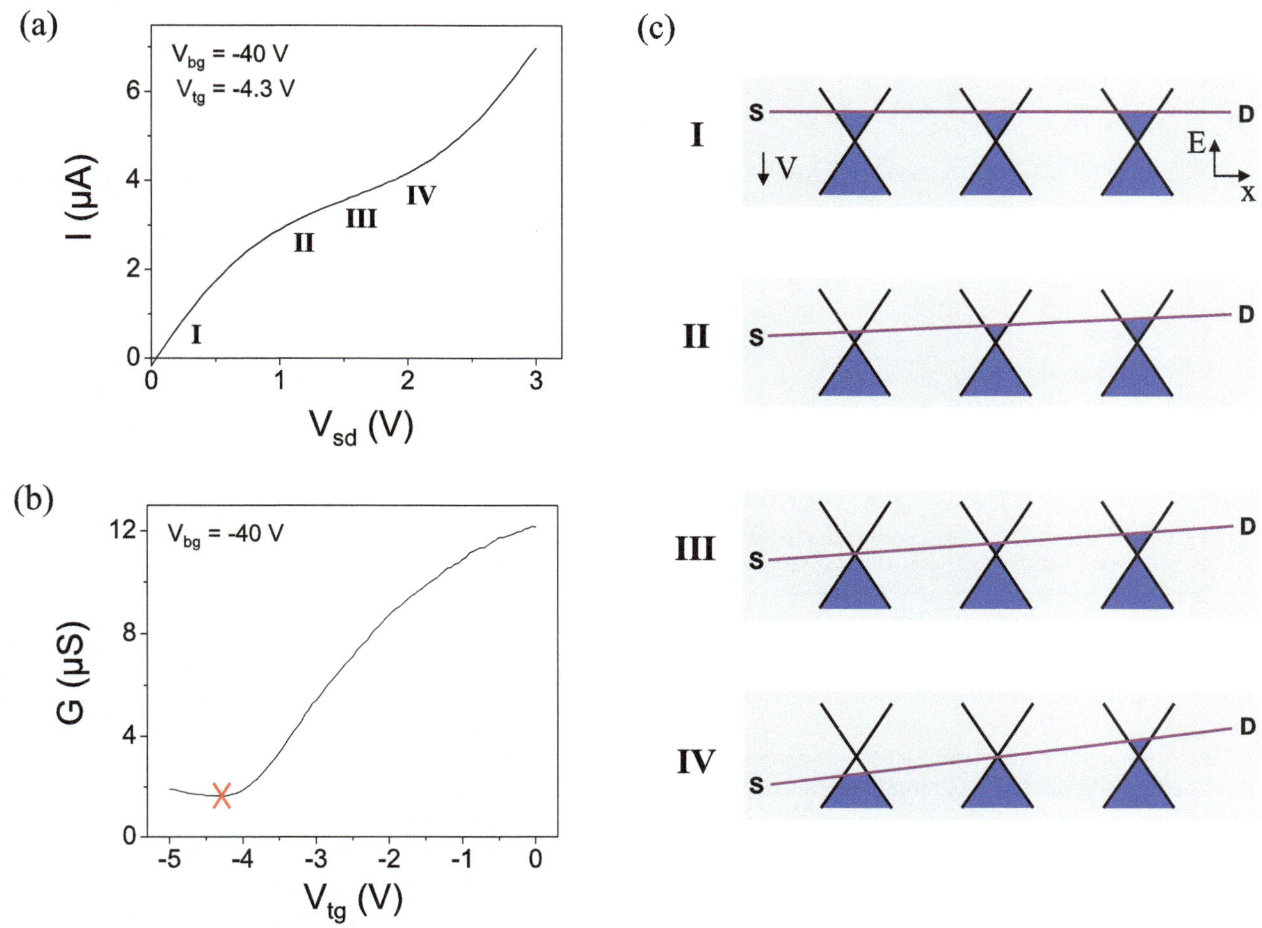

Figure 8 Kink and current saturation. (a) An I–V_{sd} curve for the device in Figure 7, highlighting the current saturation "kink" behavior. **(b)** G-V_{tg} for the same device; the red "x" highlights the low bias ($V_{sd} = 1$ meV) conditions corresponding to the curve in (a). **(c)** Cartoon schematic of the Fermi level in the channel for each condition (I–IV) marked in (a). Dirac cones with different Fermi levels along the length of the channel result from carrier density variation along the channel under the influence of a strong V_{sd}.

major differences in their performance. This result is only preliminary because the widths of the GNRs in this experiment are not well specified within the range of $W \approx$ 20–60 nm. The widths of nanoribbons lying underneath the dielectric and metal layers cannot be accurately measured in this device geometry. Estimates of the width can be made based on the expected width dependence of the low-temperature transport characteristics Δm and ΔV_b from the analysis in earlier work [20]. From these comparisons, it is estimated that the ribbons used in this experiment have $W \approx 50$ nm. Ribbons of this width are narrow enough to behave distinctly from "wide" ($W \gtrsim$ 100 nm) devices at low temperatures and low bias, but as we shall see below, they may not be narrow enough show a difference in transport characteristics at high bias. Ribbons as narrow as $W \approx 15$–20 nm are achievable by our fabrication methods, so measurements of narrower devices with larger transport gaps may still reveal distinct device behavior.

In comparing gapped graphene nanoribbons to wide graphene with no gap, there are several differences we may expect to see. First, since graphene nanoribbons have a strongly suppressed current at energies inside the gap, we may see an increased transconductance. Also, we could see larger and more fully saturated current in the "kink" region, as the presence of a gap causes the "kink" to more closely resemble pinch-off in a traditional MOSFET. We may also see the effects of edge roughness. In narrow ribbons where edge roughness constitutes a significant portion of the total ribbon width, this could lead to a decrease in maximum current carrying capabilities, or cause the devices to degrade more quickly.

Figure 9 shows I–V characteristics for graphene devices taken at two different temperatures, 77 K and 300 K. The I–V curves do not change significantly between the two temperatures. Since we expect to see thermal effects in the conductivity even away from the charge neutrality point, this suggests that the effective temperature in the device is

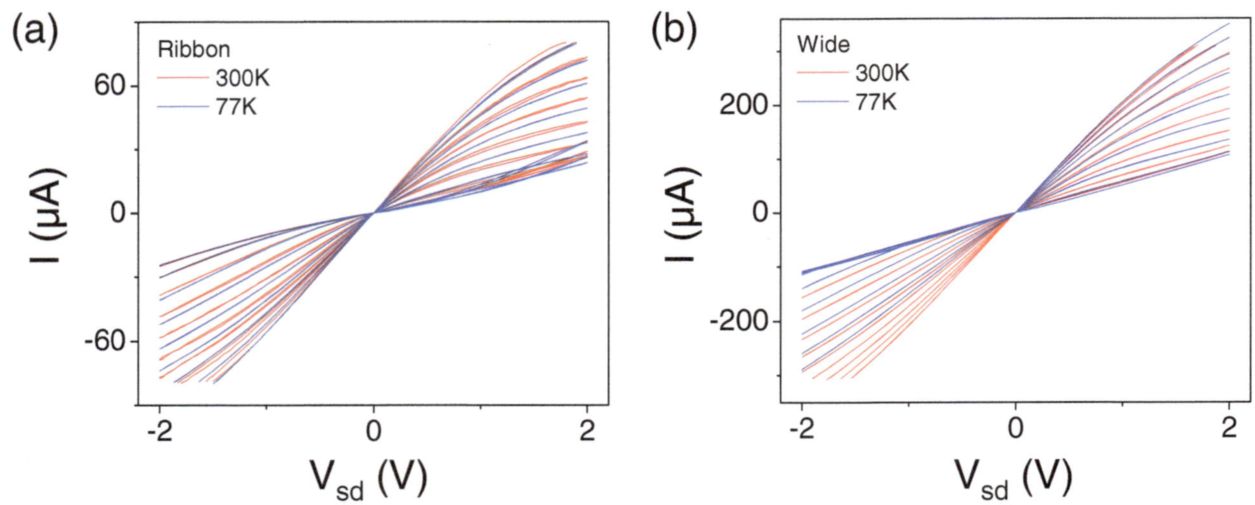

Figure 9 Top gated transfer characteristic for nanoribbons. I–V_{sd} characteristics at constant $V_{bg} = 0$ with V_{tg} varying from 0 to -8 V, measured at T=77 K and T=300 K for **(a)** a ribbon device ($W \approx 50$ nm) and **(b)** a wide graphene device ($W = 200$ nm). Both devices have $L = 500$ nm.

similar at both 77 K and 300 K, in other words, other heating in the system dominates over the ambient temperature up to 300 K.

As a straightforward method to directly compare ribbon devices with wide devices, we compare the scaled current density per width $j = I/W$ for two devices, a ribbon device with width $W \approx 50$ nm and a wide device with $W = 200$ nm, both with length $L = 500$ nm, shown in Figure 10. Here we can see that in the ribbon device, there is no difference in the size or shape of the kink behavior (Figure 10(b)), and only a minor difference in transconductance. From this data we see that a ~ 50 nm wide graphene nanoribbon shows no major differences in behavior from a wide device when operated at high bias.

To understand the similarity in behavior between 50 nm and 200 nm wide devices, we compare the gap size of the ribbon device with the relevant thermal effects in the system; if the available thermal energy in the system is larger than the gap, the effect of the gap will be washed out by thermally activated charge carriers. In earlier work [20], we found that there are three different ways to measure the size of the gap: Δ_m, from the gate voltage, ΔV_b, from the bias voltage, and E_a, from the activation energy for nearest neighbor hopping. Here we are concerned with current flow at high bias, so ΔV_b is the most relevant of these scales for distinguishing the on and off states of the device, though E_a will determine the leakage current in the off state. For the 500 nm long devices studied here, these values are similar. Since ΔV_b has a strong length dependence, if we wish to increase ΔV_b we can increase the device length L, with the trade-off of an increased the resistance and therefore a decreased current.

We consider two heating effects in this experiment. First, we compare the size of the gap with the thermal energy at room temperature, $k_B T \approx 26$ meV. Figure 11 shows how the relevant gap sizes compare with the room

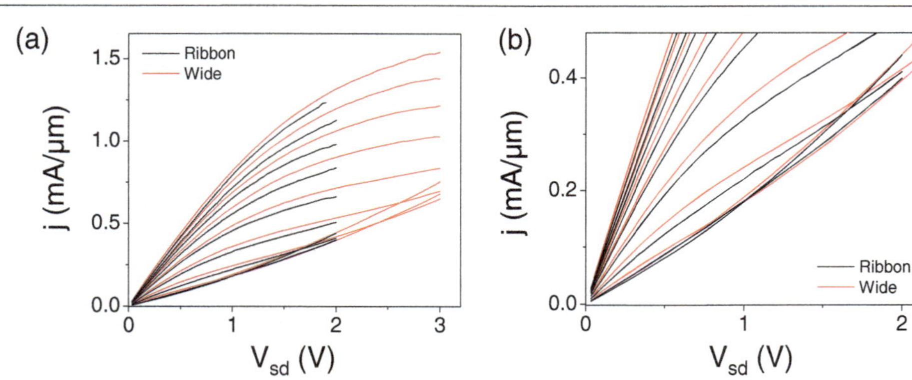

Figure 10 Current density versus bias voltage at fixed gates. (a) Current density $j = I/W$ at high V_{sd} at constant V_{bg} and varying V_{tg} for the two devices in Figure 9. **(b)** Same data as in (a), enlarged to show behavior near in the region of the "kink".

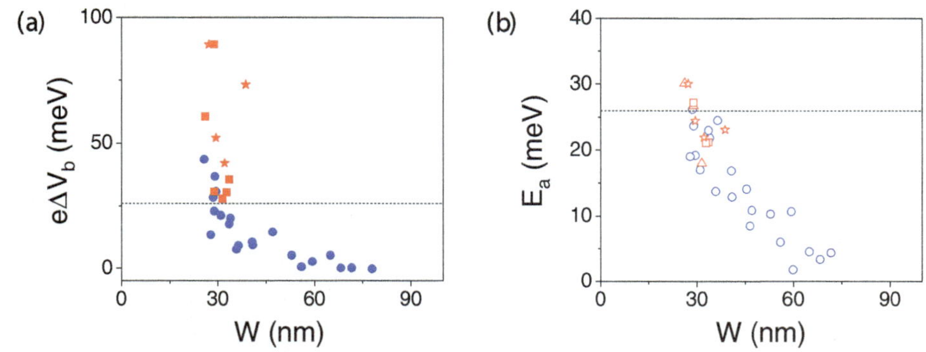

Figure 11 Bias gap and activation energy versus GNR width. (a) Measures of bias gap vs width from the data in reference [20]. **(b)** The activation erengy estimated from Arrhenius relation. The room temperature thermal energy 26 meV is highlighted with the horizontal dashed lines.

temperature thermal energy. We see that ΔV_b can be easily made greater than 26 meV by decreasing the ribbon width to below 30 nm or increasing ribbon length. However, only the narrowest ribbons shown here have a large enough E_a; narrower ribbons would be needed to ensure a low thermally activated leakage current.

If heating effects raise the device temperature above room temperature, then heating effects will be more relevant than ambient temperature effects. Several recent works [33-35] address the topic of heating in graphene at high bias. From the ribbon device data in Figure 9(a), we can expect to see dissipated electrical power $P = IV$ of up to $\approx$ 350 kW/cm^2, though power dissipation may be lower in the optimal operating regime for device applications. From the results in Reference [35] for temperature vs. power per area, this power dissipation corresponds to a temperature of 1350 K, or a thermal energy of 116 meV. From this it is clear that the thermal energy from heating greatly exceeds that from the ambient temperature, but this result was from a back-gated device. In a dual-gated device geometry, the top-gate dielectric and electrode may act as a heat sink and decrease the effect of heating.

In Figure 12, we model the heat sinking effects of a gate dielectric and top gate on a hot ribbon. This was done in the COMSOL Multiphysics finite element modeling package by assigning a heat flux to the ribbon such that maximum temperature in a back-gated device is $\sim$ 1100 K, shown in Figure 12(a). The graphene ribbon and graphene leads were assigned a thermal conductivity of 5000 K [36] and a thickness of 3.4; heat dissipation was also allowed through the 285 nm SiO_2 layer to the Si substrate below. In Figure 12(b), a 30 nm SiO_2 gate dielectric and a 30 nm gold top gate are added to the same model, again allowing heat dissipation into the gate dielectric and the top gate. Here, SiO_2 was used in place of HSQ because they are expected to have similar material properties. In this model, the maximum temperature is decreased to 825 K. If the top gate thickness is increased to 100 nm to allow for more heat sinking, the temperature decreases only slightly more, to 812 K.

The gate dielectric actually used in the experiment consists of HSQ/HfO_2 with thicknesses of 15/15 nm. Hafnium dioxide has a much higher thermal conductivity than silicon dioxide (23 W/m $\cdot$ K for HfO_2 versus 1.4 W/m $\cdot$ K for SiO_2). When the model is changed to include the proper layer thicknesses of each dielectric, the maximum nanoribbon temperature decreases to 680 K, which corresponds to an energy of 59 meV. This is the behavior we can

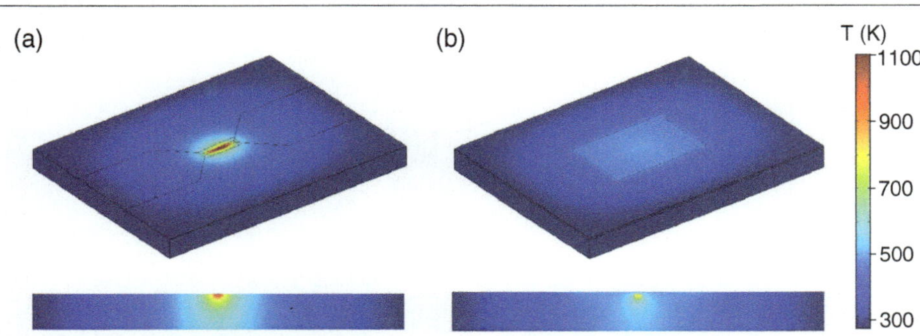

Figure 12 Thermal modeling of GNRs at high bias. Thermal modeling of **(a)** back-gated and **(b)** dual-gated graphene nanoribbon devices. Top row is a full 3D view of each device (on a 3 μm by 4 μm rectangle), bottom row is a cross-sectional slice taken midway across the ribbon (showing the oxide thickness of 285 nm), colorbar is the same for all plots.

expect to see in the actual device measured in Figures 9 and 10. From Figure 11 it is clear that E_a and ΔV_b are both far below this energy, so thermally activated carriers easily wash away any gap-related effects we might have seen in the transport at high bias.

Heat sinking could be greatly improved by removing the HSQ, such as by an hydrofluoric acid etch, and depositing $\approx$ 15 nm of hafnia only as the dielectric. In this geometry, the dielectric would be thinner and more thermally conductive, allowing for more efficient heat dissipation to the metal top gate. For the same heating conditions, this device construction would result in a maximum nanoribbon temperature of 460 K. The corresponding energy, 40 meV, is a gap size easily achievable by our nanoribbon fabrication methods. We note that from the results in our earlier work [20], the addition of a top gate tends to decrease E_a, but as the top-gated geometry provides very good heat sinking, and top gates will ultimately be needed for optimized device design, we see this as the best route for development of a graphene nanoribbon device that retains its gapped behavior at high bias.

6 Conclusions

In this review we have described a saturating I–V characteristic in graphene devices operated at high source-drain bias, and described the behavior using a model where surface phonon emission results in a carrier velocity that saturates to a Fermi energy dependent value at high applied electric field. We showed that for top-gated graphene devices have an enhanced current saturation effect at certain gate voltage combinations. This effect results from the introduction of the charge neutrality point into the channel, and is similar to pinch-off in MOSFET devices. We observe that heating effects in graphene at high bias are significant, and very narrow ribbons with a strongly heat sinking device design are required to produce a device where confinement-induced gap effects dominate over the effects of heating.

Competing interests

The authors declare that they have no competing interests.

Authors' contributions

Both authors read and approved the final manuscript.

Author details

[1]Department of Applied Physics and Applied Mathematics, Columbia University, New York, NY 10027, USA. [2]Department of Physics, Columbia University, New York, NY 10027, USA.

References

1. KS Novoselov, AK Geim, SV Morozov, D Jiang, Y Zhang, SV Dubonos, IV Grigorieva, AA Firsov, Electric field effect in atomically thin carbon films. Science. **306**(5696), 666–669 (2004)
2. KS Novoselov, AK Geim, SV Morozov, D Jiang, MI Katsnelson, IV Grigorieva, SV Dubonos, AA Firsov, Two-dimensional gas of massless dirac fermions in graphene. Nature. **438**, 197 (2005)
3. Y Zhang, Y-W Tan, HL Stormer, P Kim, Experimental observation of the quantum hall effect and berry's phase in graphene. Nature. **438**, 201 (2005)
4. MS Dresselhaus, G Dresselhaus, R Saito, Physics of carbon nanotubes. Carbon. **33**(7), 883–891 (1995)
5. K Nakada, M Fujita, G Dresselhaus, MS Dresselhaus, Edge state in graphene ribbons: Nanometer size effect and edge shape dependence. Phys. Rev. B. **54**, 17954–17961 (1996)
6. K Wakabayashi, M Fujita, H Ajiki, M Sigrist, Electronic and magnetic properties of nanographite ribbons. Phys. Rev. **59**, 8271–8282 (1999)
7. Y Son, ML Cohen, SG Louie, Energy gaps in graphene nanoribbons. Phys. Rev. Lett. **97**(21), 216803 (2006)
8. MY Han, B Özyilmaz, Y Zhang, P Kim, Energy band-gap engineering of graphene nanoribbons. Phys. Rev. Lett. **98**(20), 206805 (2007)
9. Z Chen, Y-M Lin, MJ Rooks, P Avouris, Graphene nano-ribbon electronics. Physica. E. **40**(2), 228–232 (2007)
10. X Li, X Wang, L Zhang, S Lee, H Dai, Chemically derived, ultrasmooth graphene nanoribbon semiconductors. Science. **319**(5867), 1229–1232 (2008)
11. M Ezawa, Peculiar width dependence of the electronic properties of carbon nanoribbons. Phys. Rev. **73**, 045432 (2006)
12. L Brey, HA Fertig, Electronic states of graphene nanoribbons studied with the dirac equation. Phys. Rev. B. **73**(23), 235411 (2006)
13. V Barone, O Hod, GE Scuseria, Electronic structure and stability of semiconducting graphene nanoribbons. Nano Lett. **6**(12), 2748–2754 (2006)
14. LA Ponomarenko, F Schedin, MI Katsnelson, R Yang, EW Hill, KS Novoselov, AK Geim, Chaotic dirac billiard in graphene quantum dots. Science. **320**(5874), 356–358 (2008)
15. C Stampfer, J Güttinger, S Hellmüller, F Molitor, K Ensslin, T Ihn, Energy gaps in etched graphene nanoribbons. Phys. Rev. Lett. **102**(5), 056403 (2009)
16. F Molitor, A Jacobsen, C Stampfer, J Güttinger, T Ihn, K Ensslin, Transport gap in side-gated graphene constrictions. Phys. Rev. B. **79**(7), 075462 (2009)
17. K Todd, H Chou, S Amasha, D Goldhaber-Gordon, Quantum dot behavior in graphene nanoconstrictions. Nano Lett. **9**(1), 416–421 (2009)
18. X Liu, JB Oostinga, AF Morpurgo, LMK Vandersypen, Electrostatic confinement of electrons in graphene nanoribbons. Phys. Rev. B. **80**, 121407 (2009)
19. KTP Gallagher, D Goldhaber-Gordon, Disorder-induced gap bahavior in graphene nanoribbons. Phys. Rev. B. **95**(81), 115409 (2010)
20. MY Han, JC Brant, P Kim, Electron transport in disordered graphene nanoribbons. Phys. Rev. Lett. **104**, 056801 (2010)
21. F Sols, F Guinea, AHC Neto, Coulomb blockade in graphene nanoribbons. Phys. Rev. Lett. **99**(16), 166803 (2007)
22. D Gunlycke, DA Areshkin, CT White, Semiconducting graphene nanostrips with edge disorder. Appl. Phys. Lett. **90**(14), 142104 (2007)
23. M Evaldsson, IV Zozoulenko, H Xu, T Heinzel, Edge-disorder-induced anderson localization and conduction gap in graphene nanoribbons. Phys. Rev. B. **78**(16), 161407 (2008)
24. D Querlioz, Y Apertet, A Valentin, K Huet, A Bournel, SG Retailleau, P Dollfus, Suppression of the orientation effects on bandgap in graphene nanoribbons in the presence of edge disorder. Appl. Phys. Lett. **92**(4), 042108 (2008)
25. S Adam, S Cho, MS Fuhrer, SD Sarma, Density inhomogeneity driven percolation metal-insulator transition and dimensional crossover in graphene nanoribbons. Phys. Rev. Lett. **101**(4), 046404 (2008)
26. Z Yao, CL Kane, C Dekker, High-field electrical transport in single-wall carbon nanotubes. Phys. Rev. Lett. **84**, 2941–2944 (2000)
27. Y-F Chen, MS Fuhrer, Electric-field-dependent charge-carrier velocity in semiconducting carbon nanotubes. Phys. Rev. Lett. **95**, 236803 (2005)
28. AC Ferrari, JC Meyer, V Scardaci, C Casiraghi, M Lazzeri, F Mauri, S Piscanec, D Jiang, KS Novoselov, S Roth, AK Geim, Raman spectrum of graphene and graphene layers. Phys. Rev. Lett. **97**, 187401 (2006)
29. S Fratini, F Guinea, Substrate-limited electron dynamics in graphene. Phys. Rev. B. **77**, 195415 (2008)

30. MV Fischetti, DA Neumayer, EA Cartier, Effective electron mobility in si inversion layers in metal–oxide–semiconductor systems with a high-kappa insulator: The role of remote phonon scattering. J. Appl. Phys. **90**(9), 4587–4608 (2001)
31. J-H Chen, C Jang, S Xiao, M Ishigami, MS Fuhrer, Intrinsic and extrinsic performance limits of graphene devices on SiO_2. Nat. Nanotechnol. **3**, 206–209 (2008)
32. I Meric, MY Han, AF Young, B Özyilmaz, P Kim, KL Shepard, Current saturation in zero-bandgap, top-gated graphene field-effect transistors. Nat. Nanotechnol. **3**, 654 (2008)
33. M Freitag, M Steiner, Y Martin, V Perebeinos, Z Chen, JC Tsang, P Avouris, Energy dissipation in graphene field-effect transistors. Nano Lett. **5**, 1883–1888 (2009)
34. D-H Chae, B Krauss, K von Klitzing, JH Smet, Hot phonons in an electrically biased graphene constriction. Nano Lett. **2**, 466 (2010)
35. S Berciaud, MY Han, KF Mak, LE Brus, P Kim, TF Heinz, Electron and optical phonon temperatures in electrically biased graphene. Phys. Rev. Lett. **104**, 227401 (2010)
36. AA Balandin, S Ghosh, W Bao, I Calizo, D Teweldebrhan, F Miao, CN Lau, Superior thermal conductivity of single-layer graphene. Nano Lett. **8**, 902 (2008)

Novel 3D micro- and nanofabrication method using thermally activated selective topography equilibration (TASTE) of polymers

Arne Schleunitz[2], Vitaliy A Guzenko[1], Martin Messerschmidt[2], Hakan Atasoy[2], Robert Kirchner[1] and Helmut Schift[1*]

Abstract

Micro- and nanostructures with three-dimensional (3D) shapes are needed for a variety of applications in optics and fluidics where structures with both smooth and sharp features enhance the performance and functionality. We present a novel method for the generation of true 3D surfaces based on thermally activated selective topography equilibration (TASTE). This technique allows generating almost arbitrary sloped, convex and concave profiles in the same polymer film with dimensions in micro- and nanometer scale. We describe its principal mechanism exemplified by pre-patterned poly (methyl methacrylate) resist which is exposed to high energy electrons prior to a thermal annealing step enabling the selective transformation of stepped contours into smooth surfaces. From this we conclude, that TASTE not only offers an enormous degree of freedom for future process variations, but also will advance the patterning capabilities of current standard 3D micro- and nanofabrication methods.

Keywords: Thermoplastic polymer; Thermal reflow; Three-dimensional; Surface topography; Electron beam lithography; Nanoimprint lithography; Nanofabrication

1 Background

Current applications in micro- and nanotechnology mainly use binary (i.e. two-level) surface topographies provided by state-of-the-art planar fabrication processes. However, the increasing technical requirements for new or enhanced device functionality with sophisticated 3D topographies have to be met. This includes regular sub-wavelength blazed gratings and microprisms for out-coupling of light into planar waveguides for optical application [1,2], as well as shallow tapered inlets bridging micro- to nanofluidics for life-sciences [3,4]. Currently, 3D topographies are often simply approximated by steps with adapted widths and heights which resemble provisional and quasi-3D contours. In this case, viable pattern fidelity is connected to high process complexity since non-continuous multilevel features need multiple lithographic and etching [5] or electroplating steps [6]. Alternative approaches for continuous contours using anisotropic etching of crystalline silicon in alkaline solutions are restricted to specific geometries [7,8]. Thus novel process strategies are required in order to overcome the limitations of generic micro- and nanofabrication techniques.

By implementing the TASTE method (i.e. thermally activated selective topography equilibration), we demonstrate the possibility to generate truly 3D shaped topographies with decreased technical effort while the geometrical diversity is enhanced at the same time. TASTE is a further development and generalization of an effect observed during selective thermal reflow of grayscale structures, presented in [9]. In this original process grayscale lithography and thermal reflow were combined, i.e. a positive resist was locally exposed with a modulated dose of high energy (100 keV) electrons, which enabled to create a height profile with distinct steps after development in a suitable solvent. The stepped structures could be transferred into a linear slope by selective reflow by heating to a moderate temperature. This was possible because the exposure led to the simultaneous local modification of both etch rate and glass transition temperature T_g over the entire film thickness. This led to distinct heights of remaining resist during development and selectivity during reflow, both between exposed structures which could be modified and

* Correspondence: helmut.schift@psi.ch
[1]Paul Scherrer Institut, Laboratory for Micro- and Nanotechnology, 5232 Villigen, PSI, Switzerland
Full list of author information is available at the end of the article

unexposed structures which remained unaffected. Apart from this pronounced selectivity, the basic principles of the selective reflow are:

- the high thermo-mechanical selectivity between exposed and unexposed resist
- a local reflow and redistribution of polymer of only a few steps
- the definition of the final shape due to volume conservation and pinning on the interface to the unexposed structures and to the cleared substrate.

As can be seen in the compilation of exemplary 3D structures illustrated in Figure 1, the variety of achievable structures comprises the creation of stepped, sloped, concave and convex contours, with a range of slope angles and curvatures, as well as sizes, orientations and density variations in the very same polymer surface. In all cases TASTE was applied to a 1 μm thick poly(methyl methacrylate) resist (PMMA) on a silicon substrate resulting into diverse 3D topographies exhibiting exceptionally smooth contours, enabled by the selective transformation of stepped surfaces into new continuous topographies during a single thermal annealing step. Up to now, slope angles of up to 45° in 2 μm thick resists with optically smooth surfaces with a few nm roughness were achieved [9-11].

1.1 Background: Fundamental working principle

As fundamental aspect of TASTE, we have previously identified the local variations of viscosity in the electron-beam exposed PMMA resist during the heating step, which originates from the dose dependent reduction in molecular weight M_w (here weight-averaged molecular weight) and glass transition temperature T_g of the polymer. A typical asymmetric 5-level grayscale PMMA structure is depicted in Figure 2 to which locally five different doses (D1 - D5) were applied according to a specific lateral design, prior to development in pure methyl-isobutyl-ketone (MIBK). Here, D0 represents an unexposed area, where the initial resist thickness is preserved, while D5 corresponds to the dose required to completely clear the resist upon development, i.e. dose-to-clear. Since for a stepped structure the highest step (D1) will have M_w nearest to the original M_w (D0 = 0), this results also in a reduced ability to flow. This can be reduced by using a high enough reflow temperature near to the T_g of the unexposed ridge, leading to a softening and thus slight inclination. A simplification can therefore be achieved if an already produced slope structure with a homogeneous M_w is generated, e.g. by using an imprinted thermoplastic resist structure such as PMMA which is flood exposed with a dose D > D1 over all steps. Using this, all steps can be exposed to generate low M_w, which enables a complete reflow and an equal smoothening, particularly also the uppermost step of the 5-level pattern with the lowest reduction in M_w. Furthermore it enables to do the reflow at low temperature, at which the non-exposed ridge with a high M_w is less affected by the reflow step. Aiming for a versatile and robust process, our investigations were focused on defining thresholds which enhance the selectivity in reflow behavior for a range of surface topographies, i.e. a selective transformation of structures into defined contours while others remain unaltered. For this we have analyzed the specific correlation between M_w and resulting T_g of the exemplary PMMA resist used in this work.

Figure 1 Compilation of exemplary 3D contours in a thin PMMA film enabled by the TASTE method. The SEM micrographs (angled views and cross sections) depict refined PMMA topographies after exposure of pre-patterned PMMA to high energy electrons and subsequent thermal annealing. Achievable contours comprise **(a-c)** binary, stepped, sloped, convex and concave structures, as well as **(d-f)** hybrid structures with specific pattern combinations. (scale bars: 1 μm).

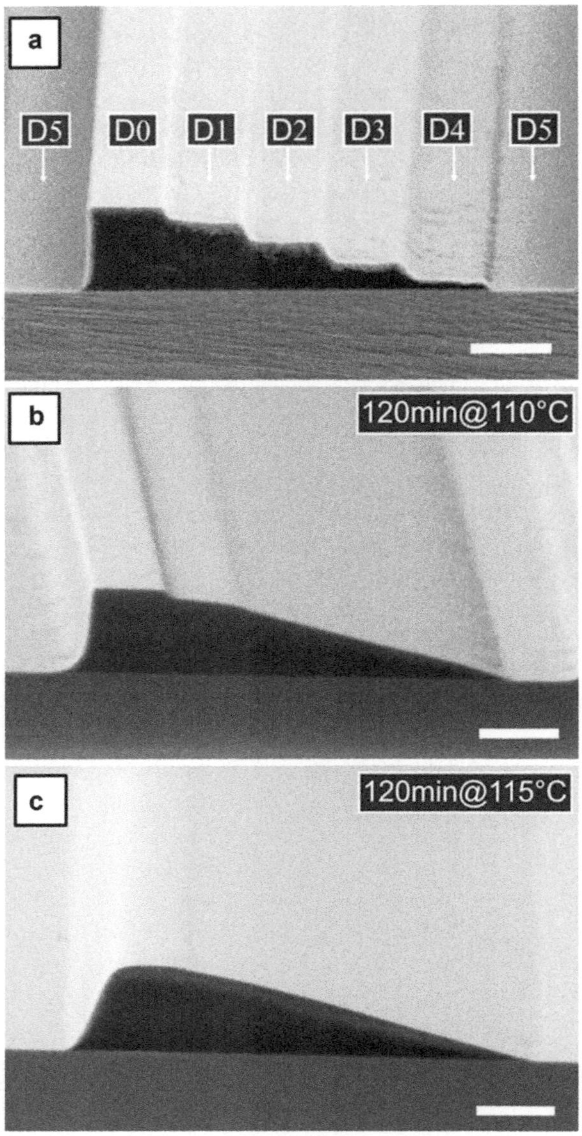

Figure 2 SEM micrographs of the selective thermal reflow of grayscale structures in PMMA: a) The stepped contour with five discrete levels (width: 1 µm; step height: ~ 200 nm) was generated in a 1 µm thick resist by local exposure to high energy (100 keV) electrons with doses D1 to D5 and wet development in pure MIBK. The annealing step was performed for 120 min at 110°C **(b)** and 115°C **(c)**. While for 110°C the step nearest to the unexposed PMMA is still visible, for 120°C it is smoothened out. (scale bars: 1 µm).

Before a detailed technical discussion is given further below, the basic process sequence of the generalized TASTE method is illustrated in Figure 3 for the generation of asymmetric 3D grating elements with linear sloped and stepped contours. Here, a thermoplastic polymer material is deposited on a solid substrate (*step 1*: material selection) and patterned with a specific stepped topography, e.g. using standard lithographic methods such as grayscale electron-beam lithography [12-15] (*step 2*: topography origination). Next, the polymer material is precisely modified over selected steps, e.g. upon irradiation to high energy electrons using defined doses (*step 3*: local modification), if not already done by grayscale lithography. This way, the thermo-mechanical properties of the polymer are locally adjusted, allowing an optimized transformation of the exposed stepped contours into smooth slopes, if the polymer is subsequently heated to an appropriate temperature (*step 4*: selective equilibration). Although the annealing step is performed to the entire film-coated substrate, only the locally exposed polymer structures are softened which allows surface tension driven topography equilibration to selectively convert steps into smooth slopes. Since the original shape of unexposed areas stays unaltered upon thermal treatment, the generation of entirely novel topographies is enabled which includes vertical, stepped and slopes contours in close vicinity to each other (see Figure 1f). This makes TASTE a unique method since it is different to traditional thermal reflow at higher temperatures which aims for the collapse into spherical or cylindrical shapes or decay of all structures [16-18].

The combination of material and lithography used in this contribution, i.e. PMMA and electron beam lithography (EBL), is only one specific example of a TASTE method which allows a precise control of resolution, dose and alignment in a well-defined material. The schematic in Figure 3 also specifies possible alternative materials and sub-processes indicating the high degree of freedom for alternative technical implementations of TASTE. This contribution has two aims: To review the achievements and possibilities of TASTE and to present an analysis of the relation exposure dose to M_w and T_g, which is the cornerstone of TASTE. By this we will demonstrate the width of process window possible by exposure with standard EBL. At the end we will give an insight about how shapes of TASTE structures could be predicted by simulation using a soapfilm-based model.

2 Methods

The PMMA for spincoating is an experimental sample provided by micro resist technology GmbH (Germany), with a specified weight averaged molecular weight M_w of 120 k (kg mol^{-1}). As it will be shown below, the actual M_w is much lower than 120 k, and for comparison, it is much more convenient to use M_n (number-averaged molecular weight) instead of M_w. In comparison to high M_w PMMA generally used for EBL, this M_w was chosen to be low enough to enable a prepatterning of resist using nanoimprint lithography (NIL) (see Figure 1e) [19,20]. This is possible since the typical temperature of imprint (160-180°C) and the T_g of the unexposed resist (122°C) are different enough to enable a selective reflow of exposed structures without affecting imprinted structures [21].

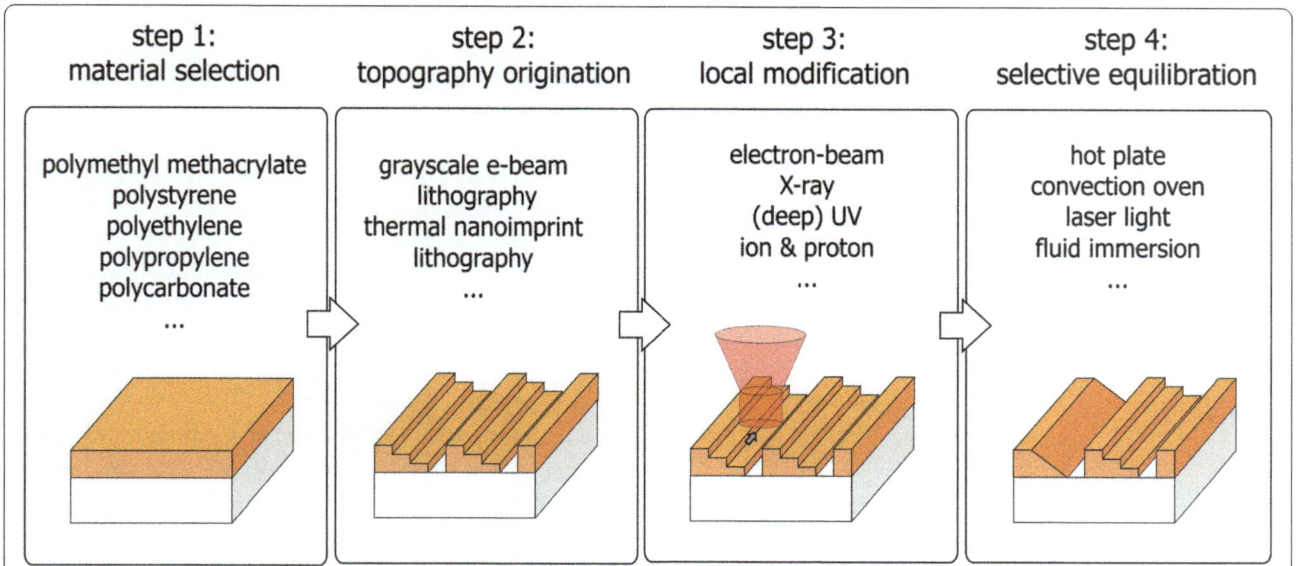

Figure 3 Schematic flow sequence of the general TASTE method for the generation of asymmetric 3D blazed grating pattern with linear sloped and stepped contours: pre-treatment of selected polymer films, generation of a specific (stepped) topography, local modification of the polymer's thermo-mechanical properties and thermal treatment for enabling a selective topography equilibration.

Furthermore the M_w was found high enough to enable a selective reflow of exposed structures with M_w below the T_g-critical threshold. The raw (i.e. unexposed) polymer material was predominantly syndiotactic PMMA (7% isotactic, 54% syndiotactiv, 39% atactic; determined by ^{1}H-NMR spectroscopy). It was dissolved in anisole enabling the spin-coating deposition of 1 µm thick layers on silicon substrate. For the PMMA resist being exposed to different doses, the distinct reduction of the molecular weight distributions was determined using gel permeation chromatography (GPC), and the T_g was analyzed using differential scanning calorimetry (DSC), by removing flakes of exposed PMMA films from the substrate. EBL exposure was performed in a Vistec EBPG 5000 Plus ES. The corresponding contrast curve can be found elsewhere [11]. After exposure, a stepped contour with five discrete levels (step width: 1 µm) was generated in a 1 µm PMMA resist after wet development in pure methyl-isobutyl-ketone (MIBK) for 30 s at 20°C. Because the aim was to create a linear slope between the top of the ridge and the substrate as two pinning points, in between a stepped slope with four intermediate steps needed to be placed. This was possible by choosing the step heights different from the average step height of ~ 200 nm, with a larger step in the center of the slope. This was done according to volume conservation over the entire slope as well as within the vicinity of each step. The annealing step was performed for 120 min at different temperatures near the T_g of the unexposed resist, on a hot plate in ambient condition. Analysis was done via scanning electron microscopy (SEM).

3 Results and discussion

3.1 Polymer analysis

In order to assess the optimum reflow conditions such as the temperature, the GPC measured M_n data were correlated to specific T_g by using DSC. The results are graphically summarized in Figure 4a and quantified by the averaged masses derived from the GPC data presented in Table 1: the exposure induced decay of M_n implicates a significant reduction of the T_g, when the dose is 225 µC cm^{-2} or higher. This results in a reduction of the T_g of unexposed PMMA (initially at $T_{g,\infty}$ = 122°C) of more than 20 K to temperatures in the range of 90 – 100°C. The grayscale steps can therefore be selectively transformed into continuous slopes, while the non-exposed ridge is less affected by the reflow step. It is particularly interesting that for PMMA exposed to doses 225 µC cm^{-2} or higher, the M_n of initially 40 kg mol^{-1} is reduced to 7 kg mol^{-1} and lower. This leads to altered thermo-mechanical properties of the exposed PMMA since the material cohesion and thus viscosity is changed in particular, when the polymer chain length is below a critical [22] M_n of about 10 kg mol^{-1}. Below this threshold, the chain entanglement is substantially reduced and thus thermally induced chain motions are less retarded, i.e. topological restriction by neighboring chains becomes less important. When the polymer is heated, the induced thermal energy is consumed by activating cooperative translation movements of the polymer molecules. For a decreased M_n, a polymer chain motion is enhanced more easily due to the increased free volume in the polymer resulting from the increased presence of chain ends. This results in a

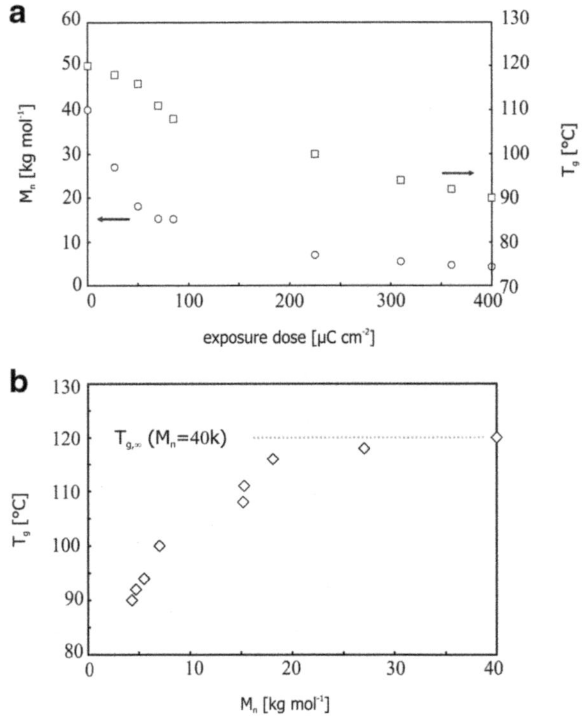

Figure 4 Dependencies of the exposure dose, the (number averaged) molecular weight M_n and the glass transition temperature T_g for electron-beam exposed PMMA films which is used in this contribution by selective topography equilibration (i.e. TASTE). When the material properties of PMMA resist are modified by exposure to high energy (100 keV) electrons **(a)**, the resulting correlation of T_g over M_n shows a smooth transition **(b)**. By creating a wide range of T_g, the selective shape transformation of the exposed (and thus T_g lowered) resist is accomplished, when the reflow temperature is adequately chosen, e.g. at least 10 K higher than the highest T_g present in the PMMA contour.

Table 1 Summary of the GPC and DSC measurements quantifying the dose dependencies of the thermo-mechanical properties of PMMA

Exp. dose [μC cm^{-2}]	M_n [kg mol^{-1}]	M_w [kg mol^{-1}]	PD [-]	T_g [°C]
unexposed	40.0	91.2	2.0	122
27	27.0	58.3	2.2	118
50	18.1	45.2	2.5	116
70	15.3	35.0	2.3	111
85	15.2	34.0	2.2	108
225	7.0	14.6	2.1	100
310	5.5	11.5	2.1	94
360	4.7	10.0	2.1	92
400	4.3	8.9	2.1	90

Upon exposure to high energy electrons, the number averaged molecular weight M_n, weight averaged molecular weight M_w, polydispersity PD (M_w/M_n) and glass transition temperature T_g. is modified according to the applied doses.

softening of exposed polymer material already at reduced temperatures which in turn corresponds to a decreased T_g [23]. The variation of a few degrees in T_g results in viscosity changes of orders of magnitude [6] for a given temperature. In the ideal case, flow only occurs between neighboring exposed steps (i.e. in "near field"), which can be accomplished when using proper reflow conditions and thus maintain a sufficient material viscosity. This prevents the formation of cylindrical or spherical pattern as are known from classical reflow processes [16-18]. Consequently, PMMA features exposed to high energy electrons can be selectively reflowed at eligible temperatures while unexposed pattern stay unaltered.

The topography transformation of the exposed areas is enabled by surface tension forces which are oriented parallel to the interface (polymer/ambience) [24] and thus are imbalanced for stepped contours with kinks (protrusions) and edges (depressions). When the PMMA resist is locally softened, the unbalanced surface tension forces initiate localized material movements until the polymer surface is in a favorable energetic state (e.g. which in high temperature reflow is based on pinning and surface minimization where the surfaces are in equilibrium). For stepped 3D resist profiles with a pinning at the unexposed ridge and the cleared substrate, this state is reached for a continuous linear slope. The resulting angle of inclination is predetermined by the pinning and initial geometry. Since the reflow process is based on kinetic effects, which takes place with different progression speeds correlated to the given viscosity variation, more than a mere smoothening into linear slopes is enabled. Convex and concave pattern of distinct shapes (e.g. non-cylindrical and -spherical, which is advantageous over standard reflow) are also possible (see Figure 1c) because the material displacement during smoothening is locally confined. This allows predefining non-linear trajectories based on contour approximation and volume conservation.

3.2 Stepped and sloped sidewalls in close vicinity

Correlating the GPC and DSC data allows drawing T_g directly as a function of M_n, which reveals a rather smooth and gradual decline of T_g (Figure 4b). Thus, an abrupt change in thermal properties for pattern with critical M_n of lower than 10 kg mol^{-1}, as anticipated previously [22], is not confirmed. This leads to an empirically found general definition of TASTE compatible process windows for the PMMA used in this work, which can be described as follows: an optimal reflow temperature is given at least 10 K below the polymers original T_g, and 10 K above the highest T_g present in the polymer pattern, respectively. For a chosen temperature of 110°C, these prerequisites are only achieved for generating PMMA patterns with a maximum M_n of 5 kg mol^{-1}, which according to Figure 4a corresponds to an exposure dose of 400 μC cm^{-2} or higher.

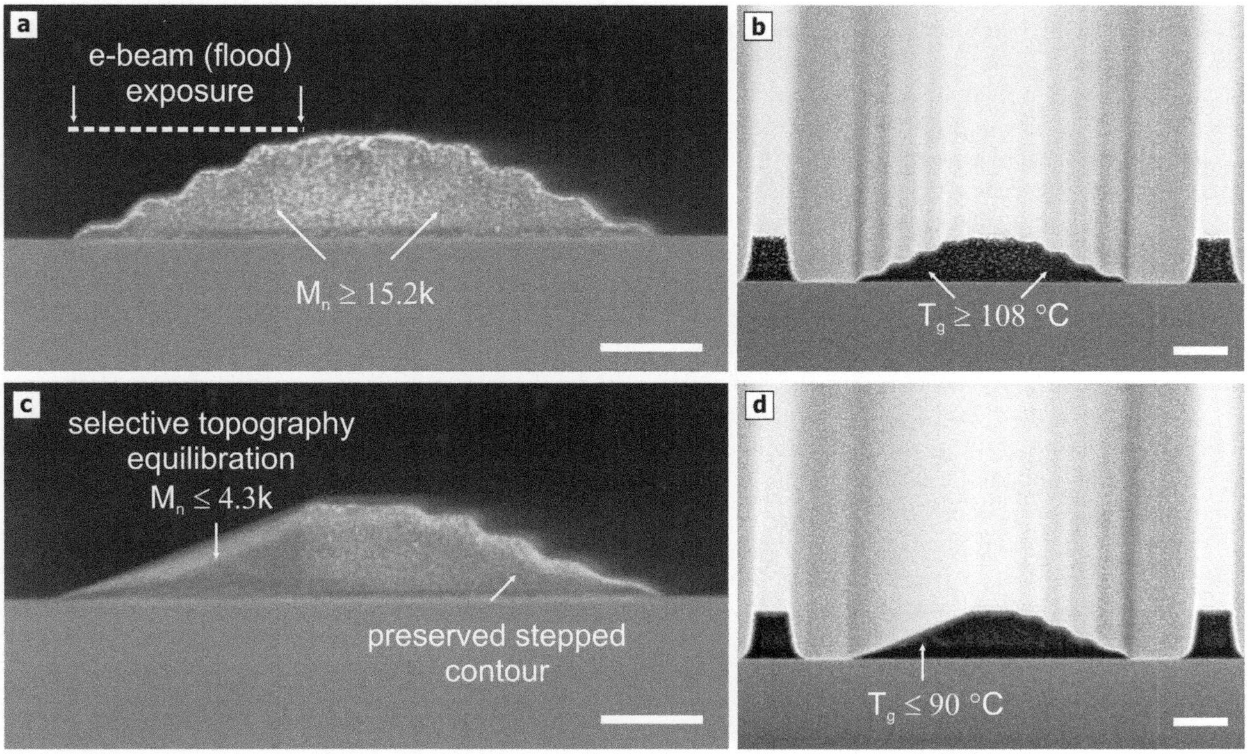

Figure 5 A partially (flood) exposing a multilevel PMMA resist contour within the TASTE process enhances the available pattern variety of 3D structures. The generation of sloped and stepped contours in very close vicinity to each other is enabled by locally exposing a stepped PMMA contours **(a-b)** with high dosage (i.e. 500 μC cm^{-2}) and thus reducing the T_g below 92°C. Applying a reflow temperature of 110°C, exposed steps are then selectively transformed into a continuous slope while other topographies are unaltered **(c-d)**. Partial flood exposure also optimizes the overall slope accuracy since the thermal properties of the exposed areas are homogenized (scale bars: 1 μm).

This way, stepped and sloped topographies can be combined in close vicinity, as demonstrated in Figure 5. Here, the stepped polymer patterns (either fabricated by grayscale EBL or 3D NIL) prior to thermal treatment, i.e. the stepped patterns, are exposed by a well aligned local flood exposure, enables the appropriate reduction of M_n into a favorable regime for reflow with sufficiently reduced T_g. In contrast to previous works by the authors, where grayscale EBL was used for topography origination which led to a rather unspecific modification of the material properties [25], this defines effective M_n values in the resist not correlated to distinct level heights. This way, typical reflow defects at upper most level height reported before are overcome and the overall slope characteristics are optimized (see also Figure 2).

The broad T_g distribution depicted in Figure 4a and 4b achieved by the electron-beam exposure also allows to further advance the available pattern variety, when two identical stepped resist contours are generated in PMMA resist, as shown in Figure 5, using a grayscale EBL with doses between 25 and 85 μC cm^{-2} and a respective development time of 15 min. Here, the comparatively low doses result in a moderate reduction of $M_n >> 10$ kg mol^{-1} for the stepped pattern and thus into marginal decrease of T_g not less than 108°C. A subsequent and locally confined flood-exposure of 500 μC cm^{-2} further reduces (and eventually homogenizes) the M_n range within the exposed steps below 5 kg mol^{-1}, which results in a reduced $T_g < 90$°C. If a thermal treatment at a fixed temperature of 110°C for 15 min is now applied, only the exposed steps are selectively transformed into a linear slope. The non-flood exposed contours, which have still relatively high T_g, stay unaltered upon reflow since the viscosity remains high. Future advancement can be achieved when an overlap of this flood exposure with the central non-exposed ridge ensures an optimal slope pinning to the upper level without any kink and consequently enhances the overall characteristics of 3D topographies.

3.3 Simulation of reflow using soap film-based model

For many applications not only the final shape, but also the pattern evolution from the initial into the reflow shape (Figure 2) depending on material and process parameters would be of interest. A simulation based on the time-temperature dependent, dynamic evolvement of the polymer surface during the reflow process was performed, based on open source software [26]. As an example, the case of a six level (five exposure-level) structure was simulated (see Figure 6), similar to that in the experiments presented in Figure 5, with an unexposed ridge on the right side and the cleared substrate on the left side. The major influencing parameters during reflow are i) the initial feature geometry after electron beam exposure, ii) the material properties of the reflown thermoplast, and iii) the reflow parameters itself. The influence of all these parameters on the shape evolution is implemented in a soapfilm-based model [26] via a dynamic contact angle between the polymer and the surface wetted by the polymer [27]. The discrete-time model allows determining precisely a point at which the relaxation should be stopped by reducing the sample temperature well below the glass transition to freeze a certain curvature. Using this, in the simulation presented in Figure 6, it is possible that the pinning at the lower end of the slope can be nearly preserved, while the polymer is able to rise up to the top of the unexposed ridge. Apart from pinned slopes, more complex structures can be predicted, including concave and convex shapes. Details can be found in [28].

4 Conclusions

The enormous degrees of freedom offered by the TASTE method brings us to conclude that the process sequence of stepped topography origination, flood exposure and reflow in a thermoplastic polymer layer needs to be specified as a new lithographic method for 3D patterning. Compared to previous works, the increased reflow selectivity and improved pattern accuracy with less geometrical defects, by introducing a flood exposure to pre-patterned polymer contours, enabled contour transformation to be performed in a pre-determined way almost entirely given by geometry. As already indicated in Figure 3, process variants could employ other polymers (e.g. polystyrene) as *step 1*, which are also known for its nanofabrication capabilities by lithographical means [29,30], or resists suitable for NIL [21]. For topography origination of stepped structures as *step 2*, instead of patterning by grayscale EBL, replication by 3D-NIL could be employed. This allows the replication of diverse preliminary (e.g. stepped, corrugated, defective) structures originating from a range of processes which can be subsequently transformed into structures with desired shape and functionality by localized flood (over entire slopes) or selected area (up to global) exposure. In future, it might also be interesting to explore the possibilities of alternative exposure techniques as *step 3* by exposure to X-rays, ions, proton or (deep) UV-light, also well known to modify the thermomechanical properties of resists [31]. As *step 4*, global heating schemes may involve a hotplate and convection oven, but can also include local treatment by laser light irradiation or immersion in heated fluids. Typically, once such complex features are fabricated, it can be used as pattern origination in mass-replication techniques like NIL or other molding processes, where 3D surface reliefs

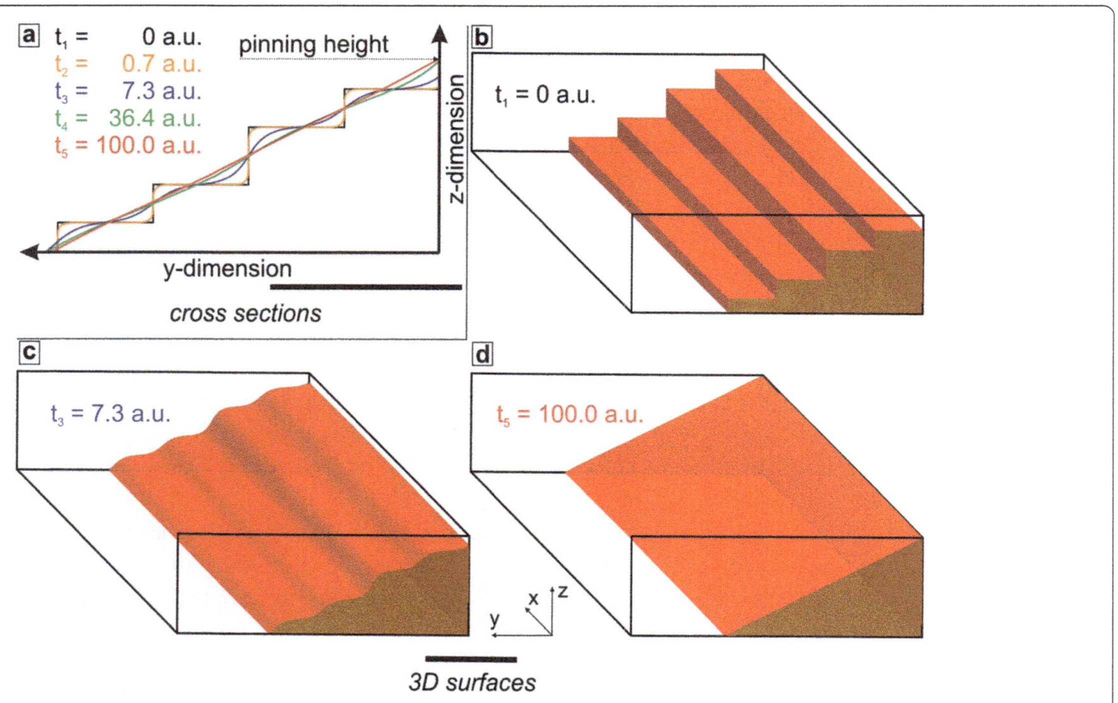

Figure 6 Simulation results for the EBL flood exposure area in Figure 5, using the true experimental geometry. The simulation visualizes the time dependent multi-step evolution from the initial shape (t_1) to the shape at the end of the reflow (t_5) by **a)** cross sections and **b-d)** 3D surface representations. The highest step is pinned during reflow at the top of the unexposed PMMA part (scale bars: 1 μm).

are repeatedly generated by spatial material displacement [19,20]. The potential of the TASTE method towards entirely novel 3D profile variants is high and will direct towards new applications where those varieties are needed.

Abbreviations
TASTE: Thermally activated selective topography equilibration; 3D: Three-dimensional; PMMA: Poly(methyl methacrylate); M_w: Weight-averaged molecular weight; M_n: Number-averaged molecular weight; T_g: glass transition temperature; MIBK: Methyl-isobutyl-ketone; EBL: Electron-beam lithography; NIL: Nanoimprint lithography; SEM: Scanning electron microscopy; GPC: Gel permeation chromatography; DSC: Differential scanning calorimetry.

Competing interests
The authors declare that they have no competing interests.

Authors' contributions
AS is the main author of this paper, who conceived and performed all reflow experiments. VAG has conceived the 3D EBL writing strategies employed in this paper. MM performed the analytics and interpretation of M_w data, HA has provided the polymer samples and contributed in the analytics of the polymer by GPC and DSC. RK has setup and performed simulations with the soapfilm based method based on the experiments. HS conceived the research project and participated in the interpretation of data; he has also written part of this paper and is responsible for the revision of the paper. All authors read and approve the final manuscript.

Authors' information
AS has been previously postdoc in the INKA group at the Paul Scherrer Institut (PSI), where all EBL and NIL experiments were performed. Since 2012 he is employed by micro resist technology GmbH.

Acknowledgements
We thank K. Vogelsang, C. David and J. Gobrecht (PSI Villigen), C. Spreu from Sensirion AG (formerly PSI), as well as J. Voskuhl, A. Kolander, M. Vogler and G. Grützner (*micro resist technology* GmbH) for their help and valuable contributions. We also thank P. Shahgaldian (Fachhochschule Nordwestschweiz) for performing the supplementing NMR measurements. The research presented here was partially funded by Swiss Federal Office for Science and Education in the framework of the EC-funded project NaPANIL (contract no. NMP 214249). The content of this work is the sole responsibility of the authors.

Author details
[1]Paul Scherrer Institut, Laboratory for Micro- and Nanotechnology, 5232 Villigen, PSI, Switzerland. [2]Micro Resist Technology GmbH, Koepenicker Str. 325, 12555 Berlin, Germany.

References
1. RH Morf, H Kiess, C Heine in: J Turunen, F Wyrowski, *Diffractive Optics* (Akademie Verlag, Berlin, 1997).
2. V Terrazzoni-Daudrix, J Guillet, X Niquille, A Shah, R Morf, A Tishchenko, V Brioude, O Parriaux, D Fischer, Mat. Res. Soc. Symp. Proc. **769**, H6.14.1–6 (2003)
3. I Fernandez-Cuesta, AL Palmarelli, X Liang, J Zhang, S Dhuey, D Olynick, S Cabrini, J. Vac. Sci. Technol. B **29**, 06F80 (2011)
4. SM Stavis, EA Strychalski, M Gaitan, Nanotechnology **20**, 165302 (2009)
5. YY Zhang, J Zhang, G Luo, X Zhou, GY Xie, T Zhu, ZF Liu, Nanotechnology **16**, 422–428 (2005)
6. E Di Fabrizio, F Romanato, M Gentili, S Cabrini, B Kaulich, J Susini, R. Barrett. Nature **401**, 895–898 (1999)
7. N Bogdanski, H Schulz, M Wissen, H-C Scheer, J Zajadacz, K Zimmer, Microelectron. Eng. **73–74**, 190–195 (2004)
8. H Seidel, L Csepregi, A Heuberger, H Baumgartel, J. Electrochem. Soc. **137**, 3612–26 (1990)
9. A Schleunitz, H Schift, J. Micromech. Microeng. **20**, 095002 (2010)
10. A Schleunitz, H Schift, Microelectron. Eng. **88**, 2736–2739 (2011)
11. A Schleunitz, VA Guzenko, A Schander, M Vogler, H Schift, J. Vac. Sci. Technol. B **29**, 06F302 (2011)
12. G Piaszenski, U Barth, A Rudzinski, A Rampe, A Fuchs, M Bender, U Plachetka, Microelectron. Eng. **84**, 945–948 (2007)
13. SY Lee, K Anbumony, J. Vac. Sci. Technol. B **25**, 2008–2012 (2007)
14. WC Cheong, WM Lee, XC Yuan, LS Zhang, K Dholakia, H Wang, Appl. Phys. Lett. **85**, 5784–5786 (2004)
15. DM Koller, A Hohenau, H Ditlbacher, N Galler, AL Baudrion, F Reil, S Schausberger, FR Aussenegg, A Leitner, JR Krenn, Microelectron. Eng. **85**, 1639–1641 (2008)
16. SY Chou, Q Xia, Nature Nanotechnology **3**, 295–300 (2008)
17. P Ruther, B Gerlach, J Göttert, M Ilie, J Mohr, A Müller, C Oßmann, Pure Appl. Opt. **6**, 643–53 (1997)
18. H Schift, C Spreu, A Schleunitz, JJ Lee, Microelectron. Eng. **88**, 87–92 (2011)
19. SY Chou, PR Krauss, PJ Renstrom, J. Vac. Sci. Technol. B **14**, 4129–4133 (1996)
20. H Schift, J. Vac. Sci. Technol. B **26**, 458–480 (2008)
21. A Schleunitz, C Spreu, M Vogler, H Atasoy, H Schift, J. Vac. Sci. Technol. B **29**, 06FC01 (2011)
22. K Ute, N Miyatake, K Hatada, Polymer **36**, 1415 (1995)
23. JMG Cowie, *Polymers: Chemistry and Physics of Modern Materials* (BlackieAcademic, Glasgow, 1991)
24. A Marchand, JH Weijs, JH Snoeijer, B Andreotti, Am. J. Phys. **79**, 999 (2011)
25. EA Dobisz, SL Brandow, R Bass, J Mitterender, J. Vac. Sci. Technol. B **18**, 107–111 (2000)
26. KA Brakke, Experimental Mathematics **1**, 141–165 (1992)
27. S Newman, J. Colloid Interface Sci. **26**, 209–213 (1968)
28. R Kirchner, A Schleunitz, H Schift, *Energy-based thermal reflow simulation for 3D polymer shape prediction using the Surface Evolver, submitted to* (J. Micromech. Microeng., 2014)
29. S Ma, V Con, M Yavuz, B Cui, Nanoscale Research Letters **6**, 446 (2011)
30. H-C Scheer, S Möllenbeck, A Mayer, K Dhima, J. Vac. Sci. Technol **28**, C6M1 (2010)
31. JO Choi, JA Moore, JC Corelli, JP Silverman, H Bakhru, J. Vac. Sci. Technol. B **6**, 2286–2289 (1988)

Transparent TiO_2 nanotube array photoelectrodes prepared via two-step anodization

Jin Young Kim[1,2*], Kai Zhu[3], Nathan R Neale[3] and Arthur J Frank[3*]

Abstract

Two-step anodization of transparent TiO_2 nanotube arrays has been demonstrated with aid of a Nb-doped TiO_2 buffer layer deposited between the Ti layer and TCO substrate. Enhanced physical adhesion and electrochemical stability provided by the buffer layer has been found to be important for successful implementation of the two-step anodization process. With the proposed approach, the morphology and thickness of NT arrays could be controlled very precisely, which in turn, influenced their optical and photoelectrochemical properties.

1. Correspondence/Findings

Aligned TiO_2 nanotube (NT) arrays have attracted considerable interest owing to their versatility in a number of optoelectronic applications, such as solar cells [1-3], photocatalysis [4,5], and electrochromics [6]. For instance, when used as electrodes in dye-sensitized solar cells (DSSCs), they were found to show higher light-harvesting and charge-collection efficiencies compared to their nanoparticle-based counterparts [7]. However, most anodized TiO_2 NT arrays have been prepared using non-transparent Ti-foil substrates, limiting the light illumination geometry to just one side of the electrode. The limited illumination geometry lowers the performance of the optoelectronic devices because the incident light is attenuated by the device components (*e.g.* electrolyte) other than the TiO_2 NT electrode [8]. Therefore, the benefits of TiO_2 NT-based electrodes can only be fully utilized by using transparent TiO_2 NT arrays. The challenges associated with preparing transparent TiO_2 NT arrays have thus far limited the numbers of routes for fabricating these electrodes [9,10]. Recently, we found that the transparent TiO_2 NT array films could be prepared via a simple and reproducible approach involving the deposition of a conducting buffer layer between the Ti layer and transparent conducting oxide (TCO) substrate [11]. The buffer layer protects the TCO from degradation through a self-terminating mechanism that arrests the TiO_2 NT growth on the Ti layer and improves the adhesion and electrical contact between the NTs and TCO.

A second challenge in TiO_2 NT array synthesis has been the ability to precisely control the microstructure dimensions, extent of alignment, pore ordering, and degree of overlayer formation. These properties have a substantial impact on the device properties. For instance, we recently showed that removing the structural disorder of TiO_2 NT arrays promoted more efficient charge transport in DSSCs [12]. Also, ordering of the NTs can affect their optical properties, in particular their ability to scatter light that affects the light-harvesting properties of the electrode. Strategies to control the NT dimensions and the extent of alignment have involved using different anodization potentials [13,14], electrolytes [15,16], drying conditions [12], and bath temperatures [4]. However, only a few studies have focused on the control of pore ordering [17] and overlayer formation [9,18]. On the other hand, it is well known that highly ordered pore structures in alumina can be achieved via a two-step process [19,20]. However, only a paucity of studies have investigated the use of the two-step anodization of TiO_2 NT arrays [14,17,21], mostly owing to the absence of proper chemical etchants that dissolve only oxide NTs without dissolving the underlying substrate. Furthermore, all of these examples used non-transparent Ti foil as the substrate for growing NTs. For the reason given above, it would be highly useful to extend this two-step anodization technique to transparent TiO_2 NT arrays.

The biggest challenge for the successful implementation of the two-step procedure for preparing transparent TiO_2

* Correspondence: kimjy@kist.re.kr; arthur.frank@nrel.gov
[1]Photo-electronic Hybrids Research Center, Korea Institute of Science and Technology (KIST), Seoul 136-791, Korea
[3]National Renewable Energy Laboratory, Golden, Colorado 80401-3393, USA
Full list of author information is available at the end of the article

NT arrays is overcoming the poor adhesion between the Ti layer and TCO substrate. During a typical two-step anodization, TiO_2 NT arrays are grown to a certain thickness and then mechanically removed either by gas evolution [14] or by adhesive tapes [17]. We find that these approaches are not applicable to TiO_2 NT arrays grown on TCO substrates. Removal of the initially grown NT layer with these approaches results in delamination of the remaining Ti film from the TCO substrate. This situation is largely attributed to the poor adhesion between the Ti film and the TCO substrate, unlike the case of NTs grown on Ti foil.

In this communication, we demonstrate the first two-step anodization procedure for growing transparent TiO_2 NT array films on TCO substrates. Key to the synthesis is the aid of a buffer layer (*i.e.* Nb-doped TiO_2; NTO), which is found to improve both the physical adhesion between the Ti layer and TCO and the electrochemical stability of the NT films during the anodization process. We also report on simple and versatile approaches to control the morphology of the NT arrays by using a post-growth pore widening procedure, which involves changing the anodization time.

A schematic procedure for the two-step anodization of transparent TiO_2 NT array films is illustrated in Figure 1a. The Ti thin film, which is deposited on the TCO substrate, is anodized to produce a sacrificial NT layer on top of the remaining Ti layer. Removal of this sacrificial layer leaves dimples on the Ti surface. These dimples serve as a template that directs pore formation during the second anodization step in which the remaining Ti is fully anodized. As described in the introduction, the strong adhesion of the Ti layer is very important for the selective removal of the sacrificial NT layer. In contrast to NT films grown directly on TCO, the presence of the NTO layer allows the first NT layer to be removed either by gas evolution or by the adhesive tape without delaminating the remaining Ti film. This indicates that the NTO layer improves the adhesion between the Ti film and TCO substrate. Figure 1b,c show typical anodization current profiles of the first and second stages of the two-step anodization process, respectively. It is interesting that the decreased current in the early stages

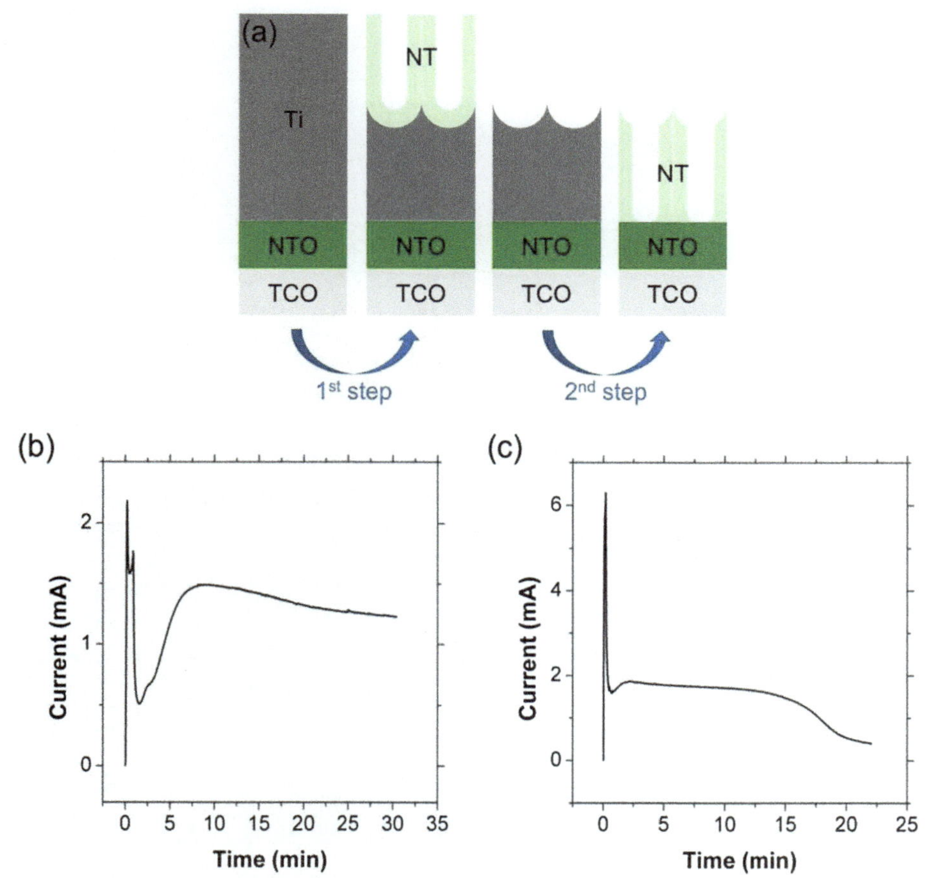

Figure 1 (a) Schematic illustration of the two-step anodization process, and the anodization current profiles of (b) first step and (c) second step anodization.

of the anodization process recovers to steady state much faster in the second anodization. The current drop and subsequent recovery are reported to be related to the respective surface roughness of the Ti metal and higher degree of ordering in the final NT film during the second anodization [17]. The electrochemical stability of the NTO layer, which was reported in our previous study [11], is also observed in the anodization current profile of the second step (Figure 1c), where the anodization current does not increase even after complete anodization.

The SEM images in Figure 2a display the surface morphology of a transparent TiO_2 NT film prepared by the single-step anodization process. It can be seen that the film surface is covered by a porous overlayer and that the individual NTs are not very discernible. Formation of the porous overlayer [18] has been ascribed to the very low chemical etching rate of the oxide in the ethylene glycol-based electrolyte [9]. We find that the overlayer can be easily removed by ion-beam etching. The inset in Figure 2a shows the surface morphology of the NT film after ion-beam etching (1.2 keV, 26 mA, and 30 min). It is clear from the SEM image that the film is composed of densely packed NT arrays with little interstitial space between adjacent tubes. On the other hand, the surface morphology after the two-step anodization (Figure 2b) looks very similar to that prepared by the two-step anodization of Ti foil [17]. The surface morphology (Figure 2b) appears more ordered than that of the single-step counterpart (Figure 2a). Better ordering of pores after the two-step anodization can be ascribed to the smoother Ti surface after removing the sacrificial NT layer and the templating effect produced by the dimples. The average pore diameter measured from the magnified image (inset in Figure 2b) is 38 (± 7) nm, and the pore-to-pore distance is 117 (± 11) nm. The NT morphology can be further modified by the post-growth pore-widening process. Figure 2c shows that the average pore diameter of the NT film increases from 38 (± 7) nm (inset in Figure 2b) to 66 (± 5) nm after the pore widening, whereas the pore-to-pore distance essentially remains the same (112 ± 13 nm). This result indicates that the pore size can be controlled by varying the pore-widening time without disrupting the ordered arrangement of the NT pores. These observations suggest that the two-step anodization and subsequent pore widening are versatile ways to control the morphology of transparent TiO_2 NT films.

Figure 3a shows a typical current profile of the single step anodization and the strategy to control the thickness of the transparent TiO_2 NT films via the two-step

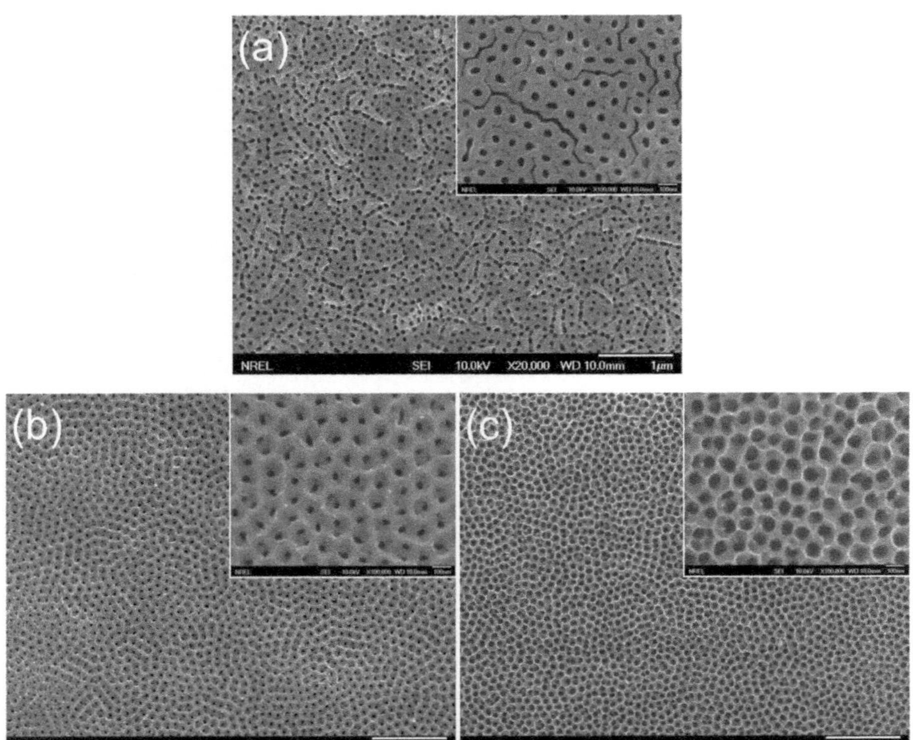

Figure 2 Top-view SEM images of TiO_2 NT arrays prepared by (a) single-step anodization process, (b) two-step anodization process, and (c) two-step anodization process and subsequent pore-widening process, where the inset in (a) shows the top surface morphology after Ar-ion etching and the insets in (b) and (c) show the images at the same magnification.

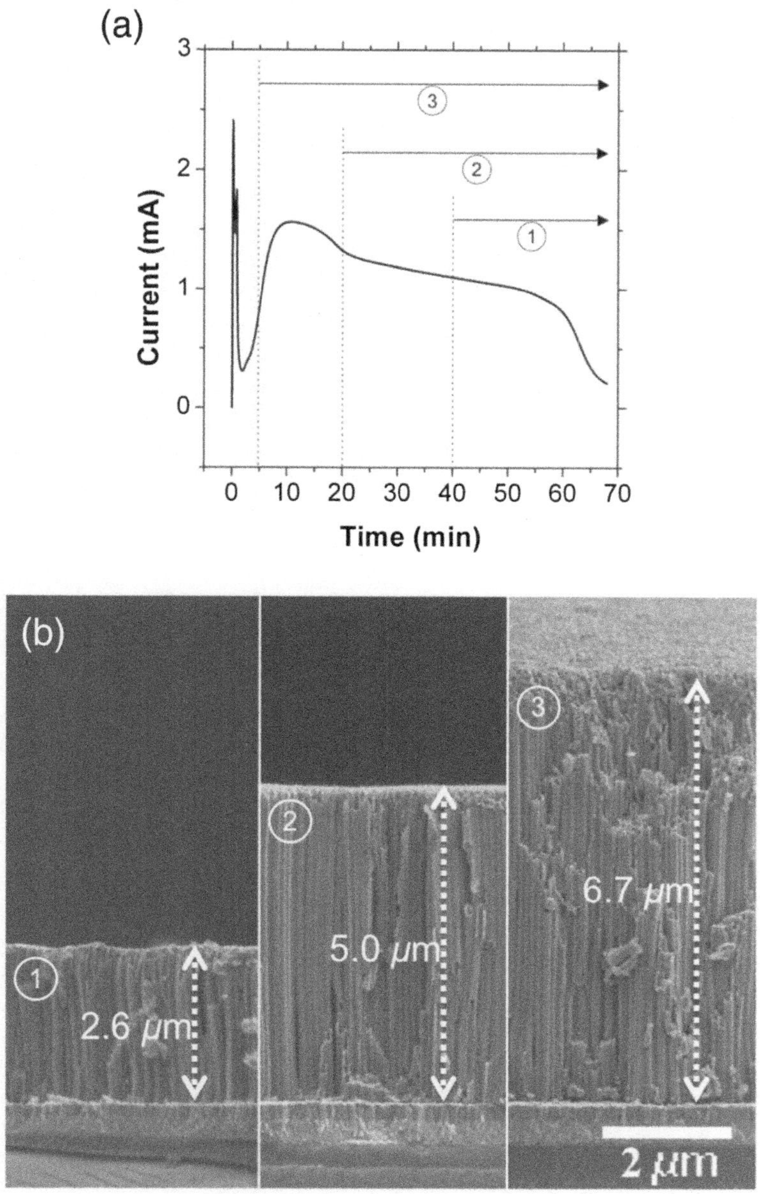

Figure 3 (a) The strategy to control the thickness of the two-step anodized transparent TiO_2 NT films, where the graph shows a typical anodization current profile of the single-step anodization and the circled numbers show the duration of the second step anodization, and (b) SEM images show NT films of various thicknesses; the circled numbers correspond to those in (a).

anodization. To control the NT thickness, we changed the reaction time between the first and second anodization processes. The numbers in the circle indicate when the first step anodization is stopped and the second step anodization is started for three different NT samples. Figure 3b compares the cross-sectional SEM images of the transparent NT samples prepared by the two-step anodization processes described in Figure 3a after the subsequent pore widening process. It can be seen that the thickness of the NT films could be varied from 6.7 μm to 2.6 μm by changing the first anodization time from 5 min to 40 min. This is a very precise approach to control the NT thickness, which is easily adaptable to other anodic oxide systems, such as the porous alumina membrane. The approximate anodization times for the second step are 25 min, 45 min, and 50 min for samples 1, 2, and 3, respectively. It should be noted that the anodization rate of the second anodization process is higher than the first anodization, resulting in a shorter anodization time compared to that shown Figure 3a. The anodization rate for the second step is calculated to be 130 ~ 150 nm/min, assuming that the final 5 min does not contribute to the film growth, which is about twice as fast as the reported anodization rate for Ti foils

[14]. The rate difference can be ascribed to the different properties of the starting Ti. Preliminary experiments show that the deposition temperature, which affects the crystallinity and preferred orientation of the Ti film, influences the anodization rate. It is noteworthy that the NT film thickness after the anodization process is significantly thicker than the starting Ti film (less than 2.6 μm depending on the first anodization time), which is consistent with literature reports [9,22]. Although the actual mechanism for the increased NT thickness is not clearly understood, it has been proposed that lateral motion of the anodized material into the walls of both the TiO_2 NTs and anodized alumina structures is responsible for the NT film being thicker than the starting Ti film [23,24]. Extremely small loss of Ti atoms owing to the re-attachment of the dissolved TiF_6^{2-} and the subsequent oxide formation has also been proposed to account for the increased thickness of the relatively thin NT films [22].

Figure 4a shows the normalized transmittance spectra of the transparent TiO_2 NT films displayed in Figure 3b. The absolute value of the transmittance (not shown) decreased with increasing NT thickness, an observation which is consistent with a previous report [9]. In order to investigate the spectral response of the NT films, the transmittance spectra were normalized with respect to the maximum spectral values. The NT films have an optical band gap (~ 3.4 eV), which is slightly larger than that of bulk anatase TiO_2 (~ 3.2 eV), regardless of the film thickness. Such a slight blue shift is attributed to the small crystallite size, which is limited by the wall thickness of the NTs [25]. Interestingly, the relative transmittance at shorter wavelengths (< 550 nm) decreases with increasing NT thickness, whereas the relative transmittance at longer wavelengths (> 550 nm) increases with increasing NT thickness. In other words, thicker NT films transmit less high-energy photons and more low-energy photons, implying that the light scattering at shorter wavelengths becomes more significant for thicker (at least up to 6.7 μm) NT arrays. The inset in Figure 4a shows the normalized absorption spectra obtained by subtracting the raw transmittance spectra of bare TiO_2 NT arrays from those of dye-adsorbed counterparts. The resulting spectra represent the absorption by just the adsorbed dye. The absorption by dye molecules exhibits a very similar behavior to the transmittance of the stained NT arrays. The amount of light absorption by dye molecules decreases with increasing NT thickness at shorter wavelengths, whereas it increases with increasing NT thickness at longer wavelengths. The higher transmittance of NT array electrodes at wavelengths > 550 nm should lead to better light absorption by dye molecules on the photoelectrodes, as more photons can reach the adsorbed dye molecules on the NT arrays due to the higher transmittance. As can be seen in Figure 4b, the thickness- and wavelength-dependent light absorption of the transparent TiO_2 NT films, in turn, affects the properties of the NT-based DSSCs. The normalized IPCE spectra show that the photocurrent response follows the same behavior as the transmittance and net absorption spectra shown in Figure 4a. For instance, the largest difference between 2.6 μm and 6.7 μm samples appears at 410 nm in both the net absorption and IPCE spectra. The IPCE is normally approximated by the product of the following three factors: light-harvesting, charge-injection, and charge-collection efficiencies. Among these factors, the charge-injection efficiency is not influenced by the thickness of the NT arrays because the NT/dye interface is identical for all samples. Also, because with the NT length is much

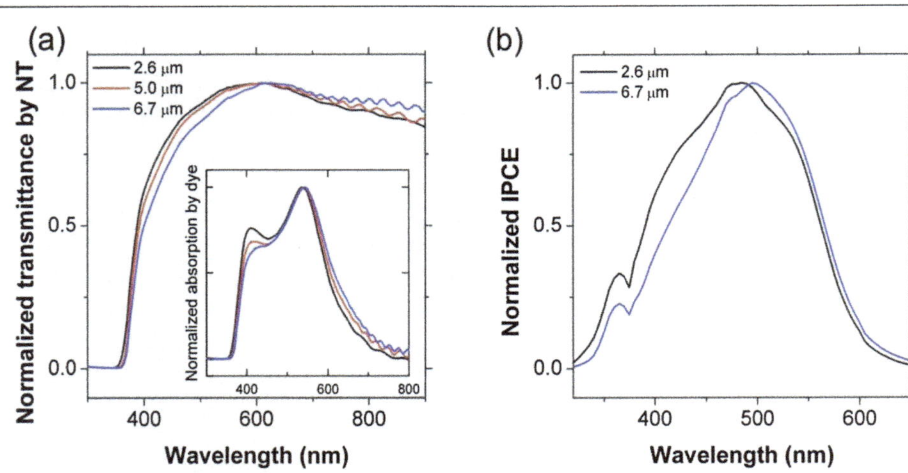

Figure 4 (a) Normalized transmittance of transparent TiO_2 NT arrays (without dye adsorption) with various thicknesses, where the inset shows the normalized absorption spectra only by the dye molecules adsorbed on the NT surfaces, and (b) Normalized IPCE spectra of DSSCs composed of transparent TiO_2 NT arrays with different thicknesses.

shorter than the electron diffusion length (~100 μm) [26] the charge-collection efficiency of TiO_2 NT-based DSSCs should be close to unity. Therefore, the IPCE spectra are influenced mainly by the light-harvesting efficiency that is represented by the amount of light absorption by the dye (*i.e.* the inset in Figure 4a). The advantage of having higher transmittance at wavelengths longer than 700 nm does not contribute substantially to the photoresponse for this particular DSSC system owing to the fact that the Z907 dye sensitizer absorbs very weakly at wavelengths above 700 nm. However, this effect will become more important for improving the performance of sensitized photoelectrodes incorporating active materials that absorb lower energy light, such as semiconductors (*i.e.* quantum dots) and IR-absorbing organic dyes.

In summary, we have demonstrated that the highly ordered transparent TiO_2 NT arrays can be prepared by a two-step anodization with the aid of a Nb-doped TiO_2 (NTO) buffer layer. Strong physical adhesion and electrochemical stability provided by the NTO layer are found to be crucial for the success of this approach. The morphology and thickness of the transparent TiO_2 NT arrays can be controlled precisely by the post-growth pore widening and the anodization time between first and second step anodization. The controlled nanostructure, in turn, affects the optical properties of the transparent TiO_2 NT arrays, which are shown to impact the photoelectrochemical properties of NT-based sensitized solar cells.

2. Experimental section

2.1 NT preparation

Transparent TiO_2 NT arrays were prepared via an electrochemical anodization process [27]. Before depositing Ti metal films, 10 at% Nb-doped TiO_2 (NTO) thin layers were deposited on the F-doped SnO_2 (TEC15, Pilkington) substrates by RF-magnetron sputtering at 400°C using a ceramic target with the same composition under Ar flow (20 sccm) with a working pressure of 5 mTorr. Ti metal films with a thickness of 2.6 μm were deposited on the NTO/TCO substrates using a Ti metal target (99.9%) under the same sputtering conditions. The Ti thin films were anodized in a two-electrode cell at 50 V (room temperature, Pt counter electrode) using 0.25 wt% NH_4F (Aldrich, 99.9%) electrolyte in ethylene glycol (Aldrich, 99%) with 1 wt% of H_2O. The anodization process was controlled by a source meter (Keithley, Model: 2425) connected to a computer. The applied voltage was increased with a ramp rate of 1 V/sec and 5 V/sec for first and second anodization processes, respectively [14]. After washing with water, the first NT layer was removed by applying a cathodic potential (−3 ~ −5 V) in 1 M H_2SO_4 aqueous solution [14] using a two-electrode cell with a platinum counter electrode, and then anodized again after washing with water and ethanol. The as-prepared samples were rinsed with water and ethanol, and dried with a gentle stream of N_2. Then, for the pore widening, the as-anodized NT films were immersed in the formamide solution consisting of 0.15 M NH_4F and 3.5 wt% H_2O at 80°C for 5 min [11]. After rinsing and drying using the same conditions, the NT films were annealed at 400°C for 1 h under an ambient atmosphere.

2.2 DSSC fabrication

For fabricating DSSCs, the films were stained with a 0.3 mM ethanolic solution of Z907 dye (cis-bis(isothiocyanato)(2,2′-bipyridyl-4,4′-dicarboxylato)(4,4′-dinonyl-2′-bipyridyl)ruthenium(II)), and sandwich type devices were fabricated with Pt-loaded TCO substrates using the similar procedure described previously [11]. The electrolyte was composed of 1.0 M 1-methyl-3-propylimidazolium iodide, 30 mM I_2, 0.5 M 1-butyl-1H-benzimidazole, and 0.1 M guanidinium thiocyanate in 3-methoxypropionitrile [28].

2.3 Characterization

The microstructure and thickness of the NT arrays were characterized by field-emission scanning electron microscopy (FE-SEM). The optical absorption/transmission spectra and external quantum efficiencies were obtained using an IPCE measurement system (PV Measurements, Inc.).

Competing interests

The authors declare that they have no competing interest.

Authors' contributions

JK and AJF designed and carried out experiments and wrote the manuscript. KZ and NRN discussed the experimental results and wrote the manuscript. All authors read and approved the final manuscript.

Acknowledgement

This work was supported by the KIST internal fund, the New & Renewable Energy Technology Development Program of the Korea Institute of Energy Technology Evaluation and Planning (KETEP) grant funded by the Ministry Of Trade, Industry & Energy (No. 20113020010040), the Nano·Material Technology Development Program through the National Research Foundation of Korea (NRF) funded by the Ministry of Science, ICT & Future Planning (2012M3A7B4049989), and the "National Agenda Project" program of Korea Research Council of Fundamental Science & Technology (KRCF). The work done at Green School was supported by the National Research Foundation of Korea Grant funded by the Korean Government (MSIP) (2013, University-Institute cooperation program). KZ, NRN, and AJF were supported by the Solar Photochemistry Program, Division of Chemical Sciences, Geosciences, and Biosciences, Office of Basic Energy Sciences, U.S. Department of Energy, under contract No. DEAC36-08GO28308.

Author details

[1]Photo-electronic Hybrids Research Center, Korea Institute of Science and Technology (KIST), Seoul 136-791, Korea. [2]Green Green School, Korea University, Seoul 136-701, Korea. [3]National Renewable Energy Laboratory, Golden, Colorado 80401-3393, USA.

References

1. JM Macak, H Tsuchiya, A Ghicov, P Schmuki, Electrochem. Commun. **7**, 1133 (2005)
2. GK Mor, K Shankar, M Paulose, OK Varghese, CA Grimes, Nano Lett. **6**, 215 (2006)
3. D Kuang, J Brillet, P Chen, M Takata, S Uchida, H Miura, K Sumioka, SM Zakeeruddin, M Grätzel, ACS Nano **2**, 1113 (2008)
4. GK Mor, K Shankar, M Paulose, OK Varghese, CA Grimes, Nano Lett. **5**, 191 (2005)
5. JH Park, S Kim, AJ Bard, Nano Lett. **6**, 24 (2006)
6. A Ghicov, H Tsuchiya, R Hahn, JM Macak, AG Muñoz, P Schmuki, Electrochem. Commun. **8**, 528 (2006)
7. K Zhu, NR Neale, A Miedaner, AJ Frank, Nano Lett. **7**, 69 (2007)
8. S Ito, NLC Ha, G Rothenberger, P Liska, P Comte, SM Zakeeruddin, P Pechy, MK Nazeeruddin, M Grätzel, Chem. Commun. **42**, 4004 (2006)
9. OK Varghese, M Paulose, CA Grimes, Nat. Nanotech. **4**, 592 (2009)
10. S Berger, A Ghicov, YC Nah, P Schmuki, Langmuir **25**, 4841 (2009)
11. JY Kim, JH Noh, K Zhu, AF Halverson, NR Neale, S Park, KS Hong, AJ Frank, ACS Nano **5**, 2647 (2010)
12. K Zhu, TB Vinzant, NR Neale, AJ Frank, Nano Lett. **7**, 3739 (2007)
13. GK Mor, OK Varghese, M Paulose, K Shankar, CA Grimes, Sol. Energy Mater. Sol. Cells **90**, 2011 (2006)
14. JM Macak, SP Albu, P Schmuki, Phys. Stat. Solidi-Rapid Res. Lett. **1**, 181 (2007)
15. JM Macak, H Tsuchiya, L Taveira, S Aldabergerova, P Schmuki, Angew. Chem. Int. Ed. **44**, 7463 (2005)
16. QY Cai, M Paulose, OK Varghese, CA Grimes, J. Mater. Res. **20**, 230 (2005)
17. Y Shin, S Lee, Nano Lett. **8**, 3171 (2008)
18. D Wang, Y Liu, B Yu, F Zhou, W Liu, Chem. Mater. **21**, 1198 (2009)
19. H Masuda, M Satoh, Jpn. J. Appl. Phys. **35**, L126 (1996)
20. W Lee, H Han, A Lotnyk, MA Schubert, S Senz, M Alexe, D Hesse, S Baik, U Gösele, Nat. Nanotech. **3**, 402 (2008)
21. D Wang, B Yu, C Wang, F Zhou, W Liu, Adv. Mater. **21**, 1964 (2009)
22. H Zheng, AZ Sadek, M Breedon, D Yao, K Latham, JD Plessis, K Kalantar-Zadeh, Electrochem. Commun. **11**, 1308 (2009)
23. HE Prakasam, K Shankar, M Paulose, OK Varghese, CA Grimes, J. Phys. Chem. C **111**, 7235 (2007)
24. P Skeldon, GE Thompson, SJ Garcia-Vergara, L Iglesias-Rubianes, CE Blanco-Pinzon, Electrochem. Solid-State Lett. **9**, B47 (2006)
25. GK Mor, OK Varghese, M Paulose, CA Grimes, Adv. Funct. Mater. **15**, 1291 (2005)
26. JR Jennings, A Ghicov, LM Peter, P Schmuki, AB Walker, J. Am. Chem. Soc. **130**, 13364 (2008)
27. Q Wang, K Zhu, NR Neale, AJ Frank, Nano Lett. **9**, 806 (2009)
28. D Kuang, C Klein, S Ito, J-E Moser, R Humphry-Baker, N Evans, F Duriaux, C Grätzel, SM Zakeeruddin, M Grätzel, Adv. Mater. **19**, 1133 (2007)

19

Study of blend composition of nano silica under the influence of neutron flux

Elchin Huseynov*, Adil Garibov and Ravan Mehdiyeva

Abstract

Nano SiO_2 compound with 160 m^2/g specific surface area and 20 nm sizes has been irradiated continuously with neutron flux up to 20 hours in various periods in TRIGA Mark II type research reactor. The initial activities of different type radionuclides defined in the result of eight day activity analysis changes between wide range of 1,5 kBq- 1,5GBq. In the result of activity analysis carried out after the irradiation, the element content of 0,5% mixture existing in nano SiO_2 compound has been defined with radionuclides of relevant element. It has been defined percentage amounts of elements in blend composition according to the performed activities.

PACS: 81.07.Wx; 61.46. + w; 61.80.Hg; 28.20.Fc; 92.20.Td

Keywords: Nanopowder; Nano silica; Nanomaterial; Neutron activation analysis; Neutron irradiation; Radioactivity; Radioisotopes

1 Introduction

Silicon and its oxide compositions are widely applied in different fields of science and technology for their unique physical, physical-chemical properties and radiation- durability [1-9]. In real application conditions the surface of silicon-based materials and devices are covered with oxide layer. The influence of the oxide layer formed on surface on physical properties of base material and effective thickness in terms of defence from subsequent oxidation process are usually in nanometer form. On the other hand nano-size silicon and SiO_2 are of great importance for their physical and surface physical-chemical properties.

Oxide compositions of silicon are widely used in electronics, radiation technology and radiochemical processes [1,2,5,9,10]. Conversion of ionizing radiation energy and transfer to surface levels is of great importance for radiative study of materials and radiation technology. From this point of view nano-size SiO_2 is a perspective and at the same time an actual model system for study of radiation defect-formation, conversion and transfer processes of ionizing radiation energy. In recent years formation of nano size SiO_2 compound and improvement of their purity degree are in the focus of attention of researchers [11-13]. At modern period the purity degree of nano silica has been brought to 99,5 – 99,9% and intensive scientific studies are being carried out in order to increase its value.

In the presented work under the influence of neutrons with 2×10^{13} n/cm^2 s intensity it has been studied the dependence of activities of radioactive nucleus, formed in the result of neutron flux in powder and extruded shaped, 99,5% purity, 20 nm size nano SiO_2 compound, on integral dose and decomposition time after irradiation. On the base of the achieved results it has been carried out the quality and quantity identification of blends in the composition of nano SiO_2 compound.

The sample used in the presented work is widespread in the nature and met in crystalline forms as quartz, rock crystal, flint, opal, etc. In these crystals silica exists with different percentages and "Obsidian" natural glass has more SiO_2 percentage (about 70-75% of SiO_2) among natural crystals existing in the nature [14,15]. Nowadays, macro silica is treated with several methods and it was available to obtain maximum macro SiO_2 (In Egypt) with 99,85% purity. However, the synthesis method of high purity SiO_2 nanopowder in nanosize differs slightly and in this case, practically obtain of SiO_2 nanopowder with perfect purity (100% purity) is almost

* Correspondence: hus.elchin@yahoo.com
Institute of Radiation Problems of Azerbaijan National Academy of Sciences, AZ 1143, B. Vahabzadeh 9, Baku, Azerbaijan

impossible [16]. Thus, if purity of nano SiO_2 powder obtained before was 75%, at present development of modern technology allowed obtaining SiO_2 nanopowder with 99,9% purity [11-13]. Nowadays, nano SiO_2 compounds with 99–99,9% purity are considered to be high purity nano SiO_2 compounds and samples with this purity are of wide application. As it was mentioned before, nano SiO_2 compound with 99,5% purity has been used during the experiment, so it can be considered as a high purity nano SiO_2 compound. However, the sample contains 0,5% of impurities and even if at first sight it can seem a small indicator, it is very large value in nano-scale and molecule compiling. Therefore, if to take into account that there is approximately 10^{22} units of particles in atomic level in 1 g of SiO_2 nanopowder, then 0,5% to be great value (approximately 5×10^{19} mixture particle) is obvious. Naturally, 0,5% mixture doesn't impact the physical parameters of sample, but it clearly manifests itself during irradiation in the reactor and is of great significance.

2 Methods

From previous studies it is known that the specific surface area of nanomaterial used in the experiment is 160 m^2/g, dimensions are 20 nm and some parameters of the used sample has been studied [17-19]. In the presented work the samples have been irradiated by neutron flux 2×10^{13} n/cm^2s in central channel (Channel A1) of TRIGA Mark II light water pool type research reactor at full power (250 kW) in "Reactor Centre" of Jozef Stefan Institute (JSI) in Ljubljana city of Slovenia. It is important to note that the JSI TRIGA reactor has been thoroughly characterized [20-25] and the computational model used for computational characterization has been thoroughly verified and validated [26,27] against several experiments. It should be mentioned that in this channel the parameters of neutron flux at full power mode are 5.107×10^{12} cm^{-2} s^{-1} (1 ± 0.0008, $E_n < 625$ eV) for thermal neutrons, $6.502x10^{12}$ cm^{-2} s^{-1} (1 ± 0.0008, $E_n \sim 625$ eV ÷ 0.1 MeV) for epithermal neutrons, $7.585x10^{12}$ cm^{-2} s^{-1} (1 ± 0.0007, $E_n > 0.1$ MeV) for fast neutrons and finally for all neutrons the flux density in central channel is 1.920×10^{13} cm^{-2} s^{-1} (1 ± 0.0005) [20,26].

The powdered nano SiO_2 has been irradiated in a special cylindrical aluminium container. Density of powdered nano SiO_2 compound is $\rho_{powder} = 0{,}1$ g/cm^3 (density in packaging is approximately ~0,3 g/cm^3), volume of the sample in cylinder-shaped radiation packaging is $V_{powder} \approx 2{,}3$ cm^3, the area of the sample which is exposed to neutron flux is $S_{powder} \approx 10$ cm^2. The sample which is approximately same amount ~0,7 g being pressed in a special press machine has been put in a special form and its parameters are: $\rho_{tablet} = 2{,}9$ g/cm^3, $V_{tablet} \approx 0{,}2$ cm^3, $S_{tablet} \approx 1{,}8$ cm^2. The prepared experiment sample was irradiated firstly for 5 minutes evaluate the final activity after 20 hour radiation. Then other 8 samples (each one apx. 0,7 g) were divided into 4 groups and each of them irradiated continuously for 5, 10, 15, 20 hours by neutron 2×10^{13} cm^{-2} s^{-1} in central channel (Channel A1) of TRIGA Mark II type research reactor at full power (250 kW). Both forms of the samples (4 powder samples and 4 tablets) have been irradiated in 2×10^{13} n/cm^2s intensity of neutron flux. Absorption dose values of the studied samples have been defined according to geometric dimensions of powdered and tablet-formed samples, intensity of irradiation, irradiation periods, density of influencing neutron flux and energy spectra of neutrons. The value of neutron flux for powdered samples is $3{,}95\times10^{18}$ ÷ $1{,}58\times10^{19}$ and for tablet-formed ones it changes within the neutron range $6{,}67\times10^{17}$ ÷ $2{,}67\times10^{18}$. The dependence of neutrons falling on the sample on irradiation period for samples of both types has been given in Figure 1. As it is seen, the interacting neutron flux and thus the amount of absorbed neutrons in powdered sample is approximately 25 times more than in tablet-formed.

Radionuclides being formed in composition of nano SiO_2 compound after the interaction with neutron have been analysed in spectrometers "Ortec HPGe detectors (Coaxial, Low and Well-Type)" and "Canberra coaxial HPGe detector". The radioactivity, isotope composition and amount of blend elements of the irradiated samples have been determined on technique [28-31].

3 Results and discussions

The main probable process of interaction of neutrons with substance is radiation capacity [28]:

$$n + {}^{A}Z \rightarrow {}^{A+1}Z^{*} \rightarrow {}^{A+1}Z + \gamma \tag{1}$$

here ${}^{A}Z$ – irradiated isotope, ${}^{A+1}Z^{*}$ - excited nucleus formed as a result of neutron capacity and γ - second gamma rays. Radioactive excited samples are formed in the result of the processes shown in irradiated samples. Their identification has been studied with the method of gamma spectroscopy. γ – ray intensities appropriate to nuclear transitions in gamma spectra are different depending on irradiation period and decay constants. One of these spectra has been given in Figure 2 as an example.

After neutron irradiation it has been studied the change of samples' activities after 192 hours. It has been determined that initial radioactivities of irradiated samples change between the ranges 1,5 kBq – 1,5GBq (powder and tablet each one total four sample approximate 3 g). Define of concentrations of elements is conducted on the base of the activities originated in appropriate energy range. The activity is determined on the following formula based on nuclear constants [28]:

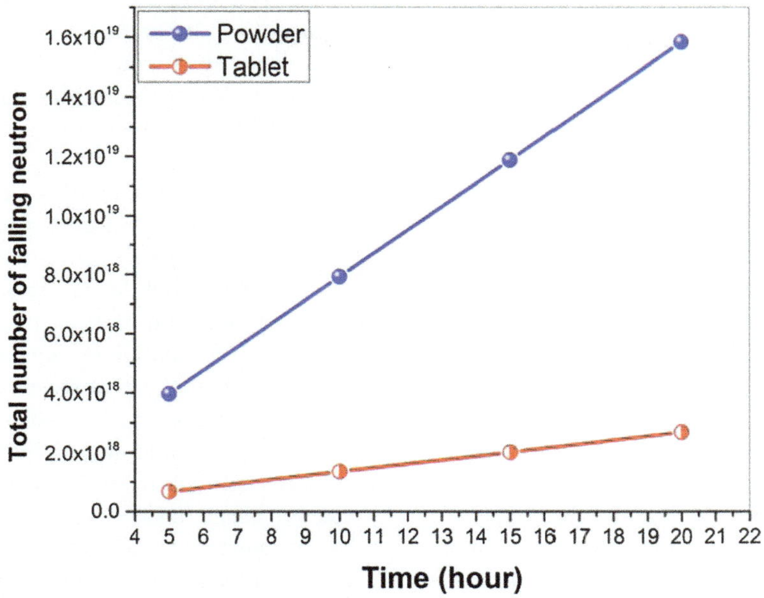

Figure 1 Dependence of neutrons falling on the sample on irradiation period.

$$A = \sigma \Phi \left(\frac{m}{M}\right) N_A \Theta P_\gamma \xi \left(1 - \exp(-\lambda \tau_{ray})\right) \times (1 - \exp(-\lambda \tau_{mes})) \exp(-\lambda \tau_{cool}) \tag{2}$$

here, A – measured activity (Bq), σ – cross section of activation of defined isotopes (cm^2), Φ – neutron flux (n/cm^2s), m – weight of defined element (g), M – atomic weight of defined element (g/mol), N_A – Avogadro number (1/mol), Θ – distribution of activated isotope, P_γ – distribution probability of E energy gamma quantum, ξ – defect efficiency according to E energy, λ – decay constants of the formed isotopes, τ_{ray}, τ_{mes}, τ_{cool}– irradiation,

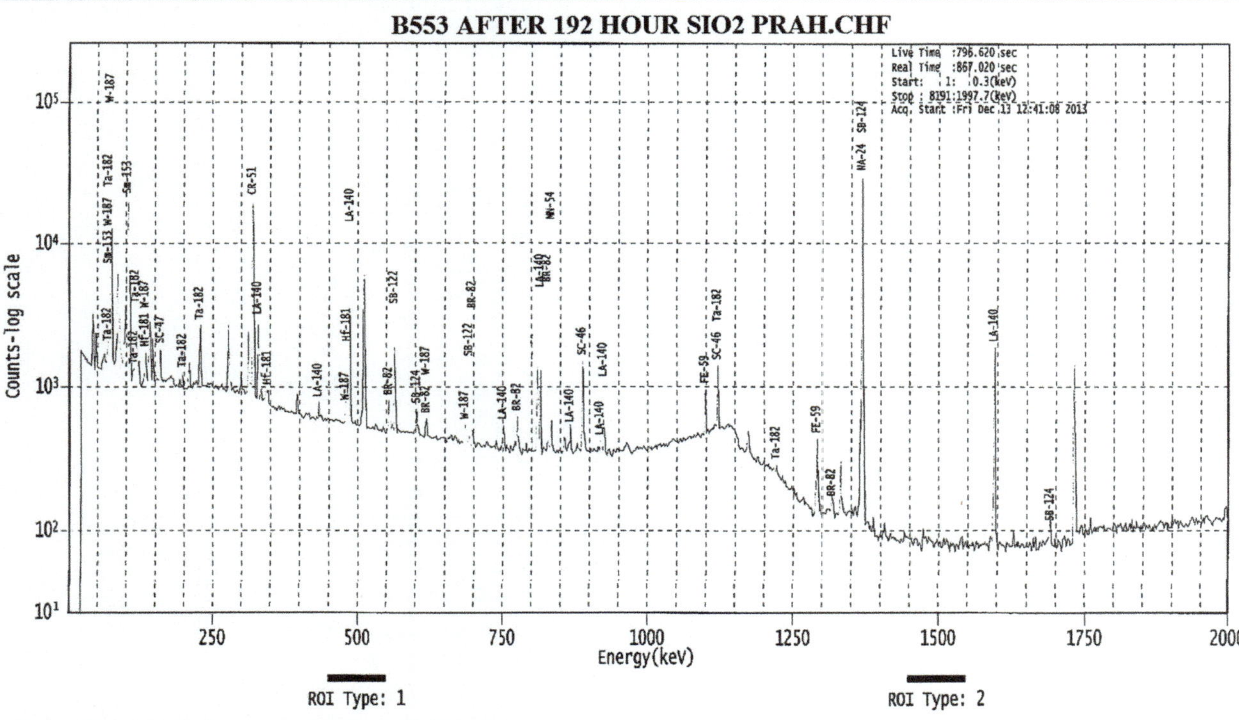

Figure 2 Gamma spectrum of nano SiO_2 sample, irradiated by neutrons with $2x10^{13}$ n/cm^2s flux for τ = 5,10,15,20 hours, in 192 hours after irradiation.

measurement and cooling time interval of samples correspondingly.

Activities of newly formed radioactive isotopes observed in irradiated samples change correspondingly to decay constants. The initial activities of different type radionuclides defined in the result of eight day activity analysis changes between wide range of 1,5 kBq- 1,5 GBq. Just for this reason we decided to divide these elements into four groups. The observed radioactive isotopes can be divided conditionally into four groups as I – A ≤ 10 kBq, II – A ≤ 100 kBq, III – A ≤ 2 MBq and IV – A ≤ 1,5 GBq for their activities. Observation time dependence of initial activities of the observed radioactive isotopes on conventional groups has been given in figures. Dependency of radioactivity on observation time has been defined for both powdered and tablet-formed nano SiO_2 (Figures 3, 4, 5 and 6). It should be noted that all used graphics cover an eight day- period and elapsed time was denominated in hours. First, let's review radionuclides with activity up to 10 kBq which is generated in nanocompound with the influence of neutron flux (Figure 3).

In general, it has been observed 7 types of radionuclides, thus their half-life changes from 80,4 hours up to 748,2 hours. For living periods, the observed "I group" radionuclides can be lined up as Sc-47 (80,4 hour), Hf-181 (1017,36 hour), Fe-59 (1067,88 hour), Sb-124 (1444,8 hour), Sc-46 (2010,96 hour), Ta-182 (2753,76 hour) and Mn-54 (7489,2 hour).

The activity of radionuclides included inside II group is up to 70 kBq (Figure 4). It has been observed two type Br-82 and Sb-122 radionuclides that their half-life are 35, 3 h for Br-82 and 64, 8 h for Sb-122 respectively. As it is seen from Figure 4, after 8 days the activity of Br-82 isotope decreased approximately down to 5 kBq, and the activity of Sb-122 down to 10 kBq.

The activity of only one of four radionuclides including to the third group is approximately up to 2 Mbq (for comparison: the activity in tablet form of relevant radionuclide is approximately up to 72 kBq, right column in the fig.) and the activity of other three radionuclides do not exceed 0,6 MBq (Figure 5). Half-lifes of observed radionuclides are 24 h (W-187), 40,32 h (La-140), 46,28 h (Sm-153) and 665 h (Cr-51) and after eight days activity for Cr-51 decreased up to approximately 0,5 MBq, and 0,1 MBq for other ones.

Only one element is included to the last IV group that its activity was approximately up to 1,5 GBq (for comparison: the activity in tablet form of relevant radionuclide is approximately 54 MBq, right column in the fig.). The half-life of observed high activity Na-24 isotope is 14,95 hours and at the end of measurement day the activity of Na-24 isotope decreased approximately down to 0,2 MBq (Figure 6). It can be said that the major part of mixture containing Na-24 isotope which is more active than other ones is the element of Na.

It is seen from the dependency given in Figure 4 that activities of isotopes with small radioactive decay

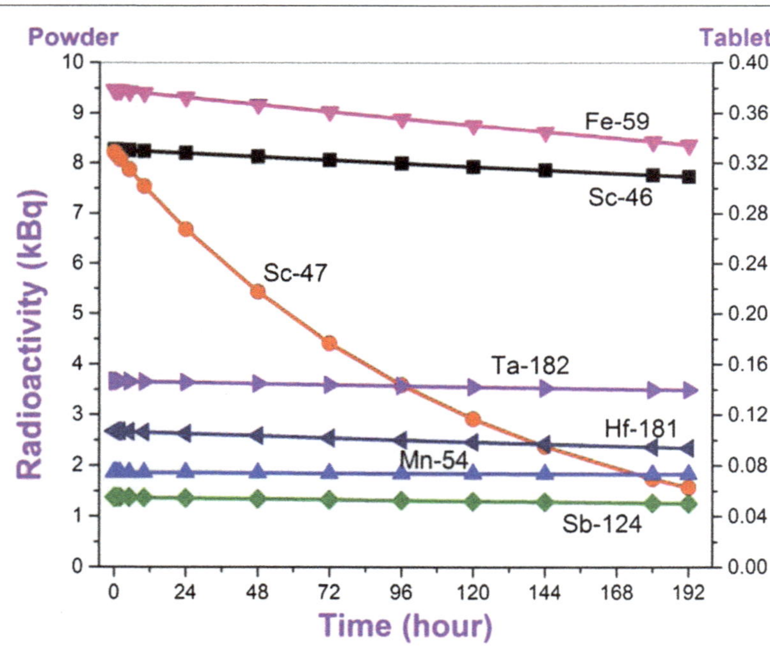

Figure 3 Dependence of activities of conditional I group radionuclides being formed in nano SiO_2 under the influence of neutron flux on measurement time.

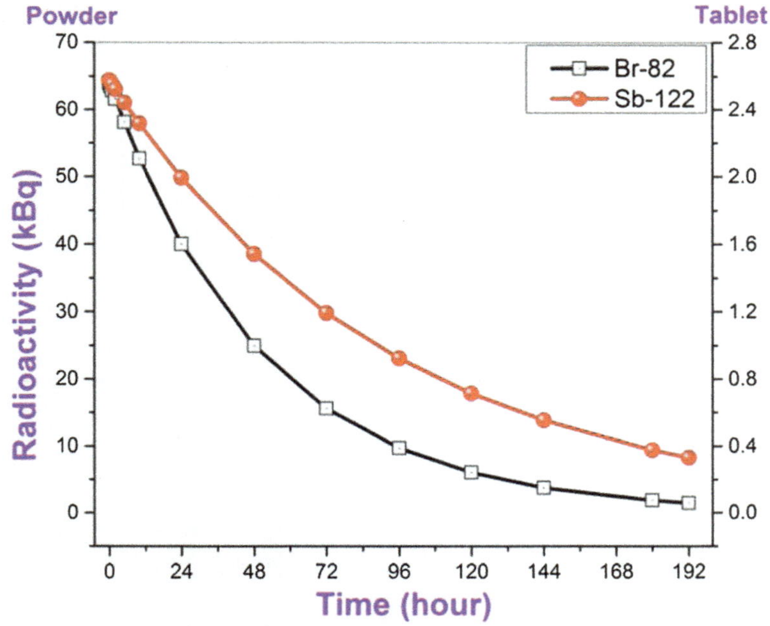

Figure 4 **Dependence of activities of conditional II group radionuclides being formed in nano SiO_2 under the influence of neutron flux on measurement time.**

constant and which are relatively high within half decay, change little during experimental observation and it is expected the influence of their radioactive decay during subsequent study of nano SiO_2 compound properties. According to comparison between activities of isotopes defined experimentally and activities of standards taken in the same condition and weight, it has been defined the amount of each element in the content of nano SiO_2 compound [28]:

$$C_{sam} = C_{st}\frac{A_{sam}}{A_{st}} \tag{3}$$

here, C_{sam} – concentration of the shown element isotope in nano SiO_2, C_{st} – concentration of comparison standard

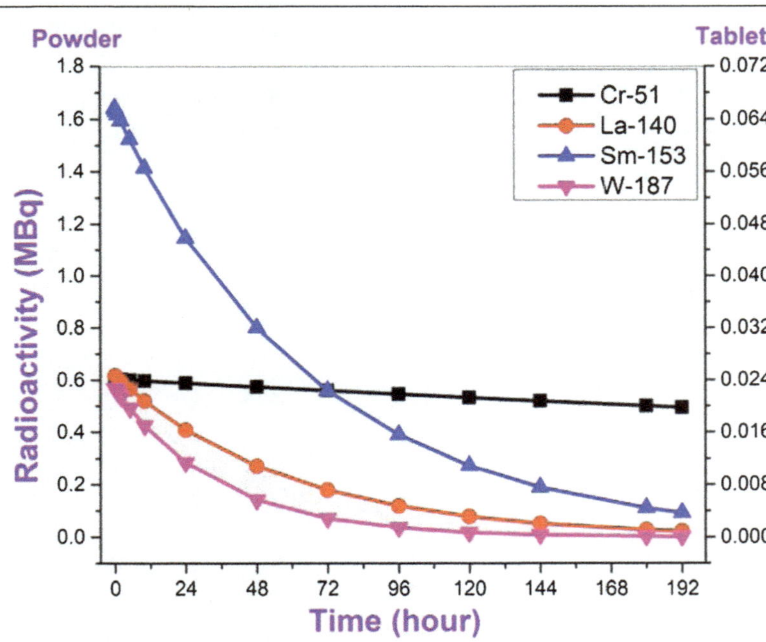

Figure 5 **Dependence of activities of conditional III group radionuclides being formed in nano SiO_2 under the influence of neutron flux on measurement time.**

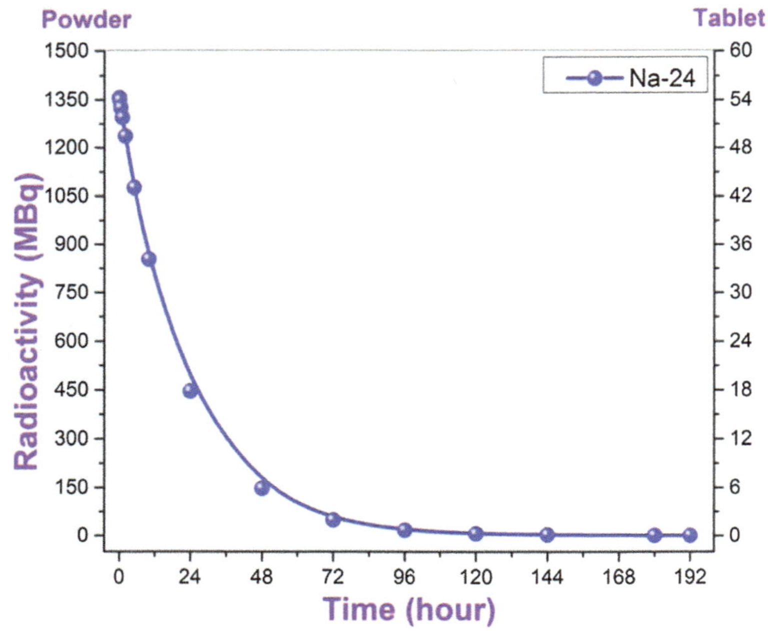

Figure 6 Dependence of activity of [24]Na isotope included to the conventional IV group originated in nano SiO_2 under the influence of neutron flux on measurement time.

in the same isotope, A_{sam} and A_{st} are activities of sample and standard, respectively. If weights of sample and standards are different, the concentration of the element [28-31] is defined on the given technique. The obtained results have been given in Table 1.

In general, it has been given in tables the different terms' radioactivity of radionuclides generated in 0,5% mixture during the influence of neutron flux to 99,5% purity nano SiO_2, their initial isotopes and approximate amount. In Table 2 III and IV group radioisotopes and their activities have been shown with MBq and lined up in the form of initial activity increase.

Table 1 Isotopes generated by neutron activation in nano SiO_2, number of neutrons spent on activation of one isotope and number of isotopes in percent in sample

Stable isotope	Radioisotope	Number of capture neutron in each nucleus	Apx. amount of blend (%)
Na-23	Na-24	+1	~0,3%
La-139	La-140	+1	
Cr-50(52)	Cr-51	+1	~0,1%
Sm-150(152)	Sm-153	+3	
W-184(186)	W-187	+3	
Sb-121	Sb-122	+1	
Sb-123	Sb-124	+1	<0,1%
Br-81	Br-82	+1	
Fe-56(57,58)	Fe-59	+3	<0,1%
Sc-45	Sc-46	+1	
Sc-45	Sc-47	+2	
Ta-181	Ta-182	+1	
Hf-178(179,180)	Hf-181	+3	
Fe-54(n,p)	Mn-54	+1	

Table 2 Conditional III and IV radioisotope groups generated in nano-compound under the influence of neutron flux

Time (hour)	Radioactivity (MBq)				
	Na-24	Sm-153	La-140	Cr-51	W-187
0	~1355	~1.6	~0.62	~0.604	~0.568
0.1	~1350	~1.6	~0.61	~0.603	~0.566
0.5	~1325	~1.6	~0.61	~0.603	~0.56
1	~1295	~1.6	~0.60	~0.602	~0.55
2	~1235	~1.59	~0.59	~0.601	~0.53
5	~1075	~1.52	~0.57	~0.6	~0.49
10	~850	~1.41	~0.52	~0.59	~0.42
24	~445	~1.14	~0.41	~0.58	~0.28
48	~146	~0.8	~0.27	~0.57	~0.14
72	~48	~0.55	~0.18	~0.56	~0.07
96	~15	~0.38	~0.12	~0.55	~0.04
120	~5	~0.27	~0.08	~0.53	~0.02
144	~1.7	~0.19	~0.05	~0.52	~0.01
180	~0.3	~0.11	~0.03	~0.5	~0.003
192	~0.2	~0.09	~0.02	~0.49	~0.002

Table 3 Conditional I and II radioisotope groups generated in nano-compound under the influence of neutron flux

Time (hour)	Radioactivity (kBq)								
	Sb-122	Br-82	Fe-59	Sc-46	Sc-47	Ta-182	Hf-181	Mn-54	Sb-124
0	~64.4	~64	~9.46	~8.25	~8.21	~3.67	~2.67	~1.86	~1.37
0.1	~64.3	~63.9	~9.46	~8.25	~8.2	~3.67	~2.67	~1.86	~1.37
0.5	~64.0	~63.44	~9.46	~8.25	~8.17	~3.66	~2.67	~1.86	~1.37
1	~63.7	~62.8	~9.45	~8.25	~8.14	~3.66	~2.66	~1.86	~1.37
2	~63	~61.6	~9.45	~8.24	~8	~3.65	~2.66	~1.86	~1.37
5	~61	~58	~9.43	~8.24	~7.87	~3.65	~2.65	~1.85	~1.36
10	~58	~53	~9.4	~8.23	~7.53	~3.65	~2.64	~1.85	~1.36
24	~50	~40	~9.32	~8.19	~6.67	~3.64	~2.62	~1.85	~1.35
48	~39	~25	~9.17	~8.12	~5.43	~3.62	~2.58	~1.85	~1.34
72	~30	~15	~9	~8.05	~4.41	~3.6	~2.54	~1.84	~1.32
96	~23	~10	~8.8	~7.98	~3.6	~3.57	~2.5	~1.84	~1.31
120	~18	~6	~8.7	~7.92	~2.9	~3.55	~2.45	~1.84	~1.29
144	~14	~4	~8.6	~7.85	~2.4	~3.53	~2.4	~1.83	~1.28
180	~9	~2	~8.4	~7.75	~1.7	~3.5	~2.35	~1.82	~1.25
192	~8	~1.5	~8.3	~7.72	~1.6	~3.48	~2.34	~1.82	~1.25

In Table 3 I and II group radioisotopes and their activities have been shown with kBq and lined up in the form of initial activity increase. It should be mentioned that radioisotopes of the elements Si and O which are the main part of the sample, which may arise, have a very small existence period and in our discussions they have not been considered. So, Si-31 has short decay-time (~2.5 h) in 20 h irradiation, there could be produced some P which is beta emitter and cannot be detected by gamma detector. In general, in Table 1 it has been depicted initial isotopes and their amounts in percent according to blend radioisotopes generated under the influence of neutron flux.

4 Conclusions

It has been carried out identification of radioactivity appeared in nano SiO_2 under the influence of neutron flux and isotopes that formed radioactivity. It has been revealed dependency of samples' activity and dose amount on irradiation time and sample dispersity. It has been defined that powdered nano SiO_2 possess an activity approximately 25 times higher than the samples made as a tablet in special press form due to the interaction field with neutron to be big. Dependencies of radio activities of the revealed isotopes on observation time and amount of blend elements in percent have been defined. In the studied nano SiO_2 samples it has been revealed the isotopes possessing relatively large half-decay time and these isotopes are suggested to be considered in explanation of physical properties of nano SiO_2 compound within the period after irradiation.

Competing interests

The authors declare that they have no competing interests.

Authors' contributions

EH make all experiment and write first version of manuscript. AG participated in the sequence alignment and gave some idea. RM participated in drafted the manuscript. All authors read and approved the final manuscript.

Acknowledgements

The work has been carried out on the base of agreement signed between the Institute of Radiation Problems of ANAS and Jozef Stefan Institute of Slovenia. We express our gratitude to colleagues of the Institute of Radiation Problems of ANAS and "Reactor Infrastructure Centre (RIC)" and "Radiation Protection Unit" laboratories of Jozef Stefan Institute of Slovenia. We wish to thank Dr. Luka Snoj and Anze Jazbec for doing experiment with TRIGA Mark II type research reactor and fruitful discussions us.

References

1. SM Jafari, DA Bradley, CA Gouldstone, PHG Sharpe, A Alalawi, TJ Jordan, CH Clark, A Nisbet, NM Spyrou, Radiat. Phys. Chem. **97**, 95–101 (2014)
2. AT Abdul Rahman, RP Hugtenburg, AIM Siti Fairus Abdul Sani, FI Alalawi, R Thomas, MA Barry, A Nisbet, DA Bradley, Appl. Radiat. Isot. **70**, 1436–1441 (2012)
3. D Guarnieri, MA Malvindi, V Belli, PP Pompa, P Netti, J. Nanoparticle Res. **16**, 2229 (2014)
4. P Buchwalter, J Rosé, B Lebeau, O Ersen, M Girleanu, P Rabu, P Braunstein, J-L Paillaud, J. Nanoparticle Res. **15**, 2132 (2013)
5. Y Ding, X Chu, X Hong, P Zou, Y Liu, Appl. Phys. Lett. **100**, 013701 (2012)
6. F Chi, L Yan, H Yan, B Jiang, H Lv, X Yuan, Opt. Lett. **37**(9), 1406–1408 (2012)

7. G Sponchia, R Marin, I Freris, M Marchiori, E Moretti, L Storaro, P Canton, A Lausi, A Benedetti, P Riello, J. Nanoparticle Res. **16**, 2245 (2014)
8. M Martini, M Montagna, M Ou, O Tillement, S Roux, P Perriat, J. Appl. Phys. **106**, 094304 (2009)
9. YP Timalsina, J Branen, DE Aston, K Noren, G Corti, R Schumacher, DN McIlroy, J. Appl. Phys. **110**, 014901 (2011)
10. T Bellunatoa, M Calvi, C Matteuzzi, M Musy, DL Perego, B Storaci, Eur. Phys. J. C **52**, 759–764 (2007)
11. IA Rahman, P Vejayakumaran, J. Nanomater. **2012**, 1–15 (2012). Article ID 132424
12. K Mohanraj, S Kannan, S Barathan, G Sivakumar, J. Optoelectron. Adv. M. **6**(3–4), 394–397 (2012)
13. A Lazaro, G Quercia, HJH Brouwers, JW Geus, World J. Nanosci. Eng. **3**, 41–51 (2013)
14. PL Barnes-Svarney, TE Svarney, *"The Handy Geology Answer Book"* (Visible Ink Press, Slovenia, 2004). ISBN 1578591562
15. M Martini, M Milazzo, M Piacentini, *"Physics Methods in Archaeometry" Società Italiana di Fisica 154* (IOS Press, Italy, 2004). ISBN 1586034243
16. K Srivastava, N Shringi, V Devra, A Rani, Int. J. Innovative Research in Science, Engineering and Technology **2**(7), 2936–2942 (2013)
17. EM Huseynov, AA Garibov, RN Mehdiyeva, Azerbaijan J. Phys. **XIX**(N1), 10–14 (2013). ISSN 1028-8546
18. RN Mehdiyeva, AA Garibov, EM Huseynov, Transactions of National Academy of Sciences of Azerbaijan, Series of Physics – Mathematical and Technical Sciences, Physics and Astronomy **XXXII**(N5), 83–88 (2012). ISSN 0002-3108
19. E Huseynov, A Garibov, R Mehdiyeva, *Physica B.* **450**, 77–83 (2014)
20. L Snoj, G Zerovnik, A Trkov, Appl. Radiat. Isot. **70**, 483–488 (2012)
21. L Snoj, A Trkov, R Jačimović, P Rogan, G Žerovnik, M Ravnik, Appl. Radiat. Isot. **69**, 136–141 (2011)
22. L Snoj, A Kavcic, G Zerovnik, M Ravnik, Ann. Nucl. Energy **37**(2), 223–229 (2010)
23. L Snoj, A Trkov, M Ravnik, G Zerovnik, Ann. Nucl. Energy **42**, 71–79 (2012)
24. L Snoj, M Ravnik, Nucl. Eng. Des. **238**(9), 2473–2479 (2008)
25. A Jazbec, G Zerovnik, L Snoj, A Trkov, Atw. Internationale Zeitschrift für Kernenergie **58**(12), 701–705 (2013)
26. V Radulović, Ž Štancar, L Snoj, A Trkov, Appl. Radiat. Isot. **84**, 57–65 (2014)
27. G Zerovnik, M Podvratnik, L Snoj, Ann. Nucl. Energy **63**, 126–128 (2014)
28. MV Frontasyeva, Physics of Elementary Particles and Atomic Nuclei **42**(2), 636–718 (2011). ISSN 0367-2026
29. E Steinnes, *"Some Neutron Activation Methods for the Determination of Minor and Trace Elements in Rocks"* (Kjeller, Norway, 1972). 3 editions published
30. G Medkour Ishak-Boushaki, K Boukeffoussa, Z Idiri, M Allab, Appl. Radiat. Isot. **70**(3), 515–519 (2012)
31. H Mommsen, J. Archaeol. Sci. **39**(3), 704–707 (2012)

Design of plasmonic cavities

Soon-Hong Kwon[1], You-Shin No[2] and Hong-Gyu Park[2*]

Abstract

In this review paper, we introduce the unique optical properties of high-quality, fully three-dimensional, subwavelength-scale plasmonic cavities. Surface-plasmon-polaritons excited at dielectric-metal interfaces are strongly confined in such cavities. The field profiles of plasmonic modes, their temperature-dependent quality factors, and subwavelength mode volumes are calculated and analyzed systematically using three-dimensional finite-difference time-domain simulations. Reasonable design of high-quality plasmonic cavities opens an opportunity to demonstrate novel plasmonic lasers enabling the further miniaturization of coherent light sources for use in ultra-compact photonic integrated circuits.

1 Introduction

Plasmonic cavities are particularly attractive for nano-scale photonic applications as their physical sizes can be smaller than the diffraction limit of light [1-9], whereas the dimensions of conventional dielectric cavities such as photonic crystals [10-12], microdisks [13-15], nanowires [16-18] and metal-cladding cavities [19-22] are limited by wavelength. The resonant wavelengths of the surface-plasmon-polaritons (SPPs) excited at the dielectric-metal interface can be shorter than the wavelengths in dielectric cavities. Therefore, SPPs can be confined in a subwavelength volume. Various high-quality (Q) plasmonic cavities were successfully demonstrated by reducing metallic absorption [8], and provided strong optical feedback for lasing as well as an accessible collection of light emission [3,4]. Lasing using plasmonic cavities could be a significant step toward an ultimate miniaturized coherent light source for nano-scale photonic integrated circuits. In this review paper, we will introduce four recently reported high-Q plasmonic cavities for strong SPP confinement and theoretically investigate their optical properties using three-dimensional (3D) finite-difference time-domain (FDTD) methods.

2 Review

2.1 Simulation Method

In our FDTD simulations, silver, which was used for the metal component in the plasmonic cavities we investigate, was modeled with a Drude model [23,24]: $\varepsilon(\omega) = \varepsilon_\infty - \omega_p^2 / (\omega^2 + i\gamma\omega)$. The Drude model fits the experimentally determined dielectric function of silver. Dielectric functions in the visible spectral range (400 to 800 nm) and the near-infrared spectral range (800 to 2000 nm) were used to fit the Drude model, respectively [25]. In the visible spectral range, the background dielectric constant ε_∞, the plasma frequency ω_p, and the collision frequency γ at room temperature are 4.1, 1.4×10^{16} s^{-1}, and 4.2×10^{13} s^{-1}, respectively. In the near-infrared spectral range, the background dielectric constant ε_∞, the plasma frequency ω_p, and the collision frequency γ at room temperature are 3.1, 1.4×10^{16} s^{-1}, and 3.1×10^{13} s^{-1}, respectively.

In order to investigate the details of photon loss channels in plasmonic cavities, we calculated corresponding Q factors separately [2-4,8,9]. The Q factor of a SPP mode is determined by contributions from both the intrinsic metal loss and optical radiation loss into free space. As the optical radiation loss is reduced, the total Q factor of a SPP mode approaches the minimum value allowed by the intrinsic properties of the metal used, the metal-loss-limited Q factor. The Q factor due to optical radiation loss can be calculated with $\gamma = 0$. To represent the dielectric function of silver at low temperature, the damping collision frequency γ, was scaled by a factor of the room-temperature conductivity divided by the low-temperature conductivity [2-5,8,9,19]. The total Q factor was calculated from the time decay of the energy of a cavity mode [10,11]. To calculate mode volume and the confinement factor, we used the effective refractive index of the metal, $\varepsilon_{eff} = d(\omega\varepsilon(\omega)) / d\omega = \varepsilon_\infty + (\omega_p / \omega)^2$, where ω is the resonant frequency of a SPP cavity mode [26]. The mode volume, V, is defined as the ratio of the total electric field

* Correspondence: hgpark@korea.ac.kr
[2]Department of Physics, Korea University, Seoul 136-701, Republic of Korea
Full list of author information is available at the end of the article

energy density of the mode to the peak energy density [10,11,26]:

$$V = \frac{1}{\max\left[\frac{1}{2}\varepsilon_{eff}\left|E(\vec{r})\right|^2\right]} \iiint_{all} \frac{1}{2}\varepsilon_{eff}\left|E(\vec{r})\right|^2 d^3\vec{r}$$

2.2 Plasmonic cavities

2.2.1 Dielectric-core/metal-shell nanowire plasmonic cavity

A plasmonic cavity based on a dielectric-core/metal-shell nanowire structure can enable deep subwavelength confinement of SPPs in all three spatial dimensions (Figure 1(a)) [2]. This plasmonic cavity is formed by introducing an axial heterostructure along the nanowire axis in a 2D plasmonic waveguide of metal-coated dielectric nanowire. The SPPs excited at the interface between the high-index dielectric nanowire core and the metal shell are confined in a cavity region with core diameter d_{NW1} and length L. Two high-index-core/low-index-shell/metal-shell nanowire structures with a smaller core diameter d_{NW2} effectively form a plasmonic mirror, sandwiching the cavity region. The refractive indices of the high-index core and low-index shell are 2.6 (e.g. CdS or InGaN in the visible wavelength [27,28]) and 1.5 (SiO_2), respectively, and the metal shell is made of silver.

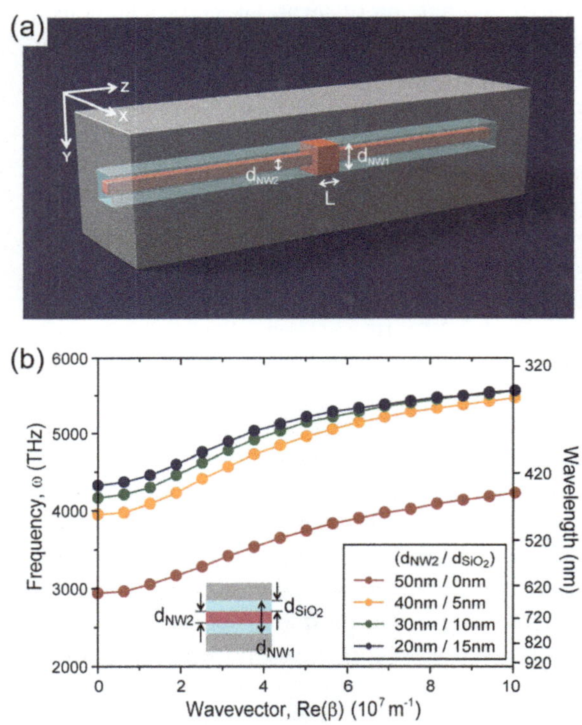

Figure 1 3D confinement of SPPs in a deep subwavelength plasmonic cavity. (a) Schematic diagram of the plasmonic cavity. d_{NW1}, d_{NW2}, and L represent the nanowire diameter of the cavity region, the diameter of the high-index core, and the cavity length, respectively. **(b)** Calculated dispersion curves of the fundamental transverse SPP modes in the square cross-sectional, infinitely long waveguide with high-index-core/low-index-shell/metal-shell structure. Refractive indices are 2.6 and 1.5 for the high-index core and the low-index shell, respectively. Silver is used as the metal shell. Inset: d_{NW1} is fixed to 50 nm, and d_{NW2}, and d_{SiO2} are the thicknesses of the high-index core, and the low-index shell, respectively. Adapted from [2].

The longitudinal confinement of SPPs in this cavity can be understood from the dispersion curves of the fundamental transverse plasmonic-guided mode in a square cross-sectional, infinitely long waveguide with a high-index-core/low-index-shell/metal-shell structure (inset of Figure 1(b)). The total diameter of the nanowire including the high-index core and the low-index shell, d_{NW1}, is fixed at 50 nm. The frequency of the SPP mode approaches a non-zero minimum as the wavevector approaches zero, which is the cutoff frequency. SPP modes with lower frequencies than this cutoff frequency cannot propagate along the waveguide. This cutoff frequency, a unique property of the proposed 2D plasmonic waveguide [29], depends significantly on the presence of a low-index shell d_{SiO2} (Figure 1(b)), explaining the large frequency gap between waveguides with and without it. We note that this frequency gap prevents the coupling of SPP modes excited in a waveguide lacking a low-index shell with those excited in a waveguide with a low-index shell. Therefore, three-dimensional (3D) confinement of SPPs is achieved in the region without the low-index shell, without modifying the metallic structure.

In order to investigate the confined SPP modes, we calculated the cavity modes for the cavities with L values of 40 and 120 nm, respectively. The electric field intensity (log E^2) mode profiles show that SPPs are strongly confined in the cavities without significant scattering (Figure 2). Consequently, strong suppression of optical radiation loss allows these cavities to achieve high Q factors (greater than 36000) approaching the metal-loss-limited value. In this case, the Q factors were calculated at the low temperature of 20 K. At this temperature, we could assume a reduced absorption loss value for silver. We note that the proposed plasmonic cavity exhibits two unique properties: the deep subwavelength-scale confinement of SPPs and extremely low optical radiation loss. The physical dimensions of the cavity are $50 \times 50 \times 40$ nm^3 at a wavelength of 533 nm. In addition, the longitudinal SPP cavity modes can be identified by the number of the electric field antinodes, m, in a similar fashion to a conventional Fabry-Perot cavity. As L increases, the value m of an excited SPP mode increases. For example, in the cavity with L = 40 nm, the longitudinal mode of m = 1 is observed in the wavelength of interest (Figure 2(a)).

Figure 3 shows the optical characteristics of the plasmonic cavity, such as the resonant wavelengths, mode volumes, Q factors, and confinement factors. The resonant wavelength can be tuned readily over a wide spectral range

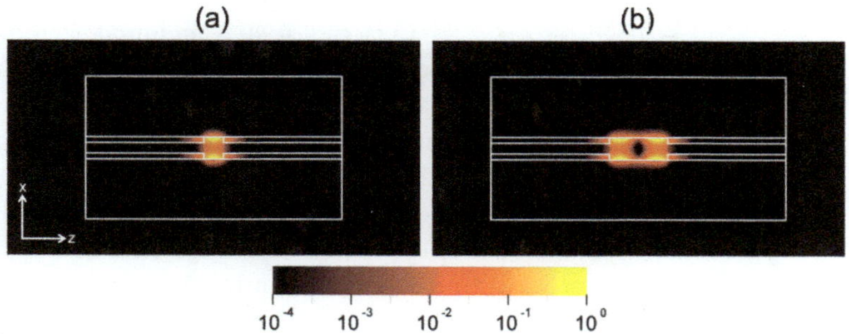

Figure 2 Electric field intensity (log E^2) mode profiles in the plasmonic cavities with (a) L = 40 nm and (b) L = 120 nm. The number of electric field antinodes, m, the resonant wavelengths, λ, and the Q factors for both cavities are as follows: **(a)** m = 1, λ = 533 nm, Q = 36100, **(b)** m = 2, λ = 529 nm, Q = 38100. Q factors are calculated at 20 K. Adapted from [2].

by varying L and thus can be well adjusted to the emission spectrum of a dielectric nanowire (left axis, Figure 3(a)). In addition, extremely small mode volumes on the order of 10^{-5} μm^3 were calculated (right axis, Figure 3(a)). These are ~ 100 times smaller than those of the smallest dielectric cavities so far reported [10-12,19-22]. The physical limit of an optical cavity's size, which is the diffraction limit of light, is overcome in the plasmonic cavity. For example, in a cavity with L = 40 nm and m = 1 (Figure 2(a)), the mode volume was calculated to be ~ 1/50 $(\lambda/2n_{NW})^3$, where n_{NW} is 2.6. Next, Q factors of the SPP cavities were calculated at a temperature of 20 K (Figure 3(b)). The high Q factor of ~ 38000 approaches the metal-loss-limited value because optical radiation loss for this cavity is negligible. Indeed, the optical Q factor is estimated at an extremely high value of 3.2×10^6 with zero collision frequency. As L and the resonant wavelength decrease, the Q factor tends to decrease in a cavity with a fixed m. Also, the high Q factor and subwavelength-scale mode volume of this SPP mode yield an extremely high λ^3Q/V value of ~ 2.6×10^8. This value is comparable to the best values of dielectric cavities such as photonic crystal or microdisk cavities [11,12,19,30]. In addition, the plasmonic cavity exhibits a large confinement factor, which is here defined as the ratio of the energy confined in the dielectric nanowire core of the cavity to the total energy of the cavity mode. It was calculated to be > ~ 0.45 (right axis, Figure 3(b)).

To investigate the effect of metallic absorption loss on Q factors, we plotted Q factors for the SPP mode with m = 1 in a 40-nm-long cavity as a function of temperature (Figure 3(c)). Since lowering temperature reduces the absorption loss of the metal and increases the metal-loss-limited Q factor [23,31], high-Q SPP modes can be obtained at low temperatures. The metal-loss-limited Q factor depends predominantly on the damping collision frequency, γ. The SPP mode's Q factor increases exponentially with decreasing temperature due to the dramatic reduction of collision frequency. Even at 80 K, a high Q factor of ~ 600 was obtained. The Q factor's inverse-proportional relationship to γ shows that the Q factor is limited by metal absorption loss, and that the optical loss of the SPP cavity mode is negligible.

2.2.2 Nanorod plasmonic cavity using cutoff mirror mechanism

Maximizing the ratio between Q factor and mode volume V, i.e., Q/V, is an effective mean to increase light-matter interaction. Instead of increasing the Q factor, which is metal-loss-limited in the plasmonic cavity, we reduced the size of the plasmonic cavity by using high-index/low-index dielectric nanorods covered with silver (Figure 4(a)) [9]. The lengths and refractive indices of the high- (low-) index dielectric nanorods are L_c (L_m) and 3.4 (1.5), respectively. Along the z-axis, the SPP mode is confined at the high-index dielectric-silver interface due to the large frequency gap between the SPP mode excited at the high-index dielectric-silver interface and the SPP mode excited at the low-index dielectric-silver interface [2]. Since the top of the nanorod is covered with silver, a subwavelength SPP cavity mode is confined along the opposite z-axis, with an extremely small mode volume. In rectangular cross-sectional dielectric waveguides covered with silver, dispersion curves calculated for the fundamental SPP waveguide modes show how the SPP confinement is built (Figure 4(b)). In the waveguides, the rectangular cross-section is defined by w = 200 nm and d = 100 nm, and the refractive indices (n) of dielectric cores are 3.4 (circles) and 1.5 (squares), respectively. In Figure 4(b), the cutoff frequencies in these 2D plasmonic waveguides strongly depend on the refractive index of the dielectric core of the waveguide, n. The cutoff frequency increases (926 to 2072 THz) with decreasing n (3.4 to 1.5). The structural parameters of the waveguides, such as w and d, also affect the cutoff frequencies.

To further reduce the cavity size while maintaining the resonant wavelength, we performed systematic FDTD simulations to obtain the cutoff frequencies of high-index dielectric-silver waveguide modes for various structural parameters (Figure 5). Here, a dipole emitter was used to excite a SPP mode, which was placed 1 nm away from

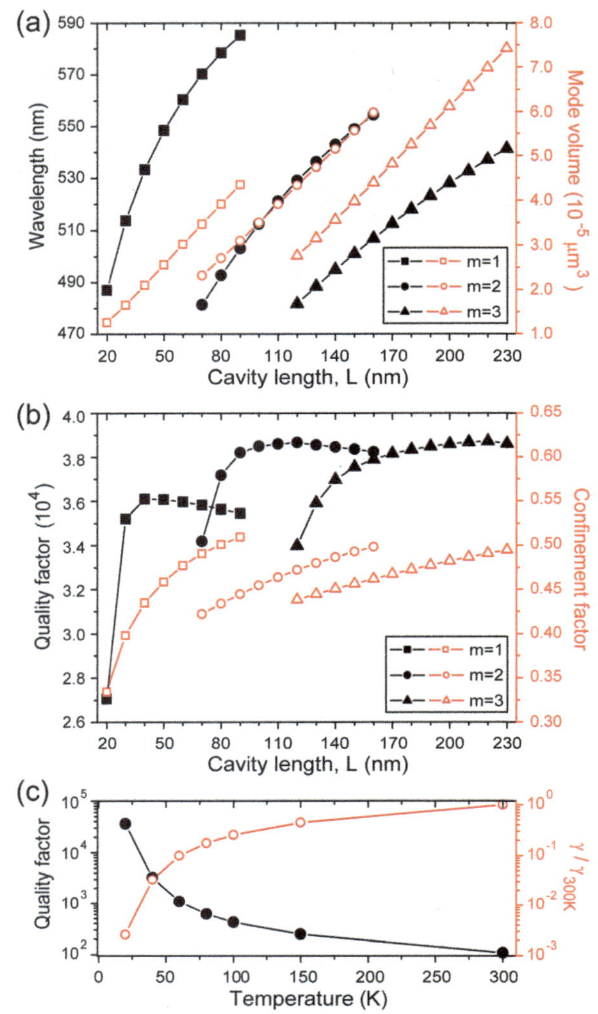

Figure 3 The optical properties of the SPP modes with m = 1, 2, and 3. (a) The resonant wavelengths and mode volumes are plotted as a function of L. An extremely small mode volume on the order of 10^{-5} μm^3 results. **(b)** The Q factors and confinement factors are plotted as a function of L at low temperature (20 K). **(c)** The Q factor of the SPP mode of m = 1 for the cavity with L = 40 nm is plotted as a function of temperature. Adapted from [2].

the sidewall of the silver surface. The SPPs are strongly confined at the high-index dielectric-silver interface (Figure 5(a)). In the color map of the cutoff frequencies plotted as a function of w and d, red (top left) and violet (bottom right) indicate higher and lower frequencies, respectively (Figure 5(b)). The resonant frequency of the SPP mode in the various cavities can be estimated based on this map, because the frequency of the cavity mode is slightly higher than the cutoff frequency. For example, if one wanted to design a cavity which has a resonant wavelength of 1550 nm, the cavities with dimensions w and d corresponding to the black dots in Figure 5(b) can be used, where the cutoff frequencies are 1550 nm.

Next, in a nanorod plasmonic cavity with $L_c = 200$ nm, the Q factors at 40 K and mode volumes were investigated as a function of the cross-sectional area of the cavity, w × d (Figure 5(c)). Since solid-state cavity quantum electrodynamics (QED) experiments studying light-matter interactions such as those involving a single photon source and strong coupling are usually performed in the temperature range from 0 to 40 K [26,32,33], the Q factor at 40 K suggests a minimum for the metal-loss-limited Q factor. The dimensional parameters, w and d, (represented as black dots in Figure 5(b) were chosen so that the resonant wavelengths remain at 1550 nm. The mode volumes (red) decrease significantly from 0.15 to 0.0038 $(\lambda/2n)^3$ with decreasing cross-sectional area from 2×10^4 to 3×10^2 nm^2, where the mode volume decreases by a factor of 40. The Q factors (black), however, decrease only slightly, from 3200 to 1500. This discrepancy is due to higher metallic absorption in the smaller cavity. We note that the mode volumes decrease much more sharply than Q factors do with decreasing physical cavity size.

The extremely small mode volume of the proposed cavity enables it to have large Purcell factors even though metallic absorption loss always occurs in a plasmonic cavity. In Figure 5(d), the Purcell factors, F, were calculated as a function of the cross-sectional area using the following equation [32]:

$$F = \frac{3}{4\pi^2}\left(\frac{\lambda}{n}\right)^3\left(\frac{Q}{V}\right)$$

Here, an emitter was assumed to be placed at the modal field maximum to calculate the maximum enhancement of Purcell factor. A large Purcell factor, greater than 2×10^5, was obtained for a cavity with dimensions $60 \times 5 \times 200$ nm^3 (Figure 5(d)). Spontaneous emission from this cavity can be enhanced considerably due to this large Purcell factor. Even if a tenfold drop in Q factor is considered at room temperature the Purcell factor remains large, ~2×10^4, owing to the tiny mode volume of 0.0038 $(\lambda/2n)^3$.

2.2.3 Room-temperature channel-waveguide plasmonic cavity

In order to reduce metallic absorption loss and increase Q factor at room temperature, a channel-waveguide plasmonic cavity is proposed by combining two silver-air channel waveguides of different widths and introducing a low-index layer on the sides and bottom walls of the silver (Figure 6) [8]. The widths of the wide and narrow waveguides are d_c and d_m, respectively. A high-index dielectric slab is added on the bottom. The thicknesses of the low-index layer and high-index slabs are t_{low} and t, respectively. In this cavity, SPPs excited at the bottom silver surface are strongly confined within the wide waveguide with length L_c.

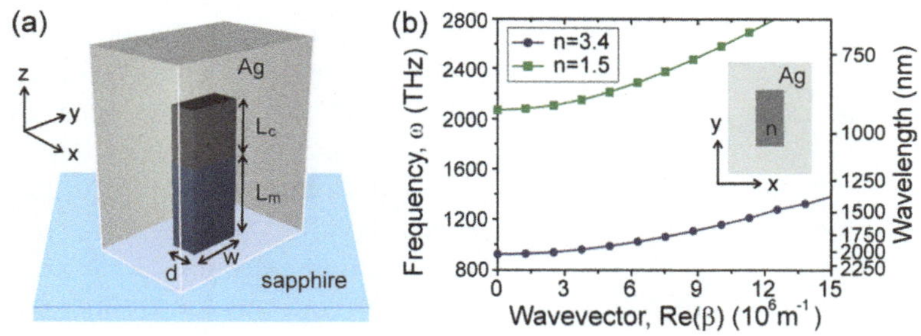

Figure 4 Plasmonic cavity with a high-index (top)/low-index (bottom) dielectric nanorod covered by silver. (a) Schematic diagram of the cavity. The cavity is placed on a sapphire substrate. **(b)** Calculated dispersion curves of the fundamental SPP mode for the core refractive index of 3.4 (blue) and 1.5 (green), with w = 200 nm and d = 100 nm. Adapted from [9].

The narrow-channel waveguides at both sides of the cavity prevent the propagation of SPP modes along the y-axis.

In a channel waveguide without a low-index layer, the SPP mode is confined at the interface of the silver and the high-index slab (left inset, Figure 7(a)). The dispersion curves significantly depend on the waveguide width, d. The cutoff frequency, appearing due to the finite waveguide width, decreases with increasing d [2,34]. The cutoff frequency of the waveguide with d = 100 nm is 1606 THz, while it is 965 THz for d = 250 nm. Therefore, the SPP mode excited in the waveguide with d = 250 nm cannot propagate into the waveguide with d = 100 nm and the narrower waveguide can act as a mode-gap mirror as long as the frequency remains in the frequency mode gap between 965 and 1606 THz. The mode-gap mirror can be used to form a channel-waveguide plasmonic cavity that consists of a wider waveguide between two narrower waveguides.

3D FDTD simulations show that a SPP cavity mode is confined efficiently in the channel-waveguide plasmonic cavity (Figure 7(b)). L_c and h are large enough ($L_c > 250$ nm and $h > 200$ nm) to achieve strong SPP confinement. In the top and side views of the electric field intensity profiles, the SPP mode is confined at the cavity's bottom dielectric-silver interface (Figure 7(c)). Along the y-axis of the waveguide, the SPP cavity mode with a resonant wavelength of 1550 nm (1216 THz) is confined within the

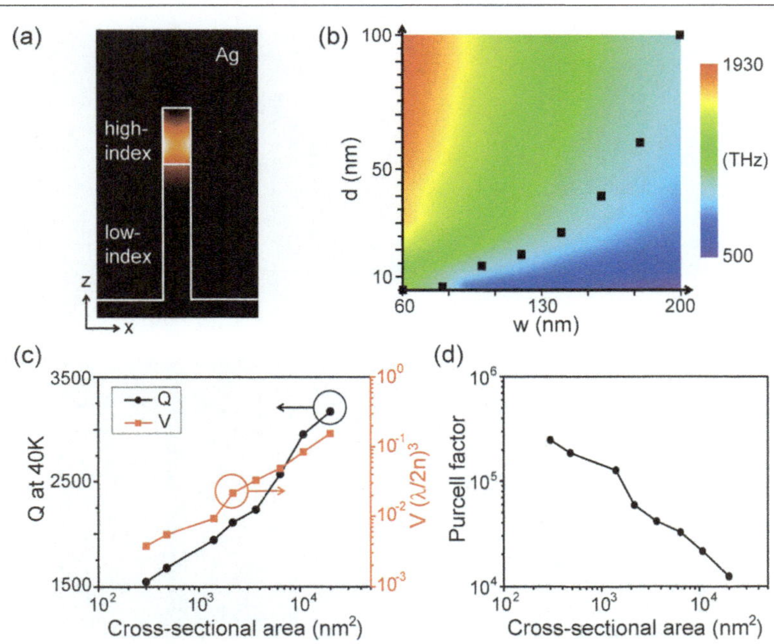

Figure 5 Optical properties of the high-index/low-index dielectric nanorod SPP cavity. (a) Electric field intensity profile of the SPP mode in the cavity. $L_C = 200$ nm, w = 200 nm, and d = 100 nm. **(b)** Cutoff frequency color map of the high-index dielectric core-silver shell SPP waveguide plotted as a function of w and d . The cutoff frequency decreases with increasing w and decreasing d. **(c)** Q factors at 40 K (black) and mode volumes (red) plotted as a function of the cross-sectional area, w × d. (w, d) ranges from (200 nm, 100 nm) to (60 nm, 5 nm). **(d)** Purcell factor plotted as a function of the cross-sectional area. Adapted from [9].

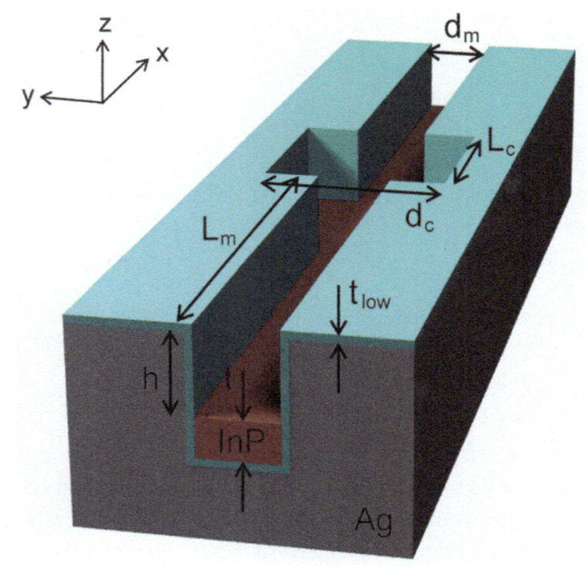

Figure 6 Channel-waveguide plasmonic cavity. SPP modes are confined in the cavity region with a width of d_c and length of L_c. Low-index and high-index layers are indicated with cyan and red colors. Adapted from [8].

wide waveguide region with dimensions $d_c \times L_c$ (250 nm × 250 nm) by the mode gap (965–1606 THz). The mode is confined along the x-axis by metallic mirrors formed with the side walls. Both of these confinement mechanisms, the mode gap and the use of metal mirrors, can be used to demonstrate the possibility of a subwavelength-scale 3D plasmonic cavity.

Losses in the metallic cavity can again be divided into optical loss and metallic absorption loss. In the channel-waveguide plasmonic cavity, radiation into free space is strongly suppressed. Therefore, optical loss can be assumed to be negligible and the cavity achieves strong optical feedback. An ultra-high optical Q factor of 1.2×10^9 was calculated by neglecting metallic absorption in the proposed channel-waveguide plasmonic cavity. The mode volume was estimated to be extremely small, $\lambda^3/10000$ or 0.0040 $(\lambda/n)^3$, where λ and n are the wavelength in free space and the refractive index of the high-index slab, respectively. As temperature increases, metallic absorption loss also increases, dominating the cavity's total losses [3,4]. The Q factor of the plasmonic cavity significantly decreases from 1.2×10^9 at 0 K to 125 at room temperature due to increased metallic absorption. The resonant wavelength and mode volume remained nearly constant across all temperatures.

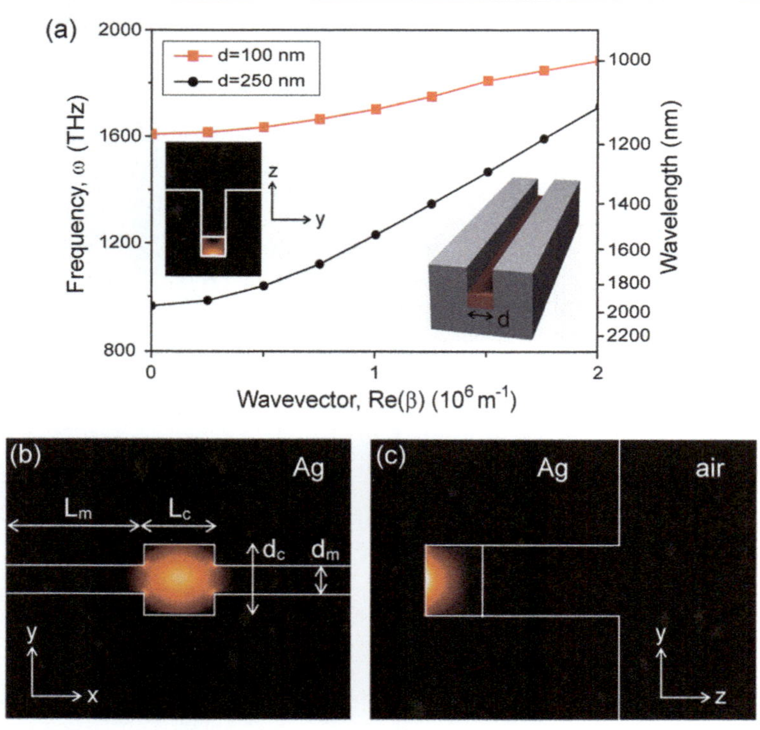

Figure 7 Field confinement in the channel-waveguide plasmonic cavity. (a) Dispersion curves of the SPP modes for the waveguides of different widths, 250 nm (black) and 100 nm (red). The right inset illustrates a channel waveguide consisting of a high-index dielectric slab and an air slot with a rectangular cross-section. The left inset shows the electric field intensity profile of the SPP waveguide mode. **(b)** Top and **(c)** cross-sectional views of the electric field intensity profile of the SPP cavity mode. d_c = 250 nm, L_c = 250 nm, d_m = 100 nm, L_m = 500 nm, t = 200 nm, and h = 500 nm. Adapted from [8].

In this case, high metallic absorption of the SPP mode is unavoidable due to the electric field overlapping with metal. Therefore, to achieve high Q at room temperature, it is necessary to minimize the amount of field energy of the SPP mode extending into metal. An efficient way to reduce metallic loss is to introduce a low-index layer at the dielectric-metal interface [35]. In the channel-waveguide plasmonic cavity, a low-index layer with a refractive index of 1.5 (e.g. SiO_2) was introduced at the silver interface (Figure 7(a)). When $t_{low} = 40$ nm, $d_c = 350$ nm, and $L_c = 350$ nm, a SPP cavity mode with a resonance wavelength of 1550 nm is strongly confined in the cavity region because of the mode gap mentioned above. Figures 8(a) and (b) show that most of the electric field intensity is strongly confined in the low-index layer, which should decrease metallic absorption loss. For this plasmonic cavity, the Q factor at room temperature is 300, which is 2.5 times larger than that of 125 for the cavity without a low index layer. The introduction of a low-index layer slightly increases the mode volume to $\lambda^3/1000$ or 0.040 $(\lambda/n)^3$, where n is the refractive index of the high-index slab.

To examine the performance of the proposed plasmonic cavity with a low-index layer, we calculated the Q (black) and confinement (red) factors at room temperature as a function of the thickness of the low-index layer, t_{low} (Figure 8(c)). Here, the confinement factor is defined by the ratio between the energy in the high-index slab and total energy of the cavity mode. The Q factor increases significantly with increasing t_{low} because the amount of mode energy overlapping into the silver decreases (Figure 8

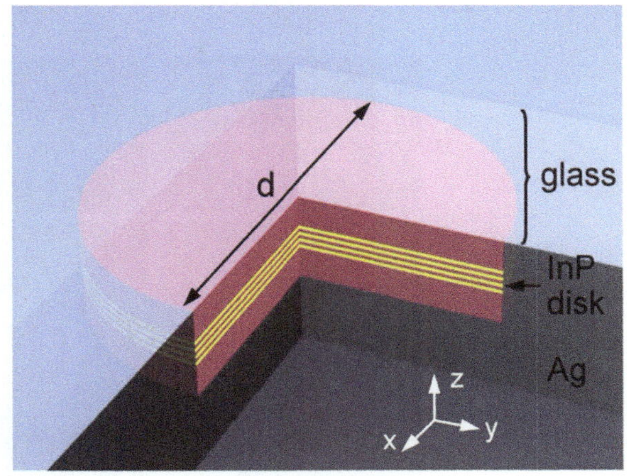

Figure 9 Schematic diagram of the nanodisk/nanopan plasmonic cavity. Adapted from [3].

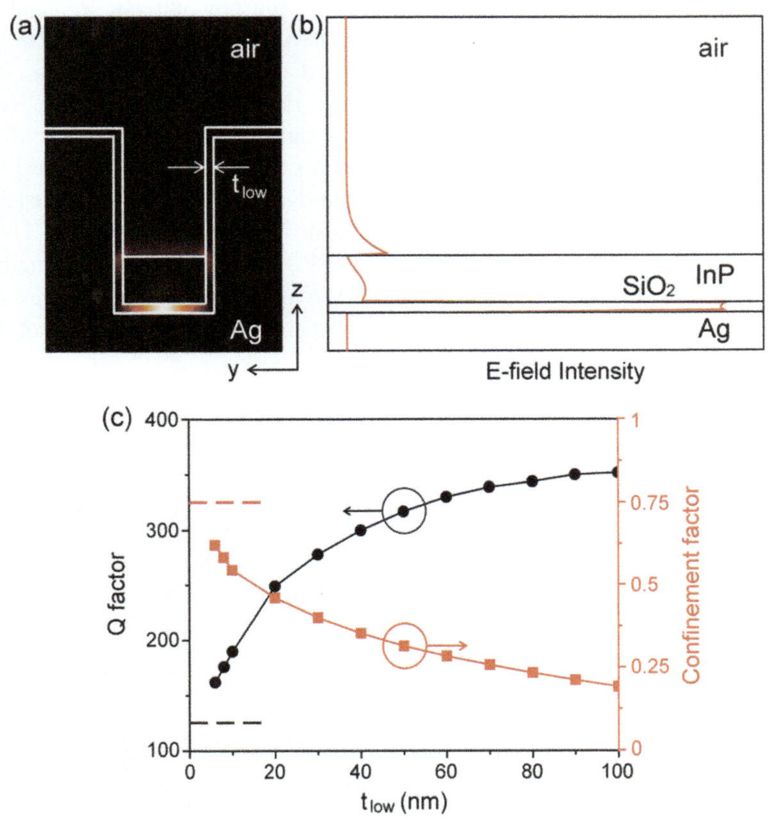

Figure 8 Plasmonic cavity with a low index layer of thickness t_{low}. (a) Cross-sectional electric field intensity profile. **(b)** Electric field intensity distribution along the z-axis at the center of the cavity. **(c)** Q factors (black) and confinement factors (red) at room temperature plotted as functions of t_{low}. Dotted lines indicate the Q and confinement factors of the cavity without a low-index layer. Adapted from [8].

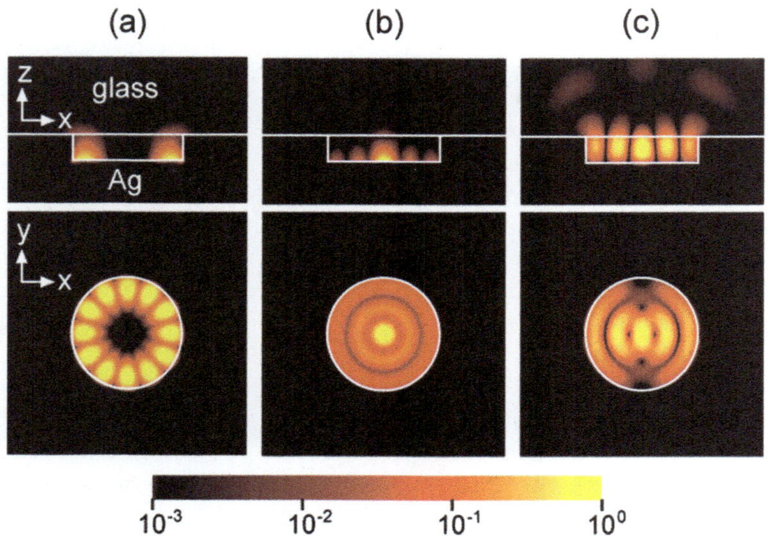

Figure 10 Side (upper panel) and top (lower panel) views of the electric field intensity profiles (log-scale) of (a) TM-like WG SPP, (b) TM-like radial SPP, and (c) dipole optical modes. Adapted from [3].

(b)). The Q factor increases up to 350, which is larger than the Q factor of 125 for the cavity without the low-index layer. On the other hand, introducing of the low-index layer decreases the confinement factor with increasing t_{low} because the energy in the low-index layer increases. Since both high Q and high confinement factors are desirable for lasing, the thickness of the low-index layer need to be optimized for the room-temperature operation of plasmonic lasers.

2.2.4 Nanodisk/nanopan plasmonic cavity

We finally examine a nanodisk/nanopan plasmonic cavity. This cavity consists of an InP nanodisk with a radius of r and a silver nanopan covering the bottom and sidewall surfaces of the nanodisk (Figure 9) [3,4]. Various SPP modes are excited at the disk-nanopan interfaces in conjunction with the suppression of optical modes. In particular, high-Q whispering-gallery (WG) modes, which are typically observed in a conventional disk cavity, appear as SPP modes in the nanodisk/nanopan structure. In addition, the silver nanopan surrounding the sidewall of the disk suppresses radial optical loss and enables the excitation of radial SPP modes, which are barely observed in a conventional disk-shaped cavity. In fact, using this nanodisk/nanopan structure, plasmonic lasing operation was experimentally demonstrated at 10 K [3,4]. Optical pumping and photoluminescence (PL) collection were performed through the glass substrate bonded to the nanodisk (Figure 9).

In order to understand the SPP modes excited in this nanodisk/nanopan structure, all possible cavity modes were modeled using FDTD simulation [3,4]. In the simulation, the refractive indices of the InP and glass were 3.2 and 1.45, respectively. Representative resonant modes are shown for a nanodisk/nanopan structure with a nanodisk radius of 500 nm in Figure 10. Among three modes, two SPP modes (transverse-magnetic-like (TM-like) radial and TM-like WG) have an electric field maximum at the disk-silver interface, whereas one optical mode (a dipole mode) has an electric field maximum at the center of the dielectric disk. The TM-like WG SPP and dipole optical modes are doubly degenerate but the TM-like radial SPP is non-degenerate. The TM-like WG and radial SPP modes have subwavelength mode volumes of 0.56 $(\lambda/2n)^3$ and 0.65 $(\lambda/2n)^3$, respectively, which are smaller than the diffraction limit of light. On the other

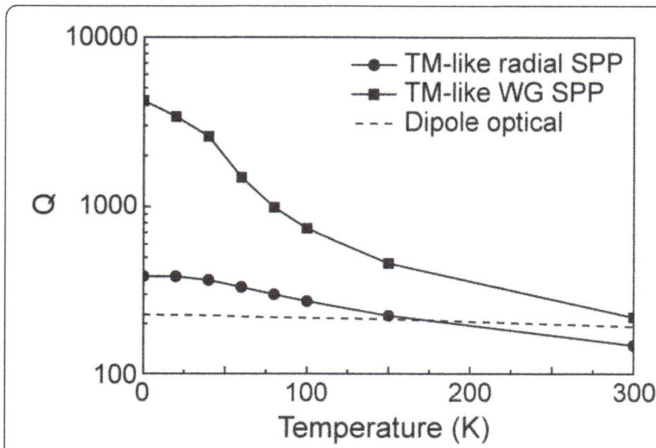

Figure 11 Q factors of TM-like WG/radial SPP modes and dipole optical mode plotted as a function of temperature. The disk radii are all 500 nm in this simulation. Adapted from [3].

hand, the mode volume of the optical dipole mode, 2.7 $(\lambda/2n)^3$, is three times larger than this limit.

The WG SPP modes derive from the conventional WG optical modes [13-15,30]. For example, the dominant electric field of conventional TM-like WG optical modes excited in a nanodisk without a silver nanopan is directed outward from the bottom surface of the disk. Introducing the silver nanopan converts these optical modes into the TM-like WG SPP modes. On the other hand, the transverse-electric-like (TE-like) WG optical modes are converted into the TE-like WG SPP modes because all of their dominant electric fields point in a perpendicular direction to the sidewall of the disk.

The Q factors of three excited modes in the nanodisk/nanopan cavity were calculated as a function of temperature using a FDTD simulation (Figure 11). The TM-like WG and radial SPP modes show significant temper ature-dependence in their Q factors. At low temperature, the Q factors of the SPP modes are higher than that of the optical mode (dipole mode) due to low metallic absorption loss and relatively low optical radiation loss, however, they seriously deteriorate with increasing temperature and absorption loss [2-4]. In contrast, the Q factors of the optical mode remained nearly constant over the calculated range of temperatures. This result agrees well with the consideration that the SPP modes' electric fields extend considerably into the metal, whereas the optical mode has relatively small overlap with the metal nanopan. Indeed, strong temperature-dependent optical loss of SPP modes is a major factor distinguishing SPP modes from conventional optical modes.

3 Conclusion

In summary, we investigated the optical properties of four subwavelength-scale plasmonic cavities: a dielectric-core/metal-shell nanowire plasmonic cavity, a nanorod plasmonic cavity using a cutoff mirror mechanism, a channel-waveguide plasmonic cavity, and a nanodisk/nanopan plasmonic cavity. Experimental demonstrations of deep subwavelength-scale photonic devices such as single photon sources, plasmonic lasers, optical memory devices, and ultrasmall biochemical sensors can be expected based on these theoretical plasmonic cavity structures with ultrasmall cavity sizes. In particular, plasmonic lasers may prove to be promising coherent light sources, as they enable the miniaturization of nanophotonic devices as well as the ultra-compact integration of photonic systems requiring minimal thermal overhead.

Competing interests

The authors declare that they have no competing interests.

Authors' contributions

S-HK, Y-SN and H-GP wrote the manuscript. All authors read and approved the final manuscript.

Acknowledgments

H.-G.P. acknowledges support by the National Research Foundation of Korea (NRF) grant funded by the Korea government (MSIP) (No. 2009–0081565). S.–H.K. acknowledges support by the Basic Science Research Program through the NRF funded by the Ministry of Education, Science and Technology (MEST) (NRF-2013R1A2A2A01014491). Y.-S.N. acknowledges the support by the TJ Park Science Fellowship.

Author details

[1]Department of Physics, Chung-Ang University, Seoul 156-756, Republic of Korea.
[2]Department of Physics, Korea University, Seoul 136-701, Republic of Korea.

References

1. M Kuttge, FJG de Abajo, A Polman, Nano Lett. **10**, 1537 (2010)
2. MK Seo, SH Kwon, HS Ee, HG Park, Nano Lett. **9**, 4078 (2009)
3. SH Kwon, JH Kang, C Seassal, SK Kim, P Regreny, YH Lee, CM Lieber, HG Park, Nano Lett. **10**, 3679 (2010)
4. SH Kwon, JH Kang, SK Kim, HG Park, IEEE J. Quantum Electron. **47**, 1346 (2011)
5. MT Hill, M Marell, ESP Leong, B Smalbrugge, Y Zhu, M Sun, PJ van Veldhoven, EJ Geluk, F Karouta, Y-S Oei, R Notzel, C-Z Ning, MK Smit, Opt. Express **17**, 11107 (2009)
6. RF Oulton, VJ Sorger, T Zentgraf, R-M Ma, C Gladden, L Dai, G Bartal, X Zhang, Nature **461**, 629 (2009)
7. R-M Ma, RF Oulton, VJ Sorger, G Bartal, X Zhang, Nat. Mater. **10**, 110 (2011)
8. JH Kang, HG Park, SH Kwon, Opt. Express **19**, 13892 (2011)
9. JH Kang, YS No, SH Kwon, HG Park, Opt. Lett. **36**, 2011 (2011)
10. O Painter, RK Lee, A Scherer, A Yariv, JD O'Brien, PD Dapkus, I Kim, Science **284**, 1819 (1999)
11. HG Park, SH Kim, SH Kwon, YG Ju, JK Yang, JH Baek, SB Kim, YH Lee, Science **305**, 1444 (2004)
12. K Nozaki, S Kita, T Baba, Opt. Express **15**, 7506 (2007)
13. L Liu, R Kumar, K Huybrechts, T Spuesens, G Roelkens, E-J Geluk, T de Vries, P Regreny, DV Thourhout, R Baets, G Morthier, Nat. Photonics **4**, 182 (2010)
14. Q Song, H Cao, ST Ho, GS Solomon, Appl. Phys. Lett. **94**, 061109 (2009)
15. YH Kim, SH Kwon, JM Lee, MS Hwang, JH Kang, WI Park, HG Park, Nat. Commun. **3**, 1123 (2012). doi:10.1038/ncomms2137
16. MH Huang, S Mao, H Feick, H Yan, Y Wu, H Kind, E Weber, R Russo, P Yang, Science **292**, 1897 (2001)
17. X Duan, Y Huang, R Agarwal, CM Lieber, Nature **421**, 241 (2003)
18. MK Seo, JK Yang, KY Jeong, HG Park, F Qian, HS Ee, YS No, YH Lee, Nano Lett. **8**, 4534 (2008)
19. MT Hill, YS Oei, B Smalbrugge, Y Zhu, T de Vries, PJ van Veldhoven, FWM van Otten, TJ Eijkemans, JP Turkiewicz, H de Waardt, EJ Geluk, SH Kwon, YH Lee, R Nötzel, MK Smit, Nat. Photonics **1**, 589 (2007)
20. K Yu, A Lakhani, MC Wu, Opt. Express **18**, 8790 (2010)
21. MP Nezhad, A Simic, O Bondarenko, B Slutsky, A Mizrahi, L Feng, V Lomakin, Y Fainman, Nat. Photonics **4**, 395 (2010)
22. CY Lu, SW Chang, SL Chuang, TD Germann, D Bimberg, Appl. Phys. Lett. **96**, 251101 (2010)
23. SA Maier, *Plasmonics: Fundamentals and Applications; Springer*, 2007
24. A Taflove, SC Hagness, *Computational Electrodynamics: The Finite-Difference Time-Domain Method*, 2nd edn. (Artech House, Boston, 2000)
25. PB Johnson, RW Christy, Phys. Rev. B **6**, 4370 (1972)
26. YY Gong, J Vuckovic, Appl. Phys. Lett. **90**, 033113 (2007)
27. ED Palik, *Handbook of Optical Constants of Solids* (Academic, New York, 1985)
28. Q Fang, Y Li, S Gradecak, H-G Park, Y Dong, ZL Wang, CM Lieber, Nat. Mater. **7**, 701 (2008)
29. JA Dionne, HJ Lezec, HA Atwater, Nano Lett. **6**, 1928 (2006)
30. K Vahala, Nature **424**, 839 (2003)
31. B Min, E Ostby, V Sorger, E Ulin-Avila, L Yang, LX Zhang, K Vahala, Nature **457**, 455 (2009)
32. D Englund, D Fattal, E Waks, B Zhang, T Nakaoka, Y Arakawa, Y Yamamoto, J Vuckovic, Phys. Rev. Lett. **95**, 013904 (2005)
33. JP Reithmaier, G Sek, A Loffler, C Hofmann, S Kuhn, S Reitzenstein, LV Keldysh, VD Kulakovskii, TL Reinecke, A Forchel, Nature **432**, 197 (2004)
34. JA Dionne, LA Sweatlock, HA Atwater, A Polman, Phys. Rev. B **73**, 035407 (2006)
35. RF Oulton, VJ Sorger, DFP Pile, DA Genov, X Zhang, Nat. Photonics **2**, 496 (2008)

21

A systematic study of core size and coating thickness on manganese-doped nanocrystals for high T2 relaxivity as magnetic resonance contrast agent

Joseph Park[1], Byunghoon Kang[1], Bongjune Kim[1], Jin-Suck Suh[2,3], Yong-Min Huh[2,3*] and Seungjoo Haam[1*]

Abstract

We describe a systematic study of coating thickness and their effect on different core sizes for the optimized preparation of highly sensitive manganese-doped magnetic nanocrystals (MnMNCs) to be served as magnetic resonance (MR) contrast agent. From these efforts, MnMNCs with 12 nm core and DA-PEG2k coating demonstrated that T2 relaxivity (r2) was increased by 7.29-fold (r2 value: 452 $mM^{-1}\ s^{-1}$) compare to conventional iron oxide (CLIO) and remarkable colloidal stability in various physiological conditions. Further *in vitro* cellular MR imaging results showed that MnMNC-PEGs were biocompatible and well suited for medical applications. This study will provide a useful synthetic strategy for the development of highly effective MR contrast agents.

Keywords: Relaxivity; Coating thickness; Manganese-doped nanocrystal; Sensitivity; MRI

1 Background

Magnetic nanocrystals (MNCs) are emerging research areas for their potential applications in biomedical sciences such as magnetic resonance imaging, cancer treatment and micro-NMR sensors [1-3]. Researchers are aiming to make high sensitive magnetic materials by means of increasing size, engineering magnetism, and surface coating variations [4-6]. Manganese-doped magnetic nanocrystals (MnMNCs) are one of the most important material because of relatively simple preparation method and increased mass magnetization value compare to conventional iron oxides [7]. However MnMNCs are coated with hydrophobic ligands and eventually they need to be encapsulated in hydrophilic ligands to disperse them in water phase. Our group reported a method to make highly water dispersible MnMNCs, they are still in suboptimal potency of sensitivity because magnetic field surrounding a magnetic nanocrystals would be fallen as the distance go far from the core material. And the coating molecules are important factor that would affect the nuclear relaxation of water protons by forming hydrogen bond. Therefore, it is necessary work to study the affect of coating materials even if their iron oxide core sizes are similar [8-11].

Herein, we report on an optimized design for manganese-doped magnetic nanocrystals (MnMNCs), capable of achieving maximal r2 and colloidal stability (Figure 1). These synthesized MnMNCs with different coating thickness can exhibit quite different r2 even if their MnMNC core sizes remain similar. These constructions could approach the maximum r2 coefficient for a given material. As a specific example in the present study, we synthesized MnMNCs for T2-weighted magnetic resonance (MR) imaging contrast agent and encapsulated into a various PEG shell. The resultant MnMNCs with 12 nm core size encapsulated with dodecanoic acid-PEG2K Da (MnMNC12-PEG2K) exhibited most high r2 relaxivity (452 $mM^{-1}\ s^{-1}$ [metal]), 7.29-fold higher compared to conventional iron oxide (CLIO), and showed excellent colloidal stability. The prepared MnMNC12-PEG2K was subsequently applied to an intra-cellular imaging system to demonstrate their use in identifying target cells *in vitro*.

* Correspondence: ymhuh@yuhs.ac; haam@yonsei.ac.kr
[2]Department of Radiology & Department of Biochemistry and Molecular Biology, Yonsei University, Seoul 120-752, South Korea
[1]Department of Chemical and Biomolecular Engineering, Yonsei University, Seoul 120-749, South Korea
Full list of author information is available at the end of the article

2 Methods

2.1 Materials

Iron(III) acetylacetonate, manganese (II) acetylacetonate, 1,2-hexadecanediol, dodecanoic acid, dodecylamine, benzyl ether, anhydrous dichloromethane, monomethylpolyethylene glycol (mPEG; Mw 1 k, 2 k, 5 k, 10 k, 20 k Da) were purchased from Sigma-Aldrich. All other chemicals and reagents were of analytical grade.

2.2 Synthesis of 6 nm $MnFe_2O_4$ Magnetic Nanocrystals

$MnFe_2O_4$ nanocrystals were synthesized by seed-mediated growth method [5]. Typically, 2 mmol iron (III) acetylacetonate, 1 mmol manganese (II) acetylacetonate, 10 mmol 1,2-hexadecanediol, 6 mmol dodecanoic acid, 6 m moldodecylamine, and 20 mL of benzyl ether were mixed under a nitrogen atmosphere. The mixture was preheated to 200°C for 120 min and then refluxed at 300°C for 60 min. After being cooled to room temperature, the products were purified with an excess of pure ethanol [12].

2.3 Synthesis of 12 nm $MnFe_2O_4$ Magnetic Nanocrystals

2 mmol iron (III) acetylacetonate, 1 mmol manganese (II) acetylacetonate, 10 mmol 1,2-hexadecanediol, 2mmoldodecanoic acid, 2 mmol dodecylamine, and 20 mL of benzyl ether were mixed and magnetically stirred under a flow of N_2. Eighty four milligram sample of pre-synthesized 6 nm $MnFe_2O_4$ nanoparticles dispersed in hexane (1 mL) was added into the mixture. The mixture was first heated to 110°C for 30 min to remove hexane, then further to 200°C for 1 h. Under a blanket of nitrogen, the mixture was further heated to reflux (300°C) for 30 min. The black-colored mixture was cooled to room temperature by removing the heat source. After being cooled to room temperature, the products were purified with an excess of pure ethanol [12].

2.4 Synthesis of DA-PEG Block Copolymers

DA-PEG block copolymer was synthesized as described previously [6,7].As reported, a solution of 30 mmol dodecanoic acid (DA) and 10 mmol mPEG dissolved in 40 mL of anhydrous dichloromethane was activated by adding 30 mmol of N,N'-dicyclohexylcarbodiimide (DCC) and 4-dimethylaminopyridine (DMAP). The reaction was carried out for 48 h at room temperature under a nitrogen atmosphere. The resulting product was filtered using a cellulose acetate syringe filter (pore size ≈ 200 nm) and dialyzed for two weeks against 10 mM sodium phosphate buffer (pH 7.4) using dialysis tube.

2.5 Surface coating of $MnFe_2O_4$nanocrystals with DA-PEG

30 mg of $MnFe_2O_4$nanocrystals was dissolved in 4 mL of chloroform. This organic phase was added to 20 mL of sodium phosphate buffer containing 200 mg of DA-PEG with various sizes of mPEG (1 K, 2 K, 5 K, 10 K and 20 K molecular weight), respectively. After mutual saturation of the organic and continuous phases, the mixture was emulsified for 15 min with an ultrasonicator (ULH700S, Ulssohitech, Korea) at 300 W. After solvent evaporation in 4 h, the product was purified 3 times with centriprep at 3,000 rpm for 30 min to remove excess DA-PEG molecules [9].

2.6 Colloidal Stability

The colloidal stability of the prepared MnMNC-PEGs was determined from their resistance to pH-induced nanoparticle aggregation. A 100 μL nanoparticle suspension (20 mg/mL) was added to 2 mL (pH 2, 4, 74, 9) at room temperature and then size of the suspension was measured using laser scattering (ELS-Z, Otsuka electronics).

2.7 Cell viability assay by MTT

The biocompatibility of the prepared MNC6-PEG1K and MNC12-PEG2K for macrophage cells was quantified by a colorimetric assay based on the mitochondrial oxidation of 3-(4,5-dimethylthiazolyl-2)-2,5-diphenyltetrazolium bromide (MTT). RAW 264.7 cells were harvested at a density of 10^4 cells/200 μL in a 96-well plate and incubated at 37°C under 5% CO_2 atmosphere. The cells were incubated for 24 h with prepared MnMNCs, rinsed with 100 μL PBS (pH 7.4, 1 mM), and then treated with freshly prepared MTT solution (10 μL) and incubated for an additional 4 h before adding 100 μL dimethylsulfoxide. After 24 h, the plates were assayed using an enzyme-linked immunosorbent assay (Spectra Max 340, Molecular Devices, USA) and the results were measured at an absorbance wavelength of 575 nm and a reference wavelength of 650 nm [9].

2.8 MR imaging

We performed MR imaging experiments with a 1.5-T clinical MRI instrument with a micro-47 surface coil (Intera; Philips Medical Systems, Best, the Netherlands). R2 relaxivities of MnMNC6-PEG(1 k ~20 k) and MnMNC12-PEG (1 k ~20 k) were measured using the Carr-Purcell-Meiboom-Gill sequence at room temperature: TR = 10 s, 32 echoes with 12 ms even echo space, number of acquisition = 1, point resolution of 156 × 156 μm, section thickness of 0.6 mm. R2 was defined as 1/T2 with units of s^{-1}. For T2-weighted MR imaging of cells *in vitro* at 1.5 T, the following parameters were used: point resolution: 156 × 156 μm, section thickness of 0.6 mm, TE = 60 ms, TR = 4000 ms, number of acquisitions = 1. For T2 mapping of cells in vitro, the following parameters were used: point resolution of 156 × 156 μm, section thickness of 0.6 mm, TE = 20, 40, 60, 80, 100, 120, 140, 160 ms, TR = 4000 ms, number of acquisitions =2.

2.9 Prussian blue stain

RAW 264.7 cells (5.0×10^5 cells/well) were seeded onto six-well plates and incubated for 24 h at 37°C. Prepared

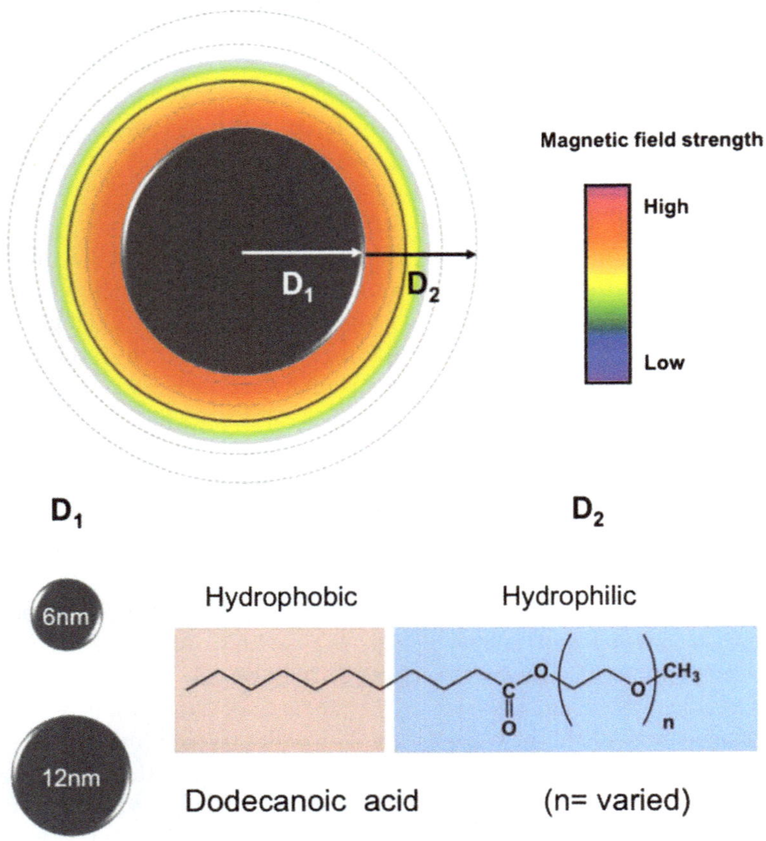

Figure 1 Schematic illustration of optimal coating thickness on different sizes of $MnFe_2O_4$ (6 nm and 12 nm). The color bar represents the magnitude of the magnetic field strength of the MnMNC cores. Solid and dashed lines represent the boundaries of PEG coating layers, respectively. Starting from the center, the lines indicate PEG1k, PEG2k, PEG5k, PEG10k, and PEG20k.

MnMNC6-PEG1K and MnMNC12-PEG2K (100 μg of MnFe/mL) were added to Dulbecco's modified eagle medium (DMEM, Gibco®, Invitrogen, USA). After incubation for 24 h at 37°C, the cells with MnMNC6-PEG1K and MnMNC12-PEG2K were detached, centrifuged and washed three times with PBS (pH 7.4, 1 mM). The detached cells were fixed and immersed in iron staining solution (20% hydrochloric acid: potassium ferrocyanate =1: 1) for 30 min at room temperature after being fixed in 95% alcohol for 5 min. Then, the samples were rinsed three times in deionized water to remove the residual staining solution. Subsequently, the samples were stained with the nuclear staining solution (Nuclear Fast Red) for 15 min, followed by three washes with deionized water, and were finally fixed in increasing concentrations of alcohol and xylene [13].

2.10 Characterization

The morphologies and the sizes of the prepared MnMNCs were analyzed using high resolution transmission electron microscopy (HR-TEM, JEM-2100 LAB6, JEOL Ltd., Japan) and laser scattering (ELS-Z, Otsuka electronics, Japan). X-ray diffraction measurement was performed by a Rigaku D/max-RB (Tokyo, Japan) powder diffractometer and image-plate photography using graphite-monochromatized Cu Kα radiation (λ =1.542 Å) to determine the lattice of the $MnFe_2O_4$. Data were collected from 20° to 80° with a step size of 0.05° and step time of 5 s. The amounts of metal ions were quantified using inductively coupled plasma atomic emission spectrometry (ICP-AES, Thermo electron corporation, USA).

3 Results and discussion

Manganese-doped magnetic nanocrystal cores ($MnFe_2O_4$; MnMNCs) exhibiting high saturation of magnetization were selected as a core material and synthesized first as reported [12]. Briefly, the particles were prepared by thermally decomposing a mixture of $Fe(acac)_3$ and $Mn(acac)_2$ in 1,2-hexadecanediol. The as-synthesized particles were then mixed and reacted with additional precursors to allow for particle growth. Using this procedure, we obtained highly monodisperse 6 nm MnMNCs (MnMNC6, size variation <5%) and 12 nm MnMNCs (MnMNC12, size variation <8.9%), respectively (Figure 2). The XRD patterns of the samples were very similar and showed characteristic reflection peaks due to the presence of a

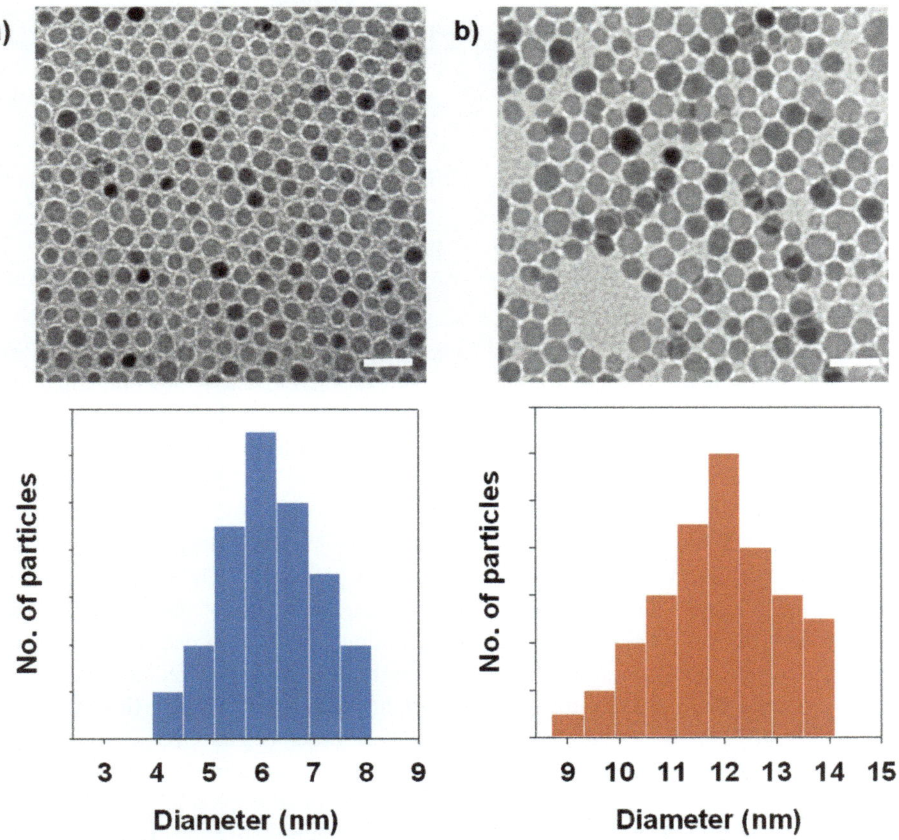

Figure 2 TEM images of $MnFe_2O_4$ magnetic nanocrystals (MnMNCs) at the same magnification. The average diameters of the MnMNC, obtained by measuring about 50 particles for each sample, **(a)** MnMNC6 **(b)** MnMNC12. All scale bars are 20 nm.

nanocrystalline inverse spinel structure (Figure 3a). The peak intensities increased from MnMNC6 to MnMNC12, indicating a progressive increase in the average particle size, also in agreement with TEM observation. The average particle sizes as determined by the Scherrer equation from line broadening of the (311) reflection (2θ =35.5) were of 5.5 and 11.6 nm for the sample of MnMNC6 and MnMNC12, respectively. The magnetic hysteresis loops of the MnMNC6 and MnMNC12 were observed at 300 K (Figure 3b) and both exhibited superparamagnetic behavior without remnant magnetic hysteresis at zero field. In addition, their saturation of magnetization values were 87.7 and 48.3 emu/g, respectively, on the basis of dried weight at 1.0 T. Due to the presence of

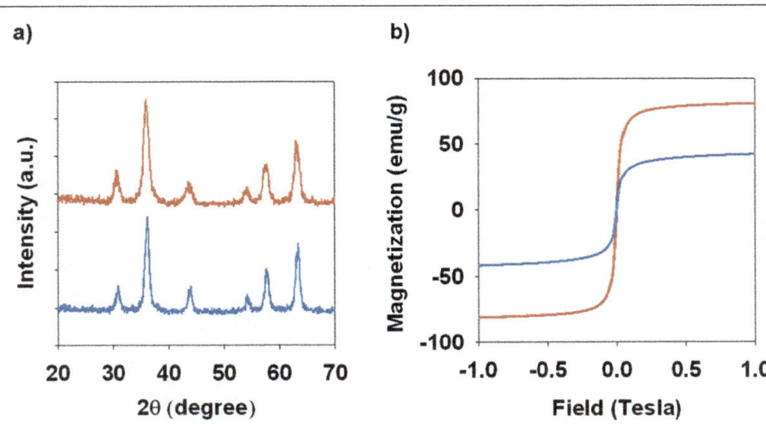

Figure 3 Characterization of MnMNCs. (a) X-ray powder diffraction patterns and **(b)** mass magnetization (M) as a function of applied external field (H) measured at 300 K for MnMNC6 (blue) and MnMNC12 (red).

Table 1 Calculated [14] and measured thickness of different core and PEG sizes

		PEG 1 K (N = 22)	PEG 2 K (N = 45)	PEG 5 K (N = 113)	PEG 10 K (N = 226)	PEG 20 K (N = 454)
MnMNC6	Total	10.8 ± 1.7	12.8 ± 1.4	18.8 ± 2.7	22.8 ± 2.1	26.8 ± 1.5
	Layer. exp	1.8 nm	3.8 nm	9.8 nm	13.8 nm	17.8 nm
	Layer. cal	1.8 nm	3.53 nm	7.4 nm	14.2 nm	24.1 nm
MnMNC12	Total	21.6 ± 2.8	23.6 ± 3.2	28.6 ± 2.7	34.6 ± 2.4	41.6 ± 2.6
	Layer. exp	2.6 nm	4.9 nm	10.3 nm	15.2 nm	22.6 nm
	Layer. cal	1.9 nm	3.8 nm	8.3 nm	11.8 nm	20.9 nm

organic components (DA-PEGs), the saturation of magnetization of MnMNC-PEGs was relatively lower than that of previously reported $MnFe_2O_4$ nanocrystals as expected.

The as-synthesized MnMNC cores were enveloped by oleic acid/oleylamine and dispersed in chloroform. For water dispersible MnMNCs, amphiphilic block copolymer was synthesized using hydrophobic dodecanoic acid (DA) and various sizes of hydrophilic monomethylpolyethyleneglycol (mPEG, 1 K, 2 K, 5 K, 10 K and 20 K) [9]. Because of the hydrophobic interaction of MnMNCs with the DA component of amphiphilic block copolymer, the PEG-coated MnMNCs were successfully dispersed in aqueous phase. As we expected, the use of PEGs with higher molecular masses resulted in larger hydrodynamic diameters. More specifically, we calculated the series of PEG layer thicknesses and compared with our experimental data. First, we hypothesize the thickness of the hydrophobic layers, formed by oleic acid/oleylamine with DA interaction, are supposed to be the same as that of the cell membrane (3 nm), the statistical length of a PEG monomer (=0.39 nm) and the density of DA-PEG on a micelle (0.59 PEG/nm^2) [14,15]. MnMNC6-PEG systems are quite well matched with calculated model systems approving the thickness of PEG variation ~ <2 nm. MnMNC12-PEG systems are slightly larger than ~ >6 nm as PEG molecular weight increase. The difference of the hydrodynamic diameters might be caused by the presence of a tiny amount of small particle clusters (e.g., dimers or trimers). This may be attributed to the strong interactions between MnMNC12 hydrophobic surfaces. The calculated and measured MnMNC-PEG layer thicknesses are listed in Table 1.

To determine the optimized thickness of various PEG coated MnMNCs listed in Table 1 for MR contrast agents, we have experimentally measured their relaxivity coefficient values of r2, defined as relaxation rates at unit concentration (1 mM metal ions, $R2 = T2^{-1}$) which were obtained from the relaxation time versus [Fe + Mn] concentration plots (Figure 4). The data clearly showed a dependency of proton relaxivity on their particle sizes (magnetic domain size). MnMNC12 exhibited consistently higher relaxation rate than the MnMNC6 for all 5-series PEG coatings. For instance, MnMNC12 with PEG2k coating showed 2.16-fold higher r2 value than that of the MnMNC6 coated with identical PEG size. As previously mentioned, this size dependency is due to the effect of surface spin anisotropy which is more pronounced for smaller magnetic domain sizes due to larger surface area to volume ratio. An interesting feature is that at the same particle core size, the measured proton relaxivity depends on the nature of PEG coating thickness,

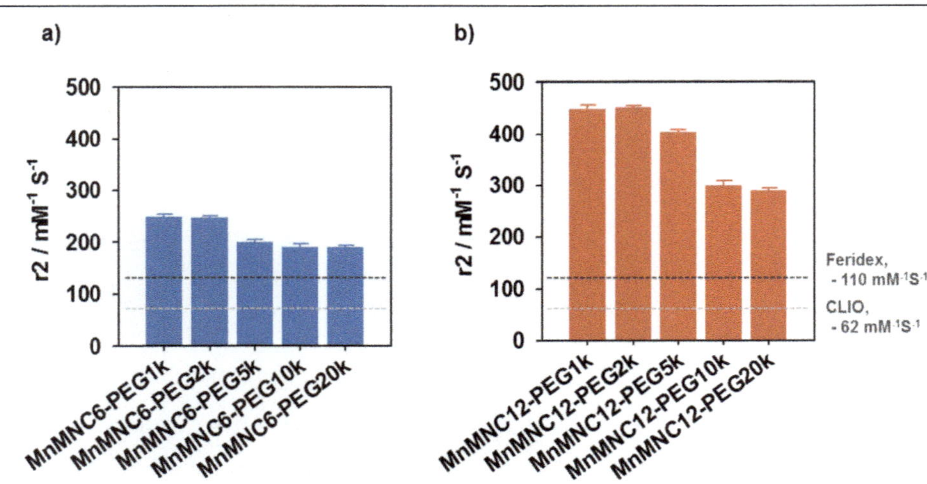

Figure 4 Dependency of relaxivity coefficient (r2, $mM^{-1}S^{-1}$) values on the core size and the PEG chain length of (a) MnMNC6, (b) MnMNC12. Comparison of r2 values of conventional iron oxide nanoparticles (dotted black line: Feridex, dotted gray line: CLIO).

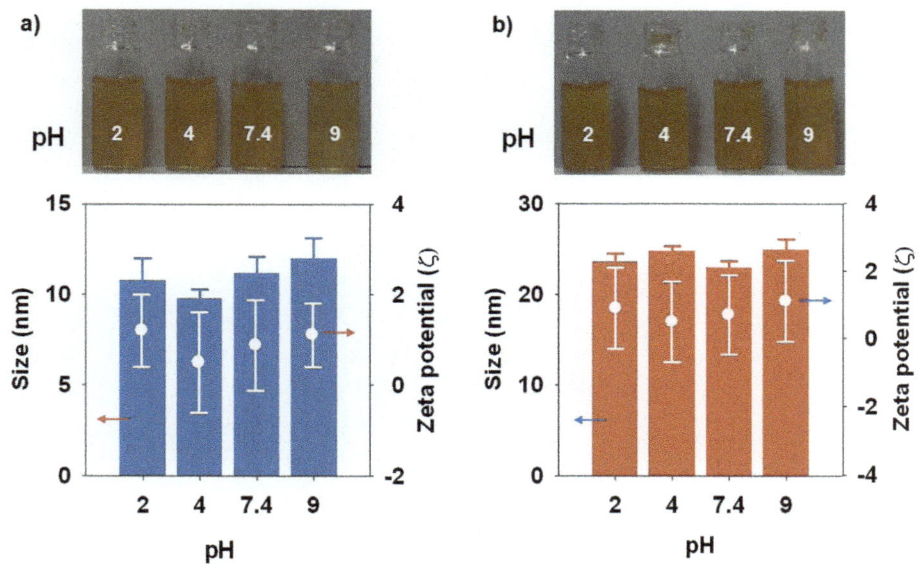

Figure 5 Photographs of colloidal stability tests and hydrodynamic sizes and zeta-potential values of (a) MnMNC6-PEG1K, (b) MnMNC12-PEG2K on various pH conditions. (pH 2, 4, 7.4 and 9).

leading to different values of relaxivity. The r2 coefficient of the MnMNC12 increased by 1.54-fold as the PEG molecular weight decreased from 20 to 2 K; however, it did not increase further as the PEG size further decreased to 2 and 1 K. MnMNC6 has a similar tendency; its r2 coefficient increased by 1.30-fold as the PEG molecular weight decreased from 20 to 1 K. Interestingly, both cores (6 and 12 nm) would have a critical PEG size, at which the r2 coefficient can be optimized through surface coating thickness. For instance, 12 nm MnMNCs core with DA-PEG2K exhibited the highest r2 coefficient of 452 ± 7.4 mM^{-1} s^{-1}, which is 7.29- and 4.10-folds higher than those of CLIO and Feridex, respectively. Consequently, from the optimal combination of the core size and coating thicknesses, r2 value maximized MnMNC design could be determined and therefore, MnMNC12-PEG2K and MnMNC6-PEG1K as optimized MnMNC candidates were selected for further physiological condition tests for MR contrast agents.

We next investigated colloidal stability of aqueous MnMNCs to be served as a MR contrast agents (Figure 5). The colloidal stabilities of MnMNCs in aqueous solution over a wide range of pH (2, 4, 7.4, and 9) are tested using laser scattering and their results were plotted against pH. We observed that their sizes were not changed indicating successful enveloping of PEG chains on the surface of MnMNCs preventing aggregation and their zeta potential values of the MnMNCs were also well maintained around at 0 ~ 2 mV (Figure 5). Although PEG coated MnMNCs exhibiting nearly neutral surface charge, excellent colloidal

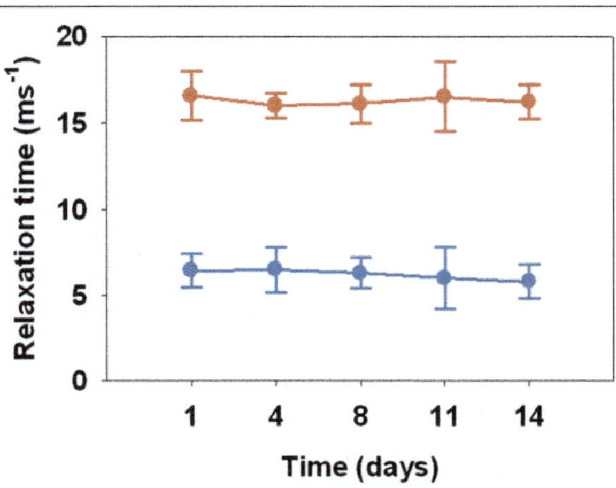

Figure 6 Time dependent (days) evaluation of transverse relaxation time (R2) of MnMNC6-PEG1K (blue) MnMNC12-PEG2K (red) in PBS ([Fe + Mn] concentration: 25 μM/ 200 μL).

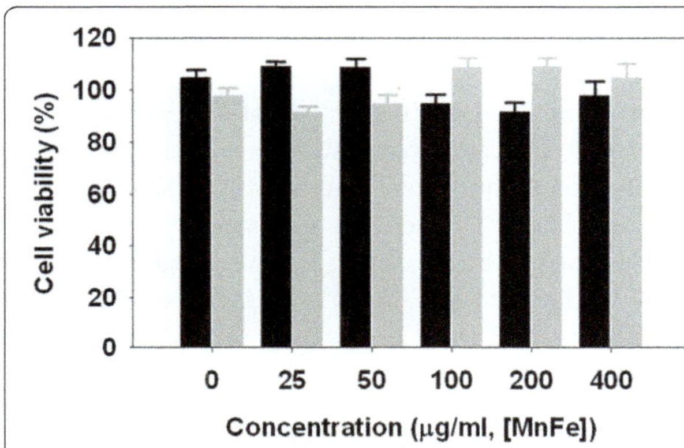

Figure 7 Cytotoxicity test of RAW 264.7 cells treated with MnMNC6-PEG1K (black) and MnMNC12-PEG2K (Gray), respectively.

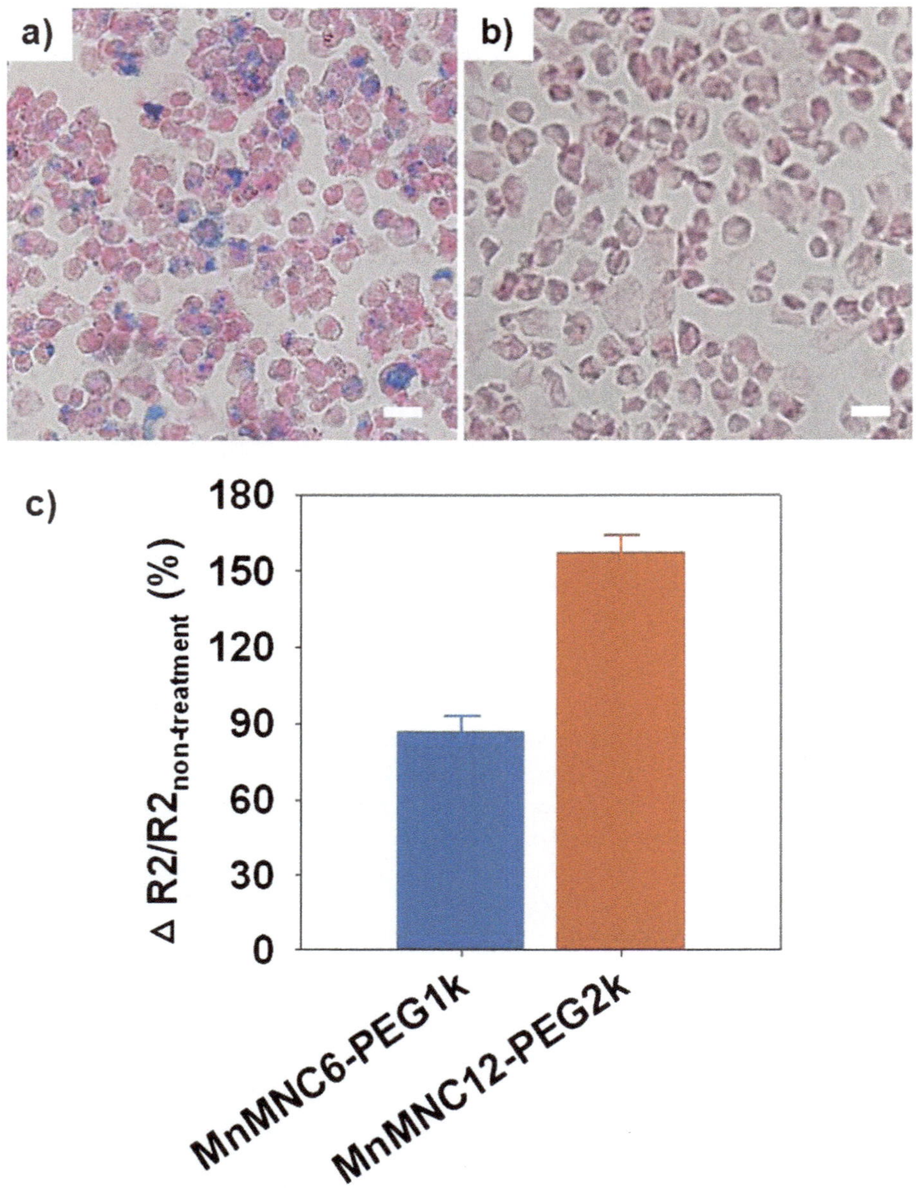

Figure 8 Prussian blue staining images of RAW 264.7 cells treated with (a) MnMNC12-PEG2K, (b) non-treatment (Scale bars: 5 μm), and (c) relative MR signal enhancement graph of $\Delta R2/R2_{non\text{-}treatment}$ value.

stability was maintained even after repeated centrifugations. This result indicates that PEG segment dispersed from the core surface into the aqueous exterior shielding overall surface charge. Further colloidal stability experiments were performed by measuring transverse relaxation time (T2) of PEG coated MnMNCs in certain period dispersed in PBS. Figure 6 showed that their relaxation time values remained unchanged and showed excellent stability for 2 week observation.

To assess the biocompatibility of MnMNCs as a MR contrast agent, cytotoxicity of MnMNCs was investigated using the MTT assay and the results showed no cytotoxicity toward RAW 264.7cells even at high concentration (400 μg/mL) (Figure 7).

Finally, we performed cellular imaging experiments with optimized MnMNC-PEG samples (MnMNC6-PEG1K and MnMNC12-PEG2K) against RAW 264.7 macrophage cells (Figure 8). Prussian blue and ferric ions of the MNCs rapidly exchange electrons thereby producing dark blue colors in the intra-cellular region. The microscope image shows that stained MnMNC12-PEG2K with RAW 264.7 cells were observed as a dotted blue color compared to non-treated cells. Moreover, the relative MR signal intensity ratio (relaxivity difference between MnMNCs treated cells and non-treated cells = $\triangle R2/R2_{non\text{-}treatment}$, where $R2 - R2_{non\text{-}treatment} = \triangle R2$ and $R2 = T2^{-1}$) showed a remarkably high MR signal sensitivity with MnMNC12-PEG2K (157.2 ± 9.2%) and less but fairly high signal enhancement was also achieved with MnMNC6-PEG1K (87.04 ± 8.5%) compared to control cells (non-treated RAW 264.7).

4 Conclusion

In summary, we developed an efficient coating method on different core size $MnFe_2O_4$ magnetic nanocrystals (MnMNCs) using dodecanoic acid (DA)-PEG amphiphilic block copolymers by solvent evaporation method, and determined r2 value maximized MnMNCs based MR probes as MRI contrast agents compared to conventional iron oxide nanoparticles (Feridex and CLIO). In addition, DA-PEG coating of MnMNCs was stable and offered an optimal shell/core to achieve successful colloidal stability as well as low cytotoxicity. The study on the combination of optimal PEG layer thickness and magnetic core size for producing maximal r2 coefficient revealed that MnMNCs with 12 nm core coated with DA-PEG2K (MnMNC12-PEG2K) provided the highest r2 coefficient for MR imaging. Consequently, these advantageous features of optimized PEG coated MnMNCs allowed us to obtain outstanding MR imaging results demonstrating the utility of this MR contrast agent design in future diagnostic MR imaging applications.

Competing interests

The authors declare that they have no competing interests.

Authors' contributions

JP, JSS, YMH and SJH were involved in all stages of design of experiments and interpretation of the result. BHK and BJK carried out synthesis of MnMNCs. JP and BHK carried out data collection by subsequent of *in vitro* cellular test with MRI. All authors read and approved the final manuscript.

Acknowledgements

This research was supported by the Bio & Medical Technology Development Program of the National Research Foundation (NRF) funded by the Korean government (MEST) (2012050077).

Author details

[1]Department of Chemical and Biomolecular Engineering, Yonsei University, Seoul 120-749, South Korea. [2]Department of Radiology & Department of Biochemistry and Molecular Biology, Yonsei University, Seoul 120-752, South Korea. [3]YUHS-KRIBB Medical Convergence Research Institute, Yonsei University, Seoul 120-752, South Korea.

References

1. JW Bulte, DL Kraitchman, Iron oxide MR contrast agents for molecular and cellular imaging. NMR Biomed. **17**(7), 484–499 (2004)
2. H Lee, E Sun, D Ham, R Weissleder, Chip-NMR biosensor for detection and molecular analysis of cells. Nat. Med. **14**(8), 869–874 (2008)
3. J-H Lee, J-T Jang, J-S Choi, SH Moon, S-H Noh, J-W Kim, J-G Kim, I-S Kim, KI Park, J Cheon, Exchange-coupled magnetic nanoparticles for efficient heat induction. Nat. Nano. **7**(6), 418–422 (2011)
4. Y-W Jun, Y-M Huh, J-S Choi, J-H Lee, H-T Song, S Kim, S Yoon, K-S Kim, J-S Shin, J-S Suh, J Cheon, Nanoscale Size Effect of Magnetic Nanocrystals and Their Utilization for Cancer Diagnosis via Magnetic Resonance Imaging. J. Am. Chem. Soc. **127**(16), 5732–5733 (2005)
5. WS Seo, JH Lee, X Sun, Y Suzuki, D Mann, Z Liu, M Terashima, PC Yang, MV McConnell, DG Nishimura, H Dai, FeCo/graphitic-shell nanocrystals as advanced magnetic-resonance-imaging and near-infrared agents. Nat. Mater. **5**(12), 971–976 (2006)
6. UI Tromsdorf, NC Bigall, MG Kaul, OT Bruns, MS Nikolic, B Mollwitz, RA Sperling, R Reimer, H Hohenberg, WJ Parak, S Forster, U Beisiegel, G Adam, H Weller, Size and surface effects on the MRI relaxivity of manganese ferrite nanoparticle contrast agents. Nano Lett. **7**(8), 2422–2427 (2007)
7. JH Lee, YM Huh, YW Jun, JW Seo, JT Jang, HT Song, S Kim, EJ Cho, HG Yoon, JS Suh, J Cheon, Artificially engineered magnetic nanoparticles for ultra-sensitive molecular imaging. Nat. Med. **13**(1), 95–99 (2007)
8. S Tong, S Hou, Z Zheng, J Zhou, G Bao, Coating Optimization of Superparamagnetic Iron Oxide Nanoparticles for High T2 Relaxivity. Nano Lett. **10**(11), 4607–4613 (2010)
9. J Yang, T Lee, J Lee, EK Lim, W Hyung, CH Lee, YJ Song, JS Suh, HG Yoon, YM Huh, SJ Haam, Synthesis of Ultrasensitive Magnetic Resonance Contrast Agents for Cancer Imaging Using PEG-Fatty Acid. Chem. Mater. **19**(16), 3870–3876 (2007)
10. M Liong, H Shao, JB Haun, H Lee, R Weissleder, Carboxymethylated Polyvinyl Alcohol Stabilizes Doped Ferrofluids for Biological Applications. Adv. Mater. **22**(45), 5168–5172 (2010)
11. T-J Yoon, KN Yu, E Kim, JS Kim, BG Kim, S-H Yun, B-H Sohn, M-H Cho, J-K Lee, SB Park, Specific Targeting, Cell Sorting, and Bioimaging with Smart Magnetic Silica Core-Shell Nanomaterials. Small **2**(2), 209–215 (2006)
12. S Sun, H Zeng, DB Robinson, S Raoux, PM Rice, SX Wang, G Li, Monodisperse MFe2O4 (M = Fe, Co, Mn) Nanoparticles. J. Am. Chem. Soc. **126**(1), 273–279 (2003)
13. E-K Lim, J Yang, CPN Dinney, J-S Suh, Y-M Huh, S Haam, Self-assembled fluorescent magnetic nanoprobes for multimode-biomedical imaging. Biomaterials **31**(35), 9310–9319 (2010)
14. M Johnsson, P Hansson, K Edwards, Spherical Micelles and Other Self-Assembled Structures in Dilute Aqueous Mixtures of Poly(Ethylene Glycol) Lipids. J. Phys. Chem. B **105**(35), 8420–8430 (2001)
15. RA Sperling, T Liedl, S Duhr, S Kudera, M Zanella, CAJ Lin, WH Chang, D Braun, WJ Parak, Size Determination of (Bio)conjugated Water-Soluble Colloidal Nanoparticles: A Comparison of Different Techniques. J. Phys. Chem. C **111**(31), 11552–11559 (2007)

Felodipine loaded PLGA nanoparticles: preparation, physicochemical characterization and *in vivo* toxicity study

Utpal Jana*, Anjan Kumar Mohanty, Sovan Lal Pal, Prabal Kumar Manna and Guru Prasad Mohanta

Abstract

Felodipine, a calcium channel blocker has been widely used for the treatment of hypertension and cardiovascular diseases; but the frequent dosing is needed for its poor solubility and variable bioavailability. In present study an attempt has been made to overcome the problems through nanoparticulate delivery system using poly (D, L-lactic-co-glycolic acid) polymer keeping in the view to get better sustainability of the formulation. The nanoparticles were prepared by single emulsion solvent evaporation technique and the physico-chemical characterization of prepared nanoparticles confirmed the particles were nanosize range with smooth and spherical morphology. Further, the compatibility of drug-polymer combination was analyzed by FTIR and DSC study. The *in vitro* drug release study of PLGA nanoparticles showed longer duration of drug release with reduced burst release compared with pure felodipine. The *in vitro* drug release data were fitted with various mathematical models to establish the drug release mechanism from the nanoparticles and found to follow mixed order kinetics. The *in vivo* toxicity study in albino mice showed no noticeable change in biochemical parameters and histopathology of organs. Hence, the developed felodipine nanoparticles were prepared, characterized and could possibly be advantageous for prolonged drug release and improving the antihypertensive effect.

Keywords: Felodipine; PLGA; Release kinetics; Biochemical parameter; Nanoparticles

1 Background

Hypertension is considered to be a major risk factor in the development of cardiovascular diseases (CVD), a leading cause of death worldwide. Hypertension is common in both developed and low- and middle-income countries [1]. The recent study reveals that half of the treated patients are not able to achieve desired blood pressure targets [2,3]. It has become a challenge to maintain the blood pressure for long time that can significantly reduce the risk of morbidity and mortality [4]. There are number of medicines used for the treatment of high blood pressure and patients might need to take their medicine for several years.

Calcium channel blockers (CCBs) are broadly used for the treatment of systemic arterial hypertension. The function of CCBs is to inhibit voltage dependent L- type calcium channels in vascular smooth muscle and heart. The antihypertensive effect of CCBs is due to arterial dilatation that causes smooth muscle relaxation. Felodipine, a dihydropyridine calcium channel blocker, has been widely used for the treatment of hypertension [4]. Clinical study revealed that felodipine has dose dependant activity for reduction of blood pressure and has antihypertensive activity similar to other antihypertensive drugs including beta blockers and angiotensin-converting enzyme (ACE) inhibitors. However, the felodipine is extensively metabolised in the gut and liver and is excreted almost entirely as metabolites [4]. Felodipine is a lipophilic crystalline powder and practically insoluble (BCS Class-II) in water (solubility-0.5 mg/l) [4]. As felodipine has poor and variable bioavailability, the multiple daily dosing is needed. The multiple daily dosing not only has the issues with fluctuation of plasma drug concentration but may also lead to poor patients' adherence.

The bioavailability of poorly water soluble drugs is dissolution rate dependant. In order to improve the bioavailability various approaches were suggested, including

* Correspondence: jana.utpal@gmail.com
Department of Pharmacy, Annamalai University, Annamalai Nagar, Tamil Nadu, India

solubilisation, use of co-solvents, salt formation, micronization and complexation with cyclodextrins [5,6]. However, more amounts of additives are required for these techniques limiting their use from the safety perspective. Another effective way to enhance the bioavailability of the drug is by making the particle size within the nano-meter range. The reduced particle size into nanometre size leads to improve dissolution rate because of increased surface area [7]. The suitable stabilizing agents are used to prevent the nano-sized particle aggregation. However, if the drug is coated with suitable polymer the crystalline nature of the drug can possibly be prevented and can be stabilized over long period of time.

In recent years, biodegradable polymeric nanoparticles have been used to improve the solubility and dissolution rate of both hydrophilic and hydrophobic drugs. The nanoparticles attracted considerable attention as potential drug delivery devices in view of their applications as sustained release of drugs, their ability to target particular organs/tissues, which helps in sustained oral drug delivery beneficial for the patients [8,9]. The sub-cellular size of the nanoparticles can improve the stability and also allows relatively higher intracellular uptake of drug than other particulate systems [10,11]. The widely used methods for nanoparticles preparation are solvent evaporation, high pressure homogenization [12], nano-precipitation and emulsion diffusion etc. [13,14].

Poly (D, L-lactic-co-glycolic acid) or PLGA is one of the most popular biodegradable and non toxic polymers used for preparation of microparticles and nanoparticles [15,16]. PLGA is randomly hydrolyzed into their biocompatible metabolite, lactic acid and glycolic acid and eliminated from the body as carbon dioxide and water [15]. PLGA is widely used for the manufacture of implants, internal sutures and also used for controlled and targeted drug delivery systems [16]. The changes in tissue distribution and pharmacokinetic profile were observed, when the drug was incorporated into PLGA nanoparticulate system. As the particle sizes of the nanoparticles are small enough they are expected to reach angiogenetic area and circulate through the capillaries [15,16]. These may lead to improve therapeutic efficacy and patient compliance. In the present study, felodipine loaded PLGA nanoparticles were prepared, characterized and evaluated for in vivo toxicity study in animal model.

2 Methods

2.1 Materials

Felodipine was a kind gift from Cadila Healthcare Limited (Ahmedabad, India). Poly (D, L-lactic-co-glycolic) acid (PLGA 50:50) (Mw = 18,000) was procured from Sigma Aldrich. Lutrol® F-68 (Poloxamer 188) was purchased from SD Fine Chemicals. Distilled- deionized water was prepared with Milli-Q plus System (Elix 10, Millipore corp. India). All other chemicals used were of the highest available grade.

2.2 Preparation of drug loaded PLGA nanoparticles

The felodipine loaded nanoparticles were prepared by single emulsion solvent evaporation in accordance with method reported with slight modification [17]. Briefly, 100 mg of PLGA (copolymer ratio 50:50) polymer and weighed quantity of felodipine (equivalent to 20% w/w dry weight of polymers) was dissolved in acetone at room temperature to make clear solution. The solution was mixed with 25 ml of aqueous phase containing 1% (w/v) of poloxamer-188 with a constant flow rate of 1 ml/min. The mixture was then homogenized using a probe homogenizer (VIRTIS, Cyclone IQ, USA), at constant agitation speeds of 15000 rpm in an ice bath to form oil-in-water (O/W) emulsion. Then the emulsion was kept at room temperature under gentle stirring for 24 h to evaporate the organic solvent. The prepared nanosuspension was then centrifuged at 30,000 rpm, for 15 min (Sorvall Ultracentrifuge, USA). The pellets were collected and washed with double distilled water to remove any unbound poloxamer-188 and free drugs. Finally, the pallets were freeze-dried (Freezone 6lt, Labconco Corp., MO) to get powdered nanoparticles and kept in freeze for further use.

2.3 Physico-chemical characterization of nanoparticles

2.3.1 Particle size and zeta potential

Dynamic laser scattering (DLS) was used to measure the hydrodynamic diameter (d nm) and Laser Doppler Anemometry (LDA) was used to determine zeta potential (mV). The DLS and LDA were analyzed using Zetasizer 3000 (Malvern Instruments, Malvern, UK). The particle size and zeta potential was determined by using a dilute suspension of nanoparticles (100 μg/ml) prepared in double distilled water and sonicated on an ice bath for 30 seconds. For each batch of sample, the mean diameter ± standard deviations of three determinations were calculated under identical conditions. The polydispersity index (PDI) was also measured to determine particle size distribution.

2.3.2 Estimation of drug entrapment efficiency and loading capacity by RP-HPLC method

The entrapment efficiency (EE) and loading capacity (LC) of felodipine loaded nanoparticles were measured by reverse phase High Performance Liquid Chromatography (RP-HPLC) method [18] with slight modification. Briefly, 1 mg/ml drug loaded nanoparticle solution was prepared in methanol and 20 μL of the sample was injected manually to HPLC equipped with Shimadzu LC-20 AD PLC pump and SPD-M20A PDA detector. The output signal was monitored and integrated using Shimadzu CLASS-VP Version 6.12 SP1 software. The chromatographic separation was carried out using analytical column Phenomenex C18 (150 × 4.6 mm, 5 μ). The measurements were made at 240 nm maintaining the flow rate at 1.0 mL/min and ambient condition using thermostat. The amount of the drug present in the sample was determined from the peak area of the chromatogram correlated with the standard curve.

2.3.3 Fourier transforms infrared spectroscopy (FTIR)

Infrared spectroscopy study was carried out using FT-IR spectrophotometer (Perkin Elmer, FT-IR Spectrometer, SPECTRUM RX I, USA) and the spectrum was recorded in the range of 4000–400 cm^{-1} with resolution of 2 cm^{-1}. Samples were mixed separately with potassium bromide (200–400 mg) and compressed by applying pressure of 200 kg/cm^2 for 2 min in hydraulic press to prepare the pellets. The pellets of the native drug, polymer and the nanoparticles were analysed by placing it on the light path and the spectrum was obtained.

2.3.4 Differential scanning calorimetry (DSC)

DSC is a thermo-analytical technique used to observe fusion and crystallization events of the drug in the prepared nanoparticles. Thermogram was obtained by the DSC analysis (DSC-60, Shimadzu, Japan). Approximately, weighed 2 mg of native drug, polymer and nanoparticles were placed separately into sealed standard aluminium pan and scanned between 25°C to 300°C with heating rate of 10°C/minute under nitrogen atmosphere. An empty aluminium pan considered as reference.

2.3.5 Scanning electron microscopy (SEM)

The shape and surface morphology of prepared nanoparticles were examined by scanning electron microscopy (SEM) (ZEISS EVO18, Carl Zeiss SMT GmbH, Germany). Moisture free lyophilised powdered samples were consigned on aluminium stubs using adhesive tapes and coated with gold using sputter coater and photomicrographs were taken at an acceleration voltage of 10–30 kV.

2.3.6 Atomic force microscopy (AFM)

Atomic force microscopy (AFM) studies were carried out to characterize the surface morphology of the drug loaded nanoparticles. The suspension of freeze dried powder was prepared with milli-Q water and dried in air on a clean glass surface for overnight. The observation was performed with AFM (JPK NanoWizard II, JPK instrument, Berlin, Germany) with silicon probes with pyramidal cantilever having force constant of 0.2 N/m. The scan speed of 2 Hz and 312 kHz resonant frequency was used to obtain the images [19].

2.3.7 In vitro drug release study and release kinetics

The in vitro drug release study was carried out by using rotating basket method with some modification [20,21]. The felodipine loaded nanoparticles (containing 5 mg felodipine) were suspended in glass bottles containing 100 ml of phosphate buffer pH 6.8. Glass bottles were placed in beaker and kept in incubator shaker (50 rpm) throughout the study, with temperature adjusted to 37°C. At specified time interval 10 ml samples were withdrawn and centrifuged at 12,000 rpm for 30 min. The precipitates were resuspended with 10 ml of fresh phosphate buffer and added to the glass bottle and supernatants were collected and analyzed by RP-HPLC. All the measurements were carried out in triplicate.

The drug release mechanism from the nanoparticulate systems were analyzed by various mathematical models. The drug release data were fitted with mathematical models including zero order kinetic [Eq. (1)], first order kinetic [Eq. (2)], Higuchi kinetic [Eq. (3)] and Korsmeyer-Peppas model [Eq. (4)].

$$Q_t = K_0 t \tag{1}$$

$$ln\,Q_t = ln\,Q_t - K_1 t \tag{2}$$

$$Q_t = K_h t^{1/2} \tag{3}$$

$$M_t / M_\alpha = K_p t^n \tag{4}$$

The plots were made: Q_t vs. t (zero order kinetic), ln (Q_0 - Q_t) vs. t (first order kinetic) and Q_t vs. $t^{1/2}$ (Higuchi model), where Q_t is the percentage of drug release at time t, Q_0 is the initial amount of drug present in the formulation and K_0, K_t and K_h are the constants of the equations. The first 60% drug release was fitted in Korsmeyer-Peppas model, where M_t/M_α are the fraction of drug release at time t, K_p is the rate constant and "n" is the release exponent. The value of "n" is calculated from the slop of the plot of log of fraction of drug released (M_t/M_α) vs. log of time to characterize the different release mechanism [20,21].

2.4 In vivo studies

2.4.1 Animals

Wistar albino mice were procured from Central Animal House, Rajah Muthiah Medical College, Annamalai University, India and housed in the Institutional animal house under standard environmental conditions (22 ± 3°C, 55 ± 5% humidity and 12 h/12 h dark/light cycle) and maintained with free access to standard diet and water ad libitum. All animal experimentations were executed in compliance with the guidelines of the Committee for the Purpose of Control and Supervision of Experimental Animals (CPCSEA), New Delhi, Government of India.

2.4.2 Acute oral toxicity study in mice

The animals were randomly selected and assigned to following four test groups (6 mice in each group) namely Group I (Control groups, treated with normal saline), Group II, III, IV and V (Test groups, treated with felodipine loaded nanoparticles equivalent to 60, 120, 240 and 480 mg/kg BW of felodipine in distilled water). The respective doses of felodipine loaded nanoparticles were freshly prepared and administered by oral gavaged in a single dose. Acute toxicity was measured by mortality and survival time and also by clinical picture of intoxication and behavioural reactions. Animals on study were observed for any adverse reaction, like changes of body weight, condition of eye, nose and motor activity. On completion of the treatment, the animals were bled via the retro orbital plexus and sacrificed by cervical dislocation and necropsied to facilitate gross pathological examination of organs, viz. size and appearance of heart, lungs, liver, spleen and kidney [22].

2.4.3 Histopathological studies

The animals were sacrificed at the end of the 14 days and the organic tissues of heart, liver, spleen and kidneys were collected and fixed in 10% formalin then subjected to histopathological examination. The tissues of organ samples were embedded in paraffin blocks, then sliced and placed onto glass slides. After histological staining the slides were observed and photos were taken using topical microscope and histopathological examination performed [23].

2.4.4 Biochemical assay

The blood samples were collected at 14th day and centrifuged at 4000 rpm for 5 min. The serum was kept at −20°C until analyzed. The levels of serum glutamate oxaloacetic transaminase (SGOT), serum alkaline phosphate (SAP), serum glutamic pyruvic transaminase (SGPT), serum creatinine, serum bilirubin and proteins were analyzed with automatic analytical instrument (Hitachi, Japan) [24,25].

2.5 Statistical analysis

All values are expressed as mean ± SD (standard deviation). Significant statistical differences were analyzed by one-way analysis of variance (ANOVA). In all comparisons, the difference was considered to be statistically significant at $p < 0.05$ (*).

3 Results

3.1 Particle size and Zeta potential (ζ)

The mean particle sizes of the drug loaded PLGA nanoparticles were measured by Photon Correlation Spectroscopy (PCS). The sizes of the nanoparticles prepared with three different ratios of drug and polymer (1:2, 1:3 and 1:4) were in the nano range. The size distributions were monodispersed (0.216 to 0.442) in all the nanoparticle formulations (Figure 1). No remarkable differences were observed between the sizes of nanoparticles obtained with different ratio of drug and polymer (Table 1). The zeta potential of the drug loaded nanoparticles exhibited negative values [26].

3.2 Surface morphology studies of nanoparticles

The surface morphology of the nanoparticles was measured using scanning electron microscopy. The shape of the particles as measured by SEM analysis revealed almost spherical in shape with relative smooth surface for all the formulations (Figure 2a). The AFM investigations showed the disc like shape of the particles (Figure 2b). The particle sizes obtained by SEM study were relatively smaller than that of the particle sizes obtained by Zetasizer [27].

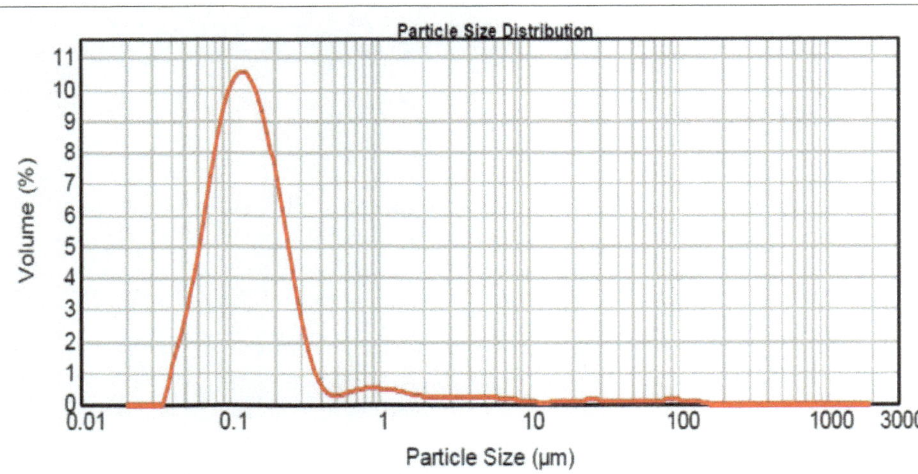

Figure 1 Mean particle size distributions of the felodipine loaded nanoparticles (F2) measured by Dynamic laser light scattering (data as mean ± S.D; n = 6).

Table 1 Particle size, polydispersity index, drug loading and entrapment efficiency of three batches of felodipine loaded nanoparticles (Data represents mean ± SD)

Batch	Drug-polymer ratio	Particle size (nm) ± SD (n = 3)	Polydispersity index	Drug loading (% w/w)	Entrapment Efficiency (% w/w)
F1	1:2	307 ± 4.51	0.442 ± 0.098	12.43 ± 0.89	68.45 ± 0.869
F2	1:3	226 ± 5.39	0.216 ± 0.009	9.47 ± 0.12	85.28 ± 0.328
F3	1:4	273 ± 6.28	0.254 ± 0.004	8.83 ± 0.04	83.23 ± 0.365

$EE\left(\%^w/_w\right) = \frac{\text{Weight of the drug in nanoparticle}}{\text{Weight of the polymer and drug added}} \times 100.$

$DL\left(\%^w/_w\right) = \frac{\text{Weight of the drug in nanoparticle}}{\text{Weight of the polymer and drug added}} \times 100.$

3.3 Drug entrapment efficiency and drug loading

The entrapment efficiency and drug loading of the nanoparticles depend on the polymer, drug, surfactant etc. The entrapment efficiency becomes high when both drug and polymer shows more affinity to the same solvent. On the other hand the low entrapment is due to the high affinity of drug and polymer to the different solvents. Drug loading and entrapment efficiency were affected by the ratio of drug and polymer used in the formulation [28]. Therefore, we prepared three different formulations of the felodipine loaded nanoparticles taking three different ratios of drug- polymer. The nanoparticle formulation (F2) having drug- polymer ratio of 1:3, shows the high entrapment efficiency of 85.28% than the other two ratio 1:2 (68.45%) and 1:4 (83.23%) (Table 1).

3.4 Fourier transform infrared spectroscopy (FTIR)

The FTIR spectral data were used to confirm chemical stability of felodipine in polymeric nanoparticles. FTIR spectra of pure felodipine, polymer and felodipine loaded nanoparticles are shown in (Figure 3). The interaction study between drug and polymer was evaluated. The absence of distinctive spectra in the felodipine loaded nanoparticles indicates that felodipine and PLGA was not involved in intermolecular interaction. However, the intensity of the peak at 3422 cm^{-1}, 2998 cm^{-1} and 2950 cm^{-1} were slightly decreased in the FTIR spectrum of the felodipine nanoparticles [29].

3.5 Differential scanning calorimetry

The physiochemical interaction of the drug encapsulated and the polymer was analyzed by differential scanning calorimetry. The study was carried out for the pure felodipine, polymer and felodipine loaded nanoparticles (Figure 4). DSC shows the characteristic peak for different compounds. The sharp endothermic peak at ~144°C represents the melting point and crystalline anhydrous nature of native felodipine [30].

3.6 In vitro drug release study

The drug-polymer composition influences the *in vitro* drug release rate from the nanoparticles. The felodipine loaded nanoparticles shows biphasic drug release pattern, as the drug encapsulated in the polymeric nanoparticles and some amount of drug adsorb on the surface due to electrostatic interaction [31]. The *in vitro* drug release

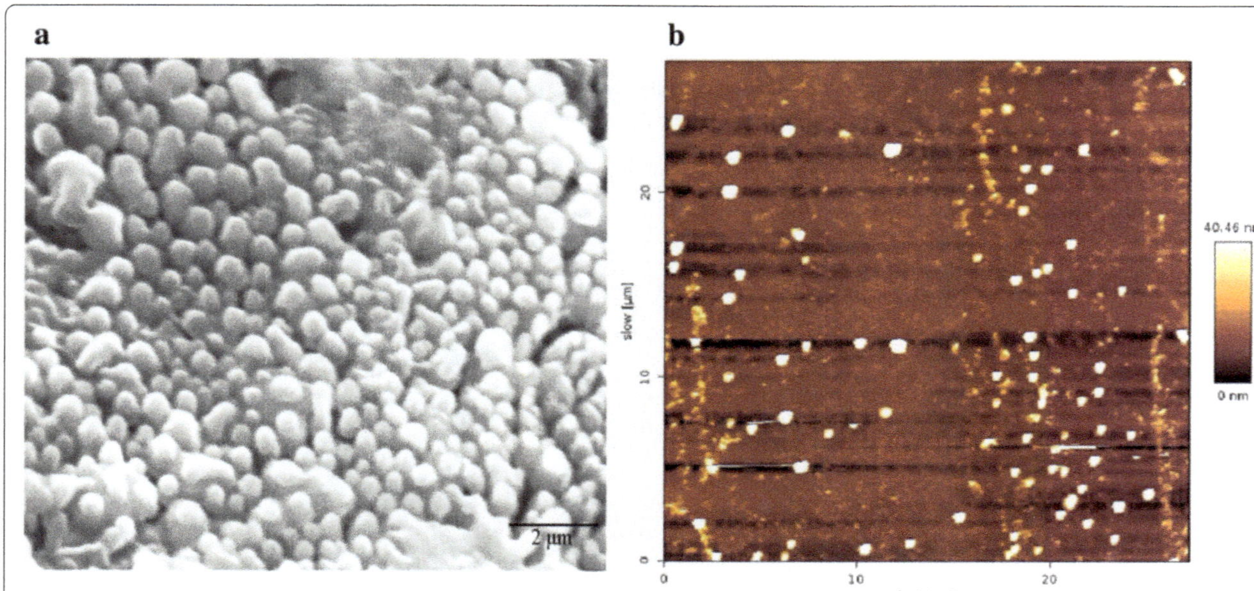

Figure 2 Surface morphological properties of nanoparticles (a) Scanning electron microscopy (SEM) photograph felodipine loaded nanoparticles (bar = 2 μm), (b) Atomic Force microscopy (AFM) images of felodipine loaded nanoparticle.

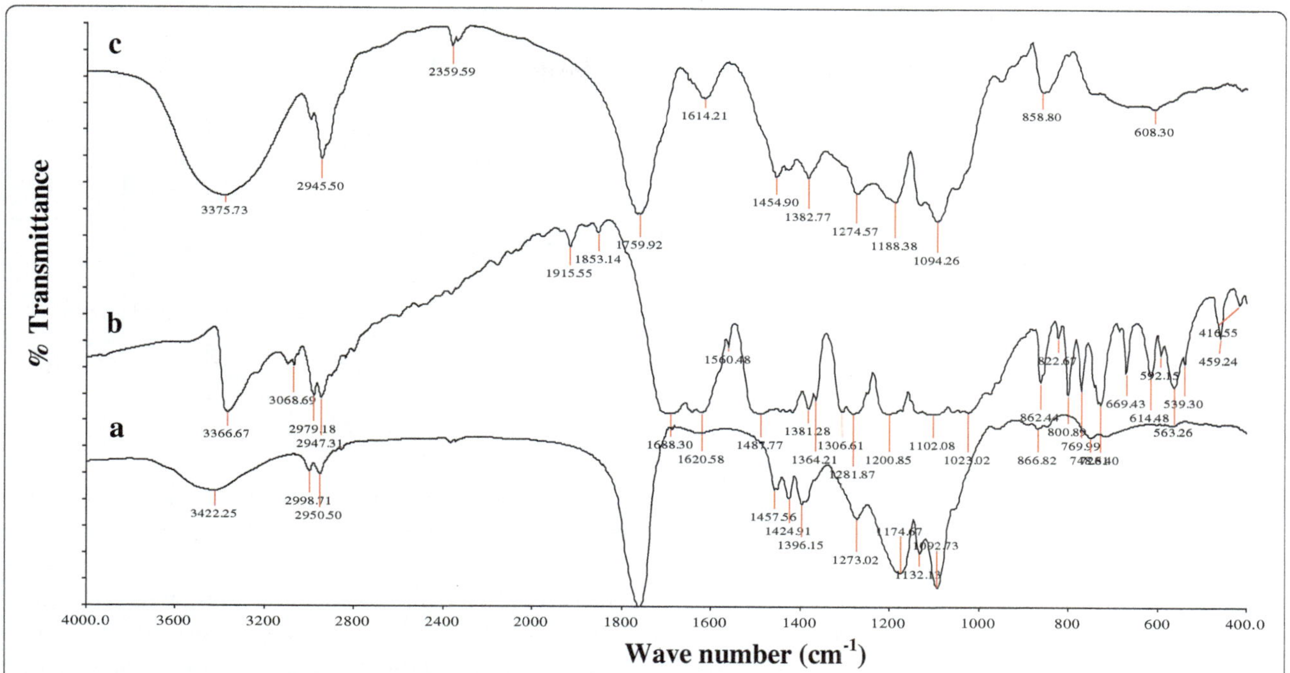

Figure 3 FTIR spectra of (a) Native felodipine, (b) PLGA polymer, (c) Felodipine loaded nanoparticles.

profile from prepared nanoparticles, intact drug powder and marketed formulation are shown in Figure 5. The felodipine loaded nanoparticles show slower drug release in comparison with intact drug powder and marketed formulation. The initial burst release of about 45% was observed for felodipine intact powder. This phenomenon was normal in the initial stage of many drug release systems [32] and it was reduced to 18% for felodipine loaded nanoparticles and also exhibited slow drug release at least 96 h due to the coating of polymer around the drug.

The drug release data were fitted with various kinetic models, data not shown. The drug release kinetics of the formulation F2 was pre-eminently explained by zero

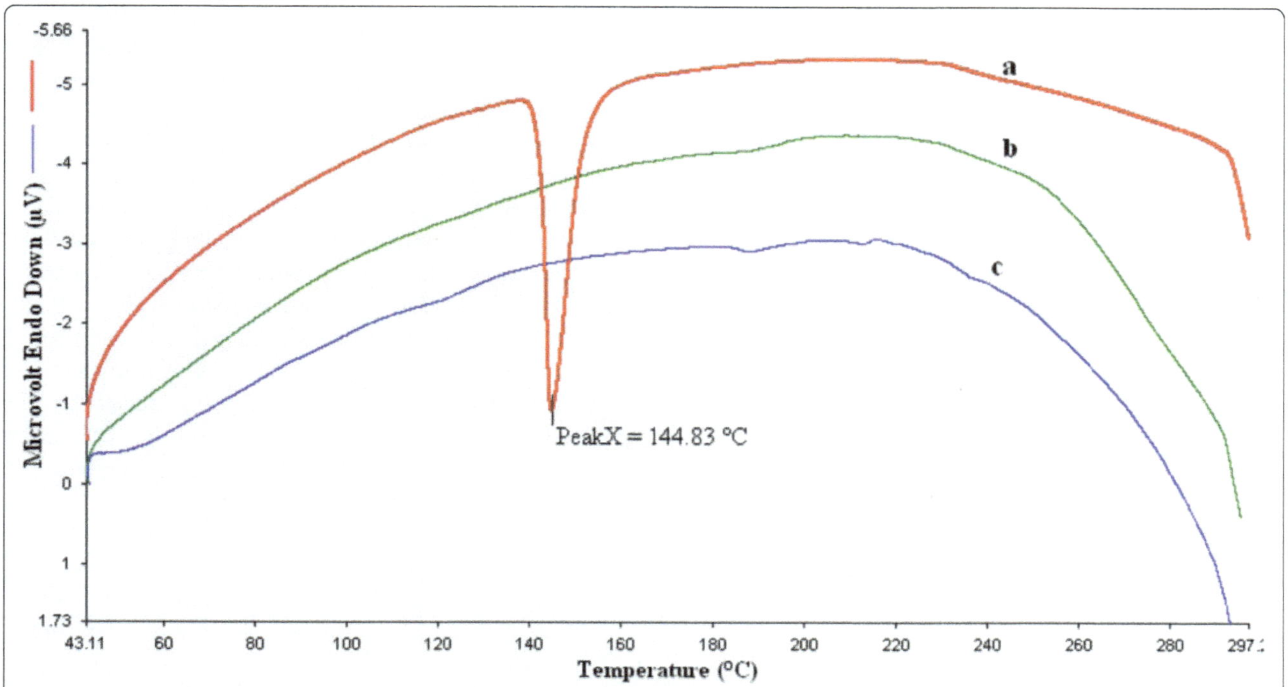

Figure 4 DSC thermograms of (a) Native felodipine, (b) PLGA polymer and (c) Felodipine loaded nanoparticles.

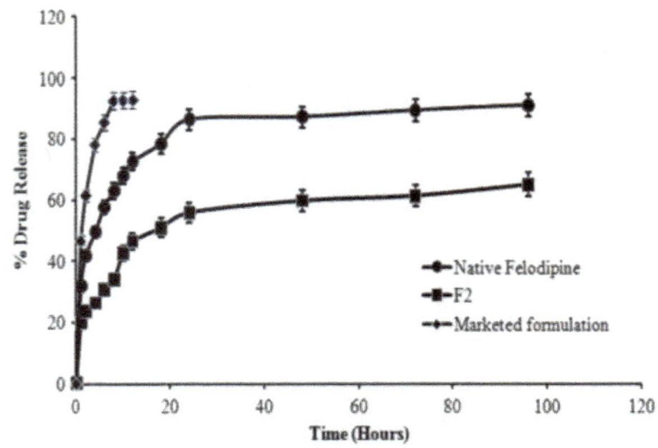

Figure 5 ***In vitro*** **drug release study of felodipine loaded nanoparticles, felodipine intact powder and marketed formulation respectively.** Values represent mean ± S.D. (n = 3).

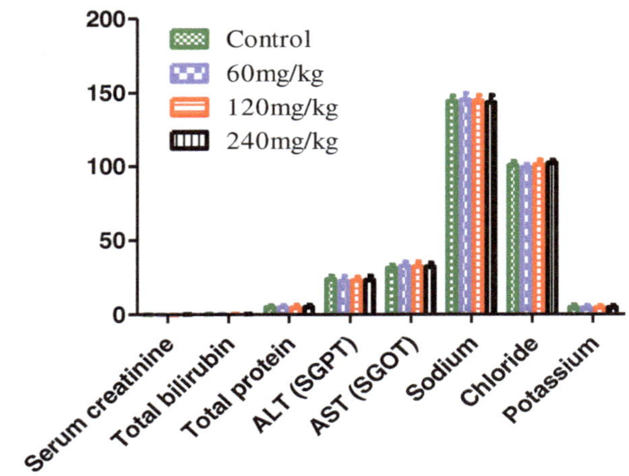

Figure 6 Biochemical parameters in control and nanoparticle treated group (I, II, III and IV) in wistar albino mice (*p < 0.05).

order and first order kinetic, followed by Higuchi model, as it showed the highest linearity. From the Peppas model the 'n' value (0.284) was calculated from the slope and found to be less than 0.5, indicating that the release mechanism from the nanoparticles was diffusion controlled.

3.7 Toxicity studies

The felodipine loaded nanoparticles can cross the small intestinal wall by persorption and distributed in various parts of the body [33]. The animals in various treatment groups as well as control group did not showed any mortality throughout the observation period except group V. In the treatment group V all the animals were died within 12 hours after the administration of the dose (480 mg/kg body weight). Throughout the observation period, the animals in group I, II, III and IV did not show any abnormality or behavioural changes (Table 2). Necroscopic examination revealed no gross changes in the vital organs of the animals in the treatment group I, II, III and IV (Figure 6).

The pathological examinations of different organs of wistar albino mice treated with 240 mg/kg body weight of felodipine loaded nanoparticles were carried out. There was no change in histopathology of glomerulus and tubules of kidney. Liver showed normal hepatocytes and absence of necrosis as compared to control group. The pathological examinations revealed that heart, kidney, liver and lungs were not exposed to nanotoxicity after oral administration of the felodipine loaded nanoparticles when compared with control groups (Figure 7).

Table 2 Behavioral parameters of drug loaded PLGA nanoparticles treated albino mice

Parameters	Results
Motor activity	Normal
Respiration	Normal
Salivation	Normal
Convulsions	Negative
Body weight	No significant change
Skin colour	Normal
Righting reflex	Positive

4 Discussion

The felodipine loaded PLGA nanoparticles were prepared by single emulsion solvent evaporation method. The particle sizes of the drug loaded nanoparticles are important parameter of the nanoparticles, which has direct impact on drug release, cellular uptake, biodistribution and stability. The drug loaded nanoparticle shows negative zeta potential and the result suggest the external localization of free drugs, adsorbed on the surface of the polymeric nanoparticles [34]. These higher negative zeta potential values reduce the particle aggregation and smaller particle size may facilitate the effective absorption of the drug loaded nanoparticles in the biological systems. The surface morphology of the nanoparticles was studied by SEM and AFM. The electron microscope allows the visualization of the nanoparticle surface, whereas zetasizer measures the particle sizes surrounded by hydrodynamic layer, so it provides larger size than the SEM images [26]. Drug loading and entrapment efficiency were measured by RP-HPLC method and the results show increased entrapment efficiency in F2 formulation having drug-polymer ratio of 1:3 [28]. The drug- polymer ratio was taken into consideration to improve the entrapment efficiency of the nanoparticle systems keeping the other parameters constant, and the result showed that, the increase in the drug-polymer ratio up to 1:4, the entrapment efficiency was almost remained constant. This may be due to the saturation level of felodipine inside the

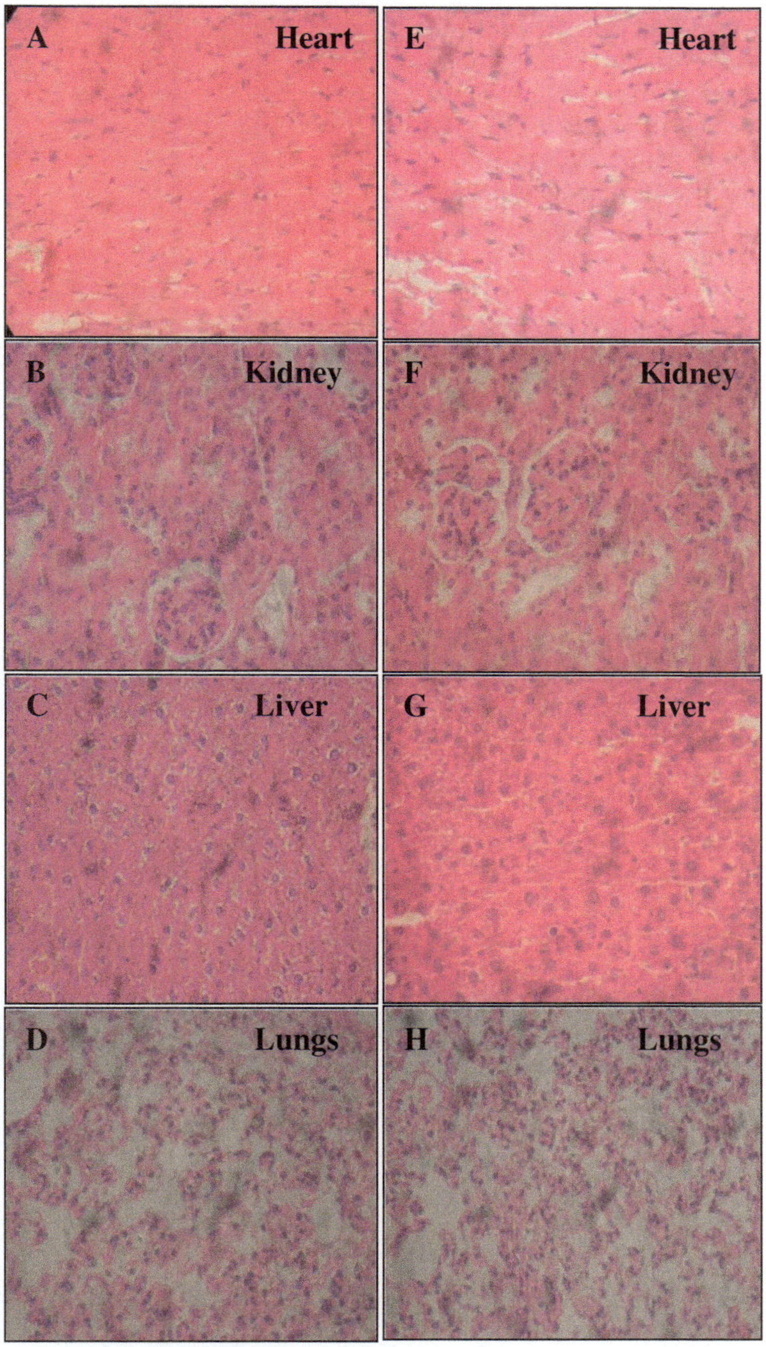

Figure 7 The histopathological slides of wistar albino mice treated with 240 mg/kg body weight of felodipine loaded nanoparticles formulation compared with control. (A-D) Slide shows the heart, kidney, liver and lungs of control mice. **(E-H)** Slide shows the heart, kidney, liver and lungs of felodipine nanoparticle treated mice.

nanoparticles. In the case of drug loading, as it increases a more porous polymeric structure forms, with large number of channels and hollow spaces. The drug can easily escape to the outer phase and decrease the drug inside the polymeric matrix of the nanoparticles [29]. Hence, the nanoparticle formulation (F2) having drug-polymer ratio of 1:3 with highest entrapment efficiency and less particle size was selected and used for the further studies. The FTIR spectrum of the felodipine nanoparticles indicates the chemical stability of the felodipine inside the nanoparticles due to the intermolecular hydrogen bonding between drug and polymer [29]. The drug loaded nanoparticles depicted no distinctive peak of the felodipine in the DSC profiles owing to the decreased crystallinity in the formulations and/or drug solvation in the amorphous carrier as well as solid state interaction induced by heating. The *in vitro* drug release signifies that the prepared nanoparticulate system possess sustained release properties. The initial burst release from the nanoparticles was due to surface adsorbed drug. After 8 h the rapid drug release was identified and this could be as a result of surface adsorbed drug released due to polymer erosion. It suggests that the combination of dissolution, diffusion and erosion are the possible mechanism to release the drug from the nanoparticles. The release kinetics of felodipine from the nanoparticles was preeminently explained by zero order and first order kinetic and it shows release of the drug was by diffusion mechanism [32]. To further explain the *in vivo* toxicity of the felodipine loaded nanoparticles, the wistar albino mice were treated with various concentration of drug loaded nanoparticles and it was found that no abnormality or behavioural changes in the group I, II, III and IV. The necroscopic examination revealed no gross changes occurred in the vital organs of the animals after the treatment. Similarly, the pathological examination of the different organ of the treated animal shows no change in histopathology. Thus the felodipine loaded nanoparticles were found to be orally safe at the single limit dose equivalent to 240 mg/kg body weight of felodipine.

5 Conclusions

Felodipine loaded nanoparticles were prepared successfully by applying emulsion-solvent evaporation technique. The nanoparticles were formulated in reproducible methodology and were able to improve the physicochemical characteristics of the native drug. The DSC study confirmed that there was no intermolecular interaction between the felodipine and PLGA. The prepared drug loaded nanoparticles showed slow release of the drug with the reduced burst release in comparison with intact drug powder. The various toxicity studies indicated the oral safety and the cellular safety of the developed felodipine loaded nanoparticles. These results may prove to be beneficial for the prolonged utilization of the formulation as an adjuvant anti-hypertensive therapy. Thus, the felodipine loaded nanoparticles may provide an effective drug delivery system for the hypertension therapy.

Competing interests

The authors declare that they have no competing interests.

Authors' contribution

UJ synthesized and chacterised the drug loaded nanoparticles, participate in animal experiments and drafted the manuscript. AKM helped in performing the characterization of nanoparticles and drafted the manuscript. SLP helps in animal experiment and biochemical parameter estimation. PKM and GPM scrutinized the results, helped in histo-pathology slide preparations and final cheek up of drafted manuscript. All authors read and approved the final manuscript.

Acknowledgments

Utpal Jana is grateful to University Grants Commission (UGC), Government of India, for providing UGC-BSR fellowship. All the authors are deeply indebted to Dr. K. Kanann, Professor and Head, Department of Pharmacy, Annama ai University and Dr. R. Manavalan, UGC-Basic Science Research Faculty Fellow (UGC-BFF), Govt. of India, for kind support.

References

1. M Ezzati, S Vander Hoorn, CM Lawes, R Leach, WP James, AD Lopez, A Rodgers, CJ Murray, Rethinking the "diseases of affluence" paradigm: global patterns of nutritional risks in relation to economic development. PLoS Med. **2**, e133–e2005 (2005)
2. MJ Brown, Science, medicine, and the future. Hypertension. Bmj. **314**, 1258–1261 (1997)
3. G Mancia, R Sega, C Milesi, G Cesana, A Zanchetti, Blood-pressure control in the hypertensive population. Lancet. **349**, 454–457 (1997)
4. JA Staessen, JG Wang, L Thijs, Cardiovascular prevention and blood pressure reduction: a quantitative overview updated until 1 March 2003. J. Hypertens. **21**, 1055–1076 (2003)
5. MJ Lawrence, GD Rees, Microemulsion-based media as novel drug delivery systems. Adv. Drug Deliv. Rev. **45**, 89–121 (2000)
6. VJ Stella, RA Rajewski, Cyclodextrins: their future in drug formulation and delivery. Pharm. Res. **14**, 556–567 (1997)
7. B Sjostrom, B Bergenstahl, B Kronberg, A method for the preparation of submicron particles of sparingly water-soluble drugs by precipitation in oil-in-water emulsions. II: influence of the emulsifier, the solvent, and the drug substance. J. Pharm. Sci. **82**, 584–589 (1993)
8. I Brigger, C Dubernet, P Couvreur, Nanoparticles in cancer therapy and diagnosis. Adv. Drug Deliv. Rev. **54**, 631–651 (2002)
9. C Vauthier, C Dubernet, C Chauvierre, I Brigger, P Couvreur, Drug delivery to resistant tumors: the potential of poly(alkyl cyanoacrylate) nanoparticles. J. Control. Release. **93**, 151–160 (2003)
10. VC Furtado, P Legrand, A Gulik, O Bourdon, R Gref, D Labarre, G Barratt, Relationship between complement activation, cellular uptake and surface physicochemical aspects of novel PEG-modified nanocapsules. Biomaterials. **22**, 2967–2979 (2001)
11. AF Ourique, AR Pohlmann, SS Guterres, RC Beck, Tretinoin-loaded nanocapsules: preparation, physicochemical characterization, and photostability study. Int. J. Pharm. **352**, 1–4 (2008)
12. C Jacobs, O Kayser, RH Muller, Production and characterisation of mucoadhesive nanosuspensions for the formulation of bupravaquone. Int. J. Pharm. **214**, 3–7 (2001)
13. F Debuigne, J Cuisenaire, L Jeunieau, B Masereel, JB Nagy, Synthesis of nimesulide nanoparticles in the Microemulsion epikuron/isopropyl myristate/ water/n-butanol (or isopropanol). J. Colloid. Interface Sci. **243**, 90–101 (2001)
14. M Trotta, M Gallarate, F Pattarino, S Morel, Emulsions containing partially water-miscible solvents for the preparation of drug nanosuspensions. J. Control. Release. **76**, 119–128 (2001)
15. K Adibkia, MR Siahi Shadbad, A Nokhodchi, A Javadzedeh, M Barzegar-Jalali, J Barar, G Mohammadi, Y Omidi, Piroxicam nanoparticles for ocular delivery: physicochemical characterization and implementation in endotoxin-induced uveitis. J. Drug Target. **15**, 407–416 (2007)
16. R Pignatello, C Bucolo, P Ferrara, A Maltese, A Puleo, G Puglisi, Eudragit RS100 nanosuspensions for the ophthalmic controlled delivery of ibuprofen. Eur. J. Pharm. Sci. **16**, 53–61 (2002)
17. RA Jain, The manufacturing techniques of various drug loaded biodegradable poly(lactide-co-glycolide) (PLGA) devices. Biomaterials. **21**, 2475–2490 (2000)
18. AK Mohanty, F Dilnawaz, C Mohanty, SK Sahoo, Etoposide-loaded biodegradable amphiphilic methoxy (poly ethylene glycol) and poly (epsilon caprolactone) copolymeric micelles as drug delivery vehicle for cancer therapy. Drug Deliv. **17**, 330–342 (2010)
19. A Trapani, J Sitterberg, U Bakowsky, T Kissel, The potential of glycol chitosan nanoparticles as carrier for low water soluble drugs. Int. J. Pharm. **375**, 97–106 (2009)
20. PV Devarajan, GS Sonavane, Preparation and in vitro/in vivo evaluation of gliclazide loaded Eudragit nanoparticles as a sustained release carriers. Drug Dev. Ind. Pharm. **33**, 101–111 (2007)

21. S Jain, S Saraf, Influence of processing variables and in vitro characterization of glipizide loaded biodegradable nanoparticles. Diabetes and Metabolic Myndrome. Clinical. Res. Rev. **3**, 113–117 (2009)
22. SE Gelperina, AS Khalansky, IN Skidan, ZS Smirnova, AI Bobruskin, SE Severin, B Turowski, FE Zanella, J Kreuter, Toxicological studies of doxorubicin bound to polysorbate 80-coated poly(butyl cyanoacrylate) nanoparticles in healthy rats and rats with intracranial glioblastoma. Toxicol Lett. **126**, 131–141 (2002)
23. UM Dhana Lekshmi, G Poovi, N Kishore, PN Reddy, In vitro characterization and invivo toxicity study of repaglinide loaded poly (methyl methacrylate) nanoparticles. Int J Pharm **396**, 194–203 (2010)
24. CW Lam, JT James, R McCluskey, RL Hunter, Pulmonary toxicity of single-wall carbon nanotubes in mice 7 and 90 days after intratracheal instillation. Toxicol Sci. **77**, 126–134 (2004)
25. G Oberdorster, E Oberdorster, J Oberdorster, Nanotoxicology: an emerging discipline evolving from studies of ultrafine particles. Environ Health Perspect. **113**, 823–839 (2005)
26. U Jana, AK Mohanty, PK Manna, GP Mohanta, Preparation and characterization of nebivolol nanoparticles using Eudragit(R) RS100. Colloids Surf B Biointerfaces. **113**, 269–275 (2014)
27. U Jana, AK Mohanty, SL Pal, PK Manna, GP Mohanta, Preparation and in vitro characterisation of felodipine loaded Eudragit® RS100 nanoparticles. Int. J. Pharm. Pharm. Sci. **6**, 564–567 (2014)
28. P Dongming, K Huang, Y Liu, S Lin, Preparation of novel polymeric microspheres for controlled release of finasteride. Int. J. Pharm. **342**, 82–86 (2007)
29. C Witschi, E Doelker, Influence of the microencapsulation method and peptide loading on poly(lactic acid) and poly(lactic-co-glycolic acid) degradation during in vitro testing. J. Control. Release. **51**, 327–341 (1998)
30. AR Tapas, PS Kawtikwar, DM Sakarkar, Enhanced dissolution rate of felodipine using spherical agglomeration with Inutec SP1 by quasi emulsion solvent diffusion method. Res Pharm Sci. **4**, 77–84 (2009)
31. SJ Douglas, SS Davis, L Illum, Nanoparticles in drug delivery. Crit. Rev. Ther. Drug Carrier Syst. **3**, 233–261 (1987)
32. E Leo, F Forni, MT Bernabei, Surface drug removal from ibuprofen-loaded PLA microspheres. Int. J. Pharm. **196**, 1–9 (2000)
33. JF Hillyer, RM Albrecht, Gastrointestinal persorption and tissue distribution of differently sized colloidal gold nanoparticles. J Pharm Sci. **90**, 1927–1936 (2001)
34. K Dillen, J Vandervoort, G Van den Mooter, A Ludwig, Evaluation of ciprofloxacin-loaded Eudragit RS100 or RL100/PLGA nanoparticles. Int. J. Pharm. **314**, 72–82 (2006)

Excellent enhancement in the device performance of nitrogen plasma treated ZnO nanorods based diodes

Nandanapalli Koteeswara Reddy[1,3], Mudusu Devika[2] and Yoon-Bong Hahn[1*]

Abstract

Impact of the plasma exposure time on the physical properties of homo-epitaxial ZnO nanorods (NRs) and their devices was investigated. Here, ZnO NRs were synthesized by chemical solution method on glass substrates and treated under high intensity nitrogen plasma at different exposure timings. The as-grown as well as treated ZnO NRs exhibited hexagonal crystal structure and (001) as a preferential orientation. While increasing the plasma exposure time from 1 to 15 min, the structural and optical quality of ZnO NRs gradually improved and above this exposure time, both the properties degraded. The devices fabricated with 15 min plasma treated ZnO NRs showed excellent diode performance than the untreated nanostructures based devices. The diodes developed with treated ZnO NRs showed a low turn-on voltage (3.3 V) than the devices developed with untreated NRs.

Keywords: ZnO Nanostructures; Chemical solution method; Plasma treatment; Structural properties; Emission properties; Diode applications

1. Background

In recent years, the development of optoelectronic, electronic and bio devices on flexible substrates has received tremendous interest due to their unique advantages over the rigid-substrates [1,2]. Zinc oxide (ZnO) is a wide band gap (~3.37 eV) semiconductor compound and exhibits high exciton binding energy (~60 meV, which is 2.5 times higher than that of room temperature thermal energy), good thermal and chemical stability [3,4]. In this aspect, different type of ZnO nanostructures including nanosheets (NSs), nanorods (NRs), nanotubes (NTs), nanoflowers (NFs), nanobelts (NBs), nanowires (NWs) and nano-cages (NCs) etc. have been synthesized using different physical and chemical methods, and also adopted for various device applications [5,6]. Among various synthetic methods, low temperature chemical solution method(s) have been widely applied for the synthesis of ZnO nanostructures due to their possible control over the growth parameters, scalability and versatility. ZnO nanostructures developed with chemical solution approach consist of pure phase and good stoichiometry since the growth occurs at atomic levels. However, these ZnO nanostructures suffer with poor structural and optical quality due to the presence of interstitial and surface defect states [7].

The plasma treatment is a versatile method for the processing of materials like surface cleaner, adhesion promotion, surface energy controlling and enhancement in bio-compatibility and device performance [8]. For example, oxygen radio frequency plasma treated multi-wall carbon nanotubes (CNT) exhibited improved uniformity in the distribution of surface defects and thereby dispersion of metallic nanoparticles [9]. Excellent electrochemical contacts between carbon nano tubes and enzymes have obtained by treating the as-grown CNTs by nitrogen plasma due to the flip of surface from hydrophobic to hydrophilic [10]. Similarly, there has been a various number of reports on plasma treated ZnO materials reported elsewhere [11-14]. By keeping in mind about the future multifunctional applications of ZnO nanostructures as flexible device and existing data, we made an attempt to produce the best quality of ZnO NRs by plasma treatment since it is safe and environmental friendly than the other traditional methods. In this present

* Correspondence: ybhahn@chonbuk.ac.kr
[1]School of Semiconductor and Chemical Engineering, Chonbuk National University, 567 Baekje-daero, Deokjin-gu, Jeonju 561 756, Republic of Korea
Full list of author information is available at the end of the article

study we have adopted nitrogen (N_2) plasma to treat the hydrothermally synthesized ZnO NRs since the ionic radius of nitrogen is close to O and it also acts as deep acceptor and good compensator [11,15]. The impact of nitrogen plasma treatment on the surface morphology, crystal structures and optical properties of ZnO NRs reported and discussed. Further, p-n junction diodes prepared with treated and untreated ZnO NRs and estimated their device performance at room temperature.

2. Methodology

Vertically aligned ZnO NRs were grown by chemical solution technique at a growth temperature of ~70°C. In this process, zinc nitrate hexahydrate and hexamethylenetetramine (from Sigma Aldrich) analytical grade chemicals were used as reagents without further purification. The ZnO NRs were grown on ZnO layer (or seed layer) coated (~50 nm) glass (bare and indium tin oxide coated; 7–8 Ω/sq resistance) substrates. A detailed description about the synthesis of ZnO NRs has been reported elsewhere [16]. The as-grown ZnO NRs structures were treated by nitrogen (4 N pure nitrogen gas with a flow rate of 100 sccm) plasma in the plasma-enhanced chemical vapor deposition system for different time spans, 0–20 min. Here, the plasma was created at a fixed gas pressure of 200 mTorr with a power of 100 W [12]. However, upon treatment we have noticed a gradual raise in temperature from 25 to 120°C. Finally, the p-n junction diodes were fabricated by spin coating of the poly(3,4-ethylenedioxythiophene) poly(styrenesulfonate) (PEDOT:PSS) polymer on ZnO NRs grown on ITO/glass substrates.

The structure and morphology of the structures were examined with X-ray diffractometer (XRD, Philips X'Pert Pro) and field emission scanning electron microscopy (FESEM, JSM-840 A). The photoluminescence (PL) measurements were carried out at room temperature with 325 nm He-Cd laser in the wavelength rage of 350–1000 nm. The electrical properties of the p-n diodes were studied at room temperature by using probe station attached with semiconductor parameter analyzer.

3. Results and discussion

3.1 Surface and structural properties

The FESEM analysis (Figure 1) reveals that the as-grown nanorods are vertically aligned and have an average length and diameter of 1.5 μm and 75 nm, respectively. It also shows that the morphology of treated ZnO NRs, even at longer exposures, remains as same as untreated structures. On the other hand, the as-grown ZnO NRs exhibited three XRD peaks diffracted at 2θ = 31.9, 34.55 and 36.4° (Figure 2), which belong to (100), (002) and (101) orientations respectively since the evaluated d-spacing values (0.2803, 02594, and 0.466 nm) exactly matched with the bulk hexagonal ZnO data (JCPDS: 36–1451) [16-18]. As compared to other diffraction peaks, (002) peak is more dominant, and therefore the as-grown ZnO NRs are preferentially oriented along <001 > direction, i.e. c-axis. While increasing plasma exposure time, the

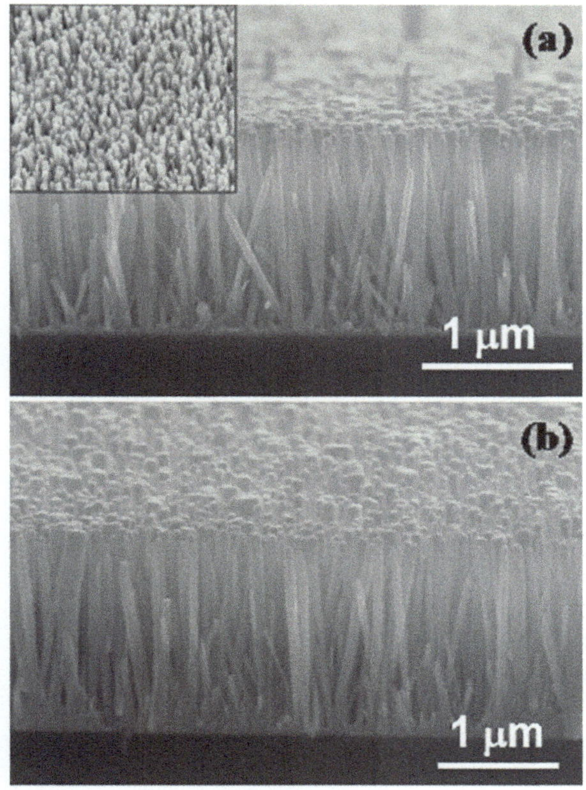

Figure 1 Cross sectional FESEM images of (a) as-grown and (b) plasma treated (20 min) ZnO NRs (inset of Figure 1a shows the surface morphology of as-grown ZnO NRs).

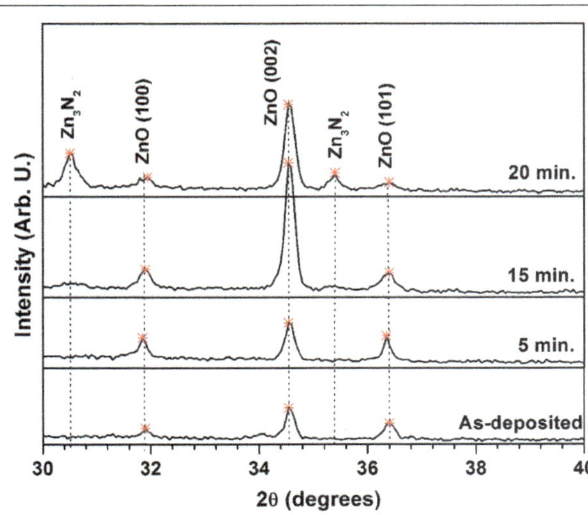

Figure 2 XRD profiles of as-grown and plasma treated ZnO NRs for the three typical exposure timings of 5, 15 and 20 min.

intensity of preferential peak gradually increased upto the exposure time of 15 min and above this, it slightly decreased. However, the crystallinity of ZnO NRs, i.e. the intensity ratio of I(002) and I(101) peaks, gradually increased with the increase of exposure time, as shown in Figure 3. Further, the evaluated full width at half maximum (FWHM) value of (002) peak decreased with the increase of exposure time, as shown in Figure 3. However, at higher exposure timings (>15 min), the plasma treated ZnO NR structures exhibited two new diffraction peaks at 30.5 (d = 0.293 nm) and 35.4° (d = 0.253 nm), which belong to Zn_3N_2 phase [19].

A considerable improvement in the crystalline quality of thermo chemically grown ZnO NRs with the treatment of high intensity plasma can be explained using the existing literature. In general, ZnO NRs grown by thermo chemical method usually consist of different surface defect states due to the presence of water and hydroxyl ions, and non-reacted Zn and O ions as interstitial defect states [20,21]. These defect states probably act as amorphous centers and as results, the as-grown structures exhibit slightly poor crystallinity. Upon plasma treatment, these surface defect states probably released due to the bombardment of energetic nitrogen ions, which also induce the re-crystallization of Zn and O ions. Further, the diffusion of Ni ions into the core-lattice of ZnO probably neutralizes the defect states present in the ZnO NRs. As result, the overall crystallinity of ZnO NRs enhanced with the increase of nitrogen plasma treatment. On the other hand, a possible reason for the formation of Zn_3N_2 phase could be unintentional raise in temperature during plasma treatment since nitrogen ions can easily replace the oxygen atoms at temperatures higher than 110°C. Therefore, these analyses clearly emphasized that the structure and phase purity of

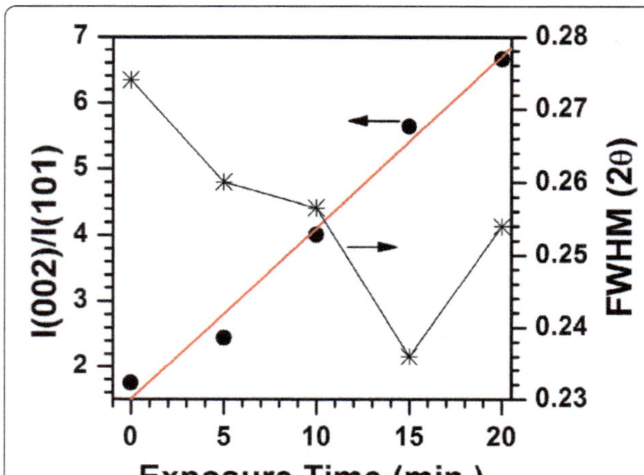

Figure 3 Variation of I(002)/I(101) ratio and FWHM of ZnO NRs as a function of plasma exposure time.

ZnO NRs remains as same as the untreated nanostructures upto the plasma exposure time of 15 min, and the crystalline quality of 15 min treated ZnO NRs improved nearly by three times than that of untreated ones.

3.2 Optical emission properties

PL studies show that the as-grown ZnO NRs have a large number of defect states that are mainly attributed to the interstitials and vacancies of zinc and oxygen atoms since the emission intensity of the broad band (BB) peak centered at 590 nm is comparatively higher than the intensity of ultra-violet (UV) peak (Figure 4) centered at 390 nm. Upon increasing the nitrogen plasma exposure time, the UV peak intensity drastically increased and BB peak intensity considerably decreased, which can be clearly seen from Figure 5. The variation of emission intensity ratio of UV and BB peaks (i.e. I_{UV}/I_{BB}) with plasma exposure time, Figure 5, reveals that while increasing exposure time the quality of ZnO NRs gradually increased upto the exposure time of 15 min. Above this, a degradation in the emission quality of NRs started. These results indicate that the structures treated for 15 min span consist of high quality optical properties since I_{UV}/I_{BB} ratio of these structures is 11 times higher than that of as-grown nanorods. An improvement in the optical quality of 15 min treated ZnO NRs is attributed to the passivation of defects states present on the surface of ZnO NRs. At higher exposure timings, the regeneration of oxygen interstitials (Oi) through the replacement of oxygen in ZnO lattice by nitrogen atoms probably lead the density of defect states to higher values and causes for the overall degradation in quality of ZnO NRs. While increasing the plasma exposure time the structures exhibited slight red-shift in the UV peak position upto the exposure time of 15 min and at higher exposure timings, the UV peak position drastically shifted towards lower wavelengths (i.e. blue-shift), Figure 6. In general, the UV peak position, represents the near band edge emission (band gap, E_g), strongly depends on the crystalline quality of ZnO NRs. As observed in XRD studies, the crystalline quality of plasma treated ZnO NRs improved upto the exposure time of 15 min that probably leads the band gap of ZnO NRs to slightly lower values due to the passivation of defect states.

It is well know that in ZnO lattice matrix nitrogen impurities act as acceptors [22,23]. In this view, various groups have adopted nitrogen as doping agent for the development of p-type ZnO films and also nanostructures [24-28]. In general, upon increasing plasma exposure time, the amount of nitrogen implantation or absorption in ZnO NRs increases. The incorporation of nitrogen into ZnO NR structures probably occurs in two ways: interstitial and substitutional doping. As interstitial doping, the nitrogen atoms neutralize the defect states present on the surface of ZnO materials, whereas in substitutional doping, nitrogen

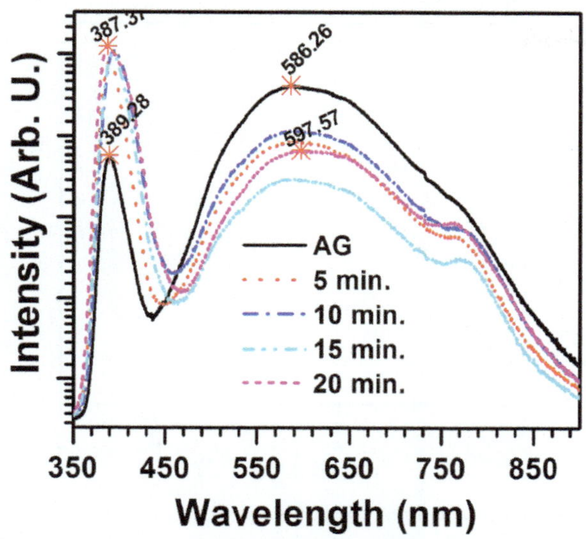

Figure 4 Photoluminescence spectra of as-grown and plasma treated ZnO NRs at different exposure timings.

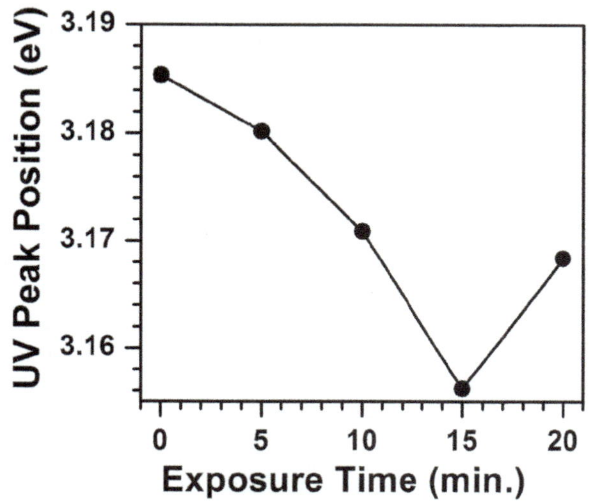

Figure 6 Shift in UV peak position of ZnO NRs with increase in plasma exposure time.

impurities generate interstitial defects (O_I) by replacing oxygen atoms. Usually, the electrical conductivity of ZnO primarily dominated by electrons generated from oxygen vacancies and zinc interstitial atoms [29,30]. In the present case, the interstitial incorporation of nitrogen atoms in place of oxygen vacancies (V_O) [11,13] diminish the existing defects states due to passivation, and leads the density of carriers to lower values. Thus, the defects related BB peak intensity strongly reduced. Further, the decrease of carrier density leads band gap of ZnO NRs to lower values since $E_g \propto n_i^{2/3}$. At higher exposure timings, there are two possible reasons for the formation of Zn_3N_2 phase: i) replacement of oxygen atoms in Zn-O lattice by entering nitrogen ions as substitutional impurity thereby release of oxygen atoms as interstitials, and/or ii) nitrification of zinc interstitials (Zn_i) under moderate temperatures [31]. The newly formed Zn-N phase and/or regenerated oxygen interstitials probably leads the band gap of ZnO NR structures to slightly higher values. Therefore, the structures exposed to nitrogen plasma for 15 min duration consist of better crystallinity as well as optical quality and thus, these structures are adopted for the development of p-n junction diodes.

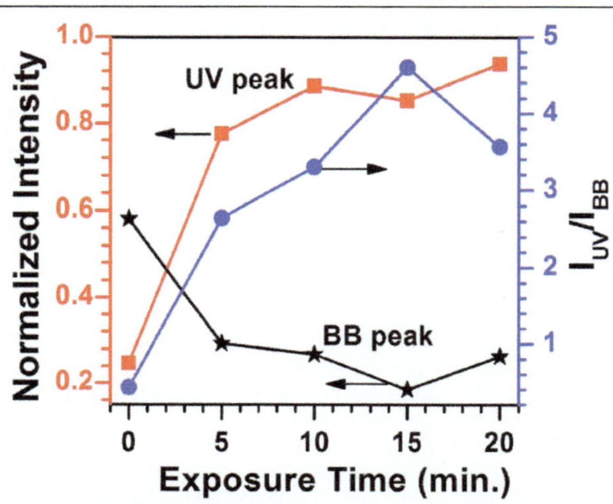

Figure 5 Variation of UV and BB peaks intensities and their ratio (I_{UV}/I_{BB}) of ZnO NRs as a function of plasma exposure time.

3.3 Devices properties

The untreated and 15 min plasma treated ZnO NRs grown on ITO substrates were coated with PEDOT:PSS using spin coater with a thickness of ~4 μm thick and allowed at room temperature for a few days. The schematic diagram and cross-section FESEM images of the device with the configuration of Glass/ITO/ZnO seed-layer/ZnO NRs/PEDOT:PSS are shown in Figure 7. Finally, a thin gold (Au) layer with a thickness of 100 nm was deposited by e-beam evaporation and the current (I) - voltage

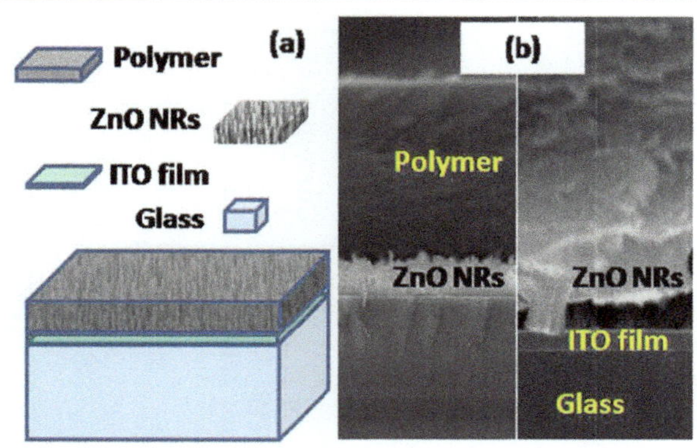

Figure 7 (a) Schematic diagram and (b) cross-section FESEM images of ZnO NRs/polymer based p-n junction diode.

(V) characteristics were measured at room temperature (Figure 8). Here, the devices prepared with untreated (D1) and treated (D2) ZnO NRs exhibited slightly similar p-n junction diode characteristics between the applied bias voltages of ± 5 V. Comparatively D2 device exhibited better I-V properties than that of D1 device [32]. For example, the maximum current through D2 device at the forward bias voltage of 5 V is about 32.3 μA, whereas in D1 device the observed maximum current is nearly 10 times low (~3.13 μA). The knee voltage (V_d) of D1 is higher (4.16 V) than that of D2 device (3.3 V) and the rectification factor (I_f/I_r) at the bias voltage of 5 V is about 78 and 408. Further, the series resistance of D1 and D2 devices is found to be 3.4×10^5 and 5.3×10^4 Ω, respectively.

From ln(I) versus V plots (Figure 9), the diode quality factor (n) of the devices was calculated using the formula: $n = (q/kT)(\Delta V/\Delta \ln(I))$, where q-charge of electron, k-Boltzmann constant and T-absolute temperature. At lower voltages (<1 V) the conduction mechanism in D1 as well as D2 devices (R1 region) dominated by diffusion of carriers since the observed diode quality factors (n) is greater than 2 [33]. Above this voltage, D1 exhibited only one region (R2), whereas D2 device exhibited three more regions (R2, R3 and R4). The ideality factor of these devices particularly at higher voltages (>2 V) is too high probably due to the presence of defects states (or space-charges) on the surface of ZnO NRs. This reveals that at higher voltages, the conduction through the devices strongly dominated by the space-charge limited current and/or its combination with series resistance. However, the space-charge limited current region (R2) for D2 device is low as compared to D1 device. A decrease in density of surface defects states in plasma treated ZnO NRs probably the cause for better device performance than the device developed with untreated ZnO NRs.

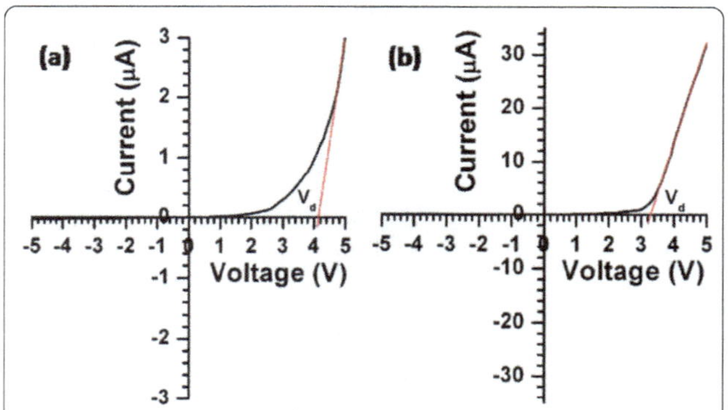

Figure 8 Current versus voltage plots of (a) untreated (D1) and (b) treated ZnO NRs (D2) based p-n junction diodes.

4. Conclusions

The ZnO NRs were synthesized using simple and low temperature chemical solution method on ZnO seeded

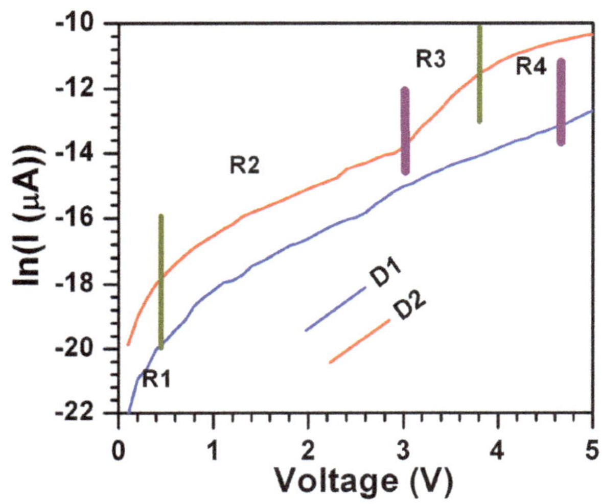

Figure 9 ln(I) versus voltage plots of p-n junctions diodes.

glass and ITO substrates. The as-grown structures were exposed to high intensity nitrogen plasma and studied their physical properties. Finally, the p-n junction diodes were fabricated using as-grown and 15 min plasma treated ZnO NRs and studied the device performance. The observed results are summarized below.

The as-grown ZnO NRs have hexagonal crystal structure and are preferentially oriented along <001 > direction. Upon exposure to nitrogen plasma, the morphology and crystal structure of ZnO NRs remained as same upto the exposure time of 15 min. The ZnO NRs exposed to plasma for 15 min duration have better crystalline and optical quality than the as-deposited and other treated nanostructures. The devices fabricated with 15 min treated ZnO NRs exhibited excellent p-n junction diode properties at room temperature. Therefore, the overall observations emphasized that the quality of ZnO NRs grown even on flexible substrates can be improved to a large extent without disturbing their morphology as well as crystal structure by treating them under nitrogen plasma for the duration of 15 min.

Competing interests
The authors declare that they have no competing interests.

Authors' contributions
NKR planned and executed the work. MD fabricated the devices and characterized. All are participated in the preparation of MS and discussion. All authors read and approved the final manuscript.

Acknowledgements
This work was supported by National Leading Research Laboratory program (2013064831) through the National Research Foundation of Korea funded by the Ministry of Science, ICT & Future Planning. N.K. Reddy wishes to acknowledge European Commission Research Executive Agency for the sanction of Marie Curie Actions - International Incoming Fellowship (No: PIIF-GA-2012-331003: NRforHF) and CSIR for the sanction of SRA fellowship under the scheme of Scientist's pool (No: 13(8525-A) 2011-Pool). M. Devika wishes to acknowledge UGC for the sanction of Dr. D.S. Kothari Postdoctoral fellowship No: F. 4-2/2006(BSR)/13-703/2012(BSR).

Author details
[1]School of Semiconductor and Chemical Engineering, Chonbuk National University, 567 Baekje-daero, Deokjin-gu, Jeonju 561 756, Republic of Korea. [2]Department of Aerospace Engineering, Indian Institute of Science, Bangalore 560 012, India. [3]Present address: Institute of Chemistry, Humboldt University, Berlin 12489, Germany.

References

1. K Zhang, JH Seo, W Zhou, Z Ma, J. Phys. D Appl. Phys. **45**, 143001 (2012)
2. DH Kim, R Ghaffari, N Lu, JA Rogers, Annu. Rev. Biomed. Eng. **14**, 113 (2012)
3. U Ozgur, YI Alivov, C Liu, A Teke, MA Reshchikov, S Doan, V Avrutin, S-J Cho, H Morkoc, J. Appl. Phys. **98**, 041301–1 (2005)
4. ZL Wang, J. Phys. Condens. Matter **16**, R829 (2004)
5. YB Hahn, J Korean, Chem. Eng. **28**, 1797 (2011)
6. U Ozgur, D Hofstetter, H Morkoc, Proc. IEEE. **98**, 1255 (2010)
7. F Oba, M Choi, A Togo, I Tanaka, Sci. Technol. Adv. Mater. **12**, 034302–1 (2011)
8. MA Lieberman, AJ Lichtenberg, *Principles of Plasma Discharges and Materials Processing*, 2nd edn. (Wiley Interscience, Hoboken, NJ: Wiley, 2005), p. 285
9. A Felten, J Ghijsen, J-J Pireaux, RL Johnson, CM Whelan, D Liang, GV Tendeloo, C Bittencourt, J. Phys. D Appl. Phys. **40**, 7379 (2007)
10. H Muguruma, T Hoshino, Y Matsui, ACS Appl. Mater. Interfaces **3**, 2445 (2011)
11. HQ Le, S Tripathy, S Chua, Appl. Phys. Lett. **92**, 141910–141911 (2008)
12. LH Quang, CS Jin, J. Electrochem. Soc. **155**, K105 (2008)
13. N Chantarat, Y-W Chen, S-Y Chen, C-C Lin, Enhanced UV photoresponse in nitrogen plasma ZnO nanotubes, Nanotechnology **20**, 395201 (2009) doi:10.1088/0957-4484/20/39/395201
14. R Huang, S Xu, W Guo, L Wang, J Song, T-W Ng, J Huang, S-T Lee, S Du, N Wang, Nitrogen deep accepters in ZnO nanowires induced by ammonia plasma, Appl. Phys. Lett. **99**, 143112 (2011) http://dx.doi.org/10.1063/1.3647773
15. CM Reddy, CC Amato-Wierda, *Gas phase analysis of $TiCl_4$ and NH_3 plasma processes by molecular beak mass spectrometry* (Electrochem. Soc. Pennington, New Jersey, USA, 2001), p. 396
16. N Koteeswara Reddy, Q Ahsanulhaq, JH Kim, YB Hahn, EPL-Europhys Lett. **81**, 38001–1 (2008)
17. N Koteeswara Reddy, Q Ahsanulhaq, JH Kim, M Devika, Y-B Hahn, Nanotechnol. **18**, 445710 (2007)
18. N Koteeswara Reddy, Q Ahsanulhaq, YB Hahn, Physica. E. **41**, 368 (2009)
19. Z Fu-Jian, MA Hong-Lei, D Wei, Z Xi-Jain, X Hong-Di, M Jin, J Feng, X Cheng-Shan, Z Hui-Zhao, Chin. Phys. Lett. **22**, 907 (2005)
20. A Toumiat, S Achour, A Harabi, N Tabet, M Boumaour, M. M Maallemi, Nanotechnol. **17**, 658 (2006)
21. H Zhou, H Alves, DM Hofmann, BK Meyer, G Kaczmarczyk, A Hoffmann, C Thomsen, Phys. stat. sol. (b) **229**, 825 (2002)
22. J Xu, R Ott, AS Sabau, Z Pan, F Xiu, J Liu, J-M Erie, DP Norton, Appl. Phys. Lett. **92**, 151112 (2008)
23. P Fons, H Tampo, AV Kolobov, M Ohkubo, S Niki, J Tominaga, R Carboni, F Boscherini, S Friedrich, Phys. Rev. Lett. **96**, 045504–1 (2006)
24. DC Look, DC Reynolds, CW Litton, RL Jones, DB Eason, G Cantwell, Appl. Phys. Lett. **81**, 1830 (2002)
25. F Gallino, C Di Valentin, G Pacchioni, M Chiesa, E Giamello, J. Mater. Chem. **20**, 689 (2010)
26. DC Look, Semicond. Sci. Technol. **20**, S55 (2005)
27. B Sieber, H Liu, G Piret, J Laureyns, P Roussel, B Gelloz, S Szunerits, R Boukherroub, J. Phys. Chem. C **113**, 13643 (2009)
28. B Yang, P Feng, A Kumar, RS Katiyar, M Achermann, J. Phys. D Appl. Phys. **42**, 195402 (2009)
29. V Khranovskyy, J Eriksson, A Lloyd-Spetz, R Yakimova, L Hultman, Thin Solid Films **517**, 2073 (2009)
30. SY Myong, KS Lim, Appl. Phys. Lett. **82**, 3026 (2003)
31. D-S Liu, C-S Sheu, C-T Lee, J. Appl. Phys. **102**, 033516–1 (2007)
32. SM Sze, *Semiconductor Devices Physics and Technology* (John Wiley & Sons, New Jersey, 1985)
33. N Koteeswara Reddy, Q Ahsanulhaq, JJ Kim, Y-B Hahn, Appl. Phys. Lett. **92**, 043127–1 (2008)

Multiplexed detection of various breast cancer cells by perfluorocarbon/quantum dot nanoemulsions conjugated with antibodies

Pan Kee Bae and Bong Hyun Chung*

Abstract

The effective targeting of cancer cell surface antigens is an attractive approach in cancer diagnosis and therapy. Multifunctional nanoprobes with cell-targeting specificity are likely to find important applications in bioanalysis, biomedicine, and clinical diagnosis. In this study, we have fabricated biocompatible perfluorocan/quantum dot nanoemulsions as bimodal imaging nanoprobes for the targeting of breast cancer cells. Perfluorocarbon/quantum dot nanoemulsions conjugated with monoclonal antibodies, as a type of bimodal imaging nanoprobe based on ^{19}F-MR and optical imaging, have been synthesized and applied for targeted imaging of three different breast cancer cells (SKBR3, MCF-7, MDA-MB 468), respectively. We have shown that the cancer-detection capabilities of antibody-conjugated PFC/QDs nanoemulsions could be successfully applied to target of various breast cancer cells. These modified PFC/QDs nanoemulsions were shown to target the cancer cell surface receptors specially. Conjugation of ligands to nanoemulsions targeting over-expressed cell surface receptors is a promising approach for targeted imaging to tumor cells. We further propose that the PFC/QDs nanoemulsions could be used in targeted imaging of breast cancer cells.

Keywords: Perfluorocarbon; Quantum dot; Breast cancer; Bimodal imaging

1 Background

Medical imaging technologies have undergone explosive growth over the past few decades and now play a central role in clinical oncology. Targeted magnetic resonance imaging (MRI) has emerged as a promising diagnostic approach offering high resolution depictions of pathological anatomy and the detection of associated disease biomarkers [1-3]. Semiconductor quantum dots (QD) or organic fluorophores as luminescence probes for many biological and biomedical applications have long been used to visualize cell biology at many levels, from molecules to complete organisms [4-8]. Biologically bifunctional nanoparticles with unique MR and optical imaging capabilities are emerging as useful probes for biolabeling, tumour-targeting, disease diagnosis, cell-based therapy, targeted molecular imaging, optical sensing and biosensors, drug delivery, and the monitoring of therapeutic effects [9-13]. The multimodal imaging will allow clinicians to not only see where a tumour is located in the body, but also to visualize the expression and activity of specific molecules and biological processes that influence tumour behaviour and/or response to therapy.

Recently, many research groups in the world have made every effort to synthesize the nanoparticles that possess both fluorescent and magnetic properties. The combination of fluorescent and magnetic properties is a powerful tool allowing manipulation by magnetic fields and visualization/detection by fluorescence. The perfluorocarbon (PFC) has been explored for various medical applications, including liquid lung ventilation [14], gastrointestinal contrast agent [15], and blood substitutes [16,17]. PFCs are a class of molecules that are highly useful for intracellular MRI tracer applications. While the ^{19}F isotope of fluorine has a natural abundance of near 100%, the biological presence is virtually zero. Therefore, the ^{19}F molecule in PFC nanoemulsions has excellent properties for MR spectroscopy and imaging without a surrounding signal from endogenous fluorine [18-20]. Because of their size-tunable emission spectra, broad absorption spectra,

* Correspondence: chungbh@kribb.re.kr
Bionanotechnology Research Center, Korea Research Institute of Bioscience and Biotechnology, Daejeon 305-806, Korea

high quantum yields, and exceptional resistance to photo and chemical degradation, QDs are currently being investigated for many biological and biomedical applications as luminescence probes [21-25].

In this context, we developed the fabrication of PFC/QD nanoemulsions as bimodal imaging nanoprobes for the targeting of breast cancer cells. Also, we have shown that the cancer-detection capabilities of antibody-conjugated PFC/QD nanoemulsions could be successfully applied to target of various breast cancer cells. We further propose that the PFC/QD nanoemulsions could be used in targeted imaging of breast cancer cells.

2 Methods

2.1. Materials

Perfluorooctylbromide (PFOB), 1H,1H,2H,2H-perfluorodecanethiol (PFDT), 3-(4,5-dimethylthiazol-2-yl)-2,5-diphenyltetrazolium bromide (MTT), and cholesterol were supplied by Sigma-Aldrich (St. Louis, MO). Perfluoro-15-crown ether (PFCE) was purchased from SynQuest Laboratories (Alachua, FL). 3-(N-succinimidyloxyglutaryl) aminopropyl polyethyleneglycol-carbamyldistearoyl phosphatidyl-ethanolamine (DSPE-PEG_{3400}-NHS) was supplied by NOF Co. (Tokyo, Japan) and CdSe/ZnS QDs was obtained from Evident Technologies (Troy, NY). L-α-phosphatidylcholine (lecithin, 95%, chicken egg) was purchased from Avanti Polar Lipids (Alabaster, AL). Antibodies were from following sources: anti-human EGF1R (AbD serotec, Oxford, UK), anti-human ErbB2 and anti-human IGF1R (R&D Systems, Inc., Minneapolis, MN). All compounds for the cell culture were supplied by Invitrogen (Carlsbad, CA).

2.2. Preparation of PFC/QD nanoemulsions

To fabricate PFC/QD nanoemulsions, two PFC materials [PFCE and PFOB] and two different colored QDs (emitting at 525, 606 nm) were used. For conventional hydrophobic ligand-capped CdSe/ZnS QDs to be compatible with PFC liquids, the hydrophobic ligands of QDs should be exchanged with PFDT. A 7 mg CdSe/ZnS QDs dispersed in toluene was mixed with 34 g PFC liquids containing 1 ml PFOT and methanol. After vigorous mixing of the two-phase solutions for 24 h, the CdSe/ZnS QDs were partitioned into the PFC phase. The QDs in PFC were then washed three times with methanol to remove the excess ligands from the solution. The PFC liquids containing CdSe/ZnS QDs were emulsified in an aqueous solution using surfactant mixtures. A surfactant mixture composed of 78.5 mol% lecithin, 20 mol % cholesterol, and 1.5 mol% DSPE-PEG_{3400}-NHS were dissolved in chloroform, and the organic solvent was evaporated using a rotary evaporator and a freezing dryer for 24 h. After dispersing the surfactant mixture into sterilized distilled water, the solution was sonicated. PFC/QD solution (40% v/v), the surfactant mixture (2% w/v) and phosphate buffered saline were mixed for 4 min using a homogenizer. The mixture was followed by microfluidization. A Microfluidizer M-110S (Microfluidics, Inc., Newton, MA) operating at a liquid pressure of approximately 20000 psi was used for all reported nanoemulsion preparations. The fabricated PFC/QDs nanoemulsions were separated using the size-exclusion column and were stored at 4°C.

2.3. Preparation of antibody conjugated PFC/QD nanoemulsions

For targeting the breast cancer cells, three different antibodies [anti-human EGF1R, anti-human ErbB2 and anti-human IGF1R] were conjugated to the PFC/QD nanoemulsions. Antibody-conjugated PFC/QD nanoemulsions were developed by linking the carboxyl groups on the surface of PFC/QD nanoemulsions with the amine groups in antibodies. To the solution containing PFC/QD nanoemulsions and antibodies, 1-ethyl-3-(3-dimethylamino propyl)carbodiimide (EDC) was added and permitted to react with the NHS ligand of PFC/QD nanoemulsions for 2 h at room temperature. To quench the reaction, add 2-mercaptoethanol to a final concentration of 10 mM. The antibody-coated PFC/QDs nanoemulsions were purified by size-exclusion with sepharose 4B column and concentrated using ultrafiltration for 30 min.

2.4. Characterization of PFC/QD nanoemulsions

The emission spectra of PFC/QD nanoemulsions were measured using a fluorescence spectrometer (LS 55, PerkinElmer Instruments, Wellesley, MA). The mean particle diameters and zeta (ζ)-potentials of the PFC/QD nanoemulsions were determined using a particle size analyzer (ELS-Z, Otsuka Electronics, Japan). Measurements were taken after diluting the nanoemulsion in water and equilibrating at room temperature for at least 30 minutes prior to each measurement. All measurements were taken at room temperature.

2.5. Cell viability assay

The cell viability was assessed by a modified MTT assay. Cell viability was measured for the following three cell lines: SKBR3, MCF-7, and MDA-MB 468 cells. These cell lines were obtained from the American Type Culture Collection (Rockville, MD). Human breast cancer cell lines SKBR3, MCF-7, and MDA-MB 468 were grown and maintained in each medium McCoy's, MEM-α, RPMI1640 supplemented with 10% heat inactivated FBS, 50 IU ml^{-1} penicillin, and 50 μg ml^{-1} streptomycin, respectively. Each cell was plated into a 96-well plate (Corning Costar, Cambridge, MA) at 1×10^4 cells/well. After incubation for 24 h, the medium was removed and the prepared PFC/QDs nanoemulsions diluted to

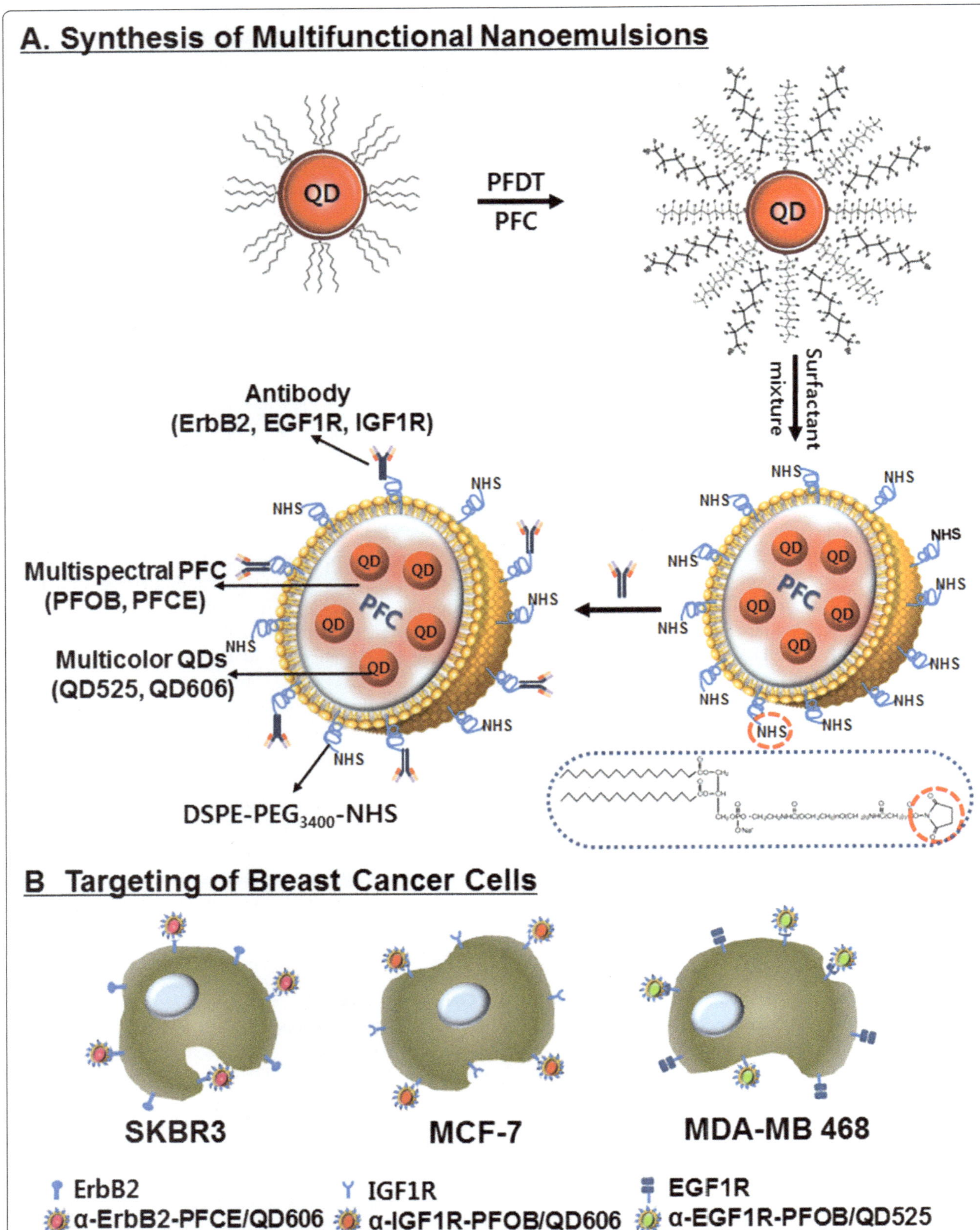

Figure 1 A schematic representation of three different liquid PFC/QDs nanoemulsions (A) and their recognition of growth factor receptor in breast cancer cells (B). The PFC core is surrounded by a lipid monolayer that is functionalized with DSPE-PEG$_{3400}$-NHS to conjugate with three different antibodies.

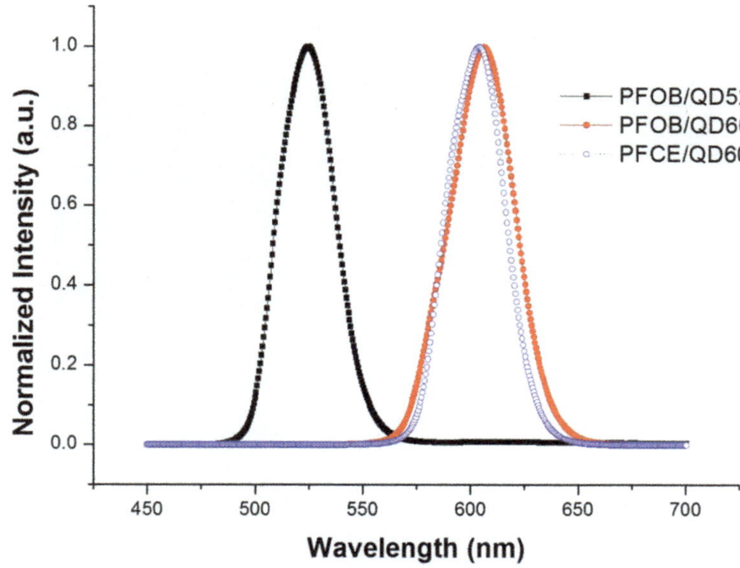

Figure 2 Fluorescence spectra of the different three PFC/QDs nanoemulsions, excited at 390 nm. Narrow emission bands (31–35 nm FWHM or full-width at half-maximum) indicate narrow particle size distributions.

several different concentrations (0, 2.5, 7.4, 22.2, 66.7, 200 μl ml^{-1}) were poured into the wells. After another 24 h incubation period, the residual nanoemulsions were removed and an MTT solution at a concentration of 2.5 mg ml^{-1} in magnesium- and calcium-free phosphate-buffered saline was added to each well. The wells were then incubated in a humidified CO_2 incubator at 37°C for 1.5 h. 100 μl of acidified isopropanol/10% Triton X-100 solution was then added and the plates were shaken to dissolve the formazan products. The absorbance at 570 nm was measured with a microplate reader (Bio-rad, Hercules, CA) The cell survival rate in the control wells without the PFC/QD solutions was considered as 100% cell survival.

2.6. In vitro fluorescence and ^{19}F-MR imaging of breast cancer cells

For fluorescence imaging, SKBR3, MCF-7, or MDA-MB 468 cells were incubated with 50 μl ml^{-1} antibody-conjugated PFC/QD nanoemulsions (ErbB2-PFCE/QD606, EGF1R-PFOB/QD525, IFG1R-PFOB/QD606) for 24 h at 37°C. After being washed with PBS, the labeled cells were fixed with Cytofix/Cytoperm solution and stained with DAPI in PBS. Fluorescence images were obtained on a Deltavision RT deconvolution microscope (Applied Precision Technologies, Issaquah, WA) using emission filters (525WB20, 600WB20, Omega Optical, Brattleboro, VT). All ^{19}F-MR imaging experiments of PFC/QD nanoemulsions were performed with a 4.7 T Bruker scanner (Biospec, Rheinstetten, Germany) using a double-tuned $^{1}H/^{19}F$ quadrature birdcage RF resonator. A ^{19}F-MR image was captured with a FLASH sequence (128 × 128 matrix; 30 × 30 mm^2 FOV; 50 ms TR; 2.6 ms TE; 12 mm slice thickness).

2.7. Statistical analysis

The statistical evaluations of the experiments were performed by ANOVA analysis followed by a Newman-Keuls multiple comparison test.

3 Results and discussion

We have fabricated multifunctional PFC/QD nanoemulsion particles for molecular imaging and targeted breast cancer cells (Figure 1). The engineered PFC/QD nanoemulsions are composed of two PFC materials (PFCE and PFOB) and two different colored QDs (CdSe/ZnS 525, 606 nm). These nanoemulsions provide both ^{19}F-based MR and optical imaging capabilities. To increase the compatibility of hydrophobic QDs with PFC, tri-*n*-octyl-phosphineoxide (TOPO) coated QDs were exchanged with PFDT [26]. After vigorous stirring of the toluene-PFC mixture, the CdSe/ZnS QDs migrated from the toluene phase to the PFC phase. For PFC liquids containing CdSe/ZnS QDs to be useful as probes, for the examinations of biological specimens, three different PFC/QD

Table 1 Summary of the optical and physicochemical properties of various PFC/QDs nanoemulsions

Sample	Emission peak (nm)	FWMH[a]	Average size (nm)	Zeta potential (mV)
PFOB/QD525	525	31	233	−21.27
PFOB/QD606	606	35	230	−20.54
PFCE/QD606	603	31	227	−18.03

[a]Full-width at half maximum.

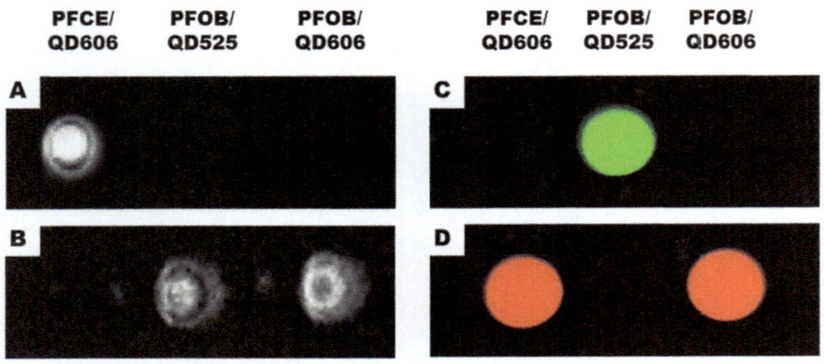

Figure 3 **^{19}F MR and fluorescence imaging of PFC/QDs nanoemulsions: A) PFCE-resonance selective MR image, B) PFOB-resonance selective MR image, C) image obtained by using a green filter, and D) image obtained by using a red filter.**

(PFCE/QD606, PFOB/QD525, PFOB/QD606) nanoemulsions were emulsified in an aqueous solutions using surfactant mixtures. The fluorescence intensity and peak wavelength of these nanoemulsions was measured using a fluorescence spectrometer (Figure 2). The full-width-at-half-maximum (FWHM) values of these nanoemulsions are 31 – 35 nm, respectively. Because of their narrow and symmetric emission spectra, high-quality PFC/QD nanoemulsions are well suited to optical multiplexing. The zeta potential and average size of different PFC/QD nanoemulsions were measured by dynamic light scattering analysis (Table 1). To investigate the dual-mode imaging capabilities of PFC/QD nanoemulsions, we measured a ^{19}F-MR image of specific PFC using a 4.7 T Bruker scanner (Figure 3A,B) and an optical image of different color QDs using suitable wavelength optical filters (Figure 3C,D). The PFC/QD nanoemulsions containing two different PFC liquids (PFCE or PFOB) are discriminated by using the distinctly different spectrum of each PFC. These results suggest that the PFC/QD nanoemulsions could serve as more effective nanoprobes for the ^{19}F-MR imaging and fluorescence imaging of various breast cancer cells.

We further investigated whether the PFC/QD nanoemulsions could be used in targeted imaging of breast cancer cells. The PFC/QD nanoemulsions were conjugated with monoclonal antibodies that can target epidermal

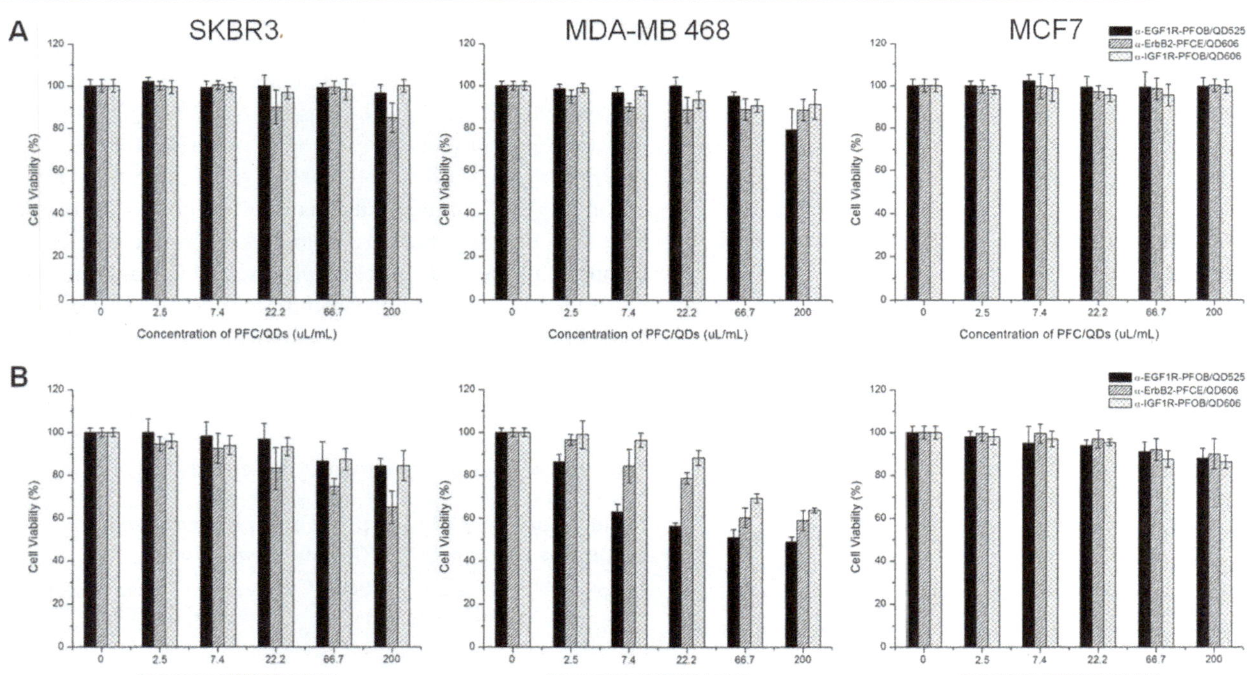

Figure 4 **Cell cytotoxicity for the different antibody-conjugated PFC/QDs nanoemulsions and different cell types, incubated at 37°C for 24 h (A) and 48 h (B).** Three different nanoemulsions are tested on the cell viability for each cell type.

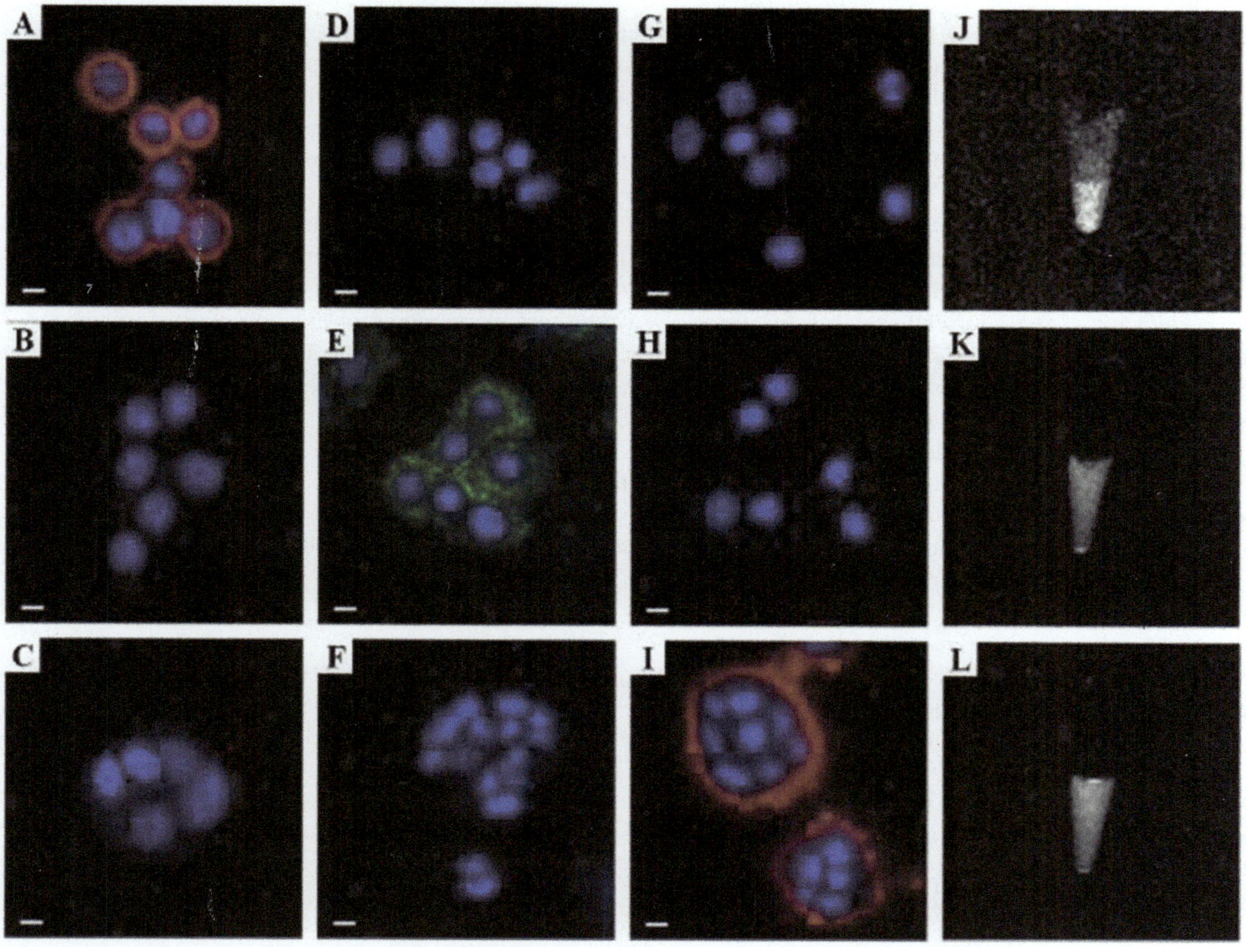

Figure 5 Luminescence (A-I) and ^{19}F MR (J-L) images of cultured SKBR3 (A, D, G, J), MDA-MB 468 (B, E, H, K), and MCF-7 (C, F, I, L) cells as incubated with α-ErbB2-PFCE/QD606 (A-C, J), α-EGF1R-PFOB/QD525 (D-F, K) and α-IGF1R-PFOB/QD606 (G-I, L). The QDs (green and red) and the DAPI-stained nuclei (blue) were recorded with Deltavision RT deconvolution microscope. The modified PFC/QDs nanoemulsions are shown in green and red, and the DAPI-stained nuclei are shown in blue. Scale Bars: 10 μm.

growth factor receptors and other growth factor receptor: EGF1R (epidermal growth factor-1 receptor), ErbB2 (also called Her2 and Neu, epidermal growth factor-2 receptor), and IGF1R (insulin-like growth factor-1 receptor) (Figure 1B). These receptors are often overexpressed on the surface of human breast cancer cell lines [SKBR3 (ErbB2-positive), MDA-MB 468 (EGF1R-positive), MCF-7 (IGF1R-positive)] [27-29]. They serve as important targets for selective cancer therapy. The expression or activation of human growth factor receptors are altered in many epithelial tumours and clinical studies indicate that they have important roles in tumour aetiology and progression. For targeting the breast cancer cells, three different monoclonal antibodies (α-EGF1R, α-ErbB2, and α-IGF1R) were reacted with NHS on the PFC/QD nanoemulsions for 2 h at room temperature, respectively. DSPE-PEG$_{3400}$-NHS lipid which has a distal NHS group at the end of PFC/QD nanoemulsions reacts with a homing molecule. The homing molecules were conjugated at the end of the lipid, and they specifically bind to their receptors on the breast cancer cell surface. The antibody-coated PFC/QD nanoemulsions purified by size-exclusion chromatography with a sepharose 4B column and the excess amount of antibodies were removed (Additional file 1: Figure S1). We evaluated the cell damage caused by the PFC/QDs nanoemulsions for biological applications. We performed cell viability assay to determine the cytotoxicity of various PFC/QD nanoemulsions based on the extracellular nanoemulsions exposure levels in cells (Figure 4). Different doses (0, 2.5, 7.4, 22.2, 66.7, 200 μl ml^{-1}) of these nanoemulsions were used for *in vitro* cytotoxicity tests. We used three cell types: SKBR3 cell, MCF-7 cell, and MDA-MB 468 cell for three PFC/QD nanoemulsions. The result showed that no differences in three type cells were observed for PFC/QD nanoemulsions at 24 h and 48 h. The tendencies of the cell viability at 24 h with PFC/QD nanoemulsions were almost the same. Treatment of SKBR3 and MDA-MB 468 with 22.2 - 200 μl ml^{-1} of antibody-conjugated PFC/QD nanoemulsions significantly decreased the cell viability with respect to control at 48 h. Within 48 h the cell viability in SKBR3 cells decreased from 92 ± 6% to 65 ± 7% at the α-ErbB2-PFCE/QD606 concentration of 7.4 - 200 μl ml^{-1}. Also, for the α-EGF1R-PFOB/QD525 concentration of 2.5 - 200 μl ml^{-1} the viability of MDA-MB 468 cells at 48 h decreased from 86 ± 3% to 49 ± 2%. There were no significant changes in cell viability for these nanoemulsions in MFC7 cells. Since QDs may slowly release the toxic Cd^{2+} or Se^{2-} ions into the solution, the particles must be as inert as possible for any in vitro application. The toxic of QDs not only depends on the concentration of free Cd^{2+} ions but also depends on whether the particles are ingested by a cell and where they are stored. The release of Cd^{2+} from the particles' surface can be reduced by employing core/shell particles or the coating of the particles with silica, polymer, or liposome.

To investigate the targeting specificity, each breast cancer cell line was incubated with three different antibody-conjugated PFC/QD nanoemulsions (α-ErbB2-PFCE/QD606, α-EGF1R-PFOB/QD525, and α-IGF1R-PFOB/QD606). Fluorescence imagings were obtained on a Deltavision RT deconvolution microscope. As shown in Figure 4, the fluorescence of α-ErbB2-PFCE/QD606 nanoemulsions was only observed in the ErbB2-positive SKBR3 breast cancer cells (Figure 5A). MDA-MB 468 and MCF-7 cells showed only slight fluorescence signals with α-ErbB2-PFCE/QD606 nanoemulsions (Figures 5B,C). The attachment of α-ErbB2-PFCE/QD606 onto the SKBR3 cells suggests that there is a specific interaction between the α-ErbB2 that bound to PFC/QDs and ErbB2. Also, α-EGF1R-PFOB/QD525 and α-IGF1R-PFOB/QD606 nanoemulsions were targeted to the MDA-MB 468 and MCF-7 cells, respectively (Figure 5D-I). Also, the ^{19}F-based MR images for the specific targeting of each antibody-conjugated PFC/QD nanoemulsion in various breast cancer cells are shown (Figure 5J-L). These results indicate that antibody-PFC/QD nanoemulsions selectively bind to the target-protein. Therefore, the modified PFC/QD can act as a useful optical and ^{19}F-MR imaging agent for the diagnosis and targeting of breast cancer cells.

4 Conclusion

In conclusion, the present study describes a novel approach for detecting the various breast cancer cells with the antibody-conjugated PFC/QD nanoemulsions as a type of bimodal imaging nanoprobe with unique MR and optical imaging capabilities. It is believed that this approach will provide a very promising tool for the diagnosis of breast cancer. Different PFC/QD nanoemulsions can be conjugated to different antibodies, each targeted to specific proteins. The specific spectra of multiple PFC/QD targeted to different tissue proteins can then be simultaneously detected and quantified on one sample. They also have enhanced photostability, allowing the emission of fluorescent light over a length of time without a brisk decrease in emission, and the strength of their fluorescence means that low-level proteins can also be detected, thereby increasing diagnostic sensitivity [30-33]. PFC/QD nanoemulsions have great capacity as an efficient nanoprobe for targeting breast cancer cells. Furthermore, these nanoprobes have potential in a wider variety of novel applications that are related to anti-receptor therapy in cancer, molecular targeted approach, and modulated drug delivery.

Additional file

Additional file 1: Figure S1. Gel permeation chromatogram of antibody-conjugated PFC/QDs nanoemulsions. The antibody-conjugated PFC/QDs nanoemulsions were separated by the sepharose 4B column. ▼, antibody-conjugated nanoemulsions; *, unreacted antibody.

Competing interests

The authors declare that they have no competing interests.

Authors' contributions

PK Bae carried out the nanoemulsion synthesis and analysis and drafted the manuscript. BH Chung conceived of the study, and participated in its design and coordination. Both authors read and approved the final manuscript.

Authors' information

Dr. Pan Kee Bae is currently a postdoctoral research associate at BioNanotechnology Research Center in KRIBB. He obtained his B.S. from ChungNam University in 2000 and Ph.D. from department of Biology, Yonsei University in 2006. His recent research is focused on the design of nanoparticles for bio-imaging against cancers and infectious disease.
Prof. Dr. Bong Hyun Chung received his PhD degree at Korea Advanced Institute of Science and Technology (KAIST) in 1987. He did his postdoctoral work at Caltech from 1990 to 1991. He is currently a director of the BioNanotechnology Research Center in KRIBB. He is also a professor of Nanobio Major at the University of Science and Technology (UST). His main research interests include (a) bio-nano interfacing science, (b) biochip/biosensor, (c) nanomedicine.

Acknowledgements

This work was supported by BioNano Health-Guard Research Center funded by the Ministry of Science, ICT & Future Planning (MSIP) of Korea as Global Frontier Project (Grant No.2013M3A6B2078950). This work was also supported by the KRIBB Initiative Research Program, Republic of Korea.

References

1. ME Juweid, BD Cheson, N. Engl. J. Med. **354**, 496 (2006)
2. R Weissleder, Nat. Rev. Cancer **2**, 11 (2002)
3. SD Caruthers, AM Neubauer, FD Hockett, R Lamerichs, PM Winter, MJ Scott, PJ Gaffney, SA Wickline, GM Lanza, Invest. Radiol. **41**, 305 (2006)
4. BC Lagerholm, M Wang, LA Ernst, DH Ly, H Liu, PM Bruchez, AS Waggoner, Nano Lett. **4**, 2019 (2004)
5. C Wu, J Hong, X Guo, C Huang, J Lai, J Zheng, J Chen, X Mu, Y Zhao, Chem. Commun. 750 (2008)
6. Y Li, YT Cu, D Luo, Nat. Biotechnol. **23**, 885 (2005)
7. B Dubertret, P Skourides, DJ Norris, V Noireaux, AH Brivanlou, A Libchaber, Science **298**, 1759 (2002)

8. JK Jaiswal, H Mattoussi, JM Mauro, SM Simon, Nat. Biotechnol. **21**, 47 (2003)
9. JH Park, G Von Maltzahn, E Ruoslahti, SN Bhatia, MJ Sailor, Angew. Chem. Int. Ed. **47**, 7284 (2008)
10. H Yang, S Santra, GA Walter, PH Holloway, Adv. Mater. **18**, 2890 (2006)
11. JR McCarthy, FA Jaffer, R Weissleder, Small **2**, 983 (2006)
12. G Beaune, B Dubertret, O Clement, C Vayssettes, V Cabuil, C Menager, Angew. Chem. Int. Ed. **46**, 5421 (2007)
13. S Wang, BR Jarrett, SM Kauzlarich, AY Louie, J. Am. Chem. Soc. **129**, 3848 (2007)
14. MA Croce, TC Fabian, JH Patton Jr, SM Melton, M Moore, LL Trenthem, J. Trauma **45**, 273 (1998)
15. JJ Brown, JR Duncan, JP Heiken, DM Balfe, AP Corr, SA Mirowitz, SS Eilengerg, JK Lee, Radiology **181**, 455 (1991)
16. AJ McGoron, R Pratt, J Zhang, Y Shiferaw, S Thomas, R Millard, Artif. Cells Blood Substit. Immobil. Biotechnol. **22**, 1243 (1994)
17. PE Keipert, S Otto, SF Flaim, JG Weers, EA Schutt, TJ Pelura, DH Klein, TL Yaksh, Artif. Cells Blood Substit. Immobil. Biotechnol. **22**, 1169 (1994)
18. GM Lanza, PM Winter, AM Neubauer, SD Caruthers, FD Hockett, SA Wickline, Curr. Top. Dev. Bio. **70**, 57 (2005)
19. ET Ahrens, R Flores, H Xu, PA Morel, Nat. Biotechnol. **23**, 983 (2005)
20. M Higuchi, N Iwata, Y Matsuba, K Sato, K Sasamoto, TC Saido, Nat. Neurosci. **8**, 527 (2005)
21. AP Alivisatos, Nat. Biotechnol. **22**, 47 (2004)
22. IL Medintz, HT Uyeda, ER Goldman, H Mattoussi, Nat. Mater. **4**, 435 (2005)
23. SW Kim, JP Zimmer, S Ohnishi, JB Tracy, JV Frangioni, MG Bawendi, J. Am. Chem. Soc. **127**, 10526 (2005)
24. X Gao, Y Cui, RM Levenson, LW Chung, S Nie, Nat. Biotechnol. **22**, 969 (2004)
25. AP Alivisatos, W Gu, C Larabell, Ann. Rev. Biomed. Eng. **7**, 55 (2005)
26. YT Lim, YW Noh, JH Cho, JH Han, BS Choi, JN Kwon, KS Hong, A Gokarna, YH Cho, BH Chung, J. Am. Chem. Soc. **131**, 17145 (2009)
27. X Huang, L Gao, S Wang, JL McManaman, AD Thor, X Yang, FJ Esteva, B Liu, Cancer Res. **70**, 1204 (2010)
28. JS Ross, JA Fletcher, Stem Cells **16**, 413 (1998)
29. D Bartusik, B Tomanek, E Lattová, H Perreault, G Fallone, J. Pharm. Biomed. Anal. **51**, 192 (2010)
30. MV Yezhelyev, X Gao, Y Xing, A Al Hajj, S Nie, RM O'Regan, Lancet Oncol. **7**, 657 (2006)
31. M Ferrari, Nat. Rev. Cancer **5**, 161 (2005)
32. X Wang, L Yang, ZG Chen, DM Shin, CA Cancer J. Clin. **58**, 97 (2008)
33. KK Jain, Clin. Chem. **53**, 2002 (2007)

Permissions

All chapters in this book were first published in NC, by Springer; hereby published with permission under the Creative Commons Attribution License or equivalent. Every chapter published in this book has been scrutinized by our experts. Their significance has been extensively debated. The topics covered herein carry significant findings which will fuel the growth of the discipline. They may even be implemented as practical applications or may be referred to as a beginning point for another development.

The contributors of this book come from diverse backgrounds, making this book a truly international effort. This book will bring forth new frontiers with its revolutionizing research information and detailed analysis of the nascent developments around the world.

We would like to thank all the contributing authors for lending their expertise to make the book truly unique. They have played a crucial role in the development of this book. Without their invaluable contributions this book wouldn't have been possible. They have made vital efforts to compile up to date information on the varied aspects of this subject to make this book a valuable addition to the collection of many professionals and students.

This book was conceptualized with the vision of imparting up-to-date information and advanced data in this field. To ensure the same, a matchless editorial board was set up. Every individual on the board went through rigorous rounds of assessment to prove their worth. After which they invested a large part of their time researching and compiling the most relevant data for our readers.

The editorial board has been involved in producing this book since its inception. They have spent rigorous hours researching and exploring the diverse topics which have resulted in the successful publishing of this book. They have passed on their knowledge of decades through this book. To expedite this challenging task, the publisher supported the team at every step. A small team of assistant editors was also appointed to further simplify the editing procedure and attain best results for the readers.

Apart from the editorial board, the designing team has also invested a significant amount of their time in understanding the subject and creating the most relevant covers. They scrutinized every image to scout for the most suitable representation of the subject and create an appropriate cover for the book.

The publishing team has been an ardent support to the editorial, designing and production team. Their endless efforts to recruit the best for this project, has resulted in the accomplishment of this book. They are a veteran in the field of academics and their pool of knowledge is as vast as their experience in printing. Their expertise and guidance has proved useful at every step. Their uncompromising quality standards have made this book an exceptional effort. Their encouragement from time to time has been an inspiration for everyone.

The publisher and the editorial board hope that this book will prove to be a valuable piece of knowledge for researchers, students, practitioners and scholars across the globe.

List of Contributors

Wounjhang Park
Department of Electrical, Computer & Energy Engineering, University of Colorado, Boulder, CO 80309-0425, USA

Pan Kee Bae
Bionanotechnology Research Center, Korea Research Institute of Bioscience and Biotechnology, Daejeon 305-806, South Korea

Juyeon Jung
Bionanotechnology Research Center, Korea Research Institute of Bioscience and Biotechnology, Daejeon 305-806, South Korea

Bong Hyun Chung
Bionanotechnology Research Center, Korea Research Institute of Bioscience and Biotechnology, Daejeon 305-806, South Korea

Sang Hoon Chae
Center for Integrated Nanostructure Physics (CINAP), Institute for Basic Science (IBS), Suwon 440-746, Republic of Korea Department of Energy Science, Department of Physics, Sungkyunkwan University (SKKU), Suwon 440-746, Republic of Korea

Young Hee Lee
Center for Integrated Nanostructure Physics (CINAP), Institute for Basic Science (IBS), Suwon 440-746, Republic of Korea Department of Energy Science, Department of Physics, Sungkyunkwan University (SKKU), Suwon 440-746, Republic of Korea

Jin-young Kim
Department of Biosystems Science and Engineering, Bioegineering Laboratory, ETH Zurich, CH-4058 Basel, Switzerland

Soo-Ik Chang
Department of Biochemistry, Chungbuk National University, Cheongju, Chungbuk 361-763, Korea

Andrew J deMello
Institute of Chemical and Bioengineering, Department of Chemistry and Applied Bioscience, ETH Zurich, CH-8093 Zurich, Switzerland

Danny O'Hare
Department of Bioengineering, Imperial College London, London SW7 2AZ, UK

Sang-Won Jee
Department of Chemical Engineering, Hanyang University, Ansan 426-791, South Korea Department of Mechanical Engineering, Texas A&M University, College Station, TX, USA

Keya Zhou
Department of Chemical Engineering, Hanyang University, Ansan 426-791, South Korea Department of Physics, Harbin Institute of Technology, Harbin, China

Dong-Wook Kim
Department of Physics, Ewha Womans University, Seoul 120-750, South Korea

Jung-Ho Lee
Department of Chemical Engineering, Hanyang University, Ansan 426-791, South Korea

Gee Yeong Kim
Department of Physics, Ewha Womans University, Seoul, Korea

Juran Kim
Department of Physics, Ewha Womans University, Seoul, Korea

William Jo
Department of Physics, Ewha Womans University, Seoul, Korea

Dae-Ho Son
Daegu Gyeongbuk Institute of Science and Technology (DGIST), Daegu, Korea

Dae-Hwan Kim
Daegu Gyeongbuk Institute of Science and Technology (DGIST), Daegu, Korea

Jin-Kyu Kang
Daegu Gyeongbuk Institute of Science and Technology (DGIST), Daegu, Korea

Young Hun Kang
Division of Advanced Materials, Korea Research Institute of Chemical Technology, Daejeon 305-600, Korea

Jae Hyung Han
Creative Research Center for Graphene Electronics, Electronics and Telecommunications Research Institute (ETRI), Daejeon 305-700, Korea

Song Yun Cho
Division of Advanced Materials, Korea Research Institute of Chemical Technology, Daejeon 305-600, Korea

Choon-Gi Choi
Creative Research Center for Graphene Electronics, Electronics and Telecommunications Research Institute (ETRI), Daejeon 305-700, Korea

Yeonwoong Jung
Department of Mechanical Engineering and Materials Science, Yale University, New Haven, CT, USA

Jie Shen
Department of Mechanical Engineering and Materials Science, Yale University, New Haven, CT, USA

Judy J Cha
Department of Mechanical Engineering and Materials Science, Yale University, New Haven, CT, USA
Energy Science Institute, Yale University West Campus, West Haven, CT, USA

Saravanan Nagappan
Department of Polymer Science and Engineering, Pusan National University, Busan 609-735, Republic of Korea

Jin Joo Park
Department of Polymer Science and Engineering, Pusan National University, Busan 609-735, Republic of Korea

Sung Soo Park
Department of Polymer Science and Engineering, Pusan National University, Busan 609-735, Republic of Korea

Chang-Sik Ha
Department of Polymer Science and Engineering, Pusan National University, Busan 609-735, Republic of Korea

Ku Youn Baik
Department of Electrical and Biological Physics, Kwangwoon University, Seoul 139-701, Korea

Sung Young Park
Center for Functional Connectomics, Korea Institute of Science and Technology, Hwarangno 14-gil 5, Seongbuk-gu, Seoul 136-791, Korea

Seon Namgung
Department of Electrical and Computer Engineering, University of Minnesota, Minneapolis, MN 55455, USA

Daesan Kim
Department of Biophysics and Chemical Biology, Seoul National University, Seoul 151-747, Korea

Dong-guk Cho
Department of Physics and Astronomy, and Institute of Applied Physics, Seoul National University, Seoul 151-747, Korea

Minju Lee
Department of Physics and Astronomy, and Institute of Applied Physics, Seoul National University, Seoul 151-747, Korea

Seunghun Hong
Department of Biophysics and Chemical Biology, Seoul National University, Seoul 151-747, Korea
Department of Physics and Astronomy, and Institute of Applied Physics, Seoul National University, Seoul 151-747, Korea

Wonjin Jo
Institute for Multidisciplinary Convergence of Matter, Korea Institute of Science and Technology, 136-791 Seoul, Republic of Korea

U Kei Cheang
Department of Mechanical Engineering & Mechanics, Drexel University, 3141 ChestnutStreet, Philadelphia, PA 19104, USA

Min Jun Kim
Department of Mechanical Engineering & Mechanics, Drexel University, 3141 ChestnutStreet, Philadelphia, PA 19104, USA

Won-Kyu Rhim
Department of Chemistry, Seoul National University, Seoul 151-747, Republic of Korea

Minho Kim
Department of Chemistry, Seoul National University, Seoul 151-747, Republic of Korea

Kevin L Hartman
Department of Chemistry, Seoul National University, Seoul 151-747, Republic of Korea

Keon Wook Kang
Department of Nuclear Medicine, Seoul National University College of Medicine, Seoul 110-744, Republic of Korea

Jwa-Min Nam
Department of Chemistry, Seoul National University, Seoul 151-747, Republic of Korea

Jae Ah Lee
Center for Bio-Artificial Muscle and Department of Biomedical Engineering, Hanyang University, Seoul 133-791 South Korea

Ray H Baughman
The Alan G. MacDiarmid NanoTech Institute, University of Texas at Dallas, Richardson, TX 75083, USA

Seon Jeong Kim
Center for Bio-Artificial Muscle and Department of Biomedical Engineering, Hanyang University, Seoul 133-791, South Korea

Richard J Parmee
Department of Engineering, Electrical Engineering Division, Cambridge University, Cambridge CB3 0FA, UK

Clare M Collins
Department of Engineering, Electrical Engineering Division, Cambridge University, Cambridge CB3 0FA, UK

William I Milne
Department of Engineering, Electrical Engineering Division, Cambridge University, Cambridge CB3 0FA, UK

Matthew T Col
Department of Engineering, Electrical Engineering Division, Cambridge University, Cambridge CB3 0FA, UK

Greta Faccio
Laboratory for Bioactive Materials, Empa, Swiss Federal Laboratories for Materials Science and Technology, Lerchenfeldstrasse 5, 9014 St. Gallen, Switzerland

Krisztina Gajda-Schrantz
Laboratory for High Performance Ceramics, Empa, Swiss Federal Laboratories for Materials Science and Technology, Überlandstrasse 129, 8600 Dübendorf, Switzerland
Department of Inorganic and Analytical Chemistry, Szeged University, Dóm tér 7, H-6720 Szeged, Hungary
Institute of Chemistry, Research Group of Environmental Chemistry, University of Szeged, Rerrich Bélatér 1, H-6720 Szeged, Hungary

Julian Ihssen
Laboratory for Bioactive Materials, Empa, Swiss Federal Laboratories for Materials Science and Technology, Lerchenfeldstrasse 5, 9014 St. Gallen, Switzerland

Florent Boudoir
Laboratory for High Performance Ceramics, Empa, Swiss Federal Laboratories for Materials Science and Technology, Überlandstrasse 129, 8600 Dübendorf, Switzerland
Department of Chemistry, University of Basel, Spitalstrasse 51, 4056 Basel, Switzerland

Yelin Hu
Laboratory for High Performance Ceramics, Empa, Swiss Federal Laboratories for Materials Science and Technology, Überlandstrasse 129, 8600 Dübendorf, Switzerland
Laboratory for Photonics and Interfaces, Ecole Polytechnique Federale de Lausanne, Rue Cantonale, 1021 Lausanne, Switzerland

Bongjin Simon Mun
Department of Physics and Photon Science, GIST School of Physics and Chemistry, GIST College, Ertl Center for Electrochemistry and Catalysis, 123 Cheomdan-gwagiro (Oryong-dong), Buk-gu, Gwangju 500-712, Korea

Debajeet K Bora
Laboratory for High Performance Ceramics, Empa, Swiss Federal Laboratories for Materials Science and Technology, Überlandstrasse 129, 8600 Dübendorf, Switzerland

Linda Thöny-Meyer
Laboratory for Bioactive Materials, Empa, Swiss Federal Laboratories for Materials Science and Technology, Lerchenfeldstrasse 5, 9014 St. Gallen, Switzerland

Artur Braun
Laboratory for High Performance Ceramics, Empa, Swiss Federal Laboratories for Materials Science and Technology, Überlandstrasse 129, 8600 Dübendorf, Switzerland

Melinda Y Han
Department of Applied Physics and Applied Mathematics, Columbia University, New York, NY 10027, USA

Philip Kim
Department of Physics, Columbia University, New York, NY 10027, USA

Arne Schleunitz
Micro Resist Technology GmbH, Koepenicker Str. 325, 12555 Berlin, Germany

Vitaliy A Guzenko
Paul Scherrer Institut, Laboratory for Micro- and Nanotechnology, 5232 Villigen, PSI, Switzerland

Martin Messerschmidt
Micro Resist Technology GmbH, Koepenicker Str. 325, 12555 Berlin, Germany

Hakan Atasoy
Micro Resist Technology GmbH, Koepenicker Str. 325, 12555 Berlin, Germany

Robert Kirchner
Paul Scherrer Institut, Laboratory for Micro- and Nanotechnology, 5232 Villigen, PSI, Switzerland

Helmut Schift
Paul Scherrer Institut, Laboratory for Micro- and Nanotechnology, 5232 Villigen, PSI, Switzerland

Jin Young Kim
Photo-electronic Hybrids Research Center, Korea Institute of Science and Technology (KIST), Seoul 136-791, Korea.
Green Green School, Korea University, Seoul 136-701, Korea

Kai Zhu
National Renewable Energy Laboratory, Golden, Colorado 80401-3393, USA

Nathan R Neale
National Renewable Energy Laboratory, Golden, Colorado 80401-3393, USA

Arthur J Frank
National Renewable Energy Laboratory, Golden, Colorado 80401-3393, USA

Elchin Huseynov
Institute of Radiation Problems of Azerbaijan National Academy of Sciences, AZ 1143, B. Vahabzadeh 9, Baku, Azerbaijan

Adil Garibov
Institute of Radiation Problems of Azerbaijan National Academy of Sciences, AZ 1143, B. Vahabzadeh 9, Baku, Azerbaijan

Ravan Mehdiyeva
Institute of Radiation Problems of Azerbaijan National Academy of Sciences, AZ 1143, B. Vahabzadeh 9, Baku, Azerbaijan

Soon-Hong Kwon
Department of Physics, Chung-Ang University, Seoul 156-756, Republic of Korea

You-Shin No
Department of Physics, Korea University, Seoul 136-701, Republic of Korea

Hong-Gyu Park
Department of Physics, Korea University, Seoul 136-701, Republic of Korea

Joseph Park
Department of Chemical and Biomolecular Engineering, Yonsei University, Seoul 120-749, South Korea

Byunghoon Kang
Department of Chemical and Biomolecular Engineering, Yonsei University, Seoul 120-749, South Korea

Bongjune Kim
Department of Chemical and Biomolecular Engineering, Yonsei University, Seoul 120-749, South Korea

Jin-Suck Suh
Department of Radiology & Department of Biochemistry and Molecular Biology, Yonsei University, Seoul 120-752, South Korea
YUHS-KRIBB Medical Convergence Research Institute, Yonsei University, Seoul 120-752, South Korea

Yong-Min Huh
Department of Radiology & Department of Biochemistry and Molecular Biology, Yonsei University, Seoul 120-752, South Korea
YUHS-KRIBB Medical Convergence Research Institute, Yonsei University, Seoul 120-752, South Korea

Seungjoo Haam
Department of Chemical and Biomolecular Engineering, Yonsei University, Seoul 120-749, South Korea

Utpal Jana
Department of Pharmacy, Annamalai University, Annamalai Nagar, Tamil Nadu, India

Anjan Kumar Mohanty
Department of Pharmacy, Annamalai University, Annamalai Nagar, Tamil Nadu, India

Sovan Lal Pal
Department of Pharmacy, Annamalai University, Annamalai Nagar, Tamil Nadu, India

Prabal Kumar Manna
Department of Pharmacy, Annamalai University, Annamalai Nagar, Tamil Nadu, India

Guru Prasad Mohanta
Department of Pharmacy, Annamalai University, Annamalai Nagar, Tamil Nadu, India

Nandanapalli Koteeswara Reddy
School of Semiconductor and Chemical Engineering, Chonbuk National University, 567 Baekje-daero, Deokjin-gu, Jeonju 561 756, Republic of Korea
Institute of Chemistry, Humboldt University, Berlin 12489, Germany

Mudusu Devika
Department of Aerospace Engineering, Indian Institute of Science, Bangalore 560 012, India

Yoon-Bong Hahn
School of Semiconductor and Chemical Engineering, Chonbuk National University, 567 Baekje-daero, Deokjin-gu, Jeonju 561 756, Republic of Korea

Pan Kee Bae
Bionanotechnology Research Center, Korea Research Institute of Bioscience and Biotechnology, Daejeon 305-806, Korea

Bong Hyun Chung
Bionanotechnology Research Center, Korea Research Institute of Bioscience and Biotechnology, Daejeon 305-806, Korea